Guo-Ping Zhang, Mingsu Si, Thomas F. George

Quantum Mechanics

Also of Interest

Quantum Mechanics
An Introduction to the Physical Background and Mathematical Structure
Gregory Naber, 2021
ISBN 978-3-11-075161-1, e-ISBN (PDF) 978-3-11-075194-9

Quantum Information Theory
Concepts and Methods
Joseph M. Renes, 2022
ISBN 978-3-11-057024-3, e-ISBN (PDF) 978-3-11-057025-0

Quantum Technologies
For Engineers
Rainer Müller, Franziska Greinert, 2024
ISBN 978-3-11-071744-0, e-ISBN (PDF) 978-3-11-071745-7

Hypersymmetry
Physics of the Isotopic Field-Charge Spin Conservation
György Darvas, 2021
ISBN 978-3-11-071317-6, e-ISBN (PDF) 978-3-11-071318-3

Data Science
Time Complexity, Inferential Uncertainty, and Spacekime Analytics
Ivo D. Dinov, Milen Velchev Velev, 2021
ISBN 978-3-11-069780-3, e-ISBN (PDF) 978-3-11-069782-7

Guo-Ping Zhang, Mingsu Si, Thomas F. George

Quantum Mechanics

DE GRUYTER

Authors
Prof. Dr. Guo-Ping Zhang
Indiana State University
Terre Haute 47809
USA
guo-ping.zhang@outlook.com

Prof. Dr. Thomas F. George
University of Missouri-St. Louis
St. Louis 63121
USA
tfgeorge@umsl.edu

Prof. Dr. Mingsu Si
Lanzhou University
Lanzhou 730000
China
sims@lzu.edu.cn

ISBN 978-3-11-067212-1
e-ISBN (PDF) 978-3-11-067215-2
e-ISBN (EPUB) 978-3-11-067228-2

Library of Congress Control Number: 2023951182

Bibliographic information published by the Deutsche Nationalbibliothek
The Deutsche Nationalbibliothek lists this publication in the Deutsche Nationalbibliografie; detailed bibliographic data are available on the Internet at http://dnb.dnb.de.

Cover image: dianaarturovna / iStock / Getty Images Plus
Typesetting: VTeX UAB, Lithuania
Printing and binding: CPI books GmbH, Leck

www.degruyter.com

To our families

Preface

Quantum mechanics introduces new physical concepts "that cannot be explained in terms of things previously known to the student, which cannot even be explained adequately in words at all," Dirac once said. This book has grown out of a quantum mechanics (QM) course that has been taught in our respective universities. The student often faces three main difficulties: logic, practice, and experimental realization.

`Logic.` Quantum mechanics introduces so many new concepts so quickly. This is partly due to the fact that materials are often organized in terms of dimensions—one-dimensional, two-dimensional, and three-dimensional systems. A prime example is that one starts from the infinite quantum well in one section and then in the next section jumps to the harmonic oscillator. A smooth transition and logical connection are missing.

In this book, we strive to gradually weave the physical concepts and ideas into the book logically, with new mathematical tools introduced as needed, lessening mathematical interruption. Each chapter is partitioned into several units. One can choose just some units for a one-semester course. To give the reader a global view, we organize materials as follows, where Chapters 5 and 7 are on mathematical methods.

Chapter	**2**	**3**	**4, 6**	**8**	**9**	**10**
matter	electron	nucleus	atom	molecules	solids	special topics
themes	tunneling	vibration	spin & orbital angular momenta	molecular orbital	energy bands	quantum computing, Berry phase

`Practice.` Practice makes perfect! Every good textbook must have lots of examples, exercises, and computer codes to guide the reader through the material. We provide many examples, and in some cases an entire section is dedicated to examples. We design three levels of exercises. The shorter ones, called exercises, 318 in total, should be worked out immediately after each section, while the longer ones at the end of each chapter, called problems, 148 in total, should be assigned after several sections have been covered. The numerical problems, which students love most, should be assigned whenever possible. We provide 16 computer source codes, so the reader can directly use them. For instance, we numerically solve the finite quantum well and the hydrogen atom. If the student cannot solve some basic problems, it is hard to say that he/she really understands the material. Once the reader is able to calculate, then the conceptual understanding becomes easier. Richard Feynman once said "What I cannot calculate, I do not understand," and "know how to solve every problem that has been solved." Computing Born's probability density for a classical object is more amenable to the student.

`Experiment.` To fully appreciate how QM works, the student needs a real experiment to see the beauty behind QM. We went to the laboratory and did some experiments

https://doi.org/10.1515/9783110672152-201

ourselves to discover a new way to teach quantum mechanics. The feeling is very different. Motivated by a sketch in the manual provided by the manufacturer, we discovered the Zeeman disk for light polarization, which never appears in any QM books.

In the following, we briefly outline each chapter and summarize the key concepts and formulas in Table 1 at the end of this Preface. The materials are designed in such a way that a one-semester course may need to cover the first few sections listed in the table without any interruption to the following chapters.

Chapter 1 is an introduction to QM. It starts with the quantization concept and ends with a LED experiment to measure the Planck constant. The very first exercise immediately reveals the insight into quantization. Asking how the photon energy changes when light enters water is the first test to see whether the student understands the photon. We find a novel way to explain Born statistics and his probability density using an airplane.

Chapter 2 is on bound and continuous states and quantum tunneling, mainly on electrons. It starts with a novel introduction to the time-dependent Schrödinger equation, which is based on the Noether theorem, and ends with insights into tunneling. Tunneling through a finite barrier is solved with a novel method, and the computer codes provide the wavefunctions with any arbitrary potential heights.

Chapter 3 is on the harmonic oscillator, mainly on nuclear vibrations. It starts with operators and ends with the explanation of blackbody radiation and the photoelectric effect. The classical probability is used as an introduction to the quantum treatment. Operators, Heisenberg uncertainty principle, and Schrödinger and Heisenberg pictures are introduced. The actual quantum oscillator is simulated by the coherent states.

Chapter 4 is on orbital angular momentum and the hydrogen atom. It combines the materials from Chapters 2 and 3, electron + nucleus, and extends it to three dimension. It starts from the Bohr model and ends with the Franck–Hertz experiment. The hydrogen code is highly recommended for the beginner.

Chapter 5 is on the time-independent method. It starts with perturbation theory and ends with the variational principle. Two entire sections are devoted to the examples.

Chapter 6 is on the electron spin. It starts from the Stern–Gerlach experiment and ends with magnetic resonance imaging. The Pauli matrices are introduced using the method of Fermi. Spin and orbital moments are introduced, together with the Zeeman energy and Lande-g factor. The Stern–Gerlach experiment is explained in both the Schrödinger and Heisenberg pictures. Covering ESR allows us to introduce the Rabi frequency and smoothly transitions to time-dependent perturbation theory.

Chapter 7 is on time-dependent perturbation theory and its applications to optics. It starts with the interaction between matter and radiation, time-dependent perturbation theory, and linear optics and ends with the laser.

Chapter 8 is the application of QM to molecules and a one-dimensional solid through the Kronig–Penny model. Going beyond most textbooks that are limited to a single atom, we introduce molecular orbitals, orbital hybridization, and charge density in atoms. We work out all the details of the Kronig–Penny model, obtain the famous equation for the energy band, and show how the band structure is constructed.

Chapter 9 is on crystalline solids. It starts with crystal structure and the Bloch theorem and ends with quantum transport and collective vibrations (phonons). It covers the nearly free-electron model, where the Brillouin zone, momentum matrix elements, and susceptibility are introduced, and the tight-binding model, where both the Berry connection and position operator are computed. The Landauer theory for quantum transport, which is missing from almost every QM book, is introduced.

Chapter 10 is on some special topics. It starts from the particle indistinguishability and ends on the Berry phase. It covers many-body states, fermions, bosons, spin singlet and triplet, magnetism, and the Heisenberg model. Also, it introduces quantum computing, in particular, qubits, gates, and hardware.

We have tested our current version in our respective universities. The overwhelming feedback from the students is that they want to have more exercises, so at least one class session in each chapter is devoted to numerical solutions. We have rederived and checked the equations in the book and adopted the best practices in the literature, which is greatly enriched by our own research on ultrafast dynamics in molecules and solids, femtomagnetism, quantum materials and information, nonlinear optics and harmonic generation, magnetic materials, and nanostructures.

In the U. S., one semester has 15 weeks, meets three times per week for 50 minutes, for a total 45 classes. A quarter system has three 10-week sessions. In China, one semester has 18 weeks, four classes per week. Each class is 45 minutes. We design the book in such a way that one can use it for either two semesters or one semester or 2–3 quarters. The table on pages XVIII–XX has a column that lists the sections of a one-semester course.

We would like to thank our students for their feedback on the draft. We also appreciate the helpful communication with Dr. Hadiseh Alaeian (Purdue University) on quantum information. We humbly recognize the formidable challenge of writing a book on a discipline as fundamental, broad and far-reaching as quantum mechanics, where we are only scratching the surface of various aspects. We sincerely welcome criticism, comments, and suggestions so that we can improve the text in the future.

Although we have cited many excellent references, we should mention that our book has been significantly influenced by

1. Books that can be used as a reference for every chapter:
   (a) D. J. Griffiths and D. F. Schroeter, Introduction to Quantum Mechanics, Third Ed., Cambridge University Press (2018).
   (b) J. Y. Zeng, Quantum Mechanics, in Chinese, Scientific Publishing, Beijing (1984).
   (c) J. J. Sakurai, Modern Quantum Mechanics, ed. by S. F. Tuan, Pearson (2005).
   (d) P. A. M. Dirac, The Principles of Quantum Mechanics, Oxford University Press (1958).
   (e) L. I. Schiff, Quantum Mechanics, McGraw-Hill, New York (1955).
2. Books that are ideal for several specific chapters:
   (a) J. S. Bell, Speakable and Unspeakable in Quantum Mechanics, Cambridge University Press, Cambridge, (1987). This is a collection of Bell's papers on quantum mechanics, including his inequality. It is ideal for Chapters 1 and 10.

(b) E. Fermi, Notes on Quantum Mechanics, The University of Chicago Press, (1961). Being able to read what Fermi wrote when he was teaching students allows one to see how detailed he was, down to real numbers in the transition rate. It is ideal for Chapters 2 and 7.
(c) (i) N. W. Ashcroft and N. D. Mermin, Solid State Physics, Cengage Learning; 1st Ed. (1976). (ii) C. Kittel, Introduction to Solid State Physics, 7th Ed., John Wiley & Sons, Inc., New York (1996). (iii) K. Huang, Solid State Physics, in Chinese, Peking University Press (2009). (iv) S. H. Simon, The Oxford Solid State Basics, Oxford (2017). These are ideal for Chapters 8 and 9.
(d) X. Sun, Solitons and Polarons in Conjugated Polymers, in Chinese, Sichuan Publishing House, (1987). This book is ideal for Chapters 8 and 9. It has a concise presentation of one-dimensional systems.
(e) M. A. Nielsen and I. L. Chuang, Quantum Computation and Quantum Information, Cambridge University Press (2000). This is arguably the best book on quantum computing and information and an a key reference for Chapter 10. Its presentation is accessible to readers of various backgrounds.
(f) Spin-first books. (i) D. H. McIntyre, Quantum Mechanics, Pearson (2012), where a correction to the Schrödinger equation is necessary. (ii) J. S. Townsend, A Modern Approach to Quantum Mechanics, University Science Books, New York (2012). These books are used for Chapter 6.
(g) Additional books. (i) L. Pauling and E. B. Wilson, Introduction to Quantum Mechanics, McGraw-Hill Book Company, Inc., New York (1935). (ii) L. D. Landau and E. M. Lifshitz, Quantum Mechanics: Non-Relativistic Theory. Vol. 3 (3rd Ed.), Pergamon Press (1958). (iii) Claude Cohen-Tannoudji, Bernard Diu and Franck Laloe, Quantum Mechanics, Wiley-VCH (2019).

3. Books with problems, solutions, and videos:
(a) F. Constantinescu and E. Magyari, Problems in Quantum Mechanics, Pergamon Press, Oxford (1971). This has lots of excellent problems.
(b) B. C. Qian and J. Y. Zeng, Problems and Analysis in Quantum Mechanics, in Chinese, China Science Publishing and Media Ltd., Beijing, China (2001).
(c) J.-L. Basdevant and J. Dalibard, The Quantum Mechanics Solver, Springer-Verlag, Berlin, Heidelberg (2006). A good option, but less systematic.
(d) S. Flügge, Practical Quantum Mechanics, Springer (1974).
(e) M. Belloni, W. Christian, and A. J. Cox, Physlet Quantum Physics, Pearson Prentice Hall, New Jersey (2006).
(f) J. F. Wang, X. H. Fang and C. J. Zhang, Concepts and Problems in Solid State Physics, in Chinese, Shangdong University Press (2010).

Terre Haute, Indiana — Guo-ping Zhang
Lanzhou — Mingsu Si
St. Louis, Missouri — Thomas F. George
December 2023

Table 1: Outline of chapters, with key concepts, formulas, experiments, and applications, Sem. denotes the one-semester course, which only needs to cover these sections.

Chapter	Key concepts	Formulas	Experiments	Sem.
1	Energy quantum, de Broglie wavelength λ_{dB}, wavefunction $\|\psi\rangle$, Born's probability density and statistics, probability amplitude, measurement, operator and Hermitian operator, expectation value.	$E_n = nh\nu, \lambda_{dB} = h/p$; $\langle\psi\|\psi\rangle = \int \psi^*(\mathbf{r})\psi(\mathbf{r})d\mathbf{r} = 1, \rho = \|\psi\|^2$, $\rho = \|\psi_1 + \psi_2\|^2$ versus $\rho = \|\psi_1\|^2 + \|\psi_2\|^2$; $\hat{\mathbf{p}} = -i\hbar\nabla, \hat{H} = \frac{\hat{\mathbf{p}}^2}{2m} + \hat{V}(\mathbf{r}), \hat{A}^\dagger = \hat{A}$, $\langle\hat{A}\rangle \equiv \langle\psi\|\hat{A}\|\psi\rangle$.	blackbody radiation, Compton effect, one-photon interference, cold atoms, LED.	
2	Electron tunneling, Schrödinger equations, Current density, Liouville equation, infinite and finite quantum wells, stationary state evolution, tunneling through a barrier.	$i\hbar\frac{\partial\psi}{\partial t} = \hat{H}\psi, \frac{\partial\rho}{\partial t} + \nabla\cdot\mathbf{J} = 0, i\hbar\frac{\partial\rho}{\partial t} = [\hat{H},\rho]$; $\hat{H}\phi_n = E_n\phi_n, \langle\phi_n\|\phi_m\rangle = \delta_{nm}$, $\sum_n \|\phi_n\rangle\langle\phi_n\| = 1, \psi(t) = \sum_n c_n e^{-iE_nt/\hbar}\phi_n$; $E_n = \frac{\hbar^2}{2m}(\frac{n\pi}{a})^2, \phi_n(x) = \sqrt{\frac{2}{a}}\sin(\frac{n\pi x}{a})$; $\eta = \pm\omega\cos\eta, E_\eta = \frac{2\hbar^2\eta^2}{ma^2}$; $\phi_I(x) = e^{ikx} + re^{-ikx}, R + T = 1$, $T \propto e^{-2\sqrt{2m(V_0-E)}a/\hbar}$.	scanning tunneling microscope.	2.1–2.6
3	Nuclear vibration, commutation relation, position, momentum, Heisenberg uncertainty principle, Schrödinger and Heisenberg pictures, coherent states, black body radiation, photoelectric effect.	$[\hat{A},\hat{B}] = \hat{A}\hat{B} - \hat{B}\hat{A}, \Delta A\Delta B \geq \|\frac{i}{2}\langle[\hat{A},\hat{B}]\rangle\|$; $[\hat{x},\hat{p}_x] = i\hbar, [\hat{a},\hat{a}^\dagger] = 1$; $\hat{H} = (\hat{a}^\dagger\hat{a} + \frac{1}{2})\hbar\omega, E_n = (n + \frac{1}{2})\hbar\omega$, $\hat{a}^\dagger\|n\rangle = \sqrt{n+1}\|n+1\rangle$, $\hat{a}\|n\rangle = \sqrt{n}\|n-1\rangle, \phi_0 = (\frac{m\omega}{\pi\hbar})^{\frac{1}{4}}e^{-\frac{m\omega}{2\hbar}x^2}$; $\frac{d\hat{A}_H(t)}{dt} = \frac{1}{i\hbar}[\hat{A}_H(t),\hat{H}] + \frac{\partial\hat{A}_H(t)}{\partial t}$; $S_\nu = \frac{2h\nu^3}{c^2}\frac{1}{e^{h\nu/k_BT}-1}$.	blackbody radiation, photoelectric effect, infrared thermometers.	3.1–3.5
4	Orbital angular momentum, hydrogen atom, Bohr model, spatial (vector) quantization, common eigenfunctions, spherical harmonics, conservation law.	$a_0 = 0.529\,\text{Å}, \upsilon_1 = \alpha c, E_f - E_i = h\nu$; $[\hat{L}_x,\hat{L}_y] = i\hbar\hat{L}_z, \hat{\mathbf{L}}\times\hat{\mathbf{L}} = i\hbar\hat{\mathbf{L}}$, $\hat{\mathbf{L}}^2Y_{lm} = l(l+1)Y_{lm}, \hat{L}_zY_{lm} = m\hbar Y_{lm}$, $\hat{L}_\pm Y_{lm} = \sqrt{l(l+1) - m(m\pm1)}\hbar Y_{lm\pm1}$; $\phi_{nlm} = R_{nl}Y_{lm}, \phi_{1s} = \frac{1}{\sqrt{\pi a_0^3}}e^{-r/a_0}$, $E_n = -\frac{13.6\,\text{eV}}{n^2}$.	spectral series, Franck–Hertz experiment.	4.1–4.6
5	Nondegenerate and degenerate perturbation theory, variational principle, molecular orbitals.	$\hat{H} = \hat{H}_0 + \hat{H}_I, E_k^{(1)} = \langle\phi_k\|\hat{H}_I\|\phi_k\rangle$, $\|\psi_k\rangle = \|\phi_k\rangle + \sum_{m\neq k}\frac{\langle\phi_m\|\hat{H}_I\|\phi_k\rangle}{\epsilon_k-\epsilon_m}\|\phi_m\rangle$, $E_k^{(2)} = \sum_{m\neq k}\frac{\|\langle\phi_k\|\hat{H}_I\|\phi_m\rangle\|^2}{\epsilon_k-\epsilon_m}$; $F = \langle\psi\|\hat{H}\|\psi\rangle - \lambda\langle\psi\|\psi\rangle, \psi_k = \sum_{i=1}^{\nu}C_i^{(0)}\phi_i$, $\begin{pmatrix}H_{AA} & H_{AB}\\ H_{BA} & H_{BB}\end{pmatrix}\begin{pmatrix}C_A\\ C_B\end{pmatrix} = E\begin{pmatrix}1 & S\\ S^* & 1\end{pmatrix}\begin{pmatrix}C_A\\ C_B\end{pmatrix}$, $\hat{H}' = \lambda^2\hat{T}' + \lambda\hat{V}', \psi(\mathbf{r}_1',\cdots,\mathbf{r}_N') =$ $\lambda^{-3N/2}\psi(\mathbf{r}_1,\cdots,\mathbf{r}_N)$.	van der Waals force, Stark effect.	5.1–5.4

Table 1 (continued)

Chapter	Key concepts	Formulas	Experiments	Sem.
6	Electron spin, Pauli matrices, spin wavefunction, total angular momentum, orbital and spin moments, Zeeman energy, Landé-g factor, Hund's rules, Schrödinger and Heisenberg pictures, Rabi frequency, nuclear resonance.	$\hat{\sigma}_x = \begin{pmatrix} 0 & 1 \\ 1 & 0 \end{pmatrix}, \hat{\sigma}_y = \begin{pmatrix} 0 & -i \\ i & 0 \end{pmatrix}, \hat{\sigma}_z = \begin{pmatrix} 1 & 0 \\ 0 & -1 \end{pmatrix}$; $\hat{\mathbf{S}} = \frac{\hbar}{2}\hat{\boldsymbol{\sigma}}, [\hat{S}_x, \hat{S}_y] = i\hbar\hat{S}_z, \hat{\mathbf{S}} \times \hat{\mathbf{S}} = i\hbar\hat{\mathbf{S}}$, $\lvert\alpha\rangle = \begin{pmatrix} 1 \\ 0 \end{pmatrix}, \lvert\beta\rangle = \begin{pmatrix} 0 \\ 1 \end{pmatrix}; [\hat{J}_x, \hat{J}_y] = i\hbar\hat{J}_z$, $\hat{\mathbf{J}} \times \hat{\mathbf{J}} = i\hbar\hat{\mathbf{J}}; {}^{2s+1}L_j, \hat{\boldsymbol{\mu}}_l = -\mu_B\frac{\hat{\mathbf{L}}}{\hbar}$, $\hat{\boldsymbol{\mu}}_s = -\mu_B\frac{2\hat{\mathbf{S}}}{\hbar}, \hat{U} = -\hat{\boldsymbol{\mu}} \cdot \mathbf{B}$; $g = 1 + \frac{j(j+1)+s(s+1)-l(l+1)}{2j(j+1)}$, $S = s_1 + s_2, \ldots, \lvert s_1 - s_2\rvert$, $M_S = S, S-1, \ldots -S, J =$ $L+S, L+S-1, \ldots, \lvert L-S\rvert$; $\hat{H}(t) = \begin{pmatrix} \hbar\omega_1 & \gamma e^{i\omega t} \\ \gamma e^{-i\omega t} & \hbar\omega_2 \end{pmatrix}, \Omega = \sqrt{\frac{\delta^2}{4} + \frac{\gamma^2}{\hbar^2}}$; $\hat{U} = -\hat{\boldsymbol{\mu}} \cdot \mathbf{B} = -\gamma\hat{\mathbf{I}} \cdot \mathbf{B}$.	Stern–Gerlach experiment, Zeeman effect, ESR, MRI.	6.1–6.4
7	Interaction between radiation and matter, dipole selection rules, mechanical momentum, canonical momentum, generalized Lorentz force, Coulomb gauge, dipole approximation, transition matrix element, $\sigma^{\pm}$ light, 1st- order wavefunction correction, constant and periodic perturbation, transition probability, linear optics, polarization and susceptibility, Beer–Lambert law, penetration depth, fluence, absorption coefficient, laser, absorption, stimulated emission, population inversion, three-level and four-level systems.	$i\hbar\frac{\partial\psi(\mathbf{r},t)}{\partial t} = [\frac{1}{2m}(\hat{\mathbf{p}} - q\mathbf{A}(\mathbf{r},t))^2 + q\varphi]\psi(\mathbf{r},t)$, $\hat{\boldsymbol{\Pi}} = \hat{\mathbf{p}} - q\mathbf{A}(\mathbf{r},t), \nabla \cdot \mathbf{A} = 0$; $\hat{H}_I = -\hat{\mathbf{D}} \cdot \mathbf{E}(t) = e\mathbf{r} \cdot \mathbf{E}, \Delta l = \pm 1$, $\Delta m_l = 0, \pm 1; \hat{H} = \hat{H}_0 + \lambda\hat{H}_I(t)$, $\psi(t) = \psi^{(0)}(t) + \lambda\psi^{(1)}(t) + \lambda^2\psi^{(2)}(t) + \cdots$, $i\hbar\dot{\psi}^{(N)} = \hat{H}_0\psi^{(N)} + \hat{H}_I\psi^{(N-1)}, C^{(1)}_{lk}(t) =$ $\frac{1}{i\hbar}\int_{t_0}^{t}\langle\phi_l\lvert\hat{H}_I(t')\rvert\phi_k\rangle e^{i(E_l-E_k)t'/\hbar}dt'$, $P_{lk}(t) = \frac{1}{\hbar^2}\lvert\int_{t_0}^{t}\langle\phi_l\lvert\hat{H}_I(t')\rvert\phi_k\rangle e^{i\omega_{lk}t}dt'\rvert^2$; $\mathbf{P} = \mathbf{P}_0 + \epsilon_0\chi_e^{(1)} : \mathbf{E} + \epsilon_0\chi_e^{(2)} :: \mathbf{EE} + \cdots$, $\mathbf{P}^{(1)}_{\text{ind}} = \epsilon_0\chi_e^{(1)}\mathbf{E}$, $\chi^{(1)}_{\alpha,\beta}(\omega) = \frac{1}{\hbar\epsilon_0\Omega}\sum_n\frac{D^{\alpha}_{kn}D^{\beta}_{nk}}{\omega_{nk}-\omega-i\Gamma_{nk}}$, $I = \frac{1}{2}c\epsilon_0 n(\omega)\lvert\mathbf{E}\rvert^2, I(z) = n(\omega)I_0e^{-\alpha z}$, $\alpha(\omega) = \frac{4\pi n_i(\omega)}{\lambda_0} = \frac{4\pi}{\lambda_0}\mathrm{Im}(\sqrt{1+\chi_e^{(1)}(\omega)})$, $I = \frac{1}{2}c\epsilon_0\lvert\mathbf{E}_{\text{tot}}\rvert^2 =$ $\frac{1}{2}c\epsilon_0(\lvert\mathbf{E}_1\rvert^2 + \lvert\mathbf{E}_2\rvert^2 + 2\mathbf{E}_1 \cdot \mathbf{E}_2), \lambda_n = \frac{L}{2n}$; $\lambda = \frac{hc}{\Delta E} = \frac{hc}{E_b - E_a}$.	Linear and nonlinear optics, laser, optical absorption.	7.1–7.3

Table 1 (continued)

Chapter	Key concepts	Formulas	Experiments	Sem.
8	Molecular orbitals, π- and σ-orbitals, bonding, hybridization, Pauli exclusion principle, conjugated polymers, polyacetylene, LCAO, overlap and hopping integrals, charge density, molecular energy, Kronig–Penny model.	$\psi_{ss\sigma} = \frac{1}{\sqrt{2}}(\phi_s^{\mathrm{at}}(A) + \phi_s^{\mathrm{at}}(B))$, $\psi_{pp\pi} = \frac{1}{\sqrt{2}}(\phi_{p_z}^{\mathrm{at}}(A) + \phi_{p_z}^{\mathrm{at}}(B))$, $\psi_{sp^2} = c_1\phi_s^{\mathrm{at}}(A) + c_2\phi_{p_x}^{\mathrm{at}}(B) + c_3\phi_{p_y}^{\mathrm{at}}(B)$, $H_{p_z,p_z}(C_1, C_2) =$ $\int \phi_{p_z}^{\mathrm{at}*}(\mathbf{r}_{C_1})\hat{H}_{\mathrm{mol}}\phi_{p_z}^{\mathrm{at}}(\mathbf{r}_{C_2})d\mathbf{r}$, $\hat{H} = \sum_i [E_s\|i\rangle\langle i\| - t(\|i\rangle\langle i+1\| + \|i+1\rangle\langle i\|)]$, $\frac{\beta^2-\alpha^2}{2\alpha\beta}\sin(\alpha a)\sinh(\beta b) +$ $\cos(\alpha a)\cosh(\beta b) = \cos[k(a+b)]$.	organic LED.	8.1–8.3
9	Crystal structure, unit cell, primitive cell, lattice constant, Block theorem, translational symmetry, Born–von Karman boundary condition, reciprocal lattice, Brillouin zone, crystal momentum, band structure, band gap, momentum, group velocity, nearly free-electron model, tight-binding model, LCAO, Pauli exclusion principle, Fermi energy, Fermi surface, Fermi velocity, chemical potential, quantum transport, Landauer's principle, density of states, transmission, I-V curve.	$\mathbf{R} = l\mathbf{a}_1 + m\mathbf{a}_2 + n\mathbf{a}_3$, $\psi_{\mathbf{k}}(\mathbf{r}+\mathbf{R}) = e^{i\mathbf{k}\cdot\mathbf{R}}\psi_{\mathbf{k}}(\mathbf{r})$, $\psi_{\mathbf{k}}(\mathbf{r}) = e^{i\mathbf{k}\cdot\mathbf{r}}u_{\mathbf{k}}(\mathbf{r}), \mathbf{G} = l\mathbf{b}_1 + m\mathbf{b}_2 + n\mathbf{b}_3$; $E_\pm = V_0 + T_n(\Delta^2 + 1 \pm \sqrt{4\Delta^2 + \frac{\|V_n\|^2}{T_n^2}})$, $\frac{1}{m^*} = \frac{1}{\hbar^2}\frac{\partial^2 E_k}{\partial k^2}$; $\psi_{\mathbf{k}}(r) = \frac{1}{\sqrt{N}}\sum_l e^{i\mathbf{k}\cdot\mathbf{R}_l}\phi_s^{\mathrm{at}}(\mathbf{r}-\mathbf{R}_l)$, $E_{\mathbf{k}} = E_s^{\mathrm{at}} - \sum_l t(\mathbf{R}_l)e^{i\mathbf{k}\cdot\mathbf{R}_l}, v_k = \frac{\partial E_k}{\hbar\partial k}, \hat{H} =$ $\sum_s t_0\|s\rangle\langle s\| - t\sum_s(\|s+1\rangle\langle s\| + \|s\rangle\langle s+1\|)$, $\langle\psi_{k_1}\|\hat{x}\|\psi_{k_2}\rangle = \frac{a}{1-e^{-i\Delta ka}} = \frac{a}{2} - \frac{ai}{2}\cot\frac{\Delta ka}{2}$; $k_F = \frac{N_e\pi}{2L}, (\frac{k_x a}{2})^2 + (\frac{k_y a}{2})^2 = 2 + \varepsilon_{\mathbf{k}}$; $f_{FD}(E_{\mathbf{k}}) = \frac{1}{e^{\beta(E_{\mathbf{k}}-\mu)}+1}, f_{BE} = \frac{1}{e^{\beta E_k}-1}$, $I = \frac{e}{h}\int_{-\infty}^{\infty} dE T(E)[f_1(E) - f_2(E)]$, $G_0 = \frac{e^2}{h}, \rho(E) = A\sum_n \frac{1}{\sqrt{(E-E_n)^2+\delta^2}}$, $\ln\frac{I(z_1)}{I(z_2)} = -S_{\mathrm{BR}}\sqrt{\phi}(z_1 - z_2)$; $\omega_k = 2\sqrt{\frac{C}{m}}\|\sin(ka/2)\|$, $\hat{H} = \sum_k(\hat{a}_k^\dagger\hat{a}_k + \frac{1}{2})\hbar\omega_k$.	X-ray diffraction, photoemission, STM, infrared and Raman spectrum	9.1–9.4
10	Many-body states, symmetric and antisymmetric wavefunctions, particle indistinguishability, bosons, fermions, Pauli exclusion principle, spin singlet and triplet, qubits, Hadmard gate, Bloch sphere, $\boxed{\mathrm{CNOT}}$, Berry phase, geometrical phase, Berry connection, exchange interaction, Heisenberg model.	$\Psi(1,2) = \pm\Psi(2,1)$, $\chi_{0,0} = \frac{1}{\sqrt{2}}[\alpha(1)\beta(2) - \alpha(2)\beta(1)]$; $E_C = \frac{e^2}{4\pi\epsilon_0}\int_1\int_2 \rho_a(1)\rho_b(2)r_{12}^{-1}d\mathbf{r}_1 d\mathbf{r}_2$, $E_X = \frac{e^2}{4\pi\epsilon_0}\int_1\int_2 \rho_{\mathrm{ex}}(1)\rho_{\mathrm{ex}}(2)r_{12}^{-1}d\mathbf{r}_1 d\mathbf{r}_2$, $\hat{H} = -\sum_{ij}J_{ij}\hat{\mathbf{S}}_i\cdot\hat{\mathbf{S}}_j$; $\|\psi(t)\rangle = \cos[\frac{\theta(t)}{2}]\|0\rangle + e^{i\varphi(t)}\sin[\frac{\theta(t)}{2}]\|1\rangle$, $\boxed{\mathrm{H}}, \boxed{\mathrm{H}}, \boxed{\mathrm{S}}, \boxed{\mathrm{X}}, \boxed{\mathrm{Y}}, \boxed{\mathrm{Z}}$; $\mathbf{A}_n(\mathbf{R}) = i\langle n(\mathbf{R})\|\nabla_{\mathbf{R}}\|n(\mathbf{R})\rangle, \gamma_n(C) =$ $\oint_C d\mathbf{R}\cdot\mathbf{A}_n(\mathbf{R}) = i\oint_C\langle n(\mathbf{R})\|\nabla_{\mathbf{R}}\|n(\mathbf{R})\rangle\cdot d\mathbf{R}$.	Magnetism, superconductivity, Josephson junction, Paul trap, quantum cryptography, photon entanglement, polarization, nonlinear optics.	10.1–10.3

Contents

Preface — VII

1 Entering the Quantum World — 1
1.1 Energy quantization — **1**
1.2 Wave-particle duality and the de Broglie wavelength — **2**
1.3 Quantum states and their wavefunctions — **4**
1.3.1 Describing a quantum state: wavefunction — **4**
1.3.2 Types of wavefunctions and Dirac notation — **5**
1.3.3 How does a wavefunction describe wave–particle duality? — **6**
1.4 Properties of a wavefunction — **6**
1.4.1 Probability density and the normalization of a wavefunction — **6**
1.4.2 Comparison between Born statistics and Maxwell–Boltzmann statistics — **7**
1.4.3 Hilbert space and the superposition principle of wavefunctions — **9**
1.5 Operators of observables and measurements — **13**
1.5.1 Expectation values of an operator — **13**
1.5.2 Hermitian operators — **13**
1.5.3 Measurements — **17**
1.6 Measuring Planck's constant using LEDs — **18**
1.7 Problems — **19**

2 Schrödinger equation: from bound and unbound states to quantum tunneling — 22
2.1 Time-dependent Schrödinger equation — **22**
2.1.1 Time symmetry and the derivation of time-dependent Schrödinger equation — **22**
2.1.2 Current density and continuity equation — **24**
2.2 Time-independent Schrödinger equation and eigenstates — **27**
2.2.1 Separation of time and space variables and properties of eigenstates — **28**
2.2.2 Stationary state evolution — **29**
2.3 Particle in a box: bound states — **31**
2.3.1 Classical treatment — **31**
2.3.2 Potential sets boundary conditions for wavefunctions — **32**
2.3.3 Finding a solution that matches the boundary conditions — **33**
2.3.4 Eigenenergies and eigenstates — **33**
2.3.5 Time-dependent state evolution: stationary-state evolution — **37**
2.4 The finite square well: bound and unbound states — **40**
2.4.1 Finding boundary conditions for wavefunctions — **41**
2.4.2 Wavefunctions in three regions — **42**
2.5 Eigenenergies and eigenstates in the finite quantum well — **43**
2.5.1 Eigenenergies and boundary conditions — **43**

2.5.2 Eigenstates and physical meanings — **45**
2.5.3 $E > V_0$ and unbound states — **48**
2.6 Tunneling through a barrier: unbound states — **48**
2.6.1 Boundary conditions — **49**
2.6.2 Wavefunctions — **50**
2.6.3 Imposing boundary conditions and transmission and reflection — **51**
2.7 Physical insights into quantum tunneling — **52**
2.7.1 Wavefunctions of continuous states and transmission — **52**
2.7.2 Quantum tunneling — **54**
2.7.3 Current — **55**
2.8 Problems — **56**

3 Harmonic oscillator and blackbody radiation — 58
3.1 Operators and Heisenberg's uncertainty principle — **58**
3.1.1 Basic properties of operators — **58**
3.1.2 Heisenberg uncertainty principle — **60**
3.1.3 Ladder operators — **62**
3.2 Harmonic oscillator — **64**
3.2.1 Classical physics — **64**
3.2.2 Quantum theory for a harmonic oscillator — **65**
3.2.3 Eigenvalues and eigenfunctions — **66**
3.3 Applications of ladder operators in position and momentum — **70**
3.3.1 Basic properties of $\hat{a}$ and $\hat{a}^\dagger$ — **70**
3.3.2 Expectation values of $\hat{x}$ and $\hat{x}^2$ — **71**
3.3.3 Expectation values of $\hat{p}_x$ and $\hat{p}_x^2$ and the uncertainty principle — **72**
3.4 Time evolution in harmonic oscillator: Schrödinger picture — **73**
3.4.1 Stationary solutions — **73**
3.4.2 Examples — **74**
3.5 Heisenberg's picture and Heisenberg equation of motion — **77**
3.5.1 Heisenberg equation of motion — **77**
3.5.2 Examples — **80**
3.6 Blackbody radiation — **81**
3.6.1 Failure of classical physics — **82**
3.6.2 Quantum theory: importance of quantization — **84**
3.6.3 Insights — **86**
3.7 Photoelectric effects — **86**
3.7.1 Experiment — **87**
3.7.2 Einstein's theory of the photoelectric effect — **88**
3.8 Problems — **90**

4 Orbital angular momentum and hydrogen atom — 93
4.1 Bohr model for the hydrogen atom: a bridge from CM to QM — **93**

4.1.1 Energy — **94**
4.1.2 Spatial quantization of angular momentum — **95**
4.1.3 Bohr radius — **95**
4.1.4 Velocity and period — **96**
4.1.5 Eigenenergies and their connections to the spectral lines — **96**
4.2 Commutable operators and their common eigenfunctions — **98**
4.2.1 Nondegenerate eigenfunctions — **98**
4.2.2 Degenerate eigenfunctions — **99**
4.3 Orbital angular momentum — **100**
4.3.1 Orbital angular momentum operators: $\hat{\mathbf{L}}$, $\hat{\mathbf{L}}^2$, $\hat{L}_z$, $\hat{L}_+$, and $\hat{L}_-$ — **100**
4.3.2 Eigenfunctions of $\hat{\mathbf{L}}^2$ are spherical harmonics Y_{lm} — **102**
4.3.3 Eigenfunction of $\hat{L}_z$ — **103**
4.3.4 Using the angular momentum ladder operators to find Y_{lm} — **104**
4.4 Understanding orbital angular momentum — **108**
4.4.1 Physical meanings: magnitude of a quantum vector — **108**
4.4.2 Matrix representations and representation theory — **110**
4.4.3 Chemical meanings: atomic orbitals — **112**
4.5 Central field potentials: symmetry and conservation law — **115**
4.5.1 Symmetry and conservation — **115**
4.5.2 Rotation of coordinates — **115**
4.5.3 Rotation of a wavefunction in the function space — **116**
4.5.4 Rotation of an operator — **116**
4.5.5 Conservation of angular momentum — **117**
4.6 Hydrogen atom — **118**
4.6.1 Spherical symmetry — **118**
4.6.2 Separation of variables and the radial Schrödinger equation — **119**
4.7 Numerical solution of the hydrogen atom: hydrogen code — **121**
4.7.1 Atomic units — **121**
4.7.2 Discretization of the eigenequation — **122**
4.7.3 Hamiltonian matrix and its eigenstates on the grid mesh — **123**
4.7.4 Hydrogen-like ions: ionic radii and Alzheimer's disease — **125**
4.8 Franck–Hertz experiment — **126**
4.8.1 Experiment — **127**
4.8.2 Verification of quantum theory — **128**
4.9 Problems — **130**

5 Time-independent approximate methods — 133
5.1 Nondegenerate perturbation theory — **133**
5.1.1 Expanding the energy and wavefunction — **133**
5.1.2 Zeroth-order perturbation — **134**
5.1.3 First-order perturbation — **134**
5.1.4 Second-order correction — **136**

5.2 Examples of nondegenerate perturbation theory — **137**
5.2.1 Charged harmonic oscillator in a weak electric field **F** — **137**
5.2.2 van der Waals force — **139**
5.3 Degenerate and nearly-degenerate perturbation theory — **142**
5.3.1 Expanding the wavefunction and energy — **143**
5.3.2 Zeroth-order correction — **144**
5.3.3 First-order correction — **144**
5.4 Examples of degenerate perturbation theory — **146**
5.4.1 A simplified approach to the H_2^+ molecule — **146**
5.4.2 Stark effect in hydrogen atom — **147**
5.5 Variational principle — **150**
5.5.1 Functional derivatives — **150**
5.5.2 Spatial scaling as a variational parameter: the virial theorem — **152**
5.6 Examples of the variational principle in the H_2^+ molecule — **153**
5.6.1 First variation — **153**
5.6.2 Second variation — **155**
5.6.3 Variational minimization of the molecular energy — **159**
5.6.4 Molecular orbitals — **160**
5.7 Problems — **162**

6 Electron spin — 165
6.1 Discovery of electron spin and representation of spin states — **165**
6.1.1 Stern–Gerlach experiment: spin is another degree of freedom — **165**
6.1.2 Pauli matrices — **166**
6.1.3 Spin matrices — **169**
6.1.4 Total wavefunction of an electron — **172**
6.2 Total angular momentum $\hat{\mathbf{J}}$ — **172**
6.2.1 Quantum number addition and spectroscopic notation — **172**
6.2.2 Spin-orbit coupling in a central field — **174**
6.2.3 Eigenfunctions and eigenvalues of $\hat{\mathbf{J}}$ and $\hat{J}_z$ — **175**
6.3 Interaction of spin and orbital moments with a magnetic field — **177**
6.3.1 Magnetic orbital and spin moments — **177**
6.3.2 Interaction with the **B** field — **179**
6.3.3 Derivation of Lande-g factor — **180**
6.4 Explanation of Stern–Gerlach experiment and measurement — **181**
6.4.1 Schrödinger picture — **183**
6.4.2 Heisenberg picture — **183**
6.4.3 Quantum measurement — **184**
6.5 Zeeman effect — **187**
6.5.1 Experiment — **187**
6.5.2 Theory and Hund's rules — **189**
6.6 Electron spin resonance (ESR) — **194**

6.6.1 Rotating wave approximation and Rabi Hamiltonian — **195**
6.6.2 Rabi model — **196**
6.6.3 Rabi frequency — **197**
6.7 Medical applications: magnetic resonance imaging — **198**
6.7.1 Interaction between a nuclear spin and magnetic field — **198**
6.7.2 Spatial resolution — **200**
6.8 Problems — **201**

7 Time-dependent perturbation theory: application to optics — 204
7.1 Interaction between radiation and matter and its selection rules — **204**
7.1.1 Quantization of interaction — **205**
7.1.2 Coulomb gauge and dipole approximation — **206**
7.1.3 Dipole selection rules — **207**
7.2 Time-dependent perturbation theory — **209**
7.2.1 Zeroth order — **210**
7.2.2 First order — **211**
7.3 Examples of time-dependent perturbation theory — **212**
7.3.1 $\hat{H}_I(t)$ is time independent — **212**
7.3.2 $\hat{H}_I(t)$ is a periodic perturbation — **213**
7.3.3 Excitation of a charged harmonic oscillator by a pulse — **215**
7.4 Linear optics: polarization and susceptibility — **216**
7.4.1 Classical optics — **216**
7.4.2 Quantum optics — **217**
7.4.3 Linear susceptibility: specialized to state $|k\rangle$ — **218**
7.4.4 Experimental connection: penetration depth and Beer–Lambert law — **220**
7.5 Laser — **222**
7.5.1 Absorption and emission — **222**
7.5.2 Population inversion and stimulated emission — **223**
7.6 Problems — **225**

8 Electrons in molecules and one-dimensional solids — 228
8.1 Molecular orbitals — **228**
8.1.1 σ-orbitals — **229**
8.1.2 π-orbitals — **230**
8.1.3 Orbital hybridization and linear combination of atomic orbitals — **230**
8.2 Molecular Hamiltonian, overlap and hopping integrals — **232**
8.3 Application to small molecules: the allyl radical — **234**
8.3.1 Occupying the molecular orbitals and many-body wavefunctions — **235**
8.3.2 Charge density — **236**
8.3.3 Dirac notation — **238**
8.4 Applications to conjugated polymers: polyacetylene — **239**
8.5 One-dimensional solid: Kronig–Penney model — **241**

8.5.1 Region 1: $-b < x < 0$ — **242**
8.5.2 Region 2: $0 < x < a$ — **242**
8.5.3 Bloch theorem — **242**
8.5.4 Region 3: $a < x \leq a + b$ — **243**
8.5.5 Derivation of the secular equation — **243**
8.5.6 Band structure — **245**
8.6 Problems — **248**

9 Electrons in crystalline solids — 250
9.1 Crystal structure — **250**
9.1.1 Unit cell and primitive cell — **250**
9.1.2 Basis and lattice points — **251**
9.1.3 Primitive lattice vectors and lattice vectors — **252**
9.1.4 Bravais lattice and lattice constant — **252**
9.2 Translational symmetry and Bloch theorem — **252**
9.2.1 Bloch theorem, Born–von Karman boundary condition and Bloch waves — **252**
9.2.2 Reciprocal lattice vectors and Brillouin zones — **255**
9.3 Nearly free-electron model — **256**
9.3.1 Nondegenerate perturbation treatment — **257**
9.3.2 Degenerate perturbation and energy band gap — **262**
9.4 Tight-binding model — **265**
9.4.1 Three-dimensional systems — **265**
9.4.2 One-dimensional ring — **270**
9.5 Occupying the energy band: Fermi surface and Fermi velocity — **272**
9.5.1 Fermi energy in metals, semiconductors and insulators — **272**
9.5.2 Fermi surface — **274**
9.5.3 Chemical potential — **276**
9.6 Transport on a nanometer scale — **277**
9.6.1 Minimum conductance and von Klitzing resistance constant — **278**
9.6.2 Landauer theory: conductance is transmission — **280**
9.6.3 Device's density of states (DOS) — **283**
9.7 Scanning tunneling microscope — **286**
9.7.1 Principle of the STM — **286**
9.7.2 How STM gains atomic resolution and the Binnig–Rohrer factor — **287**
9.8 Lattice vibrations — **288**
9.8.1 Normal modes — **288**
9.8.2 Phonons — **292**
9.9 Problems — **293**

10 Special topics: many-body systems, magnetism, and quantum information — 295
10.1 Many-body systems under various approximations — **295**
10.1.1 Born–Oppenheimer approximation — **295**
10.1.2 Independent electron approximation — **296**
10.1.3 Hartree–Fock approximation and density functional theory — **297**
10.2 Particle indistinguishability in many-body systems — **298**
10.2.1 Identical particles and entanglement — **298**
10.2.2 Bosons — **300**
10.2.3 Fermions and the Pauli exclusion principle — **300**
10.3 Spin singlet and triplet and spatial entanglement — **302**
10.3.1 Singlet and triplet — **302**
10.3.2 Entanglement between spin and spatial spaces — **303**
10.4 Kinetic, Coulomb and exchange correlation energies — **305**
10.4.1 Single-particle operators: kinetic energy — **305**
10.4.2 Coulomb and exchange correlation energies — **307**
10.5 Origin of magnetism and the Heisenberg model — **309**
10.6 Quantum information technology — **311**
10.6.1 Quantum bits or qubits — **312**
10.6.2 Single qubit and gates — **313**
10.6.3 More than one qubit and entanglement — **315**
10.6.4 Hardware and quantum bits — **317**
10.7 Berry phase — **320**
10.7.1 Dynamical phase factor with time-independent Hamiltonian — **320**
10.7.2 Time-dependent Hamiltonian and transition between states — **321**
10.7.3 Adiabatic approximation — **321**
10.7.4 Geometric phase factor: a conceptual change — **323**
10.7.5 Berry phase — **326**
10.8 Problems — **329**

11 Appendix — 332
11.1 Atomic units — **332**
11.2 Time-order operator — **332**
11.3 Spin rotation — **333**
11.3.1 Polar vectors — **333**
11.3.2 Axial vectors: spin rotation and SU(2) rotation matrices — **334**
11.3.3 $O(3)$ rotation matrices — **335**
11.3.4 What about spin eigenfunctions? — **336**
11.4 Spin matrices for photon and useful math formulas — **336**
11.5 Computer codes — **339**

Bibliography —— 361

Index —— 365

1 Entering the Quantum World

By the end of 19th century, classical mechanics, thermodynamics and statistical mechanics, and electricity and magnetism formed the three main pillars of physics. They could successfully explain the majority of phenomena observed. Classical mechanics (CM) can explain the motion of particles such as planets and mechanical waves, with continuous energies. Thermodynamics and statistical mechanics can explain physical properties of gases, liquids, and solids. Electricity and magnetism (EM) can explain electric current, light absorption, emission, and reflection. But there were some outliers that do not seem to fit any one of them. The emission spectra from atoms are found to be discrete, not continuous, as would be expected from CM. When the light intensity becomes extremely weak, EM cannot explain the intrinsic fluctuation of light. Blackbody radiation proves to be a major challenge. It is a common experience that the color of a heated iron stove transitions from red, to yellow, and to white, but never to blue or violet. Using EM theory and statistical mechanics helps to explain the increased emission at a shorter wavelength, i. e., the ultraviolet catastrophe. As a result, it is apparent that the entire physics needs a radical revolution.

This chapter is grouped into three units:

- Unit 1 consists of Sections 1.1 and 1.2, which introduces energy quantization, wave-particle duality, and the de Broglie wavelength.
- Unit 2 includes Sections 1.3 and 1.4. These two sections form the core of this chapter: quantum states and wavefunctions.
- Unit 3 includes Sections 1.5 and 1.6. It starts with operators and expectation values and concludes with an elegant experiment to measure the Planck constant.

1.1 Energy quantization

In 1900, in an effort to explain blackbody radiation (Section 3.6), Max Planck postulated that there exists a minimum energy unit, $h\nu$, which is indivisible, that he called a bundle or energy quantum. One can have only either zero energy or integer multiples of this minimum energy,

$$E_n = nh\nu, \tag{1.1.1}$$

where h is the Planck constant, $h = 6.626176\times10^{-34}$ J·s or 4.136 eV·fs, ν is the frequency of radiation, and n is the quantum number and only takes discrete integers, $n = 0, 1, 2, \ldots$. The system energy E_n has a subscript n. Quantum mechanics (QM) revolutionized 20th century science.

Different from CM, QM connects the energy with the frequency of the radiation, not the wavelength. This embodies the significance on the frequency of radiation.

https://doi.org/10.1515/9783110672152-001

Equation (1.1.1) is only for blackbody radiation. The electron in a hydrogen atom has energy of $E_n = \frac{-13.6\,\text{eV}}{n^2}$, where n takes $1, 2, 3, \ldots$, not from 0.

In the words of Planck: "And now that the way was opened, a sudden flood of new-won knowledge poured out over the whole field including the neighboring fields in physics and chemistry." For instance, to describe its electron orbitals in chemistry, we need to introduce the orbital angular momentum quantum number l to label the s, p, and d orbitals, $l = 0, 1$, and 2, respectively.

Exercise 1.1.0.

1. Max Planck in his Nobel lecture stated: "In this case the quantum of action must play a fundamental role in physics, and here was something entirely new, never before heard of, which seemed called upon to basically revise all our physical thinking, built as this was, since the establishment of the infinitesimal calculus by Leibniz and Newton, upon the acceptance of the continuity of all causative connections." To illustrate this difference between classical and quantum physics, calculate the two functions given by

$$\text{Newton}(q) = \frac{\int_{n=0}^{\infty} n\exp(-nq)dn}{\int_{n=0}^{\infty} \exp(-nq)dn}; \quad \text{Planck}(q) = \frac{\sum_{n=0}^{\infty} n\exp(-nq)}{\sum_{n=0}^{\infty} \exp(-nq)},$$

where q is positive. In Newton(q), n is continuous; in Planck(q), n is a positive integer. (a) Find Newton(q) and Planck(q). (b) Show at what value of q, Planck(q) approaches Newton(q). (c) Plot them on the same graph.

1.2 Wave-particle duality and the de Broglie wavelength

Most of us are used to viewing matter as behaving as either waves or particles. Waves interfere with each other, but particles do not. Particles can bounce off each other, but waves do not. This section introduces a radical concept: Matter has both wave and particle properties.

In 1905, in an effort to explain photoelectric effects (Section 3.7), Albert Einstein applied Planck's energy quantization idea to a light wave. He introduced a new concept, the light quantum, and in 1926 G. N. Lewis [1] coined the word "photon", a particle. Each photon has energy

$$E = h\nu \quad \text{(photon)}, \tag{1.2.1}$$

where ν is the light's frequency. According to Einstein's general relativity, the energy E of a free object is

$$E = \sqrt{p^2c^2 + m^2c^4}, \tag{1.2.2}$$

where p is the momentum, c is the speed of light, and m is the mass of the object. We apply eq. (1.2.1) to a light wave, whose static mass m is zero, so that we have

$$E = pc. \tag{1.2.3}$$

The two energies in eqs. (1.2.1) and (1.2.3) must be equal to each other, so we obtain $pc = h\nu$ and $p = \frac{h\nu}{c}$. Because the wavelength of light $\lambda = c/\nu$, we have $p = h/\lambda$, or

$$\lambda = \frac{h}{p} \quad \text{(for light wave).} \tag{1.2.4}$$

This relationship shows that photons have momentum just like regular particles, which already differs from EM. The photon concept enabled Einstein to explain the photoelectric effect, which will be addressed in Chapter 3. In 1923, A. Compton [2] demonstrated that a beam of X-ray incident on an electron, once it has collided with the electron, loses its momentum and has a longer wavelength, thus proving that light does behave like a particle.

In 1924, Louis de Broglie made a revolutionary postulate: The particle nature of a wave should be extended to all matter, not just light; a massive particle should have the wave nature. This is the wave–particle duality. The spatial extension of the quantum object is described by the de Broglie wavelength λ_{dB},

$$\lambda_{\mathrm{dB}} = \frac{h}{p} \quad \text{(for any quantum object),} \tag{1.2.5}$$

where p is the momentum. From that point on, any quantum object was understood to have dual properties. A particle must be drawn with a wavy head and tail, so the reader with the classical physics background can comprehend the concept. Figure 1.1 is a quantum illustration of a matter, a wavepacket, where the dot in the center represents the classical location. The significance of wavepackets will become clear in Section 3.1.2. This equation applies to both radiation and matter and sets a spatial limit as to whether the wave nature of a quantum object is significant. If λ_{dB} is comparable to the system size, the wave nature of an object appears. For instance, if an electron is confined in a space whose dimension is comparable to its de Broglie wavelength, the electron's wave nature, such as diffraction and interference, appears [3]. Electron diffraction experiments [4] use this property. Diffraction has been observed in C_{60} and C_{70} molecules [5, 6, 7].

In another limit, if the dimension of a system is far larger than λ_{dB}, the particle nature emerges. To get a rough idea as to how large λ_{dB} is, we look at the electron in a hydrogen atom. The electron momentum p is 1.9924×10^{-24} kgm/s, which gives $\lambda_{\mathrm{dB}} =$

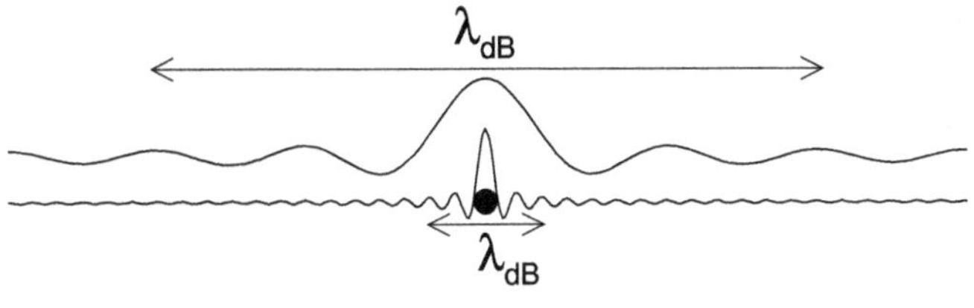

Figure 1.1: de Broglie's wave-particle duality is characterized by the de Broglie's wavelength λ_{dB}. The longer (the shorter) is λ_{dB}, the wavepacket behaves more like a wave (a particle).

3.33 Å. One can recognize that, in order to see the wave nature of an electron, the spatial dimension should be very small. In cold-atom experiments, one cools atoms down to μK so that their p is much smaller to get a measurable λ_{dB}. Consider sodium atoms at 1 μK, with velocity centered around $v_{th} = \sqrt{3k_BT/m} = 0.0329$ m/s. We find $p = mv_{th} = 1.25696 \times 10^{-27}$ kgm/s, and $\lambda_{\text{dB}} = 5271.14$ Å. This means that these Na atoms behave like waves at such a low temperature [8].

Exercise 1.2.0.

2. A baseball of mass 140 g rests on a table at room temperature. Can its de Broglie wavelength be infinite?
3. A Ti-sapphire laser has a wavelength of 800 nm. Find the photon's momentum, energy, and frequency in a vacuum.

1.3 Quantum states and their wavefunctions

In CM, position and velocity define what state the system is in. For instance, a car is traveling at 40 m/s and 1 km away from a city. QM uses the quantum state.

1.3.1 Describing a quantum state: wavefunction

In 1926, Erwin Schrödinger proposed a revolutionary concept: a mathematical entity called a wavefunction, $\psi(\mathbf{r}, t)$, to describe the quantum state of a system. $\psi(\mathbf{r}, t)$ encapsulates the de Broglie wavelength mathematically. Schrödinger introduced an equation that bears his name: the Schrödinger equation for $\psi(\mathbf{r}, t)$, which finally established the main framework of wave mechanics and put QM on a solid mathematical footing.

$\psi(\mathbf{r}, t)$ is a continuous function of the spatial coordinate $\mathbf{r}$ and time t and contains the complete information of a quantum state. In general, ψ is a complex function, $\psi^* \neq \psi$, where "*" represents a conjugation, meaning changing the sign of the imaginary part of a complex number like $\psi = a + ib \rightarrow \psi^* = a - ib$ with a and b being real. If the wavefunction is for an electron, then $\mathbf{r}$ is the electron's coordinate. If it is for a light photon, then $\mathbf{r}$ is the photon coordinate. $\psi(\mathbf{r}, t)$ is called the position $\mathbf{r}$-representation. If we use momentum $\mathbf{p}$ instead of position $\mathbf{r}$ as our spatial variable, this is called the momentum $\mathbf{p}$-representation. The wavefunction looks like $\psi(\mathbf{p}, t)$, These two representations are equivalent. The $\mathbf{p}$-representation is often used for delocalized states, with the $\mathbf{r}$ one for localized states. If we have N particles, we write down the many-body wavefunction as $\psi(\mathbf{r}_1, \mathbf{r}_2, \mathbf{r}_3, \ldots, \mathbf{r}_N, t)$, where $\mathbf{r}_i$ are the coordinates of the particles.

1.3.2 Types of wavefunctions and Dirac notation

There are two different types of wavefunctions: single-component and multi-component ones. Here, we only introduce a single-component wavefunction such as $\psi(\mathbf{r}, t)$, leaving two-component ones to Chapter 6. To describe different wavefunctions easily, Dirac introduced angle brackets: $\langle|$ and $|\rangle$ are called `bra` and `ket`, respectively. This is the Dirac notation. Here, we write $\psi(\mathbf{r}, t)$ as a ket, $|\psi\rangle \equiv \psi(\mathbf{r}, t)$. Its complex conjugate, $\psi^*(\mathbf{r}, t)$, is a bra, $\langle\psi| \equiv \psi^*(\mathbf{r}, t)$. The outer product of two wavefunctions ψ_1 and ψ_2^* is written as $|\psi_1\rangle\langle\psi_2| \equiv \psi_1(\mathbf{r}, t)\psi_2^*(\mathbf{r}, t)$. The inner product, an integral, is $\langle\psi_1|\psi_2\rangle = \int \psi_1^*(\mathbf{r}, t)\psi_2(\mathbf{r}, t)d\mathbf{r}$. However, we will put off the product of $\psi_1\psi_2$ (see eq. (8.3.7)).

We can always discretize $\mathbf{r}$ in $\psi(\mathbf{r}, t)$, as $\mathbf{r}_1, \mathbf{r}_2, \ldots$, called the mesh grid points (see Section 4.7.2), in which case $\psi(\mathbf{r}, t)$ becomes a column vector

$$\begin{pmatrix} \psi(\mathbf{r}_1, t) \\ \psi(\mathbf{r}_2, t) \\ \vdots \end{pmatrix}. \tag{1.3.1}$$

We take the first two entries $\psi(\mathbf{r}_1, t)$ and $\psi(\mathbf{r}_2, t)$ and denote them as a and b for simplicity. In Dirac notation, we write the wavefunction and its conjugation as

$$|\psi\rangle = \begin{pmatrix} a \\ b \end{pmatrix}, \quad \langle\psi| = \begin{pmatrix} a^* & b^* \end{pmatrix}. \tag{1.3.2}$$

This conjugation is a `conjugate transpose`, `Hermitian conjugate`. The outer and inner products of two wavefunctions are, respectively,

$$|\psi_1\rangle\langle\psi_2| = \begin{pmatrix} a_1 \\ b_1 \end{pmatrix} \begin{pmatrix} a_2^* & b_2^* \end{pmatrix} = \begin{pmatrix} a_1 a_2^* & a_1 b_2^* \\ b_1 a_2^* & b_1 b_2^* \end{pmatrix}, \tag{1.3.3}$$

$$\langle\psi_2|\psi_1\rangle = \begin{pmatrix} a_2^* & b_2^* \end{pmatrix} \begin{pmatrix} a_1 \\ b_1 \end{pmatrix} = a_2^* a_1 + b_2^* b_1. \tag{1.3.4}$$

Equations (1.3.1)–(1.3.4) make it clear that, once we discretize wavefunctions, the Schrödinger equation becomes an eigenequation of a matrix, such as $A\psi = \lambda\psi$, where A is a matrix and λ is an eigenvalue. This mechanics is called matrix mechanics, proposed by Heisenberg, Born, and Jordan, which is completely equivalent to the wave mechanics proposed by Schrödinger.

Exercise 1.3.2.

4. Given $|\psi_1\rangle = \frac{1}{\sqrt{2}}\binom{1}{1}$, $|\psi_2\rangle = \frac{1}{\sqrt{2}}\binom{1}{i}$, find (a) $\langle\psi_1|$ and $\langle\psi_2|$, (b) $|\psi_1\rangle\langle\psi_2|$ and $|\psi_2\rangle\langle\psi_1|$, (c) $\langle\psi_1|\psi_1\rangle$ and $\langle\psi_2|\psi_2\rangle$, (d) $\langle\psi_1|\psi_2\rangle$ and $\langle\psi_2|\psi_1\rangle$, (e) $|\psi_1\rangle\langle\psi_2|\psi_1\rangle$ and $|\psi_2\rangle\langle\psi_1|\psi_2\rangle$.

1.3.3 How does a wavefunction describe wave–particle duality?

A true wave must have a wavelength, while a true particle must have a position. As a conceptual introduction but at the risk of causing confusion, we give an example of $\psi(\mathbf{r})$ in one dimension, where we suppress the time variable for clarity. It has two parts like

$$\psi(x) = \underbrace{u(x)}_{\text{particle}} \underbrace{\exp(ikx)}_{\text{wave}}, \tag{1.3.5}$$

where $u(x)$ is a spatially localized function and represents the particle nature and $\exp(ikx)$ represents the wave nature. We consider two cases:

First, we set $u(x) = 1$, so $\psi = \exp(ikx)$. The wave vector k is related to the wavelength through $\lambda = 2\pi/k$, where λ is just λ_{dB}. This is a pure wave. Second, we choose $u(x) = \exp(-\frac{x^2}{a^2})$, so $\psi = \exp(-\frac{x^2}{a^2} + ikx)$. Now, we choose a so small that ψ peaks at $x = 0$. This is a particle. If $a \to 0$, this is a pure particle that is located at $x = 0$.

We emphasize that our separation into the particle and wave parts in eq. (1.3.5) is only approximate. In general, the wavefunction falls in between these two extremes.

1.4 Properties of a wavefunction

A wavefunction ψ allows us to describe a quantum state, but it does not have a physical meaning because it is not a physical quantity that can be measured experimentally.

1.4.1 Probability density and the normalization of a wavefunction

Schrödinger already recognized that only $\psi(\mathbf{r},t)\psi^*(\mathbf{r},t)$ has a physical meaning. Max Born provided a statistical interpretation and called it the probability density $\rho(\mathbf{r},t)$, i. e., the probability per unit length, area or volume. This means that, classically, if we have 100 % chance to find an electron in a volume of $1\,\mathrm{m}^3$, the probability density is $\rho = \frac{1}{\mathrm{m}^3}$; if 20 %, then $\rho = \frac{0.2}{\mathrm{m}^3}$. In QM, ρ is identified with $\psi(\mathbf{r},t)$ at space $\mathbf{r}$ and time t,

$$\rho(\mathbf{r},t) = \psi(\mathbf{r},t)\psi^*(\mathbf{r},t) = |\psi(\mathbf{r},t)|^2. \tag{1.4.1}$$

If we multiply ρ by the volume element *dxdydz* (the left-hand figure in Fig. 1.2), then we obtain the probability in *dxdydz*. The SI unit of ρ is 1/m, 1/m^2, and 1/m^3 in one, two, and three dimensions, respectively. Wavefunction $\psi(\mathbf{r},t)$ is now called the probability amplitude. Different from the classical statistics, "probability density" is *not stochastic or random* because, once $\psi(\mathbf{r},t)$ is given, $\rho(\mathbf{r},t)$ at every point in space and time is known without any ambiguity. For a single particle, the spatial integration over the density $\rho(\mathbf{r},t)$ must be 1 because we only have one particle,

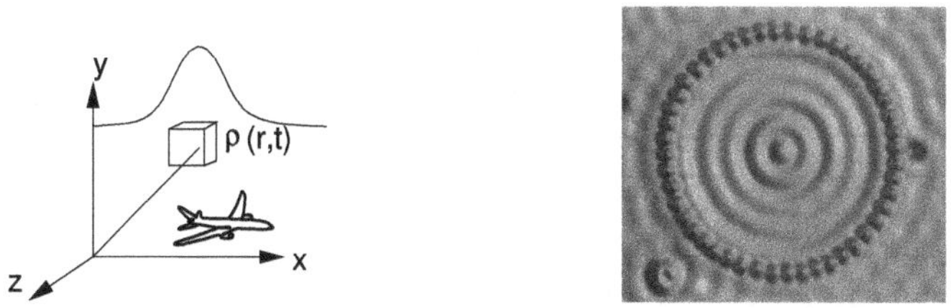

Figure 1.2: (Left) Quantum probability density $\rho(\mathbf{r},t)$. (Right) Experimental scanning tunneling microscope spatial image of the surface states of the electron in a quantum corral of 48 Fe atoms [9]. Used with permission from AAAS.

$$\int_V \rho(\mathbf{r},t)d\mathbf{r} = \int_V \psi^*(\mathbf{r},t)\psi(\mathbf{r},t)d\mathbf{r} \equiv \langle\psi|\psi\rangle = 1, \tag{1.4.2}$$

where the integration is a three-dimensional integral over a volume V and $d\mathbf{r}$ is the infinitesimal volume element, $d\mathbf{r} = dxdydz$. Equation (1.4.2) is called the normalization of the wavefunction. In Dirac notation, the integral must have the bra $\langle|$ first and then the ket $|\rangle$. This integration becomes a summation if the wavefunction is defined on real mesh grids (see eq. (1.3.1)). Suppose the wavefunction $\psi(\mathbf{r},t)$ has values at mesh points (x_i, y_j, z_k). Then the normalization is the triple summation,

$$\sum_{i,j,k} \rho_{i,j,k} = \sum_{i,j,k} \psi^*(x_i,y_j,z_k,t)\psi(x_i,y_j,z_k,t) = 1, \tag{1.4.3}$$

where x_i, y_j, z_k represent a single point in space. As a result of normalization, $\psi(\mathbf{r},t)$ multiplied by a nonzero constant c is still considered to be the same as $\psi(\mathbf{r},t)$ and represents the same state. Only properly normalized wavefunctions can be used to compute ρ and any other properties. For an unnormalizable wavefunction, see Section 2.6.

1.4.2 Comparison between Born statistics and Maxwell–Boltzmann statistics

Now, we explain the difference and similarity between Born statistics and Maxwell–Boltzmann statistics (MB statistics). The simplest way to understand Born statistics is to apply it to a classical particle, e. g., an airplane (left-hand figure in Fig. 1.2).

Consider the plane moving with a constant velocity v on a straight (1D) runway of length a. According to Born statistics, we need to find the probability density ρ in the line segment dx. We first compute the total time spent on the runway, $T = a/v$, and then find the time dt spent in dx. The ratio dt/T is the probability (not probability density),

$$\frac{dt}{T} = \frac{dx/v}{T} = \frac{dx}{vT} = \frac{dx}{a} = \rho(x)dx, \tag{1.4.4}$$

where the last equation employs the definition of probability density. So, we find $\rho(x) = 1/a$ in units of 1/m, which is constant everywhere, like a wave. This makes sense because its velocity is constant. To see whether $\rho(x)$ is indeed a correct probability density, we integrate it over space, $\int_0^a \rho(x)dx = 1$. Sure enough, it satisfies the normalization condition. This reveals the first key difference: Born statistics uses a single sample, or a single event, to find the probability density.

What about its average position? Born statistics uses the same equation as MB statistics: the integration over the product of the probability density and the position

$$\bar{x} = \int_0^a \rho(x)xdx = a/2, \tag{1.4.5}$$

which means the average position of the plane is in the middle of the runway. This also makes sense. We can do the same for the average value of x^2:

$$\bar{x^2} = \int_0^a \rho(x)x^2dx = a^2/3. \tag{1.4.6}$$

Now, we can compute the uncertainty in position using the standard deviation from classical physics, $\Delta x = \sqrt{\bar{x^2} - (\bar{x})^2} = \frac{a}{\sqrt{12}}$. The position uncertainty highlights that the plane behaves like a wave. Although our example is classical, once we accept Born statistics, we can see how the wave nature also appears through the uncertainty in position.

What if the plane stops at $x = b$ on the runway? We assume that $b < a$. Then, our probability must be 1 at $x = b$, and the rest is zero. At $x = b$, $dx = 0$, so $\rho = 1/dx = \infty$. This singular distribution is uneasy for many, so Dirac invented his δ-function,

$$\int_0^a \rho(x)dx = \int_{-\infty}^{+\infty} \delta(x-b)dx = 1, \tag{1.4.7}$$

where the second equation extends the limit to infinity and $\rho(x) = \delta(x-b)$. We can use ρ to compute both $\bar{x}$ and $\bar{x^2}$ as follows: $\int \rho(x)xdx = b$ and $\int \rho(x)x^2dx = b^2$. $\Delta x = \sqrt{\bar{x^2} - (\bar{x})^2} = 0$, i. e., zero uncertainty in position: The plane is a particle.

Now, we explain MB statistics. MB statistics needs a large number of planes, i. e., many events, so the distribution function follows the so-called MB distribution ρ_{MB}. If not, then the distribution depends on the number of planes, or events. This is where the stochasticity or randomness appears. The resultant distribution takes into account the fact that some planes may start at $x = 0$, while others may not. The runway length a "distributes" as $\rho_{\mathrm{MB}}(a)$. The final average position has an extra average over ensembles,

$$\overbrace{\int \rho_{\mathrm{MB}}(a) \underbrace{\int_0^a \rho(x)xdx}_{\text{Born}}\, da}^{\text{Maxwell–Boltzmann}}, \tag{1.4.8}$$

which is experimentally measured. We highlight the explicit difference between MB statistics and Born statistics through the underbrace and overbrace. This is the second key difference: In Born statistics, the system size and the boundary conditions must be fixed from the beginning, but, in MB statistics, they are allowed to change.

These classical examples compare and contrast Born and MB statistics. In QM, the entire procedure is similar but with two differences. Instead of using dt/T followed by a division by dx, we compute ρ from ψ. The average value is now called the expectation value (Section 1.5.1). Although the probability density provides the density value at a spatial point, it is incorrect to say, for example, that a quantum object has 20 % chance in region A, 30 % in B, and 50 % in C. We must accept the quantum object as a whole in the same way as we view a classical object. We cannot say this plane has 60 % chance being a fuselage and 15 % being a wing. The wings and fuselage are just parts of the plane. We must accept the entire spatial distribution of the quantum state as a whole.

If we have N particles, the normalization of the wavefunction is

$$\int_V \rho(\mathbf{r}_1, \mathbf{r}_2, \ldots, \mathbf{r}_N; t) d\mathbf{r}_1 d\mathbf{r}_2 \cdots d\mathbf{r}_N = N. \tag{1.4.9}$$

If we multiply the density ρ with the charge, we have the charge density. Experimentally, Crommie and coworkers probed the confined electron density from 48 Fe atoms on the Cu surface (right-hand figure in Fig. 1.2).

Exercise 1.4.2.

5. $\phi(x) = \sin(kx)$. Normalize this wavefunction within $x \in [-a/2, a/2]$.

6. If a plane starts from rest and accelerates with acceleration a on the runway of length L: (a) Find $\rho(x)$; (b) find the average position $\bar{x}$; (c) find $\overline{x^2}$; (d) find Δx.

1.4.3 Hilbert space and the superposition principle of wavefunctions

In CM, any vector **A** in Cartesian coordinates (Fig. 1.3(a)) can be written as a linear superposition of three components as

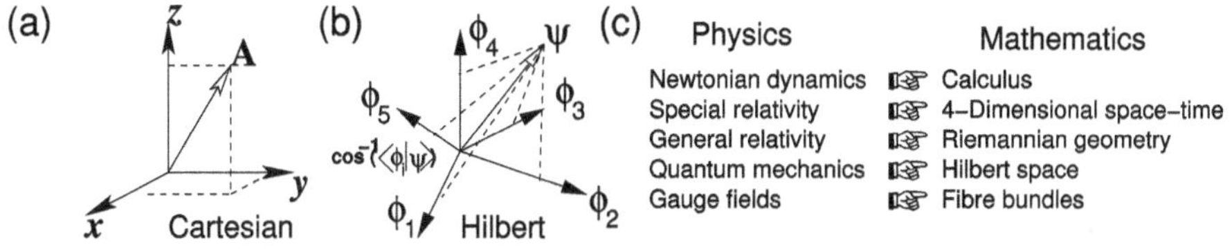

Figure 1.3: (a) Cartesian coordinates, where a vector **A** can be projected to x-, y-, and z-axis. (b) In the Hilbert space, basis functions ϕ_n are the “axes”, upon which a function ψ can be “projected” through an integration $\langle \phi_n | \psi \rangle$. (c) Physics and mathematics are connected.

$$\mathbf{A} = A_x\hat{x} + A_y\hat{y} + A_z\hat{z}, \tag{1.4.10}$$

where A_x, A_y, and A_z are three components along the x-, y-, and z-axes, respectively. This is possible as long as three unit vectors, $\hat{x}$, $\hat{y}$, and $\hat{z}$, are orthogonal to each other.

In QM, the Cartesian space is not enough, since we work with wavefunctions, not vectors alone. We need to extend it to the function space, called the Hilbert space, which are set up by a group of basis functions. Basis functions are a special group of wavefunctions and play the same role as unit vectors in the Cartesian coordinate. If two basis functions $|\phi_i\rangle$ and $|\phi_j\rangle$ have a zero inner product, i. e., $\langle\phi_i|\phi_j\rangle = 0$, they are defined as orthogonal to each other. Figure 1.3(b) schematically shows five basis functions. If we can find a group of basis functions $\{|\phi_i\rangle\}$ that are orthogonal to each other and complete (Section 2.2.1), we can expand any wavefunction $|\psi\rangle$ as

$$|\psi\rangle = \sum_i c_i|\phi_i\rangle, \tag{1.4.11}$$

which is called the superposition of wavefunctions as $\{\phi_i\}$ are linearly added up. $c_i = \langle\phi_i|\psi\rangle$ is a complex number and the coefficient just as A_x, A_y, and A_z in eq. (1.4.10). A key difference from vectors in the Cartesian coordinates is that $|\psi\rangle$ must be properly normalized. $|c_i|^2 = |\langle\phi_i|\psi\rangle|^2$ is called the probability density in basis functions $|\phi_i\rangle$. Figure 1.3(c) illustrates how the novel physics is built upon different mathematical foundations.

The superposition principle allows us to understand the one-photon interference experiment [10]. This topic often confuses the beginner: How could a single photon, a particle, interface with itself? First, we are using the wave nature of the photon to generate the interference. Because the coherent length, the length that the photon still keeps its coherence, is long, the phase of the photon wavefunction is not randomized during light propagation. Second, the interference is observed because we alter the photon wavefunction. The same thing can be said about the single electron interference. In this case, we need a device, often a beam splitter, to split its wavefunction to generate an interference pattern. Figure 1.4 schematically shows the Mach–Zehnder interferometer, where the incoming light is split into two arms and then rejoins at the exit. A single photon in state $|s\rangle$ hits on a beam splitter B_1 and splits into two paths: one along the

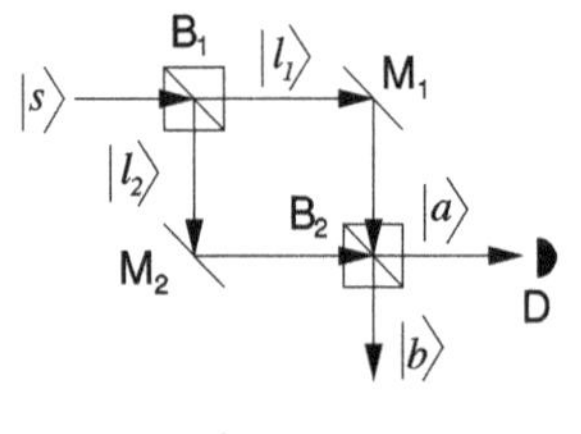

Figure 1.4: Single-photon interference with a Mach–Zehnder interferometer. A single photon emitted from the source in a state $|s\rangle$ impinges on a beam splitter B_1 and splits into two states $|l_1\rangle$ and $|l_2\rangle$. Then the photon is reflected from mirrors M_1 and M_2 before it hits on another beam splitter B_2. The photon is detected by an avalanche photodiode D in the output port a [10].

transmission direction and the other along the reflection direction, described by wavefunctions $|l_1\rangle$ and $|l_2\rangle$. Since they are spatially separated, their overlap is zero, $\langle l_1|l_2\rangle = 0$, i. e., orthogonal. $|l_1\rangle$ and $|l_2\rangle$ must be indistinguishable to the photon, so the same single photon can take "both paths", which appears surprising if one views the photon as a particle only. But this is not surprising at all because it is the single-photon wavefunction that spreads over both paths (see below for more). The level of the indistinguishability is given by the transmission probability amplitude t and the reflection probability amplitude r, so B_1 transforms the photon from $|s\rangle$ into

$$|s\rangle = t|l_1\rangle + r|l_2\rangle, \tag{1.4.12}$$

where $|s\rangle$ corresponds to $|\psi\rangle$ in eq. (1.4.11) and $|l_1\rangle$ and $|l_2\rangle$ correspond to $|\phi_i\rangle$. $|s\rangle$ is a superposition state of $|l_1\rangle$ and $|l_2\rangle$. The normalization $\langle s|s\rangle = 1$ requires $rr^* + tt^* \equiv R + T = 1$, where $T = tt^*$ and $R = rr^*$ are the probabilities in $|l_1\rangle$ and $|l_2\rangle$, respectively.

Before we proceed, we must clarify a few common confusions. Conceptually, t and $|l_1\rangle$ are both probability amplitudes as they are multiplied by each other, and the same is true for r and $|l_2\rangle$. If we left multiply eq. (1.4.12) by $\langle l_1|$, we get $\langle l_1|s\rangle = t\langle l_1|l_1\rangle + r\langle l_1|l_2\rangle = t$, which explains how one can obtain c_i from eq. (1.4.11). If, for a spatial reason or temporal reason or both, $|l_1\rangle$ and $l_2\rangle$ become inequivalent and distinguishable to the photon, the photon chooses only one path, either $|l_1\rangle$ or $|l_2\rangle$, not both, then $|s\rangle = |l_1\rangle$ or $|s\rangle = |l_2\rangle$, and there is no single-photon interference [10].

We assume that reflection from $M_{1(2)}$ does not change $|l_{1(2)}\rangle$, but the path length affects the phase because the actual photon wavefunction is approximately like e^{ikx}, where k is its wavevector and x is the position. If the path length for $|l_{1(2)}\rangle$ from B_1 to B_2 is $l_{1(2)}$, the phase is $e^{ikl_{1(2)}} \equiv e^{i\delta_{1(2)}}$. Once both have passed through B_2 into ports a and b, $|l_1\rangle$ and $|l_2\rangle$ themselves are superpositions of $|a\rangle$ and $|b\rangle$

$$|l_1\rangle = e^{i\delta_1}(r|a\rangle + t|b\rangle), \quad |l_2\rangle = e^{i\delta_2}(t|a\rangle - r|b\rangle), \tag{1.4.13}$$

where $|a\rangle$ and $|b\rangle$ are two new basis functions at ports a and b, i. e., $\langle a|b\rangle = 0$. In $|l_1\rangle$, r is before $|a\rangle$ and t is before $|b\rangle$ because the photon is reflected into port a by M_2 and is transmitted through M_2 into port b. $|l_{1(2)}\rangle$ itself now is a superposition of $|a\rangle$ and $|b\rangle$.

Substituting eqs. (1.4.13) into (1.4.12) results in

$$|s\rangle = rt(e^{i\delta_1} + e^{i\delta_2})|a\rangle + (t^2 e^{i\delta_1} - r^2 e^{i\delta_2})|b\rangle. \tag{1.4.14}$$

The avalanche photodiode D detects the signal at port a. Mathematically, we need to compute the projection of $|s\rangle$ on $|a\rangle$, $\langle a|s\rangle$, the probability amplitude. Its square, $\rho_a(\delta) = |\langle a|s\rangle|^2$, is the probability density that is measured by the detector D through $|a\rangle$,

$$\rho_a(\delta) = |\langle a|s\rangle|^2 = 2RT[1 + \cos(\delta)], \tag{1.4.15}$$

where $\delta = \delta_1 - \delta_2$. The interference emerges from $\rho_a(\delta)$ once we change the arm length of either path 1 or path 2. It disappears if two paths are distinguishable to the photon.

Table 1.1: First table: comparison between classical interference and quantum interference. Second table: terminology of wavefunction, probability density, and probability.

Physics	Minimum No. of waves for interference	Splitting
Classical	2	No
Quantum	1	Yes

Quantity	Name	Other name
$\psi(\mathbf{r}, t)$	probability density amplitude	wavefunction
$\rho(\mathbf{r}, t) = \lvert\psi(\mathbf{r}, t)\rvert^2$	probability density	
$\int \rho(\mathbf{r}, t) d\mathbf{r}$	probability	normalization
$\lvert\langle a\lvert s\rangle\rvert^2$	probability density of $\lvert s\rangle$ measured through $\lvert a\rangle$	

We can also use eq. (1.4.11) to represent a molecular wavefunction. Suppose the molecule has two atoms A and B, where the electronic wavefunction at atom A is $|\phi_{2s}\rangle$ and that at B is $|\phi_{2p_z}\rangle$. The wavefunction $|\Psi\rangle$ for the molecule is a superposition of $|\phi_{2s}\rangle$ and $|\phi_{2p_z}\rangle$

$$|\Psi\rangle = c_A|\phi_{2s}\rangle + c_B|\phi_{2p_z}\rangle. \tag{1.4.16}$$

The electron density in the Dirac notation is

$$\rho(\mathbf{r}) = |\Psi\rangle\langle\Psi| = |c_A|^2|\phi_{2s}\rangle\langle\phi_{2s}| + |c_B|^2|\phi_{2p_z}\rangle\langle\phi_{2p_z}| + 2\mathrm{Re}\left[c_B|\phi_{2p_z}\rangle c_A^*\langle\phi_{2s}|\right], \tag{1.4.17}$$

or in a regular notation is

$$\rho(\mathbf{r}) = \Psi(\mathbf{r})\Psi^*(\mathbf{r}) = \underbrace{|c_A|^2|\phi_{2s}(\mathbf{r})|^2}_{\text{density at atom 1}} + \underbrace{|c_B|^2|\phi_{2p_z}(\mathbf{r})|^2}_{\text{density at atom 2}} + \underbrace{2\mathrm{Re}\left[c_B\phi_{2p_z}(\mathbf{r})c_A^*\phi_{2s}^*(\mathbf{r})\right]}_{\text{hybridization}}. \tag{1.4.18}$$

The first and second terms on the right-hand side represent the electron densities at atoms A and B, respectively. The third term is the orbital hybridization, *sp* hybridization. The level of hybridization depends on the overlap between two wavefunctions. If the signs of c_A and c_B are the same, we have a bonding orbital, or otherwise, an antibonding orbital. Table 1.1 summarizes the key concepts that we have learned so far.

Exercise 1.4.3.

7. (a) Prove eq. (1.4.15). (b) If we place the photodector along the b port, find the probability $P_b(\delta)$.

8. Suppose $t = t_0 e^{i\alpha}$ and $r = r_0 e^{i\beta}$, where α, β, t_0, and r_0 are real. (a) Using the orthogonality relationship between $\langle l_1|l_2\rangle$ in eqs. (1.4.13), find the relationship between phases α and β. (b) Use (a) to prove $\langle s|s\rangle = 1$. (c) Prove $P_a(\delta) + P_b(\delta) = 1$. This means the number of photons must be 1 for a single-photon interference.

1.5 Operators of observables and measurements

Although ψ contains all the information about a system, ρ and the phase of ψ are the only ones that can be calculated and observed. To access other properties, one needs the operator of an observable, denoted as $\hat{O}$, where the hat on O is the unique symbol for an operator. The observable refers to the quantity that can be computed theoretically and may be measured experimentally. For instance, if we want to compute the energy of a system, we need the Hamiltonian operator $\hat{H}$. The momentum needs a momentum operator $\hat{\mathbf{p}}$, and so on. Therefore, except for the probability density and phase, a quantum measurement is composed of the wavefunction and operator. There are two types of measurements. Type I is based on an operator $\hat{O}$ and a wavefunction ψ, where ψ is not affected by $\hat{O}$. This is passive and is an ideal case. The final outcome is the so-called expectation value or something similar (such as transition matrix elements). Type II is also based on $\hat{O}$ and ψ, but ψ may be affected by $\hat{O}$ or other operators. This is related more to actual experimental measurements and causes considerable confusion. The single-photon interference in Section 1.4.3 is an example. Another example is the double-slit experiment, where the very action that one puts another camera immediately behind the slits changes ψ, so the outcome is necessarily different. At the end of this section, we will present a simple way to better understand this difference in measurements.

1.5.1 Expectation values of an operator

The expectation value is the outcome of our measurement (Type I). It is defined as

$$\langle\hat{O}\rangle \equiv \langle\psi|\hat{O}|\psi\rangle = \int \psi^* \hat{O}\psi d\mathbf{r}, \tag{1.5.1}$$

which is an integral of a product of $\psi^* \hat{O}\psi$ over the space. The order in eq. (1.5.1) matters: ψ^* is before $\hat{O}$, and ψ is after $\hat{O}$, so $\hat{O}$ operates on ψ only and their result is multiplied by ψ^*. One can think of the operator as a matrix, the wavefunction ψ as a column vector, and ψ^* as a row vector. We cannot swap $\hat{O}$ with ψ in general, but if they are, we recover the classical average value (for instance, eq. (1.4.5)). For this reason, we use the expectation value, instead of the average value, to distinguish it from CM. The particle nature of a matter, such as position, can now be revealed through the expectation value. The uncertainty in the wave nature is because $\langle\psi|\hat{O}^2|\psi\rangle \neq (\langle\psi|\hat{O}|\psi\rangle)^2$.

1.5.2 Hermitian operators

There are many operators of various kinds. We are mostly interested in Hermitian operators because they correspond to a physical quantity that can be measured experimentally. The Hermitian conjugation of an operator $\hat{O}$ consists of two operations: complex conjugation and transpose, i. e., $\hat{O}^\dagger \equiv (\hat{O}^*)^{\mathrm{T}}$, where $_{\mathrm{T}}$ indicates a transpose in the language of a matrix or an adjoint for a differential operator. The conjugation changes the sign of the imaginary part of the operator, and the transpose exchanges the upper triangle elements with the lower triangle element, while, if written as $\psi_1\hat{O}^{\mathrm{T}}\psi_2$, it means that $\hat{O}$ acts upon ψ_1, instead of ψ_2, i. e., $\psi_1\hat{O}^{\mathrm{T}}\psi_2 = (\hat{O}\psi_1)\psi_2$. If one has a product $\hat{O}|\psi\rangle$, its Hermitian conjugate is $(\hat{O}|\psi\rangle)^\dagger = \langle\psi|\hat{O}^\dagger$. If $|\psi\rangle$ is a column vector, $\langle\psi|$ is a row vector, with all the elements taking their Hermitian conjugates. If we have two operators $\hat{A}$ and $\hat{B}$, then their conjugation is $(\hat{A}\hat{B})^\dagger = \hat{B}^\dagger\hat{A}^\dagger$. If $\hat{O} = \hat{O}^\dagger$, $\hat{O}$ is called a Hermitian operator.

1.5.2.1 Momentum and position

In CM, the momentum is defined as $\mathbf{p} = m\mathbf{v}$ for a particle. We quantize it to a momentum operator in QM in the $\mathbf{r}$-representation as

$$\hat{\mathbf{p}} = -i\hbar\nabla = -i\hbar\left(\hat{x}\frac{\partial}{\partial x} + \hat{y}\frac{\partial}{\partial y} + \hat{z}\frac{\partial}{\partial z}\right), \tag{1.5.2}$$

where $\hbar$ is the reduced Planck constant $(h/2\pi)$, i is the imaginary symbol, and $\hat{x}, \hat{y}$, and $\hat{z}$ are the unit vectors along the x, y, and z directions, respectively. Its three components are $\hat{p}_x = -i\hbar\frac{\partial}{\partial x}$, $\hat{p}_y = -i\hbar\frac{\partial}{\partial y}$ and $\hat{p}_z = -i\hbar\frac{\partial}{\partial z}$. ∇ is a gradient operator, called `nabla`. Changing from classical $\mathbf{p}$ to quantum mechanical $\hat{\mathbf{p}}$ is a conceptual leap. $\hat{\mathbf{p}}$ has no mass in its operator and is the same for both electrons and photons, but its unit is exactly the same as the classical one. Different from CM, $\hat{\mathbf{p}}$ does not tell us anything about the momentum until it applies to a state or wavefunction to reveal its true value.

$\hat{\mathbf{p}}$ is a Hermitian operator, $\hat{\mathbf{p}}^\dagger = \hat{\mathbf{p}}$. Including $-i$ in $\hat{\mathbf{p}}$ is important because it ensures that the momentum expectation value is a real number (see the following example). QM does not have a specific velocity operator for all matter, but, for a massive object, the velocity operator can be written as $\hat{\mathbf{p}}/m$, where m is the mass of the object, a classical number.

The position operator $\hat{\mathbf{r}}$ is also a Hermitian operator. In the $\mathbf{r}$-representation $\hat{\mathbf{r}}$ is $\mathbf{r}$ (no hat). we can compute its expectation value similarly.

In the $\mathbf{p}$-representation, the spatial variable is $\mathbf{p}$, so $\hat{\mathbf{p}} = \mathbf{p}$, (no hat), but the position operator $\hat{\mathbf{r}}$ is changed to

$$\hat{\mathbf{r}} = i\hbar\nabla_{\mathbf{p}} = i\hbar\left(\hat{x}\frac{\partial}{\partial p_x} + \hat{y}\frac{\partial}{\partial p_y} + \hat{z}\frac{\partial}{\partial p_z}\right). \tag{1.5.3}$$

Example 1 (Expectation value of momentum). Suppose an electron moving along the z-axis, with the wavefunction normalized in a line segment within $z \in [0, L]$, $\phi(z) = \exp(ikz)/\sqrt{L}$. (a) Find the expectation value of $\hat{p}_z$. (b) Find the de Broglie wavelength.

(a) We can compute its expectation value of $\hat{p}_z$ as

$$\langle\hat{p}_z\rangle \equiv \langle\phi|\hat{p}_z|\phi\rangle = \int_0^L dz\frac{1}{\sqrt{L}}e^{-ikz}(-i\hbar)\frac{\partial}{\partial z}\frac{1}{\sqrt{L}}e^{ikz} = \hbar k.$$

(b) From eq. (1.2.5), we can compute the de Broglie wavelength $\lambda_{dB} = h/\langle \hat{p}_z \rangle = 2\pi/k$, which exactly matches our expectation, but this represents a conceptual revolution from a classical quantity p in eq. (1.2.5) to the quantum mechanical expectation value of an operator $\hat{p}_z$.

Example 2 (Expectation value of position). We have a wavefunction $\phi(x) = \delta(x-a)$, where $\delta(x)$ is a Dirac δ function.[1] Find the expectation value of the position operator.

We start from the definition of the expectation value of the position,

$$\langle \phi | \hat{x} | \phi \rangle = \int dx \delta(x-a) x \delta(x-a) = a.$$

This means that the particle is located at $x = a$.

1.5.2.2 Kinetic energy and potential energy

The classical kinetic energy is $\mathbf{p}^2/2m$. The kinetic energy operator $\hat{T}$ is

$$\hat{T} = \frac{\hat{\mathbf{p}}^2}{2m} = -\frac{\hbar^2 \nabla^2}{2m}, \tag{1.5.4}$$

where the negative sign does not mean that the kinetic energy is negative. It only means that the operator itself is negative. What matters physically is the expectation value of the kinetic energy, which must be positive. Because the kinetic energy operator is a second-order derivative with respect to the spatial coordinate, any physical wavefunction must not be a linear function of $\mathbf{r}$.

In the $\mathbf{r}$-representation, the potential energy operator is

$$\hat{V} = V(\mathbf{r}), \tag{1.5.5}$$

where $V(\mathbf{r})$ is the exact same as the classical potential. However, in the p-representation, the variable $\mathbf{r}$ is changed to $\mathbf{p}$.

1.5.2.3 Eigenfunctions, Hamiltonian and its eigenstates

If a Hermitian operator $\hat{A}$ and a wavefunction ϕ satisfy $\hat{A}\phi = a\phi$, ϕ is called the eigenfunction, or eigenvector of $\hat{A}$ in matrix mechanics, and a is its eigenvalue. The total energy operator, $\hat{H}$, called the Hamiltonian, of a system is the sum of $\hat{T}$ and $\hat{V}$,

$$\hat{H} = \hat{T} + \hat{V} = \frac{\hat{\mathbf{p}}^2}{2m} + V(\mathbf{r}).$$

Given a wavefunction $|\psi\rangle$, we can compute the expectation value of the energy,

$$\langle \psi | \hat{H} | \psi \rangle = \int_V \psi^*(\mathbf{r}) \hat{H} \psi(\mathbf{r}) d\mathbf{r}.$$

---

1 At $x = 0, \delta = \infty$. If $x \neq 0, \delta = 0$. It has some special properties: $\int_{-\infty}^{\infty} \delta(x)dx = 1$. $\int_{-\infty}^{\infty} f(x)\delta(x-a)dx = f(a)$.

If $\hat{H}$ and ϕ satisfy

$$\hat{H}\phi = E\phi, \tag{1.5.6}$$

ϕ is an eigenfunction of $\hat{H}$. Importantly, we further call ϕ an eigenstate, to distinguish it from other eigenfunctions. Only $\hat{H}$'s eigenfunction can be bestowed as an eigenstate, and E is called the eigenenergy. If $\hat{H}$ is changed, the system is changed.

1.5.2.4 Relativistic and photon Hamiltonians

For a free particle with a finite static mass, we can quantize the relativistic energy, eq. (1.2.2), to the relativistic Hamiltonian [11],

$$\hat{H} = c\sqrt{m^2c^2 + \hat{p}_x^2 + \hat{p}_y^2 + \hat{p}_z^2}. \tag{1.5.7}$$

If an object has zero static mass, one can construct a similar Hamiltonian from the classical Hamiltonian. For instance, the photon has a classical energy given by eq. (1.2.3), $E = pc$. The quantum Hamiltonian of photon [12] is

$$\hat{H} = c\boldsymbol{\alpha} \cdot \hat{\mathbf{p}}, \tag{1.5.8}$$

where $\boldsymbol{\alpha}$ is a 6×6 matrix (see Appendix Section 11.4). This dot product is necessary because $\hat{\mathbf{p}}$ is a vector, while the energy is a scalar. One can compute the eigenstate of the photon. A more rigorous approach is based on quantum field theory, where the photon polarization can be included, and the strength of the light field in classical EM, i. e., the vector potential, is replaced by the photon number operator. This enables one to describe emission, spontaneous emission, and absorption.

In summary, operators are a key feature of QM, very different from CM. Table 1.2 lists a few common operators in both **r**- and **p**-representations. Missing from the list is a force operator. Instead, in QM, we find the force operator $\hat{\mathbf{F}}$ through the time-derivative of $\hat{\mathbf{p}}$, or by taking the gradient of the Hamiltonian operator $\hat{H}$ with respect to position,

$$\hat{\mathbf{F}} = -\nabla\hat{H} = -\nabla V(\mathbf{r}, t), \tag{1.5.9}$$

Then we compute the expectation value of the force $\langle\psi|\hat{\mathbf{F}}|\psi\rangle$.

Table 1.2: Operators in the **r**- and **p**-representations.

Observable	Energy	Position	Momentum
Operator (**r**-representation)	$\hat{H} = -\frac{\hbar^2\nabla^2}{2m} + V(\mathbf{r}, t)$	$\hat{\mathbf{r}} = \mathbf{r}$	$\hat{\mathbf{p}} = -i\hbar\nabla$
Operator (**p**-representation)	$\hat{H} = -\frac{\mathbf{p}^2}{2m} + V(\mathbf{p}, t)$	$\hat{\mathbf{r}} = i\hbar\nabla_\mathbf{p}$	$\hat{\mathbf{p}} = \mathbf{p}$

Exercise 1.5.2.
9. Show that, if ψ is a real function, $\langle\psi|\hat{\mathbf{p}}|\psi\rangle = 0$.
10. Suppose $\phi(x) = \delta(x - a)$. Find $\langle x^2\rangle - (\langle x\rangle)^2$.
11. Suppose an electron moving along the x-axis, with the wavefunction normalized in a line segment within $x \in [0, L]$, $\phi(x) = \exp(ikx)/\sqrt{L}$. (a) Find the expectation value of $\hat{p}_x$. (b) In the $\mathbf{r}$-representation, the position operator is x. Find its expectation value. (c) Find the expectation value $\langle\phi|x^2|\phi\rangle$.
12. Assume the photon propagating along the z-axis. For simplicity, we assume $\boldsymbol{\alpha}$ is a unit vector $\mathbf{k}_x$ along the photon propagation x direction. Starting from eq. (1.5.8), find the eigenstate of the photon. Hint: Solve an equation like eq. (1.5.6).

1.5.3 Measurements

We take the electron single-slit experiment as an example. Figure 1.5 shows that the experiment consists of three parts: the electron source, the slit, and the screen or camera. First we consider the electron is far away from the slit. Here "far away" means that the distance between the electron and the slit is much larger than the electron's de Broglie wavelength, so our Hamiltonian is $\hat{H}_{\text{electron}}$. We can compute a physical quantity or the expectation value of an operator $\hat{O}$. This is called a type I measurement, where the measurement is passive and does not affect the quantum state of the system.

If the electron comes close to the slit, but far away from the screen, our total Hamiltonian is $\hat{H} = \hat{H}_{\text{electron}} + \hat{H}_{\text{slit}} + \hat{H}_I$, where $\hat{H}_I$ is the interaction between the electron and the slit. This is a type II measurement, where the slit affects the initial quantum state of the electron. The diffraction pattern is formed on the screen. For the *same Hamiltonian, same incoming wavefunction, and same boundary conditions*, the probability density is exactly the same, regardless of whether we send in 1 or 1,000 electrons *separately*. Caution: What we are saying is the same probability density, *not* the same diffraction pattern. Two electrons A and B may land on two different spots of the screen, but will never land in the spot with zero probability density. The probability density is the most that the quantum theory can predict. The quantum theory cannot predict where the electron lands, i. e., the diffraction pattern. This quantum indeterminacy is inexplicable from classical physics. The classical physics reader may try to narrow the slit width a to get some determinacy back, but only finds the probability density spreading even more on the screen, creating a stronger indeterminacy. This quantum indeterminacy is inherent in the wave-particle duality. That is why in Fig. 1.5, we purposely draw the wavy lines around the slit and the electron.

Does the quantum indeterminacy ever disappear at least partially? Yes, it does. For instance, we can remove the slit, and then there is no diffraction pattern. The electron probability is now determined by its initial state with a different indeterminacy. When we make a measurement, we force the electron to choose one eigenstate of the new Hamiltonian, just as the single photon interference (see eq. (1.4.12)). How can we recover the classical determinacy? This can be done by sending in lots of electrons.

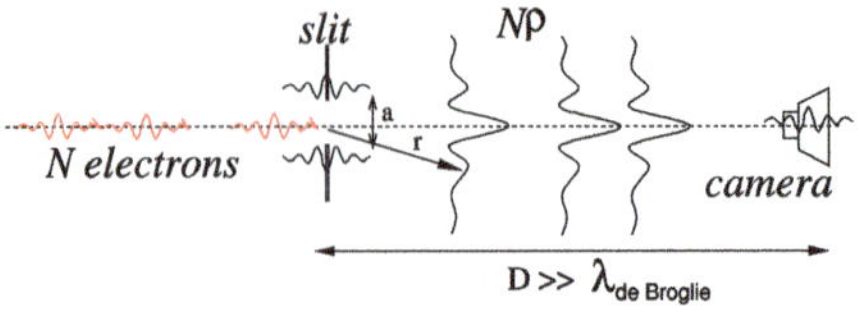

Figure 1.5: Electrons pass through a slit of width a that is comparable to the de Broglie wavelength of the electrons. The measurement, i. e., placing a slit behind the electrons, alters the property of the incident electrons. The probability density $\rho(\mathbf{r}, t)$ is detected by a camera. The density does not change its shape even with N electrons. The wavy lines around the slit and the camera denote their respective wave nature.

In summary, what QM predicts is the probability density. What is undetermined is the outcome of a single measurement. A measurement can affect the probability density. When we discuss measurement, we must first agree on the Hamiltonian. A sudden change of the Hamiltonian may constitute a transition in the view of the original Hamiltonian (see Chapter 7). There is no measurement problem.[2] More detailed discussions are presented in Section 6.4.3.

1.6 Measuring Planck's constant using LEDs

An enjoyable experiment to measure Planck constant is to use light-emitting diodes (LEDs). They are a quantum mechanical semiconductor device that can emit light by applying a voltage. The color of the emitted light directly corresponds to the energy band gap[3] of the semiconductor. The aluminum indium gallium phosphide (AlInGaP) material system produces red color, while the indium gallium nitride (InGaN) system produces blue, green, and cyan. When the externally applied voltage exceeds this gap, light is emitted, so the voltage is directly related the photon energy. The wavelength of modern LEDs ranges from 255 nm up to 2,600 nm. This allows one to measure Planck's constant quite accurately.

This experiment enters our advancd physics lab. In the following, we explain it in details. One first measures the I-V curve. Figures 1.6(a) and (b) show experimental curves for three diodes of red, green, and blue colors, with wavelengths of 660, 560 and 465 nm. The emitted photon energy is related to the threshold voltage V_d as

$$eV_d = nh\nu = nh\frac{c}{\lambda}, \tag{1.6.1}$$

where one sees that, if $n = 1$, we can compute h. Therefore, we need to find V_d. Although the I-V curve appears exponential, we cannot use an exponential function to fit the curve because doing so would lead to $V_d \to -\infty$, which is unphysical. On the other

2 David Mermin accurately explains this issue in Physics Today, June, 2022.

3 The details will be discussed in Chapter 9.

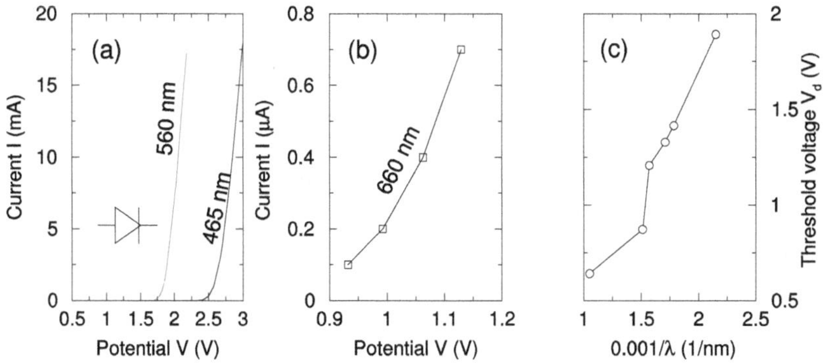

Figure 1.6: (a,b) I-V curve for three light-emitting diodes (LEDs) of wavelengths of 560, 465 and 660 nm. (b) A zoom-in view of the last four points. (c) The threshold voltage (V_d) as a function of the inverse wavelength.

hand, we cannot rely on our ammeter reading because we are limited to the accuracy of the meter. A reliable method is to choose three data points at the smallest possible current that each fall on a straight line. Then, we extrapolate this line to 0 mA. The intercept on the voltage axis is V_d. This also ensures that we have one photon emitted, so $n = 1$. Figure 1.6(b) shows a zoom-in view of the last four points for the red LED. The threshold V_d is found from the last two points. We can find the rest of V_d, where real values are given in the following exercise. Figure 1.6(c) plots the the threshold voltage as a function of inverse wavelength. Equation (1.6.1) enables us to measure $\frac{h}{e}$ accurately. We find our experimental value of Planck's constant $h = 6.26 \times 10^{-34}$ Js, with a 6 % deviation from the standard value $h = 6.626 \times 10^{-34}$ Js.

Exercise 1.6.0.

13. The following table shows the measured values of the cutoff voltage V_d of seven LEDs.

Color	blue	green	yellow	orange	red	infrared
Wavelength (nm)	465	560	585	635	660	950
Threshold V_d (V)	1.892	1.415	1.330	1.208	0.873	0.641

Assuming the speed of light is 3×10^8 m/s and the elementary charge $e = 1.602 \times 10^{-19}$ C, compute Planck's constant and the percentage of error.

1.7 Problems

1. Under what condition will the Maxwell equations break down?
2. A beam of red light enters from the air to the water tank. (a) How does the light photon energy change? (b) How does the wavelength change? (c) How does the light

momentum change? (d) Based on the light momentum, determine the direction of the force due to light. (e) Where does the force come from? This problem is based on Walter Lewin's lecture and the more quantitative information presented in the next problem.

3. The index of refraction of air is about 1, and that of water is 1.33. A beam of visible light of wavelength 500 nm enters from the air to the water tank. (a) Compute the light momentum in the air. (b) Compute the light speed in water. (c) Use eq. (1.5.7) to find the momentum in water. Is it higher or lower than the momentum in air? Why? (d) Show this momentum exactly matches the momentum if we use the de Broglie wavelength to compute the momentum. Hint: See Ref. [13].
4. A radio station has a frequency of 900 MHz and emits power of 10 kW. Compute the number of photons per second.
5. An electron is accelerated from rest through the accelerating voltage V_a. Show that the de Broglie wavelength of the electron in the nonrelativistic limit is

$$\lambda_{\mathrm{dB}} = \frac{h}{\sqrt{2em_e V_a}} = \frac{1.226439\ [\mathrm{nm}]}{\sqrt{V_a}},$$

   where λ_{dB} is in units of nanometer and V_a is in volts.
6. This problem uses the concepts of Born's statistics. An electron is accelerated from rest by the accelerating voltage V_a for a distance L on the x axis. Treat the electron classically. (a) Find the distribution function $\rho(x)$. (b) Find the average position $\bar{x}$. (c) Find $\bar{x^2}$. (d) Find Δx.
7. This problem uses the concepts of Born's statistics. A classical object of mass m falls under gravity from a height of h with zero initial velocity. (a) Find the position distribution function $\rho(x)$. (b) Find the average position $\bar{x}$. (c) Find $\bar{x^2}$. (d) Find Δx. (Answer: $\Delta x = \frac{2h}{3\sqrt{5}}$). (e) Find the velocity distribution function $\rho(v)$. (f) Find δv. (Answer: $\Delta v = \frac{2\sqrt{gh^3}}{3\sqrt{30}}$). (g) Find the uncertainty relation $\Delta x\delta p$, where p is the momentum of the object. (h) Exploration. According to the Heisenberg uncertainty principle (in Chapter 3), $\Delta x\Delta p \geq \hbar/2$. This might provide a hint about a region where both the gravity and quantum effect play simultaneous roles. This problem is motivated by Prof. Sheng-Di Lin, National Chiao Tung University.
8. Young's double-slit experiment is commonly used to demonstrate the wave nature of an object. (a) Suppose we send in one photon only. Describe and explain what happens on the screen behind. (b) Now, we send in two photons. What happens? (c) We block one slit and use one photon. In comparison to (a), what are the differences? (d) If we remove the double slit and send in one photon, what do we see on the screen? (e) From (a) to (d), how does the system Hamiltonian change?
9. A beam of red light has a wavelength of 740 nm. (a) If a free electron has the same wavelength, what is its kinetic energy in eV? (b) How do you explain the huge difference in the photon energy and electron energy?

10. In laser-cooling experiments, Rb atoms can be cooled down to μK. Suppose two atoms are separated by 1 Å. What is the temperature needed so that the wave nature of the Rb atom shows up?
11. According to eq. (1.4.14), $|s\rangle = rt(e^{i\delta_1} + e^{i\delta_2})|a\rangle + (t^2 e^{i\delta_1} - r^2 e^{i\delta_2})|b\rangle$, where $|a\rangle$ and $|b\rangle$ are indistinguishable to the photon. However, if $|a\rangle$ and $|b\rangle$ are distinguishable to the photon, we have two sets of wavefunctions, $|s_1\rangle = rt(e^{i\delta_1} + e^{i\delta_2})|a\rangle$, and $|s_2\rangle = (t^2 e^{i\delta_1} - r^2 e^{i\delta_2})|b\rangle$. (a) Find the probability densities $\rho_1 = |\langle a|s_1\rangle|$ detected at $|a\rangle$ and $\rho_2 = |\langle b|s_2\rangle|^2$ detected at $|b\rangle$. (b) Compare them with $\rho_a = |\langle a|s\rangle|^2$ and $\rho_b = |\langle b|s\rangle|^2$. (c) What can you conclude from (b)?
12. A wavefunction in one dimension $\phi(x) = \exp(ikx)$ is confined in one segment $x \in (0, L)$. (a) Normalize it. (b) Compute the expectation values of $\hat{p}_x, \hat{p}_x^2$. (c) Find the uncertainty $\Delta p_x = \sqrt{\langle \hat{p}_x^2\rangle - \langle \hat{p}_x\rangle^2}$.
13. The momentum operator also allows us to compute transitions between two states that are needed to describe optical spectra. Suppose we have two states with wavefunctions $|\phi_1\rangle$ and $|\phi_2\rangle$. The strength of the transition from $|\phi_2\rangle$ to $|\phi_1\rangle$ is determined by the transition matrix element, $\langle \phi_1|\hat{\mathbf{p}}|\phi_2\rangle$. Show $\langle \phi_1|\hat{\mathbf{p}}|\phi_2\rangle$ can be complex.
14. (a) Show $x\frac{d(\delta(x))}{dx} = -\delta(x)$. (b) Suppose $\phi(x) = \delta(x - a)$. Find $\langle \hat{p}_x^2\rangle - (\langle \hat{p}_x\rangle)^2$.
15. We have an operator, $\hat{O} = \frac{d}{dx}$. (a) Find its eigenfunctions. (b) Find its eigenvalues.

2 Schrödinger equation: from bound and unbound states to quantum tunneling

Classical physics is based on Newton's laws of motion. Quantum physics is based on two famous Schrödinger equations: the time-dependent Schrödinger equation (TDSE) and the time-independent Schrödinger equation (TIDSE). This chapter introduces them and their applications to several simple systems, with emphasis on the bound, continuous states and quantum tunneling.

This chapter is divided into four units.

- Unit 1 consists of two sections. Section 2.1 introduces TDSE using the Noether theorem and the current density operator and continuity equation. Section 2.2 introduces TIDSE as a special case of TDSE, where basic properties of eigenstates are discussed.
- Unit 2 consists of Section 2.3 which solves the Schrödinger equation for the particle in a box, where all the eigenstates are bound states. The emphasis is on the completeness, expectation values, and stationary-state evolution.
- Unit 3 includes Sections 2.4 and 2.5, both on the finite square well. Section 2.4 sets up the framework, while Section 2.5 solves it. Different from Section 2.3, some eigenstates are bound, while some are continuous. Our method is new and significantly simpler. The codes allow the reader to explore more.
- Unit 4 includes Sections 2.6 and 2.7 on tunneling through a barrier, where all eigenstates are continuous. Section 2.6 shows the details how the wavefunctions, transmission and current are computed, while Section 2.7 provides additional insights.

2.1 Time-dependent Schrödinger equation

In contrast to classical physics, where an object can be completely static or in motion, quantum mechanics does not have statics. Regardless of whether the Hamiltonian is time dependent or not, a quantum state always evolves with time.

2.1.1 Time symmetry and the derivation of time-dependent Schrödinger equation

This derivation only serves as an introduction to, instead of proof of, TDSE for the beginner. Our method is based on the Noether theorem [14], which ensures every invariance (symmetry) corresponds to a conservation law. Rotational invariance leads to the angular momentum conservation (Section 4.5.5), and spatial invariance leads to crystal momentum conservation (Section 9.2). Here, we use time invariance to derive the energy conservation. We start from Born's probability density, $\rho(\mathbf{r}, t) = |\psi(\mathbf{r}, t)|^2$. After time Δt, the density becomes $\rho(\mathbf{r}, t + \Delta t)$. Now, we assume the following two densities are the same:

https://doi.org/10.1515/9783110672152-002

$$\rho(\mathbf{r},t) = \rho(\mathbf{r},t+\Delta t), \tag{2.1.1}$$

just like the stationary state in classical physics. This property is formally called time translational symmetry, where, at any two arbitrary times t_1 and t_2, $\rho(\mathbf{r},t_1) = \rho(\mathbf{r},t_2)$. Equation (2.1.1) leads to

$$|\psi(\mathbf{r},t)|^2 = |\psi(\mathbf{r},t+\Delta t)|^2 \to \psi(\mathbf{r},t+\Delta t) = e^{-i\hat{\theta}\Delta t}\psi(\mathbf{r},t), \tag{2.1.2}$$

where $\hat{\theta}$ is an unknown operator with units of 1/s, but must be a Hermitian operator, $\hat{\theta}^\dagger = \hat{\theta}$, in order to satisfy eq. (2.1.1). One can rigorously prove the exponential form in eq. (2.1.2) is the only acceptable expression (see problem No. 1 at the end of the chapter). If we take $\hat{\theta}$ to zero or a constant, then the solution is trivial. Next, we subtract $\psi(\mathbf{r},t)$ from both sides of the second equation in eq. (2.1.2) and then divide by Δt,

$$\frac{\psi(\mathbf{r},t+\Delta t) - \psi(\mathbf{r},t)}{\Delta t} = \frac{(e^{-i\hat{\theta}\Delta t} - 1)\psi(\mathbf{r},t)}{\Delta t}. \tag{2.1.3}$$

We take the limit $\Delta t \to 0$ to obtain

$$\frac{\partial\psi(\mathbf{r},t)}{\partial t} = -i\hat{\theta}\psi(\mathbf{r},t). \tag{2.1.4}$$

Because $\hat{\theta}$ is in units of 1/s, we can replace it by the ratio of the Hamiltonian over $\hbar$, $\hat{H}/\hbar$, and then we obtain the time-dependent Schrödinger equation:

$$i\hbar\frac{\partial\psi(\mathbf{r},t)}{\partial t} = \hat{H}\psi(\mathbf{r},t) = \left[-\frac{\hbar^2\nabla^2}{2m} + V(\mathbf{r},t)\right]\psi(\mathbf{r},t). \tag{2.1.5}$$

In our derivation, $\hat{\theta}$ or $\hat{H}$, can be time-dependent. Our time translational symmetry is only imposed on the density, but in turn it requires the Hamiltonian to be real (see eq. (2.1.12)).

Example 1 (Origin of wavefunction superposition). If ψ_1 and ψ_2 satisfy eq. (2.1.5), prove that a linear superposition of $\psi_3 = c_1\psi_1 + c_2\psi_2$ is also a solution of eq. (2.1.5), where c_1 and c_2 are constants, but $\psi_4 = \psi_1\psi_2$ is not.

Since both ψ_1 and ψ_2 are solutions of eq. (2.1.5), we have

$$i\hbar\frac{\partial\psi_1(\mathbf{r},t)}{\partial t} = \hat{H}\psi_1(\mathbf{r},t) = \left[-\frac{\hbar^2\nabla^2}{2m} + V(\mathbf{r},t)\right]\psi_1(\mathbf{r},t),$$

and

$$i\hbar\frac{\partial\psi_2(\mathbf{r},t)}{\partial t} = \hat{H}\psi_2(\mathbf{r},t) = \left[-\frac{\hbar^2\nabla^2}{2m} + V(\mathbf{r},t)\right]\psi_2(\mathbf{r},t).$$

Next, we multiply both of the previous equations by c_1 and c_2, respectively, and add up these two resultant equations. We find

$$i\hbar\frac{\partial\psi_3(\mathbf{r},t)}{\partial t} = \hat{H}\psi_3(\mathbf{r},t) = \left[-\frac{\hbar^2\nabla^2}{2m} + V(\mathbf{r},t)\right]\psi_3(\mathbf{r},t).$$

Conversely, if we substitute ψ_4 into eq. (2.1.5), we find immediately that it cannot satisfy the equation. This is the origin of wavefunction superposition. It must be a linear superposition, not a multiplication.

Example 2 (Normalization of wavefunction is time-independent). Prove that if the Hamiltonian is hermitian, $\mathcal{N} = \int \psi^*(\mathbf{r},t)\psi(\mathbf{r},t)d\mathbf{r}$ is time-independent and can be taken to be 1.

We take the time derivative of $\mathcal{N}$ to get

$$\frac{d\mathcal{N}}{dt} = \int\left(\frac{\partial\psi^*}{\partial t}\psi + \frac{\partial\psi}{\partial t}\psi^*\right)d\mathbf{r}.$$

Because $i\hbar\frac{\partial\psi}{\partial t} = \hat{H}\psi = (-\frac{\hbar^2\nabla^2}{2m} + V)\psi$, we have $-i\hbar\frac{\partial\psi^*}{\partial t} = (-\frac{\hbar^2\nabla^2}{2m} + V)\psi^*$. Then we replace both partial derivatives to find

$$\begin{aligned}\frac{d\mathcal{N}}{dt} &= \frac{\hbar^2}{2i\hbar m}\int[(\nabla^2\psi^*)\psi - (\nabla^2\psi)\psi^*]d\mathbf{r} = \frac{\hbar^2}{2i\hbar m}\int\nabla\cdot[(\nabla\psi^*)\psi - (\nabla\psi)\psi^*]d\mathbf{r}\\ &= \frac{\hbar^2}{2i\hbar m}\oint[(\nabla\psi^*)\psi - (\nabla\psi)\psi^*]d\mathbf{a},\end{aligned}$$

where the last step uses the Gaussian theorem to convert a volume integral to a surface integral over $d\mathbf{a}$. The integral kernel $(\nabla\psi^*)\psi - (\nabla\psi)\psi^*$ depends on ψ. Because ψ must be normalizable, that is ψ at $\mathbf{r} \to \pm\infty$ must be zero, the integral is zero and $\mathcal{N}$ is a constant and can be chosen to be 1. The initial normalized wavefunction will stay normalized at any t.

Exercise 2.1.1.

1. Why do we say that quantum dynamics is a fundamental process of a quantum system?
2. (a) Show that, if V_1 and V_2 produce the same wavefunction ψ, then $V_1 = V_2$. (b) Show it is possible that two different wavefunctions correspond to the same V.
3. Find the condition that $\int \psi_1^*(\mathbf{r},t)\psi_2(\mathbf{r},t)d\mathbf{r}$ is time dependent, where ψ_1 and ψ_2 are two different wavefunctions.

2.1.2 Current density and continuity equation

If we take the complex conjugate of both sides of eq. (2.1.5), we have

$$-i\hbar\frac{\partial\psi^*(\mathbf{r},t)}{\partial t} = \hat{H}^*\psi^*(\mathbf{r},t) = \left[-\frac{\hbar^2\nabla^2}{2m} + V^*(\mathbf{r},t)\right]\psi^*(\mathbf{r},t). \tag{2.1.6}$$

We multiply eq. (2.1.5) by $\psi^*(\mathbf{r},t)$ and eq. (2.1.6) by $\psi(\mathbf{r},t)$ from the left. Then, subtracting $\psi^* \times$ eq. (2.1.5) by $\psi \times$ eq. (2.1.6) yields

$$i\hbar\left(\psi^*\frac{\partial\psi}{\partial t} + \psi\frac{\partial\psi^*}{\partial t}\right) = -\frac{\hbar^2}{2m}(\psi^*\nabla^2\psi - \psi\nabla^2\psi^*) + \psi^*V\psi - \psi V^*\psi^*. \tag{2.1.7}$$

Recall that the density $\rho(\mathbf{r},t) = \psi(\mathbf{r},t)\psi^*(\mathbf{r},t)$, so that the left side is $i\hbar\frac{\partial\rho}{\partial t}$. Since V commutes with ψ and ψ^*, the second term on the right side can be grouped into one term, $(V - V^*)\psi^*\psi$. Using the identity $f\nabla^2 g - g\nabla^2 f = \nabla\cdot(f\nabla g - g\nabla f)$, the first term on the right can be converted to a divergence,

$$-\frac{\hbar^2}{2m}(\psi^*\nabla^2\psi - \psi\nabla^2\psi^*) = -\frac{\hbar^2}{2m}\nabla\cdot(\psi^*\nabla\psi - \psi\nabla\psi^*). \tag{2.1.8}$$

Dividing all the terms on both sides of eq. (2.1.7) by $i\hbar$, we obtain

$$\frac{\partial\rho}{\partial t} = -\nabla\cdot\underbrace{\left[-\frac{i\hbar}{2m}(\psi^*\nabla\psi - \psi\nabla\psi^*)\right]}_{\mathbf{J}\text{ current density}} + \frac{V - V^*}{i\hbar}\rho. \tag{2.1.9}$$

In the first term on the right, we purposely introduce a negative sign, so the term inside the square bracket has another negative sign, the reason for which will soon become clear. One can check the unit for the term in the square bracket, where $\hbar$ has units of Js, m is the mass in kg, and ∇ has unit of m^{-1}, i. e., Js/(kgm) or m/s. This is a velocity, representing the flow of particles. If we multiply $\partial\rho/\partial t$ by charge q, $q\partial\rho/\partial t$, $\mathbf{J}$ is related to the electric current density. Because ψ and ψ^* have units of $1/\sqrt{m^3}$, $\mathbf{J}$ is the volume current density and is defined through the term in the square bracket as

$$\mathbf{J} = -\frac{i\hbar}{2m}(\psi^*\nabla\psi - \psi\nabla\psi^*) = \frac{1}{2m}[\psi^*\hat{\mathbf{p}}\psi - \psi\hat{\mathbf{p}}\psi^*] \tag{2.1.10}$$

$$= \frac{1}{2m}[\psi^*\hat{\mathbf{p}}\psi + (\psi^*\hat{\mathbf{p}}\psi)^*] = \mathrm{Re}\left(\psi^*\frac{\hat{\mathbf{p}}}{m}\psi\right) = \mathrm{Re}(\psi^*\hat{\mathbf{v}}\psi), \tag{2.1.11}$$

where $\hat{\mathbf{v}}$ is the velocity operator. In the second half of eq. (2.1.11), we have used $\hat{\mathbf{p}}^* = -\hat{\mathbf{p}}$. Caution: $\hat{\mathbf{p}}^\dagger = +\hat{\mathbf{p}}$. $\hat{\mathbf{v}}$ in eq. (2.1.11) is our velocity operator. If we did not introduce a negative sign in eq. (2.1.9), we could have an opposite sign for the current density with respect to the velocity. This also shows that the current density is related to $\hat{\mathbf{p}}$, but contains a mixed term (see eq. (2.1.10)). $\mathbf{J}$ in eqs. (2.1.10) and (2.1.11) is still not an expectation value since no integration over space is taken yet.

We rewrite eq. (2.1.9) as

$$\frac{\partial\rho}{\partial t} = -\nabla\cdot\mathbf{J} + \frac{V - V^*}{i\hbar}\rho. \tag{2.1.12}$$

In a closed system, the potential V must be real, so the second term is zero. The remainder gives us the current continuity equation, the same as the continuity equation in electromagnetism (EM) theory. In order to compare them directly, one has to multiple charge e on both sides of the equation. In QM, the wavefunction is the fundamental quantity, but in EM the density is the fundamental quantity. However, true physical properties are revealed by density in both QM and EM. This is the same foundation that density functional theory is built upon. It is worth noting the difference between eqs. (2.1.12) and (2.1.1): In (2.1.1), the spatial variable $\mathbf{r}$ does not change, but in (2.1.12), it does change.

The density change is equal to the negative divergence of current density,

$$\frac{\partial \rho}{\partial t} = -\nabla \cdot \mathbf{J}, \tag{2.1.13}$$

whose integration form is

$$\int \frac{\partial \rho}{\partial t} d\mathbf{r} = -\int \nabla \cdot \mathbf{J} d\mathbf{r} = -\oint \mathbf{J} \cdot d\mathbf{a}, \tag{2.1.14}$$

where $d\mathbf{r}$ and $d\mathbf{a}$ are the volume and surface infinitesimal elements, respectively. This expression guarantees the conservation of the particle number because $\int \rho d\mathbf{r} = 1$, so its time-derivative is zero, with no change in the total particle numbers. The second equation is obtained using the Gauss law, i. e., converting a volume integral to a surface integral.

Example 3 (Current and momentum conjugation). Suppose a particle of mass m propagates along the $+x$-axis, with wavefunction $\phi(x) = \exp(ikx)/\sqrt{L}$, where $1/\sqrt{L}$ is the normalization constant. Compute its current.

The expectation value of the current density operator is

$$-\frac{i\hbar}{2m}\int_0^L dx\left[\frac{\exp(-ikx)}{\sqrt{L}}\frac{\partial}{\partial x}\frac{\exp(ikx)}{\sqrt{L}} - \frac{\exp(ikx)}{\sqrt{L}}\frac{\partial}{\partial x}\frac{\exp(-ikx)}{\sqrt{L}}\right] = -\frac{i\hbar}{2m}(2ik)\frac{1}{L}\int_0^L dx = -\frac{i\hbar}{2m}(2ik) = \frac{\hbar k}{m},$$

which gives us the current.

Example 4 (Show $\hat{\mathbf{p}}^\dagger = +\hat{\mathbf{p}}$). $\hat{\mathbf{p}}^\dagger = [(\hat{\mathbf{p}})^*]^\mathrm{T}$ means a complex conjugate followed by a transpose $_\mathrm{T}$. Because $\hat{\mathbf{p}} = -i\hbar\nabla$, we have $\hat{\mathbf{p}}^* = -\hat{\mathbf{p}}$. For a matrix, the transpose means that we exchange the column and row indices, i. e., element A_{ij} becoming A_{ji}. Now, we want to compute the transpose for an operator, not for a matrix. For an operator, the transpose means that we are going to apply the operator on the function that is in front of the operator, not behind the operator. We use $\frac{\partial^\mathrm{T}}{\partial x}$ as an example. We introduce two well-defined arbitrary functions $\phi_1(x)$ and $\phi_2(x)$ that decay to zero at $x = \pm\infty$, and compute the integral as

$$\int \phi_1(x)\frac{\partial^\mathrm{T}}{\partial x}\phi_2(x)dx = \int \phi_2(x)\frac{\partial}{\partial x}\phi_1(x)dx = \underbrace{\overbrace{\phi_2(x)\phi_1(x)}^{\text{go to zero}} - \int \phi_1(x)\frac{\partial}{\partial x}\phi_2(x)dx}_{\text{chain rule}}.$$

Because ϕ_1 and ϕ_2 are arbitrary, if we compare the first and last terms, we identify $\frac{\partial^\mathrm{T}}{\partial x} = -\frac{\partial}{\partial x}$, i. e., $\hat{p}_x^\mathrm{T} = -\hat{p}_x$. So $(\hat{p}_x^\mathrm{T})^* = -(\hat{p}_x)^* = -(-\hat{p}_x) = \hat{p}_x$. Using the same method for the y and z components, we can find that $\hat{\mathbf{p}} = \hat{\mathbf{p}}^\dagger$.

Example 5 (Time-dependent Liouville equation). Prove $i\hbar\frac{\partial \rho}{\partial t} = [\hat{H}, \rho]$, the density matrix ρ whose element is defined as $\rho_{nm} = \psi_n\psi_m^*$, where $\hat{H}$ is assumed to be hermitian.

Once we are familiar with the density ρ, the extension to the density matrix is a small step. For two states ψ_n and ψ_m, we define the density matrix as $\rho_{nm} \equiv \psi_n\psi_m^* \equiv |\psi_n\rangle\langle\psi_m|$. The reader may already notice this in eq. (1.3.4). The diagonal element, ρ_{nn}, is the density ρ_n. We start from the time-dependent Schrödinger equations,

$$i\hbar\dot{\psi}_n = \hat{H}\psi_n, \tag{2.1.15}$$

$$i\hbar\dot{\psi}_m = \hat{H}\psi_m, \tag{2.1.16}$$

where the dot is a shorthand notation for $\partial/\partial t$. We take the complex conjugate and transpose of eq. (2.1.16), i. e., the Hermitian conjugate. Note that, after we take the transpose, the order of the wavefunction and the Hamiltonian is exchanged,[1]

$$-i\hbar\dot{\psi}_m^* = \psi_m^*\hat{H}. \tag{2.1.17}$$

We multiply eq. (2.1.17) by ψ_n from the left, and multiply (2.1.15) by ψ_m^* from the right, $i\hbar\dot{\psi}_n\psi_m^* = \hat{H}\psi_n\psi_m^*$, $-i\hbar\psi_n\dot{\psi}_m^* = \psi_n\psi_m^*\hat{H}$, and we subtract these two equations,

$$i\hbar\left(\dot{\psi}_n\psi_m^* + \psi_n\dot{\psi}_m^*\right) = \hat{H}\psi_n\psi_m^* - \psi_n\psi_m^*\hat{H}. \tag{2.1.18}$$

Now, we define the density matrix as $\rho_{nm} \equiv \psi_n\psi_m^*$, so eq. (2.1.18) can be rewritten as

$$i\hbar\frac{\partial\rho_{nm}}{\partial t} = \hat{H}\rho_{nm} - \rho_{nm}\hat{H} \equiv [\hat{H}, \rho_{nm}] \underbrace{\longrightarrow}_{\text{in matrix form}} i\hbar\frac{\partial\rho}{\partial t} = [\hat{H}, \rho], \tag{2.1.19}$$

which is called the time-dependent Liouville equation. ρ in eq. (2.1.19) must be understood as a matrix. The commutator $[\hat{A}, \hat{B}]$ is defined as $[\hat{A}, \hat{B}] = \hat{A}\hat{B} - \hat{B}\hat{A}$, where $\hat{A}$ and $\hat{B}$ must be a matrix or an operator.

Exercise 2.1.2.

4. Show $\hat{\mathbf{p}}^* = -\hat{\mathbf{p}}$. Hint: this is different from our example.
5. Show that, if a wavefunction is real, its current density $\mathbf{J}$ is zero.
6. A plane wave e^{ikx} is propagating along $+x$, where $k > 0$. (a) Show $\mathbf{J} = -\frac{i\hbar}{2m}(\psi^*\nabla\psi - \psi\nabla\psi^*)$ has a correct sign. (b) Find the unit of $\mathbf{J}$.
7. Show that the operator $\frac{\partial^T}{\partial y} = -\frac{\partial}{\partial y}$. Hint: Follow the previous example.
8. Show that, if a system only has one state $n = m = 1$, eq. (2.1.19) is zero. In other words, the density is constant if a system only has one eigenstate.
9. $\hat{O}$ is a Hermitian operator. (a) Prove $\int \psi_1^*(\mathbf{r})\hat{O}\psi_2(\mathbf{r})d\mathbf{r} = (\int \psi_2^*\hat{O}\psi_1 d\mathbf{r})^* = \int(\hat{O}\psi_1)^*\psi_2 d\mathbf{r}$. (b) If $\psi_1 = \psi_2$, prove $\int \psi_1^*(\mathbf{r})\hat{O}\psi_1(\mathbf{r})d\mathbf{r}$ must be real, i. e., the expectation value of a Hermitian operator is real.

2.2 Time-independent Schrödinger equation and eigenstates

TDSE is the general equation for QM. If the Hamiltonian of a system is time-independent, we have a time-independent Schrödinger equation (TIDSE).

1 In the Dirac notation, $i\hbar|\dot{\psi}_n\rangle = \hat{H}|\psi_n\rangle$, whose the Hermitian conjugate is $-i\hbar\langle\dot{\psi}_n| = \langle\psi_n|\hat{H}$.

We rewrite TDSE as

$$i\hbar\frac{\partial\psi}{\partial t} = \hat{H}\psi \rightarrow \dot{\psi} = \frac{\hat{H}}{i\hbar}\psi, \tag{2.2.1}$$

where the dot on ψ denotes the time derivative. We take another time derivative to find

$$\ddot{\psi} = \frac{\dot{\hat{H}}}{i\hbar}\psi + \frac{\hat{H}}{i\hbar}\dot{\psi} = \frac{\hat{H}}{i\hbar}\frac{\hat{H}}{i\hbar}\psi = \left(\frac{\hat{H}}{i\hbar}\right)^2\psi,$$

where in the second equation we have used the fact $\hat{H}$ is time independent, so its time derivative is zero, $\dot{\hat{H}} = 0$. In addition, we have used eq. (2.2.1) to replace $\dot{\psi}$. For the nth time-derivative, we have $\psi^{(n)} = (\frac{\hat{H}}{i\hbar})^n\psi$. The only function that can produce such a nice recursion is the exponential function. Suppose $\psi = e^{i\hat{Q}t}\psi(0)$, where $\psi(0)$ is the wavefunction at time $t = 0$. Then we have $\psi^{(n)} = (i\hat{Q})^n\psi$. If we compare it with $\psi^{(n)} = (\frac{\hat{H}}{i\hbar})^n\psi$, we can identify $\hat{Q} = -\hat{H}/\hbar$, so our final wavefunction at t is linked to its value at $t = 0$ through

$$\psi(\mathbf{r}, t) = e^{-i\hat{H}t/\hbar}\psi(\mathbf{r}, 0), \tag{2.2.2}$$

which is generic as long as $\hat{H}$ is time-independent. $\psi(\mathbf{r}, 0)$ is the initial wavefunction, and in general does not allow us to separate the spatial variable from the temporal variable in $\psi(\mathbf{r}, t)$ to simplify the Schrödinger equation. However, if $\psi(\mathbf{r}, 0)$ is an eigenfunction $\phi(\mathbf{r})$ of $\hat{H}$, then it is possible to write $\psi(\mathbf{r}, t) = \phi(\mathbf{r})f(t)$.

2.2.1 Separation of time and space variables and properties of eigenstates

We substitute $\psi(\mathbf{r}, t) = \phi(\mathbf{r})f(t)$ into both sides of eq. (2.2.1) to get

$$\underbrace{i\hbar\frac{\partial\psi(\mathbf{r}, t)}{\partial t} = i\hbar\frac{\partial[f(t)\phi(\mathbf{r})]}{\partial t} = \phi(\mathbf{r})i\hbar\frac{df(t)}{dt}}_{\text{left side}} = \underbrace{\hat{H}f(t)\phi(\mathbf{r}) = f(t)\hat{H}\phi(\mathbf{r})}_{\text{right side}}. \tag{2.2.3}$$

We divide both sides by ϕf to obtain

$$\frac{i\hbar}{f(t)}\frac{df(t)}{dt} = \frac{\hat{H}\phi(\mathbf{r})}{\phi(\mathbf{r})} = E(\mathbf{r}, t). \tag{2.2.4}$$

Most textbooks assume the left side of eq. (2.2.4) to be a constant, but this is unnecessary. Here, we set it to a general function $E(\mathbf{r}, t)$. First, we start from $\frac{\hat{H}\phi(\mathbf{r})}{\phi(\mathbf{r})} = E(\mathbf{r}, t)$ and take its time derivative.[2] Since $\frac{\hat{H}\phi(\mathbf{r})}{\phi(\mathbf{r})}$ is independent of time, $\frac{\partial E}{\partial t} = 0$, i. e., $E(\mathbf{r}, t)$ is

2 This idea is from Prof. Sheng-Di Lin, National Chiao Tung University.

time-independent, $E(\mathbf{r},t) \to E(\mathbf{r})$. Next, we start from $\frac{i\hbar}{f(t)}\frac{df(t)}{dt} = E(\mathbf{r})$ and take a spatial derivative to get $0 = \frac{\partial E(\mathbf{r})}{\partial \mathbf{r}}$, i. e., E must be independent of space, $E(\mathbf{r}) \to E$, a constant. So, we have $\frac{\hat{H}\phi(\mathbf{r})}{\phi(\mathbf{r})} = E$, which yields the time-independent Schrödinger equation,

$$\hat{H}\phi(\mathbf{r}) = \left[\frac{\hat{\mathbf{p}}^2}{2m} + V(\mathbf{r})\right]\phi(\mathbf{r}) = E\phi(\mathbf{r}) \to \hat{H}\phi_n = E_n\phi_n \quad \text{(space)}, \tag{2.2.5}$$

where E is the eigenenergy and $\phi(\mathbf{r})$ is an eigenfunction of $\hat{H}$, called the eigenstate of the system. Only the eigenfunction of $\hat{H}$ can be called an eigenstate. E and $\phi(\mathbf{r})$ form a pair. Because this is an eigenequation, i. e., an equation has two unknowns, E and $\phi(\mathbf{r})$, there are many possible solutions to eq. (2.2.5), so we label them as E_n and $\phi_n(\mathbf{r})$.

The first property of eigenstates is normalization, where all eigenstates must be normalized as

$$\langle\phi_n|\phi_n\rangle = \int \phi_n^*(\mathbf{r})\phi_n(\mathbf{r})d\mathbf{r} = 1, \tag{2.2.6}$$

to ensure the conservation of the number of particles. The integration is over the entire space. If ϕ is discrete, the equation is changed to $\sum_i \phi_n^*(i)\phi_n(i) = 1$.

The second property is orthogonalization. For two states ϕ_n and ϕ_m, $n \neq m$, their overlap integral must be zero,

$$\langle\phi_n|\phi_m\rangle = \int \phi_n^*(\mathbf{r})\phi_m(\mathbf{r})d\mathbf{r} = 0. \tag{2.2.7}$$

Eigenstates are linearly independent in the language of linear algebra.

The third property is completeness.

$$\sum_n |\phi_n\rangle\langle\phi_n| = \sum_n \phi_n(\mathbf{r})\phi_n^*(\mathbf{r}) = 1, \tag{2.2.8}$$

where the summation is now over the eigenstate index n and the final answer is independent of the variable $\mathbf{r}$.

2.2.2 Stationary state evolution

For a fixed E_n and $\phi_n(\mathbf{r})$, eq. (2.2.4) has a solution of $f(t-t_0) = f_0e^{-iE_n(t-t_0)/\hbar}$, where $e^{-iE_n(t-t_0)/\hbar}$ is called the dynamical phase factor and f_0 is a coefficient set by the initial condition at the initial time t_0. Because f_0 does not change, the time-evolution is a stationary state evolution, with a fixed frequency $E_n/\hbar$,

$$\psi(\mathbf{r},t) = e^{-iE_n(t-t_0)/\hbar}\psi(\mathbf{r},t_0) = e^{-iE_n(t-t_0)/\hbar}\phi_n(\mathbf{r}). \tag{2.2.9}$$

Equation 2.2.9 is a special case of eq. (2.2.2) when the initial state is a single eigenstate. In the following, we set t_0 to zero, so at any moment of time t, the wavefunction is

$$|\psi(\mathbf{r},t)\rangle = e^{-iE_n t/\hbar}|\psi(\mathbf{r},t_0)\rangle = e^{-iE_n t/\hbar}|\phi_n(\mathbf{r})\rangle. \tag{2.2.10}$$

If $\psi(\mathbf{r},0) = \sum_n c_n\phi_n(\mathbf{r})$,

$$\psi(\mathbf{r},t) = e^{-\frac{i}{\hbar}\hat{H}t}\psi(0) = \sum_n c_n e^{-\frac{i}{\hbar}\hat{H}t}\phi_n = \sum_n c_n e^{-iE_n t/\hbar}\phi_n, \tag{2.2.11}$$

where c_n is the amplitude of each eigenstate and must be found in the beginning. This result will be used repeatedly. But, one must remember this result is conditional. The condition is that the original Hamiltonian is time-independent and the system initially is in an eigenstate. Every eigenstate has this property. The time evolution of ψ is called the stationary state evolution.

If $\psi(\mathbf{r},t) = e^{-iE_n t/\hbar}|\phi_n(\mathbf{r})\rangle$, $\langle\psi|\hat{H}|\psi\rangle$ is equal to the eigenenergy,

$$\langle\psi(\mathbf{r},t)|\hat{H}|\psi(\mathbf{r},t)\rangle = \langle\phi_n|\hat{H}|\phi_n\rangle = \int \phi_n^*(\mathbf{r})\hat{H}\phi_n(\mathbf{r})d\mathbf{r} = E_n. \tag{2.2.12}$$

One can see that the energy expectation value remains constant at the eigenenergy E_n and has no time dependence, which is the reason why we call the state a stationary state.

Example 6 (Expand a function and compute the expectation value of $\hat{H}$). $|f(\mathbf{r})\rangle$ is a smooth function. (a) Expand it in terms of ϕ_n. (b) Compute $\langle f(\mathbf{r})|\hat{H}|f(\mathbf{r})\rangle$.

(a) We write $|f(\mathbf{r})\rangle$ as

$$|f(\mathbf{r})\rangle = 1|f(\mathbf{r})\rangle = \sum_n |\phi_n\rangle\langle\phi_n|f(\mathbf{r})\rangle = \sum_n \langle\phi_n|f(\mathbf{r})\rangle|\phi_n\rangle, \tag{2.2.13}$$

where $\langle\phi_n|f(\mathbf{r})\rangle = \int \phi_n^*(\mathbf{r})f(\mathbf{r})d\mathbf{r}$ is the expansion coefficient. ϕ_n is not necessarily an eigenstate, and, in fact, any function that is complete and orthonormalized can be used, just as the Fourier transform does. For this reason, these functions are often called basis functions [15]. This expansion is conditional. First the spatial domain of $\phi_n(\mathbf{r})$ must be equal or larger than that of $f(\mathbf{r})$. For instance, if f is defined between $x = 0$ and $x = 1$ Å, then ϕ must be defined to be at least between $x = 0$ and $x = 1$ Å. Second, $f(\mathbf{r})$ and $\phi_n(\mathbf{r})$ must match the same boundary conditions. For instance, if $f = 1$ at $x = 1$ Å, but all ϕ_n at the same location is zero, then there is no way to expand f in terms of ϕ_n.

(b) We can use eq. (2.2.13) to compute the energy expectation value,

$$\langle f|\hat{H}|f\rangle = \langle f(\mathbf{r})\overbrace{\sum_m |\phi_m\rangle\langle\phi_m|}^{=1}\hat{H}\overbrace{\sum_n |\phi_n\rangle\langle\phi_n|}^{=1}f(\mathbf{r})\rangle = \sum_n E_n\langle f|\phi_n\rangle\langle\phi_n|f\rangle = \sum_n |\langle f|\phi_n\rangle|^2 E_n,$$

where we have used eqs. (2.2.5), (2.2.6), and (2.2.7). If f, E_n and ϕ_n are given, we can compute $\langle f|\hat{H}|f\rangle$.

Exercise 2.2.2.

10. What are the major differences between the time-dependent Schrödinger equation and time-independent Schrödinger equation?
11. What are three major properties of eigenstates?
12. Some textbooks list $\psi(x) = x$ as a wavefunction. Prove this wavefunction does not satisfy TIDSE.

2.3 Particle in a box: bound states

Consider a particle of mass m sealed inside a box, or a line segment between $x \in [0, a]$, see Fig. 2.1(a). This is the particle-in-a-box (PIB) problem, also called the infinite quantum well. The wall of the box is heavy, so the particle with nonzero initial velocity bounces back and forth from the walls at $x = 0$ and $x = a$. This spatial restriction is the reason why all the eigenstates are bound states in PIB, where the energy of the particle never exceeds the infinite potential barrier. In the following, we will start with a classical approach and then introduce the quantum mechanical approach.

2.3.1 Classical treatment

Suppose the particle moves classically with velocity v. The probability for the particle appearing in a short segment dx is given by the time dt spent in dx over the total time T,

$$\frac{dt}{T} = \frac{\frac{dx}{v}}{T} = \frac{dx}{vT} = \rho(x)dx, \tag{2.3.1}$$

where $\rho(x) = \frac{1}{vT}$ is the probability density (the probability density per unit length since we have one dimension). Because the total time is $T = a/v$, $\rho(x) = 1/a$.

With $\rho(x)$, the average position $\bar{x}$ (similar to the expectation value in QM) is just

$$\bar{x} = \int_0^a x\rho(x)dx = \frac{a}{2}, \tag{2.3.2}$$

which says the average position is at the center of the box. And the average x^2 is

$$\overline{x^2} = \int_0^a x^2\rho(x)dx = \frac{a^2}{3}, \tag{2.3.3}$$

so the uncertainty in position is given by

$$\Delta x = \sqrt{\overline{x^2} - (\bar{x})^2} = \frac{a}{\sqrt{12}}, \tag{2.3.4}$$

both of which will be amazingly reproduced in QM.

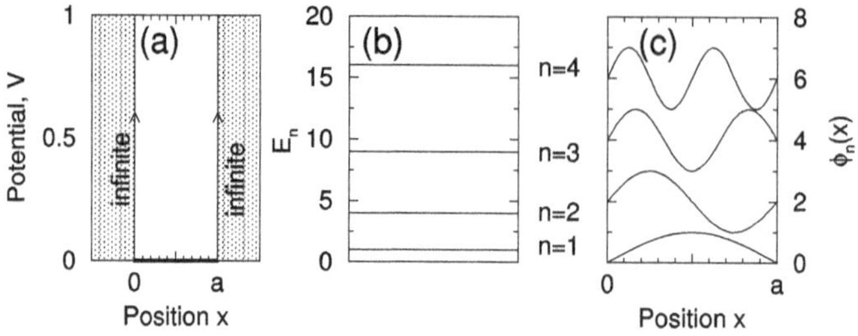

Figure 2.1: (a) Our potential. (b) Eigenvalues E_n in units of $\frac{\hbar^2\pi^2}{2ma^2}$. (c) Four eigenfunctions ϕ_n (vertically shifted) with $n = 1, 2, 3, 4$. As n increases, the number of nodes (wavefunction crossing the zero line) increases, and the energy becomes larger.

2.3.2 Potential sets boundary conditions for wavefunctions

The QM analogy for PIB is to set the potential to infinity for the wall, called the infinite potential barrier. But inside the box, the potential is zero. Figure 2.1(a) sketches such a potential, $V(x) = \infty$ if $x < 0$ or $x > a$; and $V(x) = 0$ if $x \in [0, a]$. Its shape is like a well, so one also calls it an infinite quantum well.

The time-independent Schrödinger equation is an eigenequation, i. e., one equation with two unknowns. In one dimension, eq. (2.2.5) is reduced to

$$-\frac{\hbar^2}{2m}\frac{d^2\phi}{dx^2} + V(x)\phi = E\phi \rightarrow \frac{d^2\phi}{dx^2} = -\frac{2m[E - V(x)]}{\hbar^2}\phi. \tag{2.3.5}$$

It is helpful to examine this equation closely. We notice that $\frac{d^2\phi}{dx^2}$ depends on $E - V(x)$. A different E leads to a different ϕ. If $E > V(x)$ *everywhere*, then $\frac{d^2\phi}{dx^2}$ is negative, meaning that ϕ must change sign after the second derivative. Possible solutions are $\sin kx$, $\cos kx$ or e^{ikx}, where k is real, which corresponds to a continuous state. E is continuous and not quantized or discrete. If $E < V(x)$ *everywhere*, $\phi'' > 0$, so ϕ must like $e^{\pm kx}$, where k is real. They are localized in space. These states are called bound states, where E is discretized or quantized. We emphasize *everywhere* because the solution is for the entire system.

In our case, $V(x) = \infty$ if $x < 0$ or $x > a$. E is always smaller than V, so we only have bound states. We have $\frac{d^2\phi}{dx^2} = -\frac{2m(E-\infty)}{\hbar^2}\phi = \infty\phi$, which is unphysical for any nonzero ϕ. The only meaningful solution is to set ϕ to zero for $x \le 0$ or $x \ge a$. These two wavefunctions are our boundary conditions at $x = 0$ and $x = a$,

$$\phi(x = 0) = 0, \tag{2.3.6}$$

$$\phi(x = a) = 0. \tag{2.3.7}$$

Because the particle density $\rho(x) = |\phi(x)|^2$, $\rho(x < 0) = 0$, and $\rho(x > a) = 0$. Thus, no particle cannot exist beyond the wall. These conditions define the spatial extension of our system.

2.3.3 Finding a solution that matches the boundary conditions

Our Hamiltonian within the box is $\hat{H} = \frac{\hat{p}_x^2}{2m} + V(x)$. Because $V(x) = 0$, our Schrödinger equation is simplified to

$$-\frac{\hbar^2}{2m}\frac{d^2\phi}{dx^2} = E\phi, \tag{2.3.8}$$

where E is the eigenenergy to be found. Since ϕ only depends on x, we have changed the partial derivative to a total derivative.

The solution depends on the sign of E. If E is negative, $\phi(x)$ must have an expression like $\exp(\alpha x)$, where α is a real number. Such an exponential function cannot satisfy boundary conditions (eqs. (2.3.6) and (2.3.7)) because it would increase or decrease with x indefinitely. Any solution with a negative E must be discarded.

If E is positive, $\phi(x) \sim e^{ikx} = \cos(kx) + i\sin(kx)$, where k is a real number to be determined. However, we cannot use $\cos(kx)$ because it violates BC. So the only solution is $\sin(kx)$, where we drop an irrelevant "i." Our trial wavefunction is $\phi(x) = A\sin(kx)$. Substituting ϕ into eq. (2.3.8), we have

$$+\frac{\hbar^2}{2m}k^2 A\sin(kx) = EA\sin(kx) \to E = \frac{\hbar^2 k^2}{2m}. \tag{2.3.9}$$

But, we do not know k. This is where the boundary condition comes in. Equation (2.3.6) is of no use because $x = 0$. Instead, we substitute $\phi(x) = A\sin kx$ into eq. (2.3.7) to find

$$\sin(ka) = 0 \quad \text{(quantization condition)}, \tag{2.3.10}$$

whose roots $ka = n\pi$. To distinguish a different k, we purposely add a subscript n to k, e. g., $k_n = \frac{n\pi}{a}$, where $n = 1, 2, 3, \ldots$ is called the quantum number. Similarly, we add a subscript n to ϕ_n and E_n. $n = 0$ is excluded because this corresponds to an infinite wavelength and not possible to fit within the well. The key insight is that *the spatial restriction leads to the energy quantization.*

2.3.4 Eigenenergies and eigenstates

We plug k_n into eq. (2.3.9) to find the eigenenergy

$$E_n = \frac{\hbar^2}{2m}\left(\frac{n\pi}{a}\right)^2 \quad \text{(eigenenergy)}, \tag{2.3.11}$$

which increases as n^2. Figure 2.1(b) shows four such energies. If n takes $-1, -2, -3, \ldots$, this does not produce a different solution since the energy depends on n as n^2. In the wavefunction, the negative sign can be absorbed into the coefficient A. The coefficient A is fixed by requiring the wavefunction normalized to 1.

$$\langle\phi_n|\phi_n\rangle = \int_0^a [A\sin(k_n x)][A\sin(k_n x)]dx = 1 \quad \text{(normalization)} \to A = \sqrt{\frac{2}{a}}.$$

Finally, our wavefunction is

$$\phi_n(x) = \sqrt{\frac{2}{a}}\sin\left(\frac{n\pi x}{a}\right) \quad \text{(eigenstate)}, \tag{2.3.12}$$

four of which are shown in Fig. 2.1(c). One notices that, when n increases, the number of nodes, the point that crosses the zero line, increases. The more the number of nodes is, the higher the energy becomes. The location of nodes can be found by setting $\phi_n(x) = 0$, i. e., $k_n x = (n\pi/a)x = j\pi$, so we have $x = ja/n$, where j is less than or equal to n and a larger n allows more j's. The probability density is

$$\rho_n = |\phi_n\rangle\langle\phi_n| = \phi_n(x)\phi_n^*(x) = \frac{2}{a}\sin^2\left(\frac{n\pi}{a}x\right). \tag{2.3.13}$$

All the eigenstates must be orthogonal to each other and form the Hilbert space of functions, just as x-, y- and z- axes form a three-dimensional Cartesian space. Combining normalization with orthogonalization, we have an orthonormalization condition

$$\langle\phi_n|\phi_m\rangle = \int \phi_n^*(x)\phi_m(x)dx = \delta_{nm}, \tag{2.3.14}$$

where δ is the Kronecker delta, i. e., $\delta = 1$ if $n = m$, and 0 otherwise. The orthogonalization can be understood from Fig. 2.1(c). Take $n = 1$ and 2 wavefunctions as an example. One sees that the multiplication of $\phi_1(x)\phi_2(x)$ changes sign at the midway, so the first part and second parts cancel each other out. For other pairs, the cancellation appears in different parts of the position space.

Example 7 (Completeness). Few systems can have all analytic eigenstates to test the completeness. (a) Use PIB as an example to demonstrate the completeness of eigenstates. (b) Show that $f(x) = x$ cannot be expanded by ϕ_n.

(a) We show the completeness of eigenstates through the following equation:

$$\text{Completeness} = \frac{a}{N}\sum_{n=1}^{N}\rho_n = \frac{a}{N}\sum_{n=1}^{N}|\phi_n\rangle\langle\phi_n| = \frac{a}{N}\sum_{n=1}^{N}\phi_n(x)\phi_n^*(x), \tag{2.3.15}$$

where the summation is over the number of states from $n = 1$ to N. Here, the factor $\frac{a}{N}$ is to ensure the normalization and cancel the unit of the probability density which is $1/a$. This can be easily calculated using our code `completeness` in the Appendix listing 11.1.

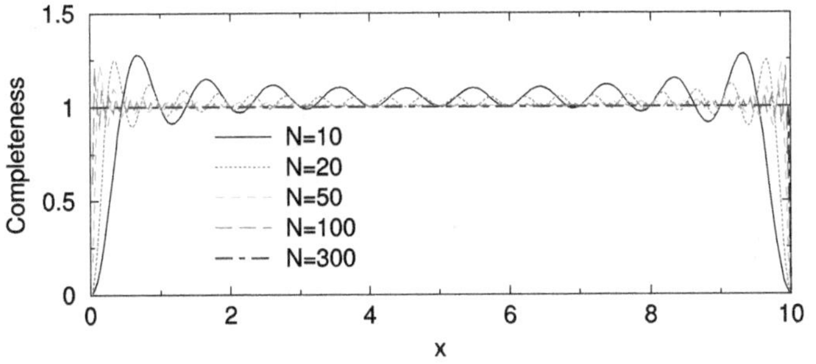

Figure 2.2: Completeness of eigenstates. Here, $a = 10$. The figure uses the code `completeness` in the Appendix listing 11.1.

Figure 2.2 shows that the completeness as a function of x for $N = 10, 20, 50$, and 100. One sees that, when N is small, the completeness deviates significantly from 1, but, as N increases, the central part approaches 1. At $N = 300$, the entire line is at 1.

(b) Because at $x = a, f = a$, but $\phi_n(x = a) = 0$, it is not possible to expand f in terms of ϕ_n.

Example 8 (Expectation values of position and momentum). Compute the expectation values of position and momentum.

For position, its expectation value is

$$\langle \phi_n(x)|\hat{x}|\phi_n(x)\rangle = \int \phi_n^*(x)x\phi_n(x)dx = \frac{2}{a}\int_0^a \sin\left(\frac{n\pi x}{a}\right)x\sin\left(\frac{n\pi x}{a}\right) = \frac{a}{2},$$

which is the same as our classical value (eq. (2.3.2)). For $\langle \phi_n(x)|\hat{x}^2|\phi_n(x)\rangle$, we find (see the following exercise)

$$\langle \phi_n(x)|\hat{x}^2|\phi_n(x)\rangle = a^2\left(\frac{1}{3} - \frac{1}{2(n\pi)^2}\right), \quad \Delta x = a\sqrt{\frac{1}{12} - \frac{1}{2(n\pi)^2}}.$$

Figures 2.3(a) and (b) illustrate that both x^2 and Δx approach the classical ones (eq. (2.3.3)) as n increases.

Similarly, we can find the momentum's expectation value $\langle \phi_n|\hat{p}_x|\phi_n\rangle$. One can prove it is always zero in PIB. We take $n = 1$ as an example to see what happens. We rewrite ϕ_1 as

$$\phi_1 = \sqrt{\frac{2}{a}}\sin\left(\frac{\pi x}{a}\right) = \sqrt{\frac{2}{a}}\frac{1}{2i}\left(e^{i\frac{\pi x}{a}} - e^{-i\frac{\pi x}{a}}\right), \tag{2.3.16}$$

$$\langle \phi_1|\hat{p}_x|\phi_1\rangle = \frac{2}{a}\frac{1}{4}\int_0^a \left(e^{-i\frac{\pi x}{a}} - e^{i\frac{\pi x}{a}}\right)\left(-i\hbar\frac{d}{dx}\right)\left(e^{i\frac{\pi x}{a}} - e^{-i\frac{\pi x}{a}}\right)dx = \frac{1}{2a}\int_0^a \left(e^{-i\frac{\pi x}{a}} - e^{i\frac{\pi x}{a}}\right)\left(\frac{\hbar\pi}{a}e^{i\frac{\pi x}{a}} + \frac{\hbar\pi}{a}e^{-i\frac{\pi x}{a}}\right)dx$$

$$= \frac{\hbar\pi}{2a^2}\int_0^a \left(1 - 1 + e^{-2i\frac{\pi x}{a}} - e^{2i\frac{\pi x}{a}}\right)dx.$$

Integration over the last two terms produces zero, and the first two terms yield

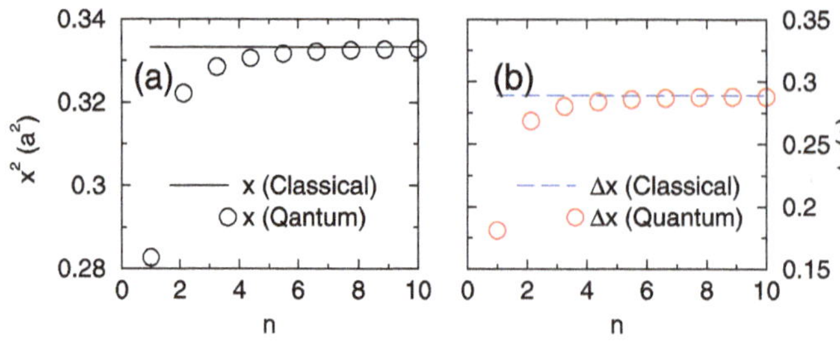

Figure 2.3: (a) The expectation value of x^2 (circles) is compared with the classical average value (solid line). (b) The uncertainty in x (circles) is compared with the classical one (dashed line).

$$\frac{\hbar\pi}{2a^2}\int_0^a 1dx = \frac{\hbar\pi}{2a}, \quad -\frac{\hbar\pi}{2a^2}\int_0^a 1dx = -\frac{\hbar\pi}{2a}. \tag{2.3.17}$$

Had we only the previous first term, our momentum expectation value would be $\frac{\hbar\pi}{2a^2}a = \frac{\hbar\pi}{2a} = \frac{\hbar k}{2}$. The momentum is half the momentum of the plane wave because eq. (2.3.16) only includes half the plane wave. If we only have the second term of eq. (2.3.17), the momentum is $-\frac{\hbar k}{2}$. The momentum for the right-moving wave cancels that for the left-moving wave. In QM, when we compute the expectation value, we sum over both terms, so the momentum is canceled out. Physically, our solution corresponds to a standing wave.

Example 9 (Parity). The parity describes how the Hamiltonian operator $\hat{H}$, or, more precisely, the potential energy operator $V(\mathbf{r})$, changes under the spatial inversion from $\mathbf{r}$ to $-\mathbf{r}$. If $V(-\mathbf{r}) = V(\mathbf{r})$, then the system has inversion symmetry or parity. If $V(-\mathbf{r}) \neq V(\mathbf{r})$, the system has no parity. The parity determines many properties often without a detailed calculation. Figure 2.4(a) shows three potentials for the one-dimensional system with x as its variable. The thin dotted line with $V(x) = x^2 + 0.2x^3$ has no parity because, if we change x to $-x$, the potential changes. Our potential for PIB starts from 0 to a. This is convenient for us to find a solution, but it misses a crucial property: *parity*. We can recover its parity by shifting the potential by $-\frac{a}{2}$, see Fig. 2.4(b). (a) Find the eigenstates. (b) Find the eigenvalues.

Because the kinetic energy operator $\hat{T} = -\frac{\hbar^2}{2m}\frac{d^2}{dx^2}$ remains symmetric under the spatial inversion from x to $-x$, our Hamiltonian is symmetric with respect to the origin of the coordinate, and has a definite parity, $\hat{H}(x) = \hat{H}(-x)$.

If a system has parity, its eigenstates have two possible parities: even and odd parities. If $\phi_n(x) = \phi_n(-x)$, the eigenstate is said to have an even parity, or gerade in German. If $\phi_n(x) = -\phi_n(-x)$, it has an odd parity, or ungerade. This property is generic, not restricted to our problem, nor in a one-dimensional system. For PIB, the even parity eigenstates are $\phi_n^e(x) = \sqrt{\frac{2}{a}}\cos(k_n^e x)$, and the odd ones are $\phi_l^o(x) = \sqrt{\frac{2}{a}}\sin(k_l^o x)$. In order for these eigenstates to match the boundary conditions at $x = \pm\frac{a}{2}$, k_n^e and k_l^o take different values. $k_n^e = \frac{n\pi}{a}$, where n is odd, and $k_l^o = \frac{l\pi}{a}$, where l is even.

A direct consequence of parity is that the matrix element of the position x, $\langle\phi_i|x|\phi_j\rangle$, is zero if state i and state j have the same parity,

$$\langle\phi_{n_1}^e|x|\phi_{n_2}^e\rangle = 0, \quad \langle\phi_{l_1}^o|x|\phi_{l_2}^o\rangle = 0.$$

Nonzero elements are only between states of opposite parities such as $\langle\phi_n^e|x|\phi_l^o\rangle$. In optics, this is called the selection rule.

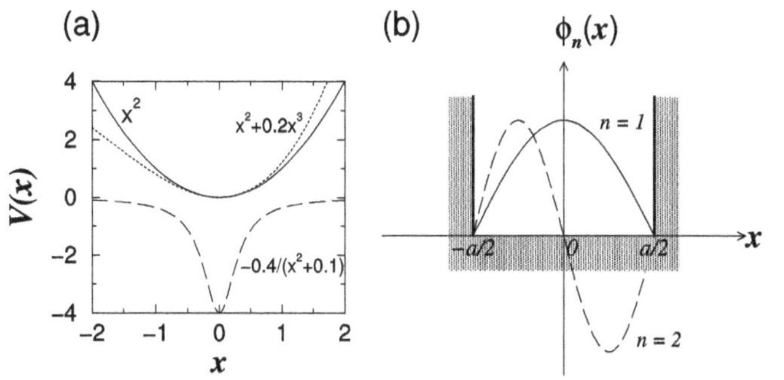

Figure 2.4: (a) Parity is determined by the symmetry of the potential energy of a system with respect to the origin of coordinate. The thick solid line denotes a potential of $V(x) = x^2$ and the long-dashed line $V(x) = -0.4/(x^2 + 0.1)$, both of which have parity. But the dotted thin line ($V(x) = x^2 + 0.2x^3$) does not have. (b) The parity of the PIB is restored when we center the infinite quantum well at $x = 0$. It accommodates eigenstates of two types: One is even and the other is odd. $\phi_{n=1}(x)$ is even, while $\phi_{n=2}$ is odd.

The eigenenergies are given by

$$E_n = \frac{\hbar^2 k^2}{2m} = \frac{\hbar^2 (n\pi)^2}{2ma^2}, \tag{2.3.18}$$

which matches eq. (2.3.11). Here, n can take even or odd integers.

Exercise 2.3.4.

13. Use an `Excel spreadsheet` to test the completeness of eigenstates of PIB for the potential (a) without parity and (b) with parity.

14. (a) Compute the expectation value $\langle \phi_n | \hat{x}^2 | \phi_n \rangle$. (b) Compute the uncertainty in position, $\Delta x = \sqrt{\langle \phi_n(x) | \hat{x}^2 | \phi_n(x) \rangle - \langle \phi_n(x) | \hat{x} | \phi_n(x) \rangle^2}$, and compare it with the classical approach.

15. Compute (a) the expectation value $\langle \phi_n | \hat{p}_x^2 | \phi_n \rangle$, and (b) the uncertainty in momentum Δp_x.

16. Prove (a) if $n = m$, $\langle \phi_n | \phi_m \rangle = 1$, which is called the normalization, and (b) if $n \neq m$, $\langle \phi_n | \phi_m \rangle = 0$, which is called the orthogonalization.

17. Use the eigenfunctions with parity to prove (a) $\langle \phi^e_{n_1} | x | \phi^e_{n_2} \rangle = 0$ and (b) $\langle \phi^o_{l_1} | x | \phi^o_{l_2} \rangle = 0$. (c) Find $\langle \phi^o_{l_1} | x | \phi^e_{n_2} \rangle$.

18. From $x = 0$ to $a/2, f(x) = Ax$, and, from $x = a/2$ to $a, f(x) = A(-x + a)$. (a) Normalize $f(x)$ for $x \in (0, a)$. (b) Expand $f(x)$ in terms of ϕ_n up to $n = 3$. (c) Based on (b), find the expectation value of $\langle f | \hat{H} | f \rangle$.

2.3.5 Time-dependent state evolution: stationary-state evolution

In the following, we show two examples of the time-evolution of a state as we briefly mentioned in the previous section. This time evolution is a stationary-state evolution, and the energy of the system is unchanged from the beginning.

2.3.5.1 Pure state

First, we assume that the system is in a single eigenstate $\phi_n(\mathbf{r})$. This is called the pure state evolution. At any instant of time t, the wavefunction is $\psi_n(t) = e^{-iE_n t/\hbar}\phi_n$. Essentially, we multiply the phase factor $e^{-iE_n t/\hbar}$ with the original eigenstate to find a state at time t. The coefficient $e^{-iE_n t/\hbar}$ only provides a phase and oscillates periodically with time. The larger E_n is, the shorter the period becomes. If $E_n = 1\,\text{eV}$, we find the period of the state oscillation, $\tau_n = h/|E_n|$, to be 4.14 fs. If we compute the energy using this wavefunction, we have

$$\langle\psi_n(t)|\hat{H}|\psi_n(t)\rangle = \langle\phi_n e^{+iE_n t/\hbar}|\hat{H}|e^{-iE_n t/\hbar}\phi_n\rangle = E_n, \tag{2.3.19}$$

which is independent of t. For other quantities, their expectation values are also time independent as long as their operators are time independent. For instance, $\langle\psi_n(t)|\hat{x}|\psi_n(t)\rangle = \langle\phi_n|\hat{x}|\phi\rangle = \frac{a}{2}$. However, this is not necessarily true for a mixed state. It depends on whether the operator commutes with the Hamiltonian (see Chapter 3).

2.3.5.2 Mixed states with known coefficients

Second, our system is in a mixed state, a state which consists of more than one eigenstate. Suppose our system at $t = 0$ is in a state $\psi(0) = c_1\phi_1 + c_2\phi_2$, where c_1 and c_2 are coefficients of these two states and $|c_1|^2$ and $|c_2|^2$ represent the probabilities in these two states, $|c_1|^2 + |c_2|^2 = 1$. At time t, $\psi(t) = c_1 e^{-iE_1 t/\hbar}\phi_1 + c_2 e^{-iE_2 t/\hbar}\phi_2$.

If we compute the expectation value of the energy, $\langle\psi(t)|\hat{H}|\psi\rangle = |c_1^2|E_1 + |c_2^2|E_2$, again independent of time (see Fig. 2.5(a)). This is a character of the stationary state. For other quantities, it may or may not depend on time.[3] For instance, the position of the particle, $\langle\psi(t)|x|\psi(t)\rangle$, depends on time as $\langle\psi(t)|x|\psi(t)\rangle = |c_1^2|\langle\phi_1|x|\phi_1\rangle + |c_2^2|\langle\phi_2|x|\phi_2\rangle + c_1^* c_2 e^{i(E_1-E_2)t/\hbar}\langle\phi_1|x|\phi_2\rangle + c_2^* c_1 e^{i(E_2-E_1)t/\hbar}\langle\phi_2|x|\phi_1\rangle = \frac{a}{2} + \frac{40a}{9\pi^2}\mathrm{Re}(c_1^* c_2 e^{i\omega_{12}t})$, where $\omega_{12} =$

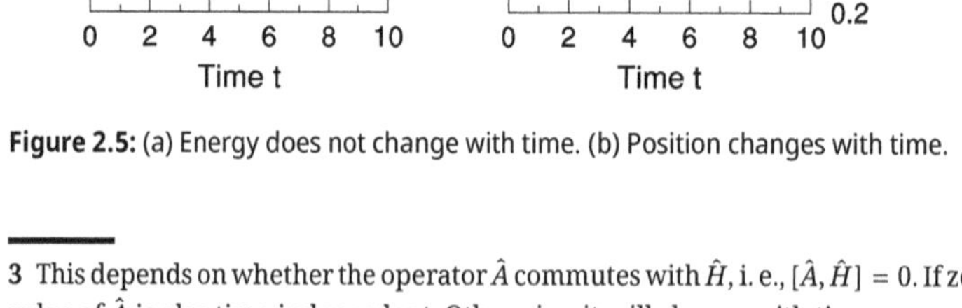

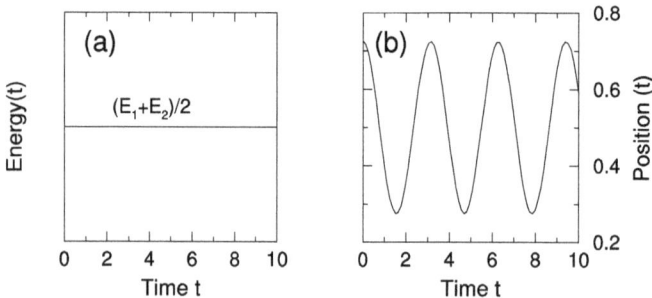

Figure 2.5: (a) Energy does not change with time. (b) Position changes with time.

3 This depends on whether the operator $\hat{A}$ commutes with $\hat{H}$, i. e., $[\hat{A},\hat{H}] = 0$. If zero, then the expectation value of $\hat{A}$ is also time-independent. Otherwise, it will change with time.

$(E_1 - E_2)/\hbar$. This shows that the position oscillates with time. Suppose $c_1 = c_2 = \frac{1}{\sqrt{2}}$ and $a = 1$. Figure 2.5(b) illustrates an example.

2.3.5.3 Mixed states with unknown coefficients

The previous coefficients of eigenstates for the initial state $\psi(\mathbf{r}, 0)$ are given, but, if not, we must find them first by expanding $\psi(x, 0)$ (in 1D) in terms of eigenstates ϕ_n like,

$$\psi(x,0) = \sum_n c_n \phi_n(x), \tag{2.3.20}$$

where $c_n = \int \psi(x,0)\phi_n^*(x)dx$ (see eq. (2.2.13)). Then, at time t, the wavefunction is

$$\psi(x,t) = \sum_n c_n \phi_n(x) e^{-iE_n t/\hbar}. \tag{2.3.21}$$

Caution must be taken that this expansion is conditional. First, $\{\phi_n(x)\}$ must be complete, $\sum_n \phi_n^* \phi_n = 1$. This is the reason why we emphasize the Hilbert space for functions. Second, $\psi(x, 0)$ must have the same boundary condition as ϕ_n and exist in the spatial domain as ϕ_n. For instance, if $\psi(x, 0) = x^2$, we cannot expand in the eigenstates of PIB ϕ_n because at $x = a$, $\psi = a^2$, not 0.

Example 10 (Find the coefficient). Given $\psi(x, 0) = bx(a - x)$, where b is a normalization constant, find c_n.
First, we multiply eq. (2.3.20) by $\phi_m(x)^*$ from the left and integrate both sides,

$$\int_0^a \phi_m^*(x)\psi(x,0)dx = \sum_n c_n \int_0^a \phi_m^*(x)\phi_n(x)dx. \tag{2.3.22}$$

Then, we use the orthogonality $\langle \phi_m | \phi_n \rangle = \delta_{nm}$ to simplify the right side,

$$\int_0^a \phi_m^*(x)\psi(x,0)dx = \sum_n c_n \delta_{nm} = c_m \rightarrow c_m = \int_0^a \phi_m^*(x)\psi(x,0)dx.$$

After integration, we find $c_m = 8\sqrt{15}/(m\pi)^3$, where m is an odd number.

Exercise 2.3.5.

19. At $t = 0$, our system is in a mixed state with the wavefunction $\psi(t = 0) = 1/\sqrt{3}\phi_2 + \sqrt{2/3}\phi_3$, where ϕ_2 and ϕ_3 are the eigenstates of PIB. (a) Write down $\psi(t \neq 0)$. (b) Find the expectation value of energy $\langle \psi(t)|\hat{H}|\psi(t)\rangle$, where $\hat{H}$ is the Hamiltonian of PIB. (c) Find $\langle \psi(t)|\hat{x}|\psi(t)\rangle$ and $\langle \psi(t)|\hat{p}_x|\psi(t)\rangle$ and sketch a diagram of $x - p_x$. (d) Why is the momentum no longer zero?

20. $\psi(x, 0) = bx(a - x)$. (a) Using the normalization condition to find b. (b) Find c_n for ϕ_n. (c) Compute $\psi(x, t)$, and find $\langle \psi(x,t)|\hat{x}|\psi(x,t)\rangle$ and $\langle \psi(x,t)|\hat{p}_x|\psi(x,t)\rangle$.

21. At $t = 0$, the particle is in a state $|\psi(x, 0)\rangle$, where, from $x = 0$ to $a/2$, $\psi(x) = Ax$, and, from $x = a/2$ to a, $\psi(x, 0) = A(-x + a)$. (a) Normalize $\psi(x, 0)$ for $x \in (0, a)$. (b) Compute $\langle \psi(x, 0)|\phi_n\rangle$ up to $n = 3$. (c) Based on (b), find $\psi(x, t)$.

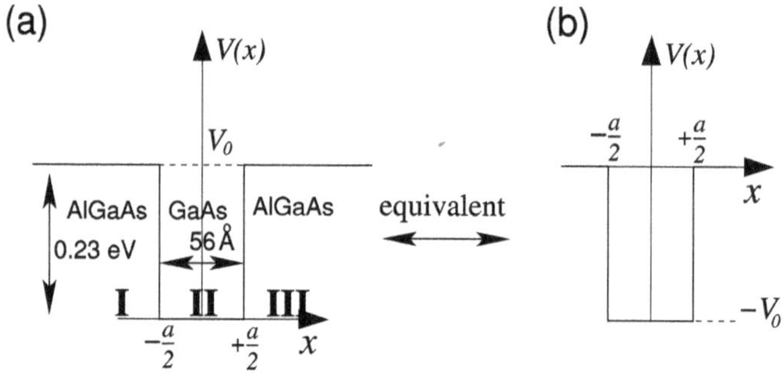

Figure 2.6: Finite quantum well. (a) The potential is zero in the well, but is V_0 outside the well. The entire potential is separated into three regions. Region I is from $x = -\infty$ to $-\frac{a}{2}$. II is between $x = -\frac{a}{2}$ and $\frac{a}{2}$. III is from $x = \frac{a}{2}$ to ∞. We can choose an energy window which is either higher than the well height V_0 or lower than V_0. For the former, we will get unbound states; for the latter, we will get bound states. The actual parameters shown are taken from Ref. [16]. (b) The equivalent well as in (a) but the potential inside the well is $-V_0$ and is zero outside the well.

2.4 The finite square well: bound and unbound states

In PIB, the potential outside the well is infinite and inside the well is zero, where all the eigenstates are bound states. A more practical case is that the potential outside the well is finite, where both bound and unbound states appear, depending on whether the particle energy is above or below the barrier height V_0. Semiconductor quantum wells are examples. These wells are built from two types of semiconductors, such as GaAs and $Al_{0.2}Ga_{0.8}As$ [16]. They are used in semiconductor and other optoelectronic devices. Figure 2.6(a) illustrates such a potential for our finite square well, where the potential inside the well is zero, but outside the well is a positive constant, $V_0 > 0$. If we rigidly shift the origin of the potential by $-V_0$, we end up with Fig. 2.6(b). These two are equivalent physically, and both are used in textbooks.

The finite square well introduces several important differences from PIB. First of all, the potential is now

$$
\begin{aligned}
V &= 0, \quad \text{if} \quad -\frac{a}{2} \le x \le \frac{a}{2}, \\
V &= V_0, \quad \text{if} \quad x < -\frac{a}{2} \quad \text{or} \quad x > \frac{a}{2}.
\end{aligned}
$$

This potential separates the space into three regions, so our wavefunction consists of three subwavefunctions in three regions, $\phi_{\mathrm{I}}(x)$, $\phi_{\mathrm{II}}(x)$ and $\phi_{\mathrm{III}}(x)$ for Regions I, II, and III. In the end, three subwavefunctions must be stitched together at the boundaries smoothly to form the final wavefunction. In PIB, we only have two boundaries, but here we have four, $x = \pm\infty$ and $x = \pm\frac{a}{2}$.

2.4.1 Finding boundary conditions for wavefunctions

When we have boundaries, a part of the wavefunction in one region must be mathematically connected to another part in a different region. Boundary conditions (BCs) are the relationships between different parts of the same wavefunction, and they put a constraint on the wavefunction and build the physics into the Schrödinger equation. Take $x = \pm\infty$ as an example. If we want to describe a traveling wave, which is spatially delocalized, then $\phi(\pm\infty)$ must be finite. If we want to describe a bound state, spatially localized in the finite region, $\phi(\pm\infty)$ must be zero. In the following, we seek a bound state solution, so we choose $\phi(\pm\infty) = 0$. We will explain why only ϕ and $\frac{d\phi}{dx}$ can be used as the boundary conditions.

The beginner may wonder why we cannot use the second-order derivative ϕ'' to set up the BCs. We take $x = -\frac{a}{2}$ as an example, which is the boundary between Regions I and II. The Schrödinger equation in Region I is

$$-\frac{\hbar^2}{2m}\frac{d^2\phi_\mathrm{I}}{dx^2} + V_0\phi_\mathrm{I} = E\phi_\mathrm{I}, \rightarrow -\frac{\hbar^2 d^2\phi_\mathrm{I}}{2mdx^2} = (E - V_0)\phi_\mathrm{I}, \tag{2.4.1}$$

but in Region II the wavefunction ϕ_II satisfies

$$-\frac{\hbar^2 d^2\phi_\mathrm{II}}{2mdx^2} = (E - 0)\phi_\mathrm{II}. \tag{2.4.2}$$

As far as V_0 differs from 0, $\frac{d^2\phi_\mathrm{I}}{dx^2} \neq \frac{d^2\phi_\mathrm{II}}{dx^2}$. This means that the second-order derivative of the wavefunction at the BCs cannot be used.

Next, we consider ϕ'. The generic Schrödinger equation with one-dimensional potential $V(x)$ is

$$-\frac{\hbar^2 d^2\phi(x)}{2mdx^2} + V(x)\phi(x) = E\phi(x) \rightarrow \frac{d^2\phi(x)}{dx^2} = -\frac{2m[E - V(x)]}{\hbar^2}\phi(x). \tag{2.4.3}$$

We integrate eq. (2.4.3) to get

$$\frac{d\phi}{dx} = -\frac{2m}{\hbar^2}\int_{x_1}^{x_2}[E - V(x)]\phi(x)dx = -\frac{2m}{\hbar^2}\left[\int_{x_1}^{x_2}E\phi(x)dx - \int_{x_1}^{x_2}V(x)\phi(x)dx\right], \tag{2.4.4}$$

where the boundary falls between x_1 and x_2. As x_1 approaches x_2, the first integral is always zero,[4] but the second integral depends on $V(x)$. If $V(x)$ is not singular, such as our current case, then the integral must be the same as we approach $-\frac{a}{2}$ from Region I or from Region II, which is also true between Regions II and III:

4 Because two limits become the same, the integral is zero.

$$\frac{d\phi_{\mathrm{I}}}{dx}\left(x=-\frac{a}{2}\right)=\frac{d\phi_{\mathrm{II}}}{dx}\left(x=-\frac{a}{2}\right),\quad \frac{d\phi_{\mathrm{II}}}{dx}\left(x=\frac{a}{2}\right)=\frac{d\phi_{\mathrm{III}}}{dx}\left(x=\frac{a}{2}\right).$$

But, if the potential has a singularity (see problems 9 and 10 at the end of the chapter), such as $V(x) = V_0\delta(x-c)$, where c is a constant,

$$\frac{d\phi_{\mathrm{I}}}{dx}-\frac{d\phi_{\mathrm{II}}}{dx}=\frac{2m}{\hbar^2}\int_{x_1}^{x_2} V_0\delta(x-c)\phi(x)dx=\frac{2m}{\hbar^2}V_0\phi(c).$$

Finally, we consider ϕ. We integrate eq. (2.4.4) to get $\phi = \int \frac{d\phi}{dx}dx$. Since $\frac{d\phi}{dx}$ is always finite, ϕ is unconditionally continuous, $\phi_{\mathrm{I}}(x=-\frac{a}{2}) = \phi_{\mathrm{II}}(x=-\frac{a}{2})$ and $\phi_{\mathrm{II}}(x=\frac{a}{2}) = \phi_{\mathrm{III}}(x=\frac{a}{2})$. The following table summarizes our BCs.

boundary	wavefunction	first-order derivative
$x=-\infty$	$\phi_{\mathrm{I}}(-\infty)=0$	$\phi'_{\mathrm{I}}(-\infty)=0$
$x=\infty$	$\phi_{\mathrm{III}}(\infty)=0$	$\phi'_{\mathrm{III}}(\infty)=0$
$x=-a/2$	$\phi_{\mathrm{I}}(-a/2)=\phi_{\mathrm{II}}(-a/2)$	$\phi'_{\mathrm{I}}(-a/2)=\phi'_{\mathrm{II}}(-a/2)$
$x=a/2$	$\phi_{\mathrm{III}}(a/2)=\phi_{\mathrm{II}}(a/2)$	$\phi'_{\mathrm{III}}(a/2)=\phi'_{\mathrm{II}}(a/2)$

2.4.2 Wavefunctions in three regions

We start to solve the Schrödinger equation in I and III since they have the same potential,

$$-\frac{\hbar^2}{2m}\frac{d^2\phi}{dx^2}+V_0\phi=E\phi \rightarrow \frac{d^2\phi}{dx^2}=-\frac{2m(E-V_0)}{\hbar^2}\phi=k^2\phi, \tag{2.4.5}$$

where we define k through the last equation, $k=\sqrt{2m(V_0-E)}/\hbar$. We are interested in a bound state solution, so we choose $E < V_0$; then, k is a real and positive number. This differential equation has two possible solutions, $\phi(x) = Ae^{\pm kx}$. To determine which one is valid, we now impose our boundary condition to $\phi(x)$. For I, we only choose $\phi_{\mathrm{I}}(x) = A_{\mathrm{I}}e^{kx}$ so at $x \rightarrow -\infty$, $\phi_{\mathrm{I}} = 0$. For Region III, $\phi_{\mathrm{III}}(x) = A_{\mathrm{III}}e^{-kx}$, since as $x \rightarrow \infty$, $\phi_{\mathrm{III}} \rightarrow 0$.

In Region II, with $V = 0$, our Schrödinger equation is

$$-\frac{\hbar^2}{2m}\frac{d^2\phi}{dx^2}=E\phi \rightarrow \frac{d^2\phi}{dx^2}=-\frac{2mE}{\hbar^2}\phi=-q^2\phi,$$

where $q=\sqrt{2mE}/\hbar$ is defined through the last equation as previously. We choose $E > 0$. This equation has two independent roots, $\phi_{\mathrm{II}}(x) = A_{\mathrm{II}}\cos(qx)$ and $A_{\mathrm{II}}\sin(qx)$. In the following, we choose $\phi_{\mathrm{II}}(x) = A_{\mathrm{II}}\cos(qx)$, and the calculation for $\phi_{\mathrm{II}}(x) = A_{\mathrm{II}}\sin(qx)$ is left as an exercise.

2.5 Eigenenergies and eigenstates in the finite quantum well

In order to find $\phi_{\mathrm{I}}(x)$, $\phi_{\mathrm{II}}(x)$ and $\phi_{\mathrm{III}}(x)$, we must find k and q. In PIB, k is related to the width of the well (eq. (2.3.10)), but this is not the case for the finite quantum well, because in Regions I and III, x goes to $\pm\infty$. To reduce the number of unknowns, we can write k in terms of E and V_0

$$k = \sqrt{2m(V_0 - E)}/\hbar, \tag{2.5.1}$$

and combine it with q through

$$k^2 + q^2 = \frac{2mV_0}{\hbar^2}, \tag{2.5.2}$$

which shows that k and q are dependent on each other.

2.5.1 Eigenenergies and boundary conditions

The BCs build in physics and deliver k and q. We choose $x = -\frac{a}{2}$ between Regions I and II. Since $\phi_{\mathrm{I}}(x) = A_{\mathrm{I}}e^{kx}$ and $\phi_{\mathrm{II}}(x) = A_{\mathrm{II}}\cos qx$, their values and derivatives satisfy BCs,

$$A_{\mathrm{I}}e^{-ka/2} = A_{\mathrm{II}}\cos(-qa/2) = A_{\mathrm{II}}\cos(qa/2), \tag{2.5.3}$$

$$A_{\mathrm{I}}ke^{-ka/2} = A_{\mathrm{II}}(-q)\sin(-qa/2) = A_{\mathrm{II}}q\sin(qa/2). \tag{2.5.4}$$

Introducing two dimensionless quantities ξ and η, $\xi = \frac{ka}{2}$ and $\eta = \frac{qa}{2}$, we divide eq. (2.5.4) by eq. (2.5.3) and multiply both sides with $a/2$. The result is

$$\xi = \eta\tan\eta. \tag{2.5.5}$$

Equation (2.5.2) leads to

$$k^2 + q^2 = \frac{2mV_0}{\hbar^2} \to \xi^2 + \eta^2 = \frac{2mV_0a^2}{4\hbar^2} = \frac{mV_0a^2}{2\hbar^2} \equiv \omega^2, \tag{2.5.6}$$

where $\omega = \sqrt{\frac{mV_0a^2}{2\hbar^2}}$ is a dimensionless quantity.

Most textbooks [17, 18] stop here and resort to the numerical solution. In fact, one can simplify it further by substituting eq. (2.5.5) into eq. (2.5.6), so the equation becomes

$$\eta^2\tan^2\eta + \eta^2 = \omega^2, \to \eta^2\sec^2\eta = \omega^2, \to \eta = \pm\omega\cos\eta \quad \text{(energy quantization)}, \tag{2.5.7}$$

which finally constrains our solution and leads to energy quantization, just as eq. (2.3.10) does to PIB. These two equations are elementary. For a given ω (given V_0 and a), one can find a set of η. This equation shows that, regardless of V_0, there is always at least one root.

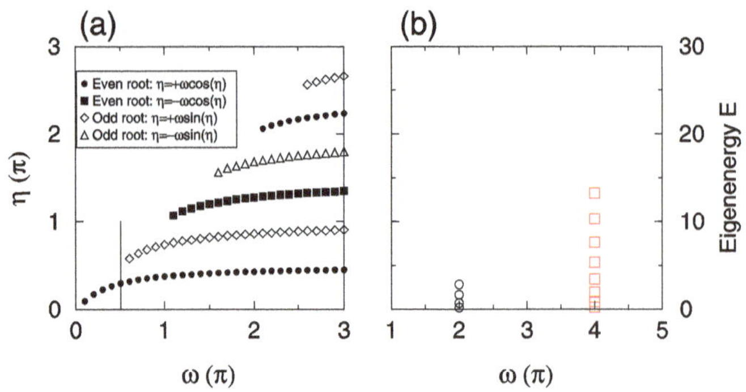

Figure 2.7: (a) Roots of $\eta = \pm\omega\cos\eta$ and $\eta = \pm\omega\sin\eta$. Odd states only appear after $\pi/2$ (the vertical line). These roots are computed from two codes `evenroot.f` and `oddroot.f` in the Appendix listings 11.2 and 11.4, respectively. (b) Eigenenergies at $\omega = 2\pi$ and 4π, in units of $\frac{2\hbar^2}{ma^2}$. As ω increases, the number of roots increases. At $\omega = 5\pi$, there are ten bound states (Tab. 2.1), whose eigenfunctions are plotted in Fig. 2.8.

This is because $\cos\eta = 1 - \eta^2/2 + \cdots$. If η is very small, $\cos\eta \approx 1$, so we have $\eta \approx \omega$, at least one bound state. In the another limit, if V_0 is large, i. e., a large ω, there are multiple bound-state solutions. We add two comments: η must be expressed in radians, and only the positive roots of η and ξ are kept.

Figure 2.7(a) shows the numerical roots for $\omega = 0$ to 3π for both even and odd eigenstates. It is interesting to note that the odd roots do not start until $\omega = \frac{\pi}{2}$. Below this, all roots are even. The programs to find these roots are in the Appendix listings 11.2 and 11.4. We find that these roots are very easily missed. In the code, we first roughly screen where roots appear and then use a middle-point method to find their exact values. The negative roots are discarded because we assume both k and q, or η and ξ, are positive. The reader is strongly encouraged to try them out.

ξ is found through $\xi = \sqrt{\omega^2 - \eta^2} = \sqrt{\omega^2 - \omega^2\cos^2\eta} = \omega\sin\eta$. Once η and ξ are found, the eigenenergy can be calculated from (note $\eta = \frac{qa}{2}$)

$$E = \frac{\hbar^2 q^2}{2m} = \frac{2\hbar^2\eta^2}{ma^2}. \tag{2.5.8}$$

These eigenenergies are discrete, though we cannot assign a quantum number n to it because it depends on η.[5] Figure 2.7(b) shows two sets of eigenenergies E at $\omega = 2\pi$ and 4π, in units of $\frac{2\hbar^2}{ma^2}$. k and q also give us de Broglie wavelengths, $\frac{2\pi}{k}$ in Regions I and III and $\frac{2\pi}{q}$ in II. They are different, even though the energy is the same. This further demonstrates that the wavelength can change, but the energy does not, which is why the light photon energy stays the same in water and air (see the problems in Chapter 1). Table 2.1 lists the numerical eigenenergies with $\omega = 5\pi$.

5 We could write E as E_η.

Table 2.1: Eigenenergies E in units of $(\frac{2\hbar^2}{ma^2})$ at $\omega = 5\pi$. The parity alternates with E.

No.	1	2	3	4	5	6	7	8	9	10
E	0.221	0.883	1.985	3.524	5.497	7.896	10.71	13.92	17.50	21.38
parity	even	odd	even	odd	even	odd	even	odd	even	odd

Our energy quantization condition (eq. (2.5.7)) is very convenient. It allows us to recover all the prior results in PIB. We take $V_0 \to \infty$, i. e., $\omega \to \infty$, in eq. (2.5.7) to get $\cos\eta = 0$, $\eta = (2n+1)\frac{\pi}{2}$, where n is an integer. Then the energy is

$$E_n = \frac{\hbar^2\pi^2}{2ma^2}(2n+1)^2,$$

which matches the eigenenergies of an odd n in eq. (2.3.18). The eigenenergies in eq. (2.3.18) with an even n are from the odd root $\phi_{\rm II} = A_{\rm II}\sin(qx)$.

Example 11 (Compute an even root with a calculator). For $\omega = 1$, use a calculator to find (a) η, (b) ξ, and (c) E.

We start from $\eta = \pm\omega\cos\eta$. Since $\omega = 1$, our equation is $\eta = \cos\eta$. We start from $\eta = 0$ to see whether it satisfies both sides of the equation. Clearly, $\eta = 0$ is too small. We gradually increase in small steps up to $\eta = 1$. For $\eta = 0.1$, we get $\cos\eta = 0.995$. At $\eta = 0.7$, we have 0.7648, close to the exact value of 0.739095133214, which is computed using Code `evenroot.f` in the Appendix listing 11.2. We find $\xi = 1 \times \sin 0.739095133214 = 0.673612029$, which is positive and is a valid root. One must check this, though we do not need ξ to compute the eigenenergy. Since $\omega = 1$, $V_0 = \frac{2\hbar^2}{ma^2}$. We can find the eigenenergy $E = \frac{\hbar^2 4\eta^2}{2ma^2} = \eta^2 V_0 = 0.453753166V_0 < V_0$. This means that the particle can exist even if its energy is below the potential barrier, a finding that classical physics cannot explain.

Exercise 2.5.1.

22. Use `evenroot.f` and `oddroot.f` to find all the roots shown in Fig. 2.7(a).
23. Compute the energy levels for $\omega = 3\pi$.
24. (a) Show that, if we start from the odd root $\phi_{\rm II} = A_{\rm II}\sin(qx)$, the energy quantization condition is $\eta = \pm\omega\sin\eta$. (b) In the limit $\omega \to \infty$, $E_n = \frac{\hbar^2}{2ma^2}(2n)^2$.

2.5.2 Eigenstates and physical meanings

To determine the eigenfunctions, we need to find the coefficients $A_{\rm I}$, $A_{\rm II}$, and $A_{\rm III}$. The symmetry in the potential leads to $A_{\rm I} = A_{\rm III}$.[6] Equations (2.5.3) and (2.5.4) provide $A_{\rm I}e^{-\xi} = A_{\rm II}\cos\eta$ and $A_{\rm II} = A_{\rm I}e^{-\xi}/\cos\eta$, where $\xi = \frac{ka}{2}$ and $\eta = \frac{qa}{2}$ are used.

6 For odd eigenstates, they may differ by a sign.

The eigenfunction has three parts for three regions I, II, and III,

$$\phi_{\mathrm{I}}(x) = A_{\mathrm{I}}e^{kx}, \quad \text{if} \quad x < -a/2,$$
$$\phi_{\mathrm{II}}(x) = A_{\mathrm{II}}\cos qx, \quad \text{if} \quad -a/2 < x < a/2,$$
$$\phi_{\mathrm{III}}(x) = A_{\mathrm{III}}e^{-kx}, \quad \text{if} \quad x > a/2.$$

We use the normalization condition to find (see the details in the following exercise)

$$A_{\mathrm{I}} = \sqrt{\frac{2}{a}}\frac{e^{\xi}}{\sqrt{\frac{1}{\xi} + \frac{2\eta+\sin 2\eta}{2\eta\cos^2\eta}}}, \tag{2.5.9}$$

which can be used to find A_{II} and A_{III}.

As an example, we choose $\omega = 5\pi$ and compute 10 eigenfunctions, with their eigenenergies given in Table 2.1. Figure 2.8(a) shows five even eigenfunctions as a function of x with η increasing from the bottom to top. They are symmetric with respect to $x = 0$, $\phi(-x) = \phi(x)$. Within the well, the wavefunction is a cosine function. To have

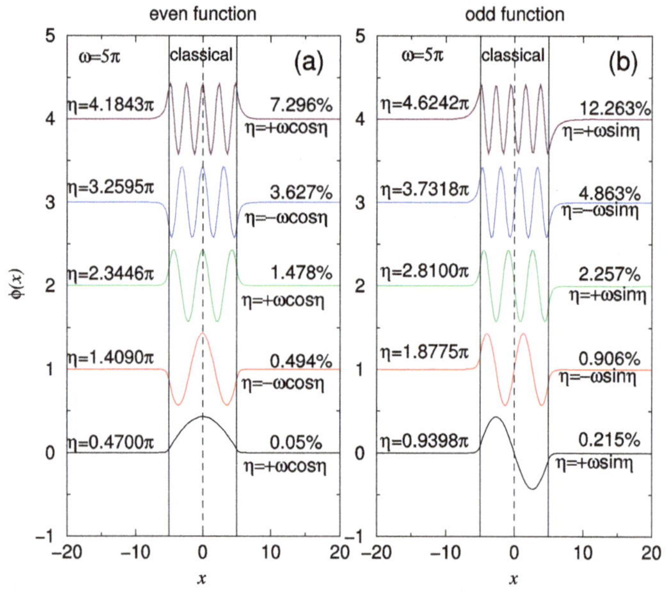

Figure 2.8: (a) Even eigenfunctions at $\omega = 5\pi$ as a function of x. All the wavefunctions are normalized and are shifted vertically by one unit, except the lowest one. The classical region is denoted by two vertical solid lines. The percentage denotes the probability for the particle found outside the classical region. η is from Fig. 2.7. (b) Odd eigenfunctions. Their energies are higher, so the probability for the particle existing outside the well is larger. The codes used to find these eigenstates are `finite.well.odd.wavefunction.f` and `finite.well.even.wavefunction.f` in Appendix listings 11.3 and 11.5, whose input parameters are read from `odd.w.eta.used` and `even.w.eta.used`.

an easy view, we vertically shift all the wavefunctions by 1, except the first one. We first note that the number of nodes in the wavefunction is 0, 2, 4, 6, and 8, increasing in energy. The ground state has $\eta = 0.4700\pi$, with no node.

Figure 2.8(b) shows the odd eigenfunctions. They are antisymmetric, $\phi(-x) = -\phi(x)$, with odd numbers of nodes, 1, 3, 5, 7, and 9, increasing in energy. Within the well, the wavefunction is a sine function.

The most striking feature of the eigenfunction is that, although $E < V_0$, i. e., the particle's energy is lower than the barrier height, its has a finite probability density being outside the well, increasing from only 0.05 % to 7.296 % for the even eigenfunctions (Fig. 2.8(a)) and from 0.215 % to 12.163 % for the odd eigenfunctions (Fig. 2.8(b)). This is purely quantum mechanical: Due to the wave nature of the particle, even for bound states, a quantum particle can appear in the classically forbidden region.

It is also interesting to see how the particle can manage to appear in a nonclassical region even if its energy is below the potential barrier. It turns out that the local energy in Region II (the well where the potential is zero), $E_{\mathrm{II}} = \langle\phi_{\mathrm{II}}|\hat{H}|\phi_{\mathrm{II}}\rangle$, where the integration is over Region II only, is far below V_0. The particle has a larger probability, which leaves some room for the particle in Regions I and III to have a higher local energy, higher than V_0, but with a much lower probability. This ensures the resultant total energy is still below V_0. Essentially, we have $E = |c_{\mathrm{I}}|^2 E_{\mathrm{I}} + |c_{\mathrm{II}}|^2 E_{\mathrm{II}} + |c_{\mathrm{III}}|^2 E_{\mathrm{III}} < V_0$, where $|c_{\mathrm{I}}|^2$, $|c_{\mathrm{II}}|^2$ and $|c_{\mathrm{III}}|^2$ represent the probabilities in three regions, respectively.

To this end, all the eigenfunctions are even or gerade because we have chosen an even symmetry solution. If we choose an odd function for Region II, the eigenfunction is $\phi_{\mathrm{II}}(x) = A_{\mathrm{II}} \sin qx$. Similarly, we can use the boundary conditions at $x = -\frac{a}{2}$, $\phi_{\mathrm{I}}(x = -\frac{a}{2}) = \phi_{\mathrm{II}}(x = -\frac{a}{2})$ and $\phi_{\mathrm{I}}'(x = -\frac{a}{2}) = \phi_{\mathrm{II}}'(x = -\frac{a}{2})$:

$$A_{\mathrm{I}} e^{-ka/2} = A_{\mathrm{II}} \sin(-qa/2) = -A_{\mathrm{II}} \sin(qa/2),$$
$$kA_{\mathrm{I}} e^{-ka/2} = qA_{\mathrm{II}} \cos(-qa/2) = qA_{\mathrm{II}} \cos(qa/2).$$

Dividing the first equation by the second one leads to $\frac{1}{k} = -\frac{\tan(qa/2)}{q}$. Then, we again introduce $\xi = \frac{ka}{2}$ and $\eta = \frac{qa}{2}$, which gives us the similar equation $\xi = -\eta \cot \eta$.

Exercise 2.5.2.

25. Use the normalization condition to find A_{I} in eq. (2.5.9). Hint: Compute three separate integrals, and then add up to 1.
26. Compute the local energy E_{I}, E_{II}, and E_{III}.
27. This needs a computer. Use Code `evenroot.f` in Appendix listing 11.2 to find all the roots and plot them as Fig. 2.7. The roots are stored in files `evenroot.eta-omegaxcoseta` and `evenroot.eta+omegaxcoseta`.

2.5.3 $E > V_0$ and unbound states

The states that we have computed so far have the eigenenergy $0 < E < V_0$. These are bound states, whose eigenenergies are discrete. This conclusion is generic, even if our potential is not a constant. In other words, we get a bound state if the eigenenergy is below the potential everywhere in space. In the other limit, if the eigenenergy is above the potential everywhere in space, we get an unbound state, whose eigenenergies may or may not be discrete (see Chapters 8 and 9). If $E > V_0$, our states are not bound states and are delocalized continuous states across the entire space. For instance, Region I has the Schrödinger equation,

$$\frac{d^2\phi}{dx^2} + \frac{2m(E - V_0)}{\hbar^2}\phi = 0,$$

which has two roots, e^{+ik_1x} and e^{-ik_1x}. The general solution is

$$\phi_I = A_I e^{ik_1x} + B_I e^{-ik_1x} \quad \text{for Region I.} \tag{2.5.10}$$

Additional information about the system is necessary to decide whether we want to keep both terms A_I and B_I. For instance, if we are only interested in a particle moving to the right ($+x$), then we only keep the A_I term. This situation occurs for a voltage bias on a device. If we are interested in both the right and left propagating waves, then both terms are kept. This occurs in scattering from a potential barrier. In solids, both terms are kept.

Once we choose a solution like eq. (2.5.10) for Region I, the solution for Region II is similar, $\phi_{II} = A_{II}e^{ik_2x} + B_{II}e^{-ik_2x}$. We use the boundary condition at $x = -\frac{a}{2}$, $\phi_I(x = -\frac{a}{2}) = \phi_{II}(x = -\frac{a}{2})$, which leads to $A_I e^{-ik_1a/2} + B_I e^{ik_1a/2} = A_{II}e^{-ik_2a/2} + B_{II}e^{ik_2a/2}$, and $\phi_I'(x = -\frac{a}{2}) = \phi_{II}''(x = -\frac{a}{2})$, which leads to $ik_1A_Ie^{-ik_1a/2} - ik_1B_Ie^{ik_1a/2} = ik_2A_{II}e^{-ik_2a/2} - ik_2B_{II}e^{ik_2a/2}$. We have the normalization condition $\int dx|\phi(x)|^2 = 1$.

In total, we have three equations but we have four unknowns, A_I, B_I, A_{II}, and B_{II}. This shows that we have to put more constraints on our wavefunctions in order to describe a physical process, which is the goal of the next section. What we should emphasize is that the energy spectrum for $E > V_0$ is continuous. This means, for any E, we always can find a wavefunction. E is not quantized any more, similar to CM, but not exactly the same, as will be seen in the next section.

2.6 Tunneling through a barrier: unbound states

The previous section concerns a particle confined in a finite quantum well, which is an eigenvalue problem of bound states with the boundary conditions. This section deals with a different problem: tunneling through a quantum barrier. This is not an eigenvalue problem. Instead, it describes that a particle or a wave propagates toward a barrier, then reflects from and transmits through the barrier. Importantly, all the states are unbound

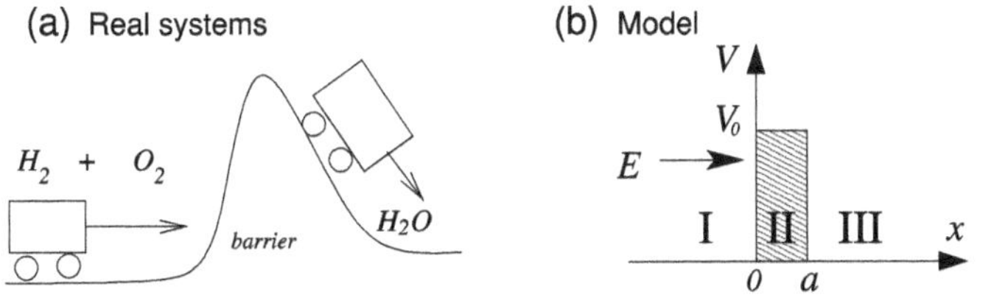

Figure 2.9: (a) Quantum scattering and tunneling are a quantum mechanical way to describe a chemical reaction or a matter passing through a barrier. (b) Our model potential barrier, with height V_0, has a finite width a, whose solution is sought in Regions I, II, and III.

states, i. e., continuous states. We know the incident energy of the particle, but we need to find the reflection and transmission coefficients. This simulates a variety of phenomena in physics and chemistry. The chemical reactants is such a process, where two reactions pass through a barrier to form a new product. Figure 2.9(a) shows such a process. The ultimate goal of our model is to reveal that even if a particle's initial energy is lower than the barrier height, it can still have a finite probability to tunnel through a barrier, i. e., quantum tunneling. This has a practical application in the quantum transport in a device.

2.6.1 Boundary conditions

We consider a particle traveling toward a square potential barrier (Fig. 2.9(b)),

$$V = 0, \quad \text{if} \quad x < 0, \quad \text{or} \quad x > a,$$
$$V = V_0 > 0, \quad \text{if} \quad 0 \leq x \leq a.$$

We have three regions and two boundary conditions: one at $x = 0$ and the other at $x = a$, where the wavefunction itself and its derivatives must match. The following table lists these conditions:

position	wavefunction	first derivative
$x = 0$	$\phi_{\rm I}(0) = \phi_{\rm II}(0)$	$\phi'_{\rm I}(0) = \phi'_{\rm II}(0)$
$x = a$	$\phi_{\rm II}(x = a) = \phi_{\rm III}(x = a)$	$\phi'_{\rm II}(x = a) = \phi'_{\rm III}(x = a)$

We want to know how much it is reflected and transmitted through the barrier. In contrast to the square well, where eigenfunctions are bound states, here we look for a traveling state for a given incident energy E of the wave. For this reason, E is commonly not called eigenenergy because it is the energy of an incident particle. Another difference is that here our wavefunction is delocalized, meaning that the wavefunction at $x = \pm\infty$ is not zero and is not normalized because we only know where the barrier is but do not know the system size. We are only interested in reflection and transmission of the incident particle. These differences require us to adopt a different strategy.

2.6.2 Wavefunctions

We first consider the solution in Region I (Fig. 2.9(b)). Here, a free particle with energy $E < V_0$ approaches the potential barrier from the left side. The potential is zero, so the Schrödinger equation is

$$-\frac{\hbar^2}{2m}\frac{d^2\phi_\mathrm{I}}{dx^2} = E\phi_\mathrm{I} \rightarrow \frac{d^2\phi_\mathrm{I}}{dx^2} = -\frac{2mE}{\hbar^2}\phi_\mathrm{I} = -k^2\phi_\mathrm{I}, \tag{2.6.1}$$

where $k = \frac{\sqrt{2mE}}{\hbar}$. We choose $V_0 > E > 0$,[7] ϕ_I has two roots of $e^{\pm ikx}$, where the positive (negative) exponent denotes which way the wave travels to the + (−) x-axis.

Before the particle hits the barrier ($x < 0$), the incident wavefunction is e^{ikx}, where we only choose a positive k, because the wave propagation direction is along the $+x$-axis. Note that $\phi(x)$ is not normalized and has the coefficient of 1. Once the particle hits the barrier, it reflects back, so its wavefunction is re^{-ikx}, where $|r|^2$ represents the probability of the reflected wavefunction. The reflected wave has the same wavevector with the direction flipped since both the incident and reflected waves are in the same medium and their wavelength and wavevector must be the same.[8] Our total wavefunction is a superposition of the incident and reflected wavefunctions,

$$\phi_\mathrm{I}(x) = e^{ikx} + re^{-ikx}, \quad (x < 0), \text{ Region I.} \tag{2.6.2}$$

Similarly, in Region III with $x > a$, the wave, after it tunnels through the barrier (Region II), enters the same medium as that in Region I with $x < 0$. The wavefunction has a similar form but has no reflected wave,

$$\phi_\mathrm{III} = te^{ikx}, \quad (x > a), \text{ Region III,} \tag{2.6.3}$$

where $|t|^2$ is the probability of the wave in region $x > a$. t and r are complex numbers.

Next, we consider II, $0 \le x \le a$, where the potential is V_0,

$$-\frac{\hbar^2}{2m}\frac{d^2\phi_\mathrm{II}}{dx^2} + V_0\phi_\mathrm{II} = E\phi_\mathrm{II} \rightarrow \frac{d^2\phi_\mathrm{II}}{dx^2} = \frac{2m(V_0 - E)}{\hbar^2}\phi_\mathrm{II} \rightarrow \frac{d^2\phi_\mathrm{II}}{dx^2} = q^2\phi_\mathrm{II}, \tag{2.6.4}$$

---

7 Here E must be positive. For a negative E, the wavefunction would look like $\exp(\pm kx)$, and it is not the traveling wave that we want.

8 If we consider two different media, we must have a different wavelength.

where $q = \sqrt{2m(V_0 - E)}/\hbar$. Its general solution is[9]

$$\phi_{\mathrm{II}} = Ae^{+qx} + Be^{-qx}, \quad (0 \le x \le a), \text{ Region II,} \tag{2.6.5}$$

where A and B are complex coefficients.

2.6.3 Imposing boundary conditions and transmission and reflection

We have four BCs and four unknowns, r, t, A, B. To simplify our notation, we set $\eta = \frac{k}{q}$. The boundary conditions at $x = 0$ are $\phi_{\mathrm{I}}(x = 0) = \phi_{\mathrm{II}}(x = 0)$, $\phi'_{\mathrm{I}}(x = 0) = \phi'_{\mathrm{II}}(x = 0)$, which leads to

$$1 + r = A + B, \tag{2.6.6}$$

$$ik(1 - r) = q(A - B). \tag{2.6.7}$$

The boundary conditions at $x = a$ are $\phi_{\mathrm{II}}(x = a) = \phi_{\mathrm{III}}(x = a)$, $\phi'_{\mathrm{II}}(x = a) = \phi'_{\mathrm{III}}(x = a)$, which yields

$$Ae^{qa} + Be^{-qa} = te^{ika}, \tag{2.6.8}$$

$$q(Ae^{qa} - Be^{-qa}) = ikte^{ika}. \tag{2.6.9}$$

Solving these four equations (with details in the following exercise and problem at the end of the chapter), we find

$$te^{ika} = \frac{-2i\eta}{(1 - \eta^2)\sinh qa - 2i\eta \cosh qa}, \tag{2.6.10}$$

$$r = \frac{(1 + \eta^2)\sinh qa}{-(1 - \eta^2)\sinh qa + 2i\eta \cosh qa}. \tag{2.6.11}$$

By definition, the reflection coefficient R and transmission coefficient T (dimensionless) are given by

$$R \equiv rr^* = \frac{[(1 + \eta^2)\sinh qa]^2}{[(1 - \eta^2)\sinh qa]^2 + 4\eta^2(\cosh qa)^2},$$

$$T \equiv tt^* = \frac{4\eta^2}{[(1 - \eta^2)\sinh qa]^2 + 4\eta^2(\cosh qa)^2},$$

9 In contrast with Region II of the finite quantum well problem previously addressed, where the $\sin(qx)$ and $\cos(qx)$ terms can be treated separately, the transport problem does not have symmetry. We either choose the wave moving from left to right or right to left, so there is no symmetry.

and

$$R + T = \frac{[(1+\eta^2)\sinh qa]^2 + 4\eta^2}{[(1-\eta^2)\sinh qa]^2 + 4\eta^2(\cosh qa)^2}.$$

Because $[(1-\eta^2)\sinh qa]^2+4\eta^2(\cosh qa)^2 = [(1+\eta^2)(\sinh qa)]^2+4\eta^2, R+T = 1$. This is the result of the particle number conservation. The sum of the reflected and transmitted particle numbers must be equal to the incident one.

Exercise 2.6.3.
28. Show $A = [1 + r + i\eta(1 - r)]/2$, $B = [1 + r - i\eta(1 - r)]/2$.
29. In a similar manner, show $A = \frac{e^{-qa}}{2}(te^{ika} + i\eta te^{ika}) = \frac{e^{-qa+ika}}{2}t(1+i\eta)$, $B = \frac{e^{qa}}{2}(te^{ika} - i\eta te^{ika}) = \frac{e^{qa+ika}}{2}t(1-i\eta)$.
30. Show r, t, η satisfy $r(1 - i\eta) = (1 + i\eta)[te^{-qa+ika} - 1]$, $r(1 + i\eta) = (1 - i\eta)[te^{qa+ika} - 1]$. Hint: Use the previous two results.
31. Prove $[(1 - \eta^2)\sinh qa]^2 + 4\eta^2(\cosh qa)^2 = [(1 + \eta^2)(\sinh qa)]^2 + 4\eta^2$.

2.7 Physical insights into quantum tunneling

We now reveal the physics behind our equations of quantum tunneling by directly computing the wavefunction, transmission coefficient, and current.

2.7.1 Wavefunctions of continuous states and transmission

We take the electron as an example in a region from $x = -50$ Å to $+50$ Å. We choose the barrier width $a = 5$ Å, which starts at $x = 0$ and ends at 5 Å. The barrier height is 1 eV and the incident electron energy is 0.9 eV, from which we can compute k and q numerically, $k = 0.486$ Å^{-1} and $q = 0.162$ Å^{-1}. Both r and t are complex: $r = 0.554 - 0.621i$ and $t = -0.072 - 0.549i$. Although the wavefunctions are relatively simple, it is difficult to see them clearly because our states are traveling states, i. e., continuous states, and their wavefunctions are not normalized. The amplitude of the incident wave is set at 1. Figure 2.10(a) shows the incident wave from −50 to 0 Å. The solid and dashed lines are the real and imaginary parts of ϕ. The reflected wave is shown in Fig. 2.10(b), whose amplitude is less than the incident wave. One might expect a larger reduction, but this is not the case because only $|r|^2 = 0.693$ determines the amplitude. What is nontrivial is the wavefunction inside the barrier (Fig. 2.10(c)). One can see that its amplitude is close to 2. This is because the incident and reflected wavefunctions are not normalized. In order to match the boundary condition at $x = 0$, A and B become larger: $A = -0.154 + 0.358i$ and $B = 1.709 - 0.979i$. Figure 2.10(d) illustrates that the transmitted wave has a smaller amplitude, where $|t|^2 = 0.307$. The entire wavefunction across all three regions is shown in Fig. 2.10(e).

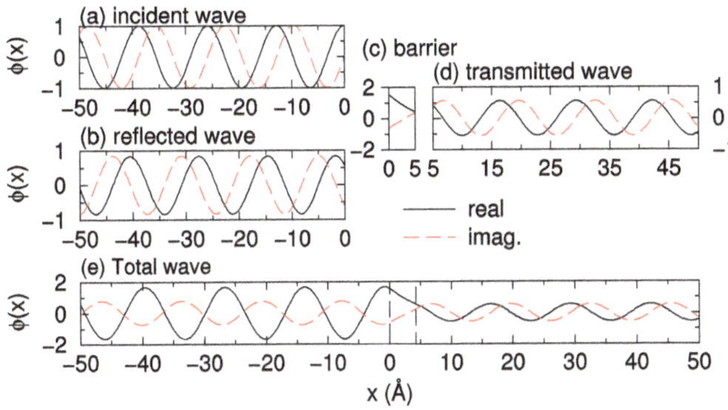

Figure 2.10: (a) Incident wave. (b) Reflected wave. (c) Wave in the barrier. (d) Transmitted wave. (e) Sum of all the waves. The dashed box represents the barrier.

Now, we systematically investigate how quantum tunneling depends on the system parameters. First, we choose the barrier width $a = 10$ Å, which is from $x = 0$ to $= 10$ Å (the two vertical lines in Fig. 2.11(a)). The barrier height is $V_0 = 1$ eV, and the incident electron energy E is 0.2 eV, $E = 0.2V_0$. Note that our wave is a traveling wave and contains both incident and reflected waves. Here the wavefunction is complex and has both real and imaginary parts: $\phi_r(x)$ and $\phi_i(x)$, denoted by the solid and dashed lines in Fig. 2.11(a), respectively. Figure 2.11(a) displays the wavefunction as a function of x. We focus on the wavefunction amplitude. In Region I, i. e., $x < 0$ Å, the majority of the electron wave reflects from the barrier, and only a small portion of the wave tunnels through the barrier (Region II) and enters Region III. Tunneling can be enhanced if we decrease the barrier width to $a = 5$ Å, while keeping the rest of the parameters unchanged (see Fig. 2.11(b)). We can also enhance tunneling by increasing the incident energy E (compare Figs. 2.11(a) with 2.11(c)).

So far, we have only investigated the wavefunction change. Next, we explore how the transmission changes with E. First, we fix $V_0 = 2$ eV but change E from 0 to 2 eV. Figure 2.11(d) shows that, when the barrier width a is 1 Å, the transmission increases sharply with E and then tapers off. If we use a larger $a = 3$ Å, the increase is much smaller (indicated by the dashed line). We stay with $a = 3$ Å and plot both T and R in the same figure (Fig. 2.11(e)) as a function of α, where α is defined through $E = \alpha V_0$. We notice that, as T becomes larger, R decreases sharply, but the sum is 1.

The dependence of T on a is shown in Fig. 2.11(f). The transmission is extremely sensitive to the barrier width a. Beyond 5 Å, the transmission is reduced to a small number, which can be analytically proven.

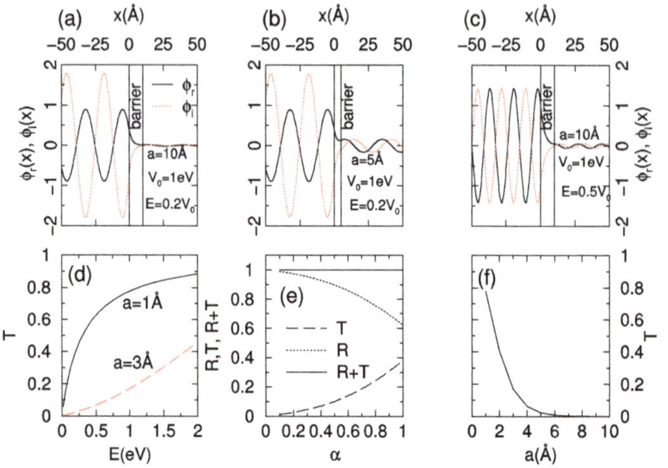

Figure 2.11: (a,b,c) The incident wave is from the left (Region I), tunnels through a barrier (Region II), and reaches the right side (Region III). (a) Unnormalized wavefunctions in three regions. The solid line is the real part of the wavefunction, while the dotted line is the imaginary part. All the parameters are given in the figure. The barrier is denoted by a box. (b) Same as (a), but with $a = 5$ Å. (c) Same as (b), but with a higher incident energy. (d) Transmission as a function of the incident particle's energy E at $a = 1$ Å and 3 Å. (e) Transmission T (dashed line) and reflection R (dotted line) coefficients as a function of the energy parameter α, where α is the proportional constant for $E = \alpha V_0$. The sum of $R + T$ is shown in the solid line. (f) Transmission T as a function of barrier width a. The codes are `transmission.f` and `tunnel.f` in Appendix listings 11.6 and 11.7, respectively.

2.7.2 Quantum tunneling

In CM, a particle cannot pass through a potential energy barrier if the particle's kinetic energy is smaller than the barrier height, i. e., the transmission coefficient $T = 0$. In QM, due to the wave nature of a particle, T can differ from zero. This phenomenon is called quantum tunneling.

In the following, we start from

$$\begin{aligned} T &= \frac{4\eta^2}{[(1-\eta^2)\sinh qa]^2 + 4\eta^2(\cosh qa)^2} \\ &= \frac{4k^2/q^2}{[(1-k^2/q^2)\sinh qa]^2 + 4(k/q)^2(\cosh qa)^2}, \end{aligned} \tag{2.7.1}$$

and then take a limit of $qa \gg 1$ to get

$$T = \frac{16E(V_0 - E)}{(V_0 - E + E)^2} e^{-2\sqrt{2m(V_0-E)}a/\hbar} = \frac{16E(V_0 - E)}{V_0^2} e^{-2\sqrt{2m(V_0-E)}a/\hbar}. \tag{2.7.2}$$

The details are relegated to the following exercise. We see that indeed the transmission decreases exponentially with the barrier width a.

Example 12 (Transmission through a barrier). An electron with kinetic energy of 0.1 eV tunnels through a potential barrier V_0 of 0.2 eV with the barrier width of 20 Å. Find the transmission.

We are going to use eq. (2.7.2), but we must convert the units in its exponent to the SI units first, so our transmission is

$$T = \frac{16 \times 0.1 \times (0.2 - 0.1)}{0.2^2} e^{-\frac{2\times 20\times 10^{-10}\times 2\pi}{6.626\times 10^{-34}}\sqrt{2\times 9.11\times 10^{-31}\times(0.2-0.1)\times 1.602\times 10^{-19}}}$$

$$= 4e^{-\frac{4\times 10^{25}\times 2\pi}{6.626}\times 10^{-25}\sqrt{2\times 9.11\times 0.1\times 1.602}} = 4e^{-6.480} = 0.00613.$$

This shows that the transmission through such a barrier is very small, which is very important to the success of scanning tunneling microscopy (see Section 9.7).

Exercise 2.7.2.

32. Start from eq. (2.7.1) to get eq. (2.7.2), $T = \frac{16E(V_0-E)}{V_0^2} e^{-2\sqrt{2m(V_0-E)}a/\hbar}$. Hint: Expand both cosh qa and sinh qa.

33. (a) Using eq. (2.7.2) with $E = 1$ eV, $V_0 = 2$ eV, and $a = 2$ Å, find T and compare it with the previous T.

34. Following (a), plot T as a function of a, with a from 1 Å to 10 Å, where the rest of the parameters are the same.

2.7.3 Current

The microscopic current density, J, can be calculated in each region. For Region I, J_I is computed from eq. (2.1.11) as $J_\text{I} = \frac{\hbar}{2mi}(\phi_\text{I}^* \frac{\partial}{\partial x}\phi_\text{I} - \phi_\text{I}\frac{\partial}{\partial x}\phi_\text{I}^*)$. Since $\phi_\text{I} = e^{ikx} + re^{-ikx}$, we have $J_\text{I} = \frac{\hbar k}{m}(1 - |r|^2)$. The net current flowing from the left to right is the difference between the incident and reflected currents.

The current in Region II is $J_\text{II} = \frac{\hbar q}{mi}(AB^* - BA^*) = \frac{\hbar k}{m}(1 - |r|^2)$, and the current in Region III is $J_\text{III} = \frac{\hbar k}{m}|t|^2 = \frac{\hbar k}{m}(1 - |r|^2)$. This shows that the currents in the three regions are exactly the same. We can use the continuity equation (eq. (2.1.9)) to find the charge density change,

$$\frac{\partial \rho}{\partial t} = -\nabla \cdot \mathbf{J} = -\frac{\partial}{\partial x}J_x = 0.$$

So we can see that the charge density is constant and is independent of time.

Exercise 2.7.3.

35. Show the currents in Regions I, II, and III are $J_\text{I} = J_\text{II} = J_\text{III} = \frac{\hbar k}{m}(1 - |r|^2)$.

36. Prove $\frac{\partial \rho}{\partial t} = 0$. What does this tell us?

2.8 Problems

1. A time-translational operator $\hat{T}_t$ will shift the time variable of a function $f(0)$ to $f(t)$, i. e., $\hat{T}_t f(0) = f(t) = C(t)f(0)$. Show $C(t)$ must have an exponential form. All translational operators have this property. Hint: Follow eq. (9.2.6) on Section 9.2.1.
2. Starting from eq. (2.1.4), show that if we replace $\hat{\theta}$ by the ratio of the relativistic Hamiltonian (eq. (1.5.7)) with respect to $\hbar$, we have a relativistic Schrödinger equation.
3. Suppose an eigenfunction of a one-dimensional potential $V(x)$ is $\phi_1(x) = \exp(-x^2/a^2)$ and the eigenvalue is $E_1 = \frac{\hbar^2}{ma^2}$, where a is a constant. Find the potential.
4. (a) In a one-dimensional system, if two eigenstates $\phi_1(x)$ and $\phi_2(x)$ have the same eigenenergy E, then $\phi_1\phi_2' - \phi_1'\phi_2 = C$ must be a constant C. (b) If both $\phi_1(x)$ and $\phi_2(x)$ are bound states, $\phi_1(x), \phi_2(x) \to 0$ at $x \to \pm\infty$, $C = 0$ and ϕ_1 differs from ϕ_2 by a constant, i. e., there is no degeneracy in nondegenerate. Adopted from [19].
5. $\phi_k(x)$ and $\phi_l(x)$ are two bound eigenstates of a one-dimensional system with potential $V(x)$, with the eigenenergies E_k and E_l. Assume that at $x \to \pm\infty$, ϕ_k and $\phi_l \to 0$. Show $0 = (E_k - E_l)\int_{-\infty}^{+\infty} \phi_k(x)\phi_l(x)dx$.
6. $\phi_n(x)$ and $\phi_m(x)$ are two eigenstates of PIB, where $x \in (0, a)$ (eq. (2.3.12)). Compute the transition matrix elements of (a) the position and (b) the momentum operators $\langle\phi_n|\hat{x}|\phi_m\rangle$ and $\langle\phi_n|\hat{p}_x|\phi_m\rangle$, where $n \neq m$. These matrix elements are called the transition matrix element, often used in optics as on Section 7.6.
7. A particle is in an eigenstate of PIB, ϕ_n. Compute the uncertainty relation $\Delta x \Delta p_x$, where $\Delta A \equiv \sqrt{\langle\phi_n|\hat{A}^2|\phi_n\rangle^2 - (\langle\phi_n|\hat{A}|\phi_n\rangle)^2}$.
8. A particle is initially in a mixed state of PIB, $\psi(t = 0) = A(\phi_1 + \phi_2)$, where ϕ_i is an eigenstate of PIB. (a) If $\langle\psi(0)|\psi(0)\rangle = 1$, find A. (b) Find $\psi(t \neq 0)$. (c) Find $\langle\psi(t)|\hat{p}_x|\psi(t)\rangle$. (d) Find $\langle\psi(t)|\hat{p}_x x|\psi(t)\rangle$. (e) Find $\langle\psi(t)|x\hat{p}_x|\psi(t)\rangle$. Hint: (d) and (e) have a different order of $\hat{p}_x$ and x. (f) Do (d) and (e) have the same results?
9. A particle is initially in $\psi(x, 0)$, which consists of two segments. From $x = 0$ to $a/2$, $\psi(x, 0) = Ax$ and from $a/2$ to a, it is $\psi(x, 0) = A(-x + a)$. (a) Use the normalization condition to find A. (b) Expand it in terms of eigenstates of PIB and find c_n. (c) Compute $\psi(x, t)$ and find $\langle\psi(x, t)|\hat{x}|\psi(x, t)\rangle$ and $\langle\psi(x, t)|\hat{p}_x|\psi(x, t)\rangle$. Adopted from [17].
10. A particle of mass m is subject to a δ potential, $V(x) = -V_0\delta(x)$, where $\delta(x)$ is the Dirac delta function

$$\delta(x) = 0, \quad \text{if } x \neq 0; \quad \delta(x) = 1, \quad \text{if } x = 0; \quad \int_{-\infty}^{+\infty} \delta(x)dx = 1.$$

(a) Find its eigenfunction of the bound state, where the wavefunction is zero at $x = \pm\infty$. Hint: This problem only has one bound state. (b) Find its eigenenergy. (c) Suppose we confine the particle in a box with x within $-a/2$ and $a/2$. The potential remains the same. Find the eigenfunctions of continuous states and normalize the wavefunction within this box.

11. A particle of mass m is inside a double-delta potential, $V(x) = V_0(\delta(x-\frac{a}{2})+\delta(x+\frac{a}{2}))$. (a) Find the bound eigenstates. (b) Find the bound eigenenergies.
12. Suppose $V(x) = 0$ if $x < -\frac{a}{2}$ or $x > \frac{a}{2}$, and $V(x) = -V_0$ for $-\frac{a}{2} < x < \frac{a}{2}$. A particle of mass m is placed inside the potential just cited. (a) Write down the wavefunctions in three regions. (b) Write down their boundary conditions. (c) Match the boundary conditions to find an equation for the eigenvalues.
13. Suppose that a particle is in the ground state of PIB. Compute the force on the wall due to the particle using two different methods: (a) $\frac{\Delta\langle p\rangle}{\Delta t}$, and (b) $-\frac{\partial E_1}{\partial a}$. (Answer: $\frac{\hbar^2 k^2}{ma}$.)
14. A one-dimensional rotator with a fixed rotation axis can simulate the photoisomerization of bacteria rhodopsin. Suppose the moment of inertia is I. The potential is $V(\theta) = -V_0\cos(\theta)$, where θ is the rotational angle and V_0 is a positive constant. (a) Write down the time-independent Schrödinger equation for this rotator. Hint: Find the rotational kinetic energy operator. (b) Solve the Schrödinger equation to find the eigenstates and eigenenergies.
15. A particle of mass m is subject to a one-dimensional potential, where $V(x) = 0$ if $-\frac{a}{2} \le x \le \frac{a}{2}$; otherwise $V(x) = V_0 > 0$. (a) Find the odd eigenstates. (b) Find their respective eigenvalues.
16. Consider tunneling through a δ potential, $V(x) = V_0\delta(x)$, where $V_0 > 0$. Suppose a particle has mass of m. (a) Find the reflection coefficient $R = rr^*$. (b) Find the transmission coefficient $T = tt^*$.
17. Code `finite.well.even.wavefunction.f` in Appendix listing 11.3 uses an input file called `even.w.eta.used` which is created from the two files. In the code, $a = 10$ Å. Compute and plot the wavefunctions as done in Fig. 2.8.
18. In the tunneling through a potential on Section 2.6.3, prove eqs. (2.6.10) and (2.6.11):

    (a)

$$te^{ika} = \frac{-2i\eta}{\sinh qa(1-\eta^2) - 2i\eta\cosh qa},$$

    (b)

$$r = \frac{(1+\eta^2)\sinh qa}{-(1-\eta^2)\sinh qa + 2i\eta\cosh qa}.$$

3 Harmonic oscillator and blackbody radiation

Starting from this chapter, we are going to apply the Schrödinger equation to a harmonic oscillator. The harmonic oscillator model historically has served as the beginning of QM, where Planck employed the energy quatum of the harmonic oscillator to simulate blackbody radiation and overcame the old theory's ultraviolet catastrophe. The harmonic oscillator has lots of applications. Nuclear vibrations can be approximated by a harmonic oscillator. Its extension to solids allows us to introduce phonons.

This chapter is grouped into four units.

- Unit 1 is Section 3.1. This introduces some basic operations of operators and Heisenberg's uncertainty principle and concludes with the ladder operators, ready for the harmonic oscillator.
- Unit 2 consists of Sections 3.2 and 3.3. Section 3.2 starts with a classical prediction of the probability and introduces the QM Hamiltonian for the harmonic oscillator. Then, it explains how to solve the Schrödinger equation for the harmonic oscillator. Section 3.3 digs into the ladder operators to finally compute the expectation values of position and momentum. This connects with the uncertainty principle in Unit 1.
- Unit 3 is on dynamic evolution and consists of Sections 3.4 and 3.5. Section 3.4 is on the Schrödinger picture and stationary state. Section 3.5 is on the Heisenberg's picture and equation of motion.
- Unit 4 addresses applications to blackbody radiation and the photoelectric effect and includes Sections 3.6 and 3.7. Section 3.6 explains why Planck's quantization is essential to blackbody radiation, and Section 3.7 presents an experiment on the photoelectric effect and Einstein's theory.

3.1 Operators and Heisenberg's uncertainty principle

In Chapter 1, we briefly discussed operators. Since operators are at the center of measurement, it is necessary to explore some crucial properties. We warn the reader this section is somewhat abstract.

3.1.1 Basic properties of operators

In CM, two quantities A and B, each of which is a function of canonical positions $\{q_i\}$ and momenta $\{p_i\}$, have the Poisson brackets

$$\{A, B\} \equiv \sum_i \left(\frac{\partial A}{\partial q_i} \frac{\partial B}{\partial p_i} - \frac{\partial A}{\partial p_i} \frac{\partial B}{\partial q_i} \right), \tag{3.1.1}$$

where the summation is over components i. Then, $\{A, A\} = 0$, $\{A, B\} = -\{B, A\}$, $\{A, B+C\} = \{A, B\} + \{A, C\}$, $\{A, BC\} = \{A, B\}C + B\{A, C\}$.

https://doi.org/10.1515/9783110672152-003

Dirac proposed a similar relation for QM but with some key differences. The right side of the equation must be divided by $i\hbar$. A and B are replaced by two operators $\hat{A}$ and $\hat{B}$, where the hat is used as a reminder for operators. Different from regular numbers but like matrices, operators are normally noncommutable, $\hat{A}\hat{B} \neq \hat{B}\hat{A}$. Dirac uses a square bracket to denote this commutation as

$$[\hat{A}, \hat{B}] \equiv \hat{A}\hat{B} - \hat{B}\hat{A}, \tag{3.1.2}$$

which is called the commutation relation of $\hat{A}$ and $\hat{B}$. It is easy to show $[\hat{A}, \hat{B}] = -[\hat{B}, \hat{A}]$, very much similar to the Poisson brackets. To find what $[\hat{A}, \hat{B}]$ is, we need to apply $\hat{A}\hat{B}$ and $\hat{B}\hat{A}$ to an auxiliary wavefunction f, which has the same variables that $\hat{A}$ and $\hat{B}$ have.

Example 1 (Commutation). Find the fundamental commutation relation $[\hat{x}, \hat{p}_x]$.

Since both $\hat{x}$ and $\hat{p}_x$ depend on x, we choose a function $f(x)$. We compute $\hat{x}\hat{p}_x(f)$ and then $\hat{p}_x(\hat{x}f)$.[1] We apply $\hat{p}_x$ first to a wavefunction f and then apply $\hat{x}$ to f,

$$\hat{x}\hat{p}_x f = x(-i\hbar)\frac{\partial f}{\partial x} = -i\hbar x\frac{\partial f}{\partial x}. \tag{3.1.3}$$

Next, if we reverse the order of application, we have

$$\hat{p}_x(\hat{x}f) = -i\hbar\frac{\partial}{\partial x}(xf) = -i\hbar f - i\hbar x\frac{\partial f}{\partial \hat{x}}. \tag{3.1.4}$$

We see the order of operators matters. We subtract (3.1.4) from (3.1.3). Since f is an arbitrary function, we have

$$\hat{x}\hat{p}_x f - \hat{p}_x\hat{x}f = i\hbar f, \rightarrow [\hat{x}, \hat{p}_x] = i\hbar.$$

In matrix mechanics, we write $[\hat{x}, \hat{p}_x] = i\hbar\hat{I}$, where $\hat{I}$ is an identity matrix.

Example 2 (Operator commutation). Prove $[\hat{A}\hat{B}, \hat{C}] = [\hat{A}, \hat{C}]\hat{B} + \hat{A}[\hat{B}, \hat{C}]$.

We use our definition of the commutation relation (eq. (3.1.2)) to write $[\hat{A}\hat{B}, \hat{C}] = \hat{A}\hat{B}\hat{C} - \hat{C}\hat{A}\hat{B} = \hat{A}\hat{B}\hat{C} - \hat{A}\hat{C}\hat{B} + \hat{A}\hat{C}\hat{B} - \hat{C}\hat{A}\hat{B} = \hat{A}(\hat{B}\hat{C} - \hat{A}\hat{C}) + (\hat{A}\hat{C} - \hat{C}\hat{A})\hat{B} = \hat{A}[\hat{B}, \hat{C}] + [\hat{A}, \hat{C}]\hat{B} = [\hat{A}, \hat{C}]\hat{B} + \hat{A}[\hat{B}, \hat{C}]$.

We would like to mention some other properties. Suppose c is a classical number. Operators also obey the distribution, $\hat{A}(c_1\psi_1 + c_2\psi_2) = c_1\hat{A}\psi_1 + c_2\hat{A}\psi_2$. A linear combination of operators, $c_1\hat{A} + c_2\hat{B}$, is a good operator. If the operator applied to a function returns to the original function, the operator is called the identity operator $\hat{I}$. $\hat{I}\psi = \psi$. A Hermitian operator is $\hat{A}^\dagger = \hat{A}$, where $\dagger$ represents complex conjugation followed by transpose. We can multiply several operators such as $\hat{A}\hat{B}\hat{C}$, $\hat{A}^2$, or exponents $\exp(-\hat{A})$. Division of the

---

1 Because $\hat{p}_x$ contains a partial derivative, it must apply to both x and f. For this reason, we use parentheses around x and f, such as (xf), to remind us about the unique feature of differential operators.

operators is also possible, such as $\hat{A}/\hat{B}$, but this is often written as $\hat{A}\hat{B}^{-1}$. A square root of $\hat{A}$ is written as $\sqrt{\hat{A}}$.

Exercise 3.1.1.

1. (a) Prove $[\hat{A}+\hat{B},\hat{C}] = [\hat{A},\hat{C}] + [\hat{B},\hat{C}]$. (b) Find $[\hat{A},\hat{B}\hat{C}]$. (c) Prove $[\hat{A},[\hat{B},\hat{C}]] + [\hat{B},[\hat{C},\hat{A}]] + [\hat{C},[\hat{A},\hat{B}]] = 0$.
2. Compute $[\hat{x},\hat{p}_x^2]$, $[\hat{x}^2,\hat{p}_x]$, $[\hat{x},\hat{y}]$, $[\hat{p}_x,\hat{y}]$, $[\hat{p}_x,\hat{p}_z]$, $[\hat{p}_x^2,\hat{x}^2]$, and $[\hat{p}_x,f(x)]$.
3. Prove $[\hat{A}-\hat{B},\hat{C}-\hat{D}] = [\hat{A},\hat{C}] + [\hat{B},\hat{D}] - [\hat{A},\hat{D}] - [\hat{B},\hat{C}]$.
4. (a) Show $[\hat{A},\hat{B}\hat{C}]$ can be written as $[\hat{A},\hat{B}]\hat{C} + \hat{B}[\hat{A},\hat{C}]$. (b) The brackets like $[\hat{A},\hat{B}]_- = \hat{A}\hat{B} - \hat{B}\hat{A}$ are called commutators, where we purposely add a subscript "−". There is another type of bracket, $[\hat{A},\hat{B}]_+ = \hat{A}\hat{B} + \hat{B}\hat{A}$, called anticommutator. Show $[\hat{A},\hat{B}\hat{C}]_- = [\hat{A},\hat{B}]_+\hat{C} - \hat{B}[\hat{A},\hat{C}]_+$.

3.1.2 Heisenberg uncertainty principle

The Heisenberg uncertainty principle constitutes a milestone for QM. It states that the expectation values of two noncommutable Hermitian operators $\hat{A}$ and $\hat{B}$, i. e., $[\hat{A},\hat{B}] \neq 0$, cannot be determined accurately and simultaneously for the same state $|\psi\rangle$. This principle has two conditions: $|\psi\rangle$ cannot be an eigenstate of $\hat{A}$ or $\hat{B}$, and $\hat{A}$ and $\hat{B}$ must be Hermitian.[2] The experimental evidence of the uncertainty principle is overwhelming. When a single beam of electrons passes through a narrower slit (a smaller position of uncertainty Δx), the wave behind the slit spreads more (a larger momentum of uncertainty Δp_x) or $\Delta x \Delta p_x \geq \hbar/2$. Time and energy form another pair of these quantities, where time and energy cannot be determined simultaneously, $\Delta t \Delta E \geq \hbar/2$, just as a Fourier transform requires. When we introduced the de Broglie wavelength in Section 1.2 through the momentum p, we invoked the wavepacket concept to avoid a conceptual difficulty. This is because, if p is known with 100 % certainty, then the position is completely undetermined, a wave, not a particle. For a wavepacket, with multiple wavelengths and various frequencies, its group momentum is used as the de Broglie's momentum p [11].

Now, we prove Heisenberg's uncertainty principle mathematically,

$$\Delta A \Delta B \geq \left| \frac{i}{2} \langle [\hat{A},\hat{B}] \rangle \right|, \tag{3.1.5}$$

where $\langle$ and $\rangle$ denote states $\langle\phi|$ and $|\phi\rangle$, respectively. The respective uncertainties of $\hat{A}$ and $\hat{B}$ are defined as $\Delta A \equiv \sqrt{\langle \hat{A}^2 \rangle - \langle \hat{A} \rangle^2}$ and $\Delta B \equiv \sqrt{\langle \hat{B}^2 \rangle - \langle \hat{B} \rangle^2}$, where the two terms under the square root are the expectation value of the operator squared and the square of the expectation value of the operator. This definition follows the uncertainty used in classical physics, and the only difference is that we use the expectation values. We caution that here $|\phi\rangle$ *cannot* be chosen as an eigenfunction of either $\hat{A}$ or $\hat{B}$ because

---

2 This condition is not stringent since operators of physical observables are Hermitian operators.

doing so is equivalent to setting one of terms on the left side of eq. (3.1.5) to zero since there is no uncertainty in its eigenfunction.

First, we form a state $|\psi\rangle$ which is a linear combination of $\hat{A}|\phi\rangle$ and $\hat{B}|\phi\rangle$,

$$|\psi\rangle = \alpha\hat{A}|\phi\rangle + i\hat{B}|\phi\rangle, \tag{3.1.6}$$

where α is a *real* number and $|\psi\rangle$ is not necessarily normalized. The i in front of $\hat{B}$ ensures that the norm $\langle\psi|\psi\rangle$ is $\hat{B}^2$, not $-\hat{B}^2$. which simplifies the derivation. For any wavefunction, its norm $\langle\psi|\psi\rangle$, the probability density, must be nonnegative,

$$\langle\psi|\psi\rangle = \int \psi^*\psi d\mathbf{r} \geq 0, \tag{3.1.7}$$

where the equality corresponds to $|\psi\rangle = 0$. If we substitute eq. (3.1.6) into eq. (3.1.7), we have

$$\langle\psi|\psi\rangle = (\langle\phi|\alpha\hat{A}^\dagger - i\langle\phi|\hat{B}^\dagger)(\alpha\hat{A}|\phi\rangle + i\hat{B}|\phi\rangle) = \alpha^2\langle\phi|\hat{A}^2|\phi\rangle + \alpha i\langle\phi|[\hat{A},\hat{B}]|\phi\rangle + \langle\phi|\hat{B}^2|\phi\rangle, \tag{3.1.8}$$

where we have used $\hat{A}^\dagger = \hat{A}$ and $\hat{B}^\dagger = \hat{B}$. We introduce a shorthand notation for $i[\hat{A},\hat{B}] = \hat{C}$, where $\hat{C}$ is also Hermitian. So, eq. (3.1.7) becomes

$$\langle\psi|\psi\rangle = \alpha^2\langle\phi|\hat{A}^2|\phi\rangle + \alpha\langle\phi|\hat{C}|\phi\rangle + \langle\phi|\hat{B}^2|\phi\rangle \geq 0. \tag{3.1.9}$$

Because α is a real number as we assumed above, to satisfy eq. (3.1.9), the discriminant of the equation must be negative or zero, $(\langle\phi|\hat{C}|\phi\rangle)^2 - 4\langle\phi|\hat{A}^2|\phi\rangle\langle\phi|\hat{B}^2|\phi\rangle \leq 0$, which can be simplified to

$$\sqrt{\langle\phi|\hat{A}^2|\phi\rangle\langle\phi|\hat{B}^2|\phi\rangle} \geq \frac{1}{2}\langle\phi|\hat{C}|\phi\rangle, \tag{3.1.10}$$

provided $\langle\phi|\hat{C}|\phi\rangle$ is positive. If it is not, we must take its absolute value, since uncertainties must be positive.

Next, we use a trick. Since $\hat{A}$ and $\hat{B}$ can be any operators, we set $\hat{A}$ to $\hat{A} - \langle\hat{A}\rangle$ and $\hat{B}$ to $\hat{B} - \langle\hat{B}\rangle$. This does not change $\hat{C}$ because we just subtract the expectation value from the original operator. But doing so changes $\langle\phi|\hat{A}^2|\phi\rangle$ in the square root of eq. (3.1.10) to $\langle\phi|\hat{A}^2|\phi\rangle - (\langle\phi|\hat{A}|\phi\rangle)^2$, and a similar expression for $\hat{B}$. According to our stated definition of uncertainty, $\langle\phi|\hat{A}^2|\phi\rangle - (\langle\phi|\hat{A}|\phi\rangle)^2$ is just ΔA^2, so eq. (3.1.10) becomes

$$\Delta A\Delta B \geq \frac{1}{2}\langle\phi|\hat{C}|\phi\rangle = \frac{i}{2}\langle\phi|[\hat{A},\hat{B}]|\phi\rangle \tag{3.1.11}$$

if $i\langle\phi|[\hat{A},\hat{B}]|\phi\rangle$ is positive. Otherwise, one has to take the absolute value of the expression as eq. (3.1.5). This concludes our proof. We note in passing that the uncertainty ΔA is a classical number, not an operator.

There is a large group of operator pairs that are subject to this relation. For instance, $[\hat{y}, \hat{p}_y]$, $[\hat{z}, \hat{p}_z]$, $[\hat{l}_x, \hat{l}_y]$, $[\hat{l}_y, \hat{l}_z]$, $[\hat{l}_z, \hat{l}_x]$, where $\hat{l}_i$ is the angular momentum operator. It is interesting to examine the lower limit of the uncertainty principle, $\Delta A \Delta B = \frac{i}{2}\langle\phi|[\hat{A}, \hat{B}]|\phi\rangle$, and in particular what wavefunction ϕ satisfies this relation.

Take $\hat{x}$ and $\hat{p}_x$ as an example, whose uncertainty relation is $\Delta x \Delta p_x \geq \frac{\hbar}{2}$. Recall the equality in eq. (3.1.7) corresponds to $|\psi\rangle = 0$, $|\psi\rangle = \alpha\hat{x}|\phi\rangle + i\hat{p}_x|\phi\rangle = 0$, whose solution ϕ is the basis of ladder operators.

Exercise 3.1.2.

5. Show that, if $\hat{A}$ and $\hat{B}$ are Hermitian, then $i[\hat{A}, \hat{B}] = \hat{C}$ is also Hermitian.
6. Prove that, if we change $\hat{A}$ to $\hat{A} - \langle\hat{A}\rangle$, $\langle\phi|\hat{A}^2|\phi\rangle$ becomes $\langle\phi|\hat{A}^2|\phi\rangle - (\langle\phi|\hat{A}|\phi\rangle)^2$.
7. If $\alpha\hat{x}|\phi\rangle + i\hat{p}_x|\phi\rangle = 0$, find an expression for ϕ.

3.1.3 Ladder operators

The previous subsection shows that operators can be added or subtracted from each other. Here we introduce two auxiliary dimensionless operators $\hat{a}$ and $\hat{a}^\dagger$, which help us find the eigenvalues and eigenfunctions of the harmonic oscillator.

In CM, the one-dimensional harmonic oscillator has the Hamiltonian

$$H = \frac{{p_x}^2}{2m} + \frac{1}{2}Kx^2 = \frac{{p_x}^2}{2m} + \frac{1}{2}m\omega^2 x^2 \quad \text{(CM)},$$

where m is the mass, ω is its angular frequency, and K is the spring constant, $K = m\omega^2$.

We obtain the QM Hamiltonian operator by quantizing the position and momentum,

$$\hat{H} = \frac{\hat{p}_x^2}{2m} + \frac{1}{2}m\omega^2\hat{x}^2 = \frac{1}{2m}[\hat{p}_x^2 + (m\omega\hat{x})^2], \tag{3.1.12}$$

where $\hat{x}$ and $\hat{p}_x$ are position and momentum operators, respectively. We consider the commutation of position and Hamiltonian $[\hat{x}, \hat{H}]$,

$$[\hat{x}, \hat{H}] = \left[\hat{x}, \frac{\hat{p}_x^2}{2m} + \frac{1}{2}m\omega^2\hat{x}^2\right] = \left[\hat{x}, \frac{\hat{p}_x^2}{2m}\right] = \frac{\hbar}{m}i\hat{p}_x. \tag{3.1.13}$$

This means the commutation of the position operator with the Hamiltonian gives us the momentum operator. Similarly, we can derive

$$[\hat{p}_x, \hat{H}] = -i\hbar m\omega^2\hat{x}, \tag{3.1.14}$$

which is equivalent to

$$[i\hat{p}_x, \hat{H}] = \hbar m\omega^2\hat{x} = \hbar\omega m\omega\hat{x}. \tag{3.1.15}$$

We multiply (3.1.13) by a constant α and (3.1.15) by β, and then add both sides to get

$$\underline{[\alpha\hat{x} + \beta i\hat{p}_x}, \hat{H}] = \hbar\omega\underline{\left(\beta m\omega\hat{x} + \frac{\alpha}{m\omega} i\hat{p}_x\right)}, \tag{3.1.16}$$

which reveals a crucial fact: Although $\hat{x}$ and $\hat{p}_x$ each do not commute with the Hamiltonian, the commutation of their linear combination with the Hamiltonian produces a similar linear combination of $\hat{x}$ and $\hat{p}_x$. We can make the underlined terms exactly the same by matching the respective coefficients of $\hat{x}$ and $\hat{p}_x$ on both sides by setting

$$\alpha = \beta m\omega, \quad \beta = \frac{\alpha}{m\omega}. \tag{3.1.17}$$

Consequently, eq. (3.1.16) becomes

$$[\alpha\hat{x} + \beta i\hat{p}_x, \hat{H}] = \hbar\omega[\alpha\hat{x} + \beta i\hat{p}_x]. \tag{3.1.18}$$

α and β can take any values so long as they satisfy (3.1.17) or $\frac{\alpha}{\beta} = m\omega$. This is where the ladder operators come in.

We define the ladder operator $\hat{a}$ as

$$\hat{a} = \frac{1}{\sqrt{2m\hbar\omega}}(m\omega\hat{x} + i\hat{p}_x). \tag{3.1.19}$$

First, we check the units of both terms. Note that both $\hat{x}$ and $\hat{p}_x$ have units. The units of the coefficient and those in the parenthesis are $\frac{1}{\sqrt{\text{kgJs/s}}}\text{kg}\frac{\text{m}}{\text{s}} = \sqrt{\frac{\text{kgm}^2}{\text{Js}^2}} = 1$, dimensionless. So $\hat{a}$ is dimensionless, and its conjugate is

$$\hat{a}^\dagger = \frac{1}{\sqrt{2m\hbar\omega}}(m\omega\hat{x} - i\hat{p}_x), \tag{3.1.20}$$

where we have used the properties of $\hat{x}^\dagger = \hat{x}$ and $\hat{p}_x^\dagger = \hat{p}_x$. $\hat{a}$ and $\hat{a}^\dagger$ form a pair of ladder operators. Since $\hat{a} \neq \hat{a}^\dagger$, $\hat{a}$ is not a Hermitian operator. This means that $\hat{a}$ or $\hat{a}^\dagger$ alone does not represent a physical quantity, but $\hat{a}^\dagger\hat{a}$ and $\hat{a}\hat{a}^\dagger$ do. Let's compute $\hat{a}^\dagger\hat{a}$.

$$\begin{aligned}
\hat{a}^\dagger\hat{a} &= \frac{1}{2m\hbar\omega}(m\omega\hat{x} - i\hat{p}_x)(m\omega\hat{x} + i\hat{p}_x) = \frac{1}{2m\hbar\omega}[(m\omega\hat{x})^2 + \hat{p}_x^2 + im\omega\underbrace{(\hat{x}\hat{p}_x - \hat{p}_x\hat{x})}_{[\hat{x},\hat{p}_x]=i\hbar}] \\
&= \frac{1}{2m\hbar\omega}[(m\omega\hat{x})^2 + \hat{p}_x^2 - m\hbar\omega] = +\frac{1}{\hbar\omega}\underbrace{\left[\frac{1}{2}m\omega^2\hat{x}^2 + \frac{\hat{p}_x^2}{2m}\right]}_{\hat{H}} - \frac{1}{2}.
\end{aligned} \tag{3.1.21}$$

The first term in eq. (3.1.21) contains the harmonic oscillator Hamiltonian. So, we can rewrite the Hamiltonian in terms of $\hat{a}$ and $\hat{a}^\dagger$ as

$$\hat{H} = \hbar\omega\left(\hat{a}^\dagger\hat{a} + \frac{1}{2}\right). \tag{3.1.22}$$

Exercise 3.1.3.
8. Using the definitions of $\hat{a}$ and $\hat{a}^\dagger$, compute the following commutations: (a) $[\hat{a}, \hat{a}^\dagger]$ and (b) $[\hat{a}^\dagger, \hat{a}]$.
9. Use $[\hat{a}, \hat{a}^\dagger]$ to compute $[\hat{a}^\dagger, \hat{a}^2]$.

3.2 Harmonic oscillator

In molecules, atoms vibrate with respect to their equilibrium positions. Figure 3.1(a) shows the Lennard–Jones potential around x_{eq}. When atoms are not far from their equilibrium positions, the potential can be simulated by a harmonic potential (dashed line) as it in a pendulum. The harmonic oscillator is a model for molecular vibrations. The energy scale is on the order of meV, i. e., the infrared region. Although the model is quite simple, it helps us understand the coherent motion of iodine in I_2 under ultrafast laser excitation. The harmonic oscillator is analytically solvable. In the following, we will consider a one-dimensional system and shift the equilibrium position x_{eq} to 0.

3.2.1 Classical physics

According to classical Newtonian dynamics, the one-dimensional equation of motion of a particle of mass m in a harmonic potential is

$$m\frac{d^2x}{dt^2} = -Kx, \quad \text{or} \quad \frac{d^2x}{dt^2} = -\omega^2 x, \tag{3.2.1}$$

where K is the spring constant and the angular frequency is $\omega = \sqrt{\frac{K}{m}}$. The period is $T = 2\pi/\omega$. The solution is

$$x = x_0 \sin(\omega t + \phi), \tag{3.2.2}$$

where x_0 is the maximum amplitude of position and ϕ is the initial phase, not to be confused with wavefunctions. The velocity and momentum are $v_x = \omega x_0 \cos(\omega t + \phi)$ and $p_x = m\omega x_0 \cos(\omega t + \phi)$. x-p_x forms the Lissajous orbit.

We can find the probability of the particle within the line segment dx by computing the time dt that it needs to transverse dx,

$$\rho(x)dx = \frac{2dt}{T} = \frac{2}{2\pi/\omega}\frac{dx}{v_x} = \frac{\omega}{\pi}\frac{dx}{\omega x_0 \cos(\omega t + \phi)} = \frac{dx}{\pi\sqrt{x_0^2 - x^2}}, \tag{3.2.3}$$

where in the last step we have used eq. (3.2.2) to get $\sqrt{x_0^2 - x^2}$. Here, a factor of 2 is present because the bob passes through the same segment twice for a single period $T = 2\pi/\omega$. We find the classical probability density to be

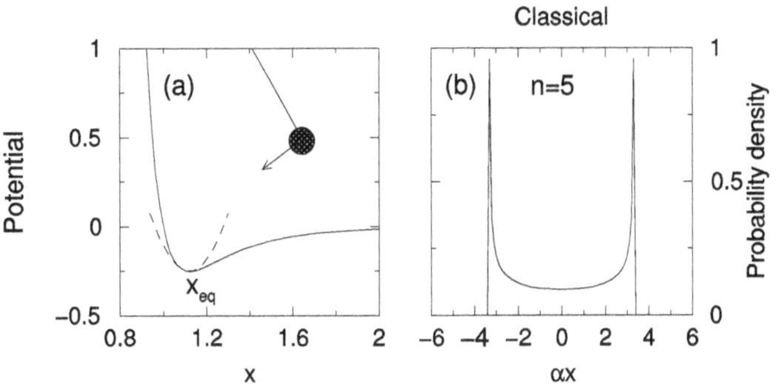

Figure 3.1: (a) The harmonic potential (dashed line), $V(x) = V_0 + \frac{1}{2}k(x - x_{eq})^2$, is an approximation to the Lennard–Jones potential (solid line), $V(r) = 4\epsilon[(\frac{\sigma}{r})^{12} - (\frac{\sigma}{r})^6]$. (b) Classical probability density $p(x)$ as a function of αx (dimensionless), where $\alpha = \sqrt{\frac{m\omega}{\hbar}}$. The largest $p(x)$ is infinite at its position maximum x_0, which is computed by matching the classical energy to the quantum one through $\frac{1}{2}\omega^2 x_0^2 = (n + \frac{1}{2})\hbar\omega$, i. e., $x_0^2 = \frac{(2n+1)\hbar}{m\omega}$. Here $n = 5$.

$$\rho(x) = \frac{1}{\pi\sqrt{x_0^2 - x^2}} \tag{3.2.4}$$

for a particle with the fixed energy. Figure 3.1(b) shows the probability as a function of displacement. In classical physics, x_0 can take any values, but, in QM, this is not possible. In the figure, our maximum displacement ξ_0 ($\xi_0 = \alpha x_0 = \sqrt{\frac{m\omega}{\hbar}} x_0$) is computed by equating the classical energy to the quantum energy with $n = 5$.

Exercise 3.2.1.

10. (a) A vibrational mode in fcc Ni has frequency 1 THz ($\nu = 10^{12}$ Hz) at room temperature $T = 300$ K. According to the equipartition theorem in classical statistics, each degree of freedom has the same energy of $\frac{1}{2}k_B T$. Assuming that at one instance all the energy is the potential energy $\frac{1}{2}m\omega^2 x^2$, where m is the mass of Ni at 58.699 amu ($58.699 \times 1.66 \times 10^{-27}$ kg) and $\omega = 2\pi\nu$, find $x_{\text{classical}}$. (b) According to QM, even at $T = 0$ K, an oscillator has energy of $\hbar\omega/2$. Assuming that the energy is all stored in the potential energy, find x_{quantum} for Ni atom.

11. Using the classical probability density (eq. (3.2.4)), (a) prove $p(x)$ is normalized to 1, and find (b) $\bar{x}$, (c) $\bar{x^2}$, (d) $\bar{p_x}$, (e) $\bar{p_x^2}$, (f) Δx, (g) Δp_x, and (h) $\Delta x \Delta p_x$. Note that the quantum treatment must approach them in the large quantum number n limit.

3.2.2 Quantum theory for a harmonic oscillator

The quantum mechanical Hamiltonian of a one-dimensional harmonic oscillator is

$$\hat{H} = -\frac{\hbar^2}{2m}\frac{\partial^2}{\partial x^2} + V(x), \tag{3.2.5}$$

where the first term is the kinetic energy operator $\hat{T}$ and $V(x)$ is the potential energy operator. In the x-representation, $V(x)$ is just a regular function and contains no partial derivative. Specifically, $V(x)$ is $\frac{1}{2}m\omega^2x^2$, where m is the mass, ω is the angular frequency and x is the position. Since this is a one-dimensional problem, we can change the partial derivative to a full derivative, so the Schrödinger equation becomes

$$\left(-\frac{\hbar^2}{2m}\frac{d^2}{dx^2}+\frac{1}{2}m\omega^2x^2\right)\phi = E\phi, \tag{3.2.6}$$

where E and ϕ are the eigenvalues and eigenfunctions to be found.

Before we solve eq. (3.2.6), we make some observations about ϕ. To ensure that the wavefunction remains finite at infinity, the boundary condition for ϕ is

$$x \to \pm\infty, \quad \phi \to 0.$$

Second, E cannot be negative for a harmonic oscillator. To see why this is the case, we rewrite eq. (3.2.6) as $\frac{\hbar^2}{2m}\frac{d^2\phi}{dx^2} = (-E+\frac{1}{2}m\omega^2x^2)\phi$. If E is negative, then $-E+\frac{1}{2}m\omega^2x^2$ must be positive, forcing $\frac{d^2\phi}{dx^2}$ to grow with ϕ infinitely, which is unphysical. Therefore, the eigenenergies of a harmonic oscillator must be positive.

There are two ways to solve this differential equation. One is to employ the Taylor expansion, and the other is to use ladder operators. We pursue the second method, which is much more elegant. There are three steps. First, we rewrite the Hamiltonian in terms of $\hat{a}$ and $\hat{a}^\dagger$. Then, we find the eigenenergy and finally the eigenfunction.

With the help of eq. (3.1.21), we rewrite the Hamiltonian as

$$\hat{a}^\dagger\hat{a} = -\frac{1}{2}+\frac{1}{\hbar\omega}\hat{H} \to \hat{H} = \hbar\omega\left(\hat{a}^\dagger\hat{a}+\frac{1}{2}\right). \tag{3.2.7}$$

This shows that, once we convert the kinetic energy operator and potential energy operator to $\hat{a}$ and $\hat{a}^\dagger$, the Hamiltonian becomes "diagonal" in $\hat{a}$ and $\hat{a}^\dagger$ if we consider $\hat{a}$ as a row vector and $\hat{a}^\dagger$ as a column vector. If we could find the expectation values of the diagonal elements $\hat{a}^\dagger\hat{a}$, we would find the eigenvalues of the harmonic oscillator, even though we have not started our calculation. This is the beauty of lowering and raising operators.

3.2.3 Eigenvalues and eigenfunctions

In the following, we first find the eigenvalues and then the eigenfunctions. Suppose that we find an eigenfunction ϕ and eigenenergy E of $\hat{H}$, i. e., $\hat{H}\phi = E\phi$, where we do not need to know or assume what E and ϕ are. This assumption is always possible for any $\hat{H}$. For our problem, this means that

$$\hat{H}\phi = E\phi \rightarrow \hbar\omega\left(\hat{a}^\dagger\hat{a} + \frac{1}{2}\right)\phi = E\phi, \tag{3.2.8}$$

where we have used eq. (3.2.7). Next, we shall show $\hat{a}^\dagger\phi$ is also an eigenstate of $\hat{H}$, if ϕ is an eigenstate, i. e., $\hat{H}(\hat{a}^\dagger\phi) = E'(\hat{a}^\dagger\phi)$, where E' is the eigenenergy corresponding to the eigenstate $\hat{a}^\dagger\phi$. The same can be said about $\hat{a}\phi$.

First, we apply $\hat{H}$ to $\hat{a}^\dagger\phi$,

$$\hat{H}\hat{a}^\dagger\phi = \hbar\omega\left(\hat{a}^\dagger\hat{a} + \frac{1}{2}\right)\hat{a}^\dagger\phi = \hbar\omega\left(\hat{a}^\dagger\hat{a}\hat{a}^\dagger + \frac{1}{2}\hat{a}^\dagger\right)\phi. \tag{3.2.9}$$

In order to use our assumption (eq. (3.2.8)), we need to exchange $\hat{a}$ with $\hat{a}^\dagger$ in $\hat{a}\hat{a}^\dagger$. We use the relation proven in the above exercise $[\hat{a}, \hat{a}^\dagger] = 1$, $\hat{a}\hat{a}^\dagger - \hat{a}^\dagger\hat{a} = 1$, so $\hat{a}\hat{a}^\dagger = 1 + \hat{a}^\dagger\hat{a}$. Then $\hat{a}^\dagger\hat{a}\hat{a}^\dagger = \hat{a}^\dagger(1 + \hat{a}^\dagger\hat{a})$, so the terms before ϕ in eq. (3.2.9) can be rewritten as

$$\hbar\omega\left[\hat{a}^\dagger(1 + \hat{a}^\dagger\hat{a}) + \frac{1}{2}\hat{a}^\dagger\right] = \hbar\omega\hat{a}^\dagger\left(\hat{a}^\dagger\hat{a} + \frac{1}{2} + 1\right) = \hat{a}^\dagger\Bigg[\underbrace{\hbar\omega\left(\hat{a}^\dagger\hat{a} + \frac{1}{2}\right)}_{\hat{H}} + \hbar\omega\Bigg] = \hat{a}^\dagger[\hat{H} + \hbar\omega].$$

We apply this expression to ϕ and use (3.2.8) to get

$$\hat{a}^\dagger[\hat{H} + \hbar\omega]\phi = \hat{a}^\dagger[E + \hbar\omega]\phi = [E + \hbar\omega]\hat{a}^\dagger\phi = E'\hat{a}^\dagger\phi \rightarrow \hat{H}\hat{a}^\dagger\phi = E'\hat{a}^\dagger\phi,$$

which proves $\hat{a}^\dagger\phi$ is also an eigenstate, but with the eigenenergy E' increased by $\hbar\omega$, i. e., $E' = E + \hbar\omega$. Similarly, we can show $\hat{H}\hat{a}\phi = [E - \hbar\omega]\hat{a}\phi$, with the eigenenergy decreased by $\hbar\omega$.

This concludes that, if ϕ is an eigenstate, both $\hat{a}\phi$ and $\hat{a}^\dagger\phi$ are eigenstates, with the eigenenergy decreased or increased by $\hbar\omega$, i. e., $(E - \hbar\omega)$ and $(E + \hbar\omega)$, respectively. The energy gap between the subsequent two levels is always $\hbar\omega$ (see Figure 3.2(a)). If we successively apply $(\hat{a}^\dagger)^n$ to ϕ n times, we get $n\hbar\omega$. On the other hand, $\hat{a}\phi$ reduces the energy by $-\hbar\omega$, and in the end, we get $n = 0$.[3] Since $\hat{H} = \hbar\omega(\hat{a}^\dagger\hat{a} + \frac{1}{2})$ has an extra $\frac{1}{2}$, when $n = 0$, the minimum eigenenergy is $+\frac{1}{2}\hbar\omega$. For $n \neq 0$, the subsequent eigenenergies must be $(n + \frac{1}{2})\hbar\omega$.

In other words, our final Schrödinger equation is

$$\hat{H}\phi_n = \hbar\omega\left(\hat{a}^\dagger\hat{a} + \frac{1}{2}\right)\phi_n = \left(n + \frac{1}{2}\right)\hbar\omega\phi_n = E_n\phi_n, \tag{3.2.10}$$

whose eigenvalues E_n are

$$E_n = \left(n + \frac{1}{2}\right)\hbar\omega. \tag{3.2.11}$$

3 For this reason, we call these operators ladder operators, just like climbing up or down a ladder.

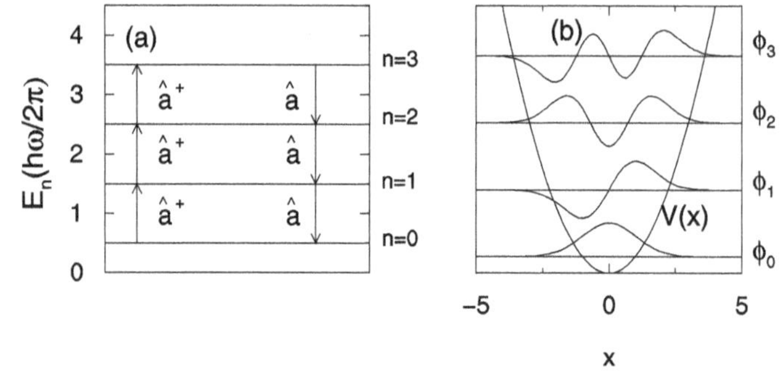

Figure 3.2: (a) Eigenvalues from $n = 0$ to 3. (b) Harmonic potential and eigenfunctions $\phi_n(x)$ which are vertically shifted for clarity.

Here, n is called quantum number and only takes nonnegative integers. This expression is similar to what Planck used (see Section 1.1.1), except that we use $\hbar$ and ω instead of h and ν, but the main difference is an extra term $\frac{1}{2}\hbar\omega$, which is called the zero-point energy. Physically, this means that even if the atom is at a temperature of 0 K, its energy cannot be zero. This quantum feature is rooted in the Heisenberg uncertainty principle.

Next, we search for eigenfunctions. We employ the fact that every application of the lowering operator $\hat{a}$ to an eigenfunction ϕ_n produces another eigenfunction with the eigenenergy lowered by $\hbar\omega$. However, when we hit the ground state ϕ_0, further application must yield zero, i. e., $\hat{a}\phi_0 = 0$, whose explicit expression is

$$\hat{a}\phi_0 = \frac{1}{\sqrt{2}}\left(\sqrt{\frac{m\omega}{\hbar}}x + \frac{\hbar}{\sqrt{m\hbar\omega}}\frac{\partial}{\partial x}\right)\phi_0 = 0. \tag{3.2.12}$$

The subscript 0 denotes the ground state. This differential equation has a root,

$$\phi_0 = Ae^{-\frac{m\omega}{2\hbar}x^2}, \tag{3.2.13}$$

where A is the normalization coefficient to be determined through

$$\int \phi_0^*\phi_0 dx = A^2 \int_{-\infty}^{\infty} e^{-\frac{m\omega}{\hbar}x^2} dx = 1, \rightarrow A = \left(\frac{m\omega}{\pi\hbar}\right)^{\frac{1}{4}}, \tag{3.2.14}$$

whose detailed derivation is left as an exercise. Then, the ground state has the following eigenfunction and eigenvalue:

$$\phi_0 = \left(\frac{m\omega}{\pi\hbar}\right)^{\frac{1}{4}} e^{-\frac{m\omega}{2\hbar}x^2}; \quad E_0 = \frac{1}{2}\hbar\omega. \tag{3.2.15}$$

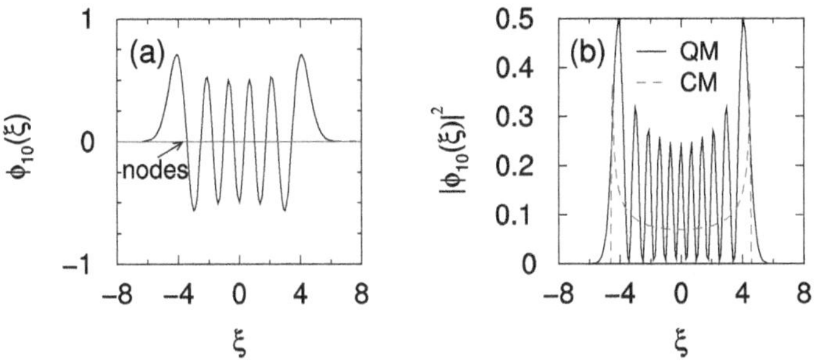

Figure 3.3: (a) ϕ_{10} as a function of ξ. (b) Probability (wavefunction squared). The dashed line is the classical result from eq. (3.2.4), whose x_0 is computed by setting the classical maximum energy to the quantum energy, i. e., $E_{\text{classical}} = \frac{1}{2}m\omega^2 x_0^2 = (n + \frac{1}{2})\hbar\omega$.

The remaining eigenstates, denoted as ϕ_n, can be found by $(\hat{a}^\dagger)^n \phi_0$,

$$\phi_n(x) = (\hat{a}^\dagger)^n \phi_0/\sqrt{n!} = \left(\frac{m\omega}{\pi\hbar}\right)^{\frac{1}{4}} \frac{1}{\sqrt{2^n n!}} H_n(\xi) e^{-\xi^2/2}, \tag{3.2.16}$$

where $1/\sqrt{n!}$ is to ensure the normalization, ξ is a dimensionless quantity, $\xi = \sqrt{\frac{m\omega}{\hbar}}x$ and $H_n(\xi)$ is the so-called Hermite polynomial, with a first few terms $H_0 = 1$, $H_1 = 2\xi$, $H_2 = 4\xi^2 - 2$. More generally, one can find them through the Rodrigues formula

$$H_n(\xi) = (-1)^n e^{\xi^2} \left(\frac{d}{d\xi}\right)^n e^{-\xi^2}, \tag{3.2.17}$$

or through the recursion relation $H_n(\xi) = 2\xi H_{n-1}(\xi) - 2(n-1)H_{n-2}(\xi)$.

Figure 3.2(a) shows four eigenvalues, whose eigenfunctions from $n = 0$ to 3 are displayed in Fig. 3.2(b) on top of the harmonic potential. As n increases, the wavefunction spreads out, and the number of nodes in $\phi_n(x)$ crossing zero increases. The more nodes a wavefunction has, the higher energy it has.

Recall from Fig. 3.1(b) that the classical prediction has the largest probability at the position maximum. We can make a comparison with this classical prediction. Figure 3.2(b) shows that, at $n = 0$, the maximum of the wavefunction $\phi_0(x)$ is situated at $x = 0$, and the probability $|\phi(x)|^2$ has the maximum at $x = 0$, which is very different from the classical prediction (see eq. (3.2.4)). But, as n, i. e., the eigenenergy, increases, we notice the wavefunction starts to spread away from $x = 0$. We choose a state with a large quantum number, here $n = 10$, and compute the probability $|\phi_{10}|^2$ using the recursion relation with code `Hermite.f` as listed in the Appendix 11.8. Figure 3.3(a) shows ϕ_{10} as a function of ξ. There are 10 nodes crossing 0. Figure 3.3(b) compares the quantum (solid line) with the classical results (dashed line). As n becomes larger, the quantum results approach the classical limit. There is a small probability for the quantum oscillator to appear in the classically forbidden region, i. e., beyond the dashed line. Note that, in order to make a proper comparison, the classical result x_0 is computed by setting the classical maximum energy to the quantum energy, i. e., $E_{\text{classical}} = \frac{1}{2}m\omega^2 x_0^2 = (n + \frac{1}{2})\hbar\omega$.

Exercise 3.2.3.

12. Suppose ϕ is an eigenstate of the harmonic oscillator Hamiltonian $\hat{H}$, where $\hat{H}\phi = E\phi$. Show $\hat{a}\phi$ is also an eigenstate of $\hat{H}$, i. e., $\hat{H}\hat{a}\phi = (E - \hbar\omega)\hat{a}\phi$.

13. The Hermite polynomial is used to construct the wavefunctions for the harmonic oscillator under code `Hermite.f` in Appendix listing 11.8. (a) Based on the code, how does the code compute the classical analogue of x_0? (b) Plot three wavefunctions, at $n = 2, 6$, and 9, and denote which one represents the wavefunctions, probability, and classical probability.

3.3 Applications of ladder operators in position and momentum

In this section we are going to demonstrate the power of ladder operators through the expectation values of position and momentum. These operators also allow us to compute the time dependence of position and momentum.

3.3.1 Basic properties of $\hat{a}$ and $\hat{a}^\dagger$

We start from eq. (3.2.10), $\hbar\omega(\hat{a}^\dagger\hat{a} + \frac{1}{2})\phi_n = (n + \frac{1}{2})\hbar\omega\phi_n$, so $\hat{a}^\dagger\hat{a}\phi_n = n\phi_n$. In the Dirac notation, we denote ϕ_n as $|n\rangle$, so $\hat{a}^\dagger\hat{a}|n\rangle = n|n\rangle$, where the n in front of $|n\rangle$ is an integer, $0, 1, 2, \ldots$ and the n in $|n\rangle$ must be understood as a label for the nth eigenstate.

From the previous section, we know that $\hat{a}^\dagger|n\rangle$ is also an eigenstate,

$$|n + 1\rangle = c_{n+1}\hat{a}^\dagger|n\rangle, \tag{3.3.1}$$

where the normalization coefficient c_{n+1} is determined by requiring

$$\langle n + 1|n + 1\rangle = \langle n|\hat{a}\hat{a}^\dagger|n\rangle|c_{n+1}|^2 = 1. \tag{3.3.2}$$

Here, we have used the complex conjugate of $|n + 1\rangle$, $\langle n + 1| = \langle n|ac^*_{n+1}$. Then, we use $\hat{a}\hat{a}^\dagger = \hat{a}^\dagger\hat{a} + 1$ and $\hat{a}^\dagger\hat{a}|n\rangle = n|n\rangle$ to reduce eq. (3.3.2) to

$$\langle n|(n + 1)|n\rangle|c_{n+1}|^2 = 1 \to (n + 1)|c_{n+1}|^2 = 1.$$

We take a real value for $c_{n+1} = \frac{1}{\sqrt{n+1}}$ and substitute it into eq. (3.3.1) to find

$$\hat{a}^\dagger|n\rangle = \sqrt{n + 1}|n + 1\rangle. \tag{3.3.3}$$

Similarly, one can find

$$\hat{a}|n\rangle = \sqrt{n}|n - 1\rangle, \quad \text{if } n \geq 1. \tag{3.3.4}$$

If $n = 0$, $\hat{a}|n\rangle = 0$. Since $\hat{a}^\dagger$ and $\hat{a}$ are related to $\hat{x}$ and $\hat{p}_x$, we are going to show that the expectation values of $\hat{x}$ and $\hat{p}_x$ can be computed using these two relations.

Exercise 3.3.1.

14. (a) Show $\hat{a}|n\rangle = \sqrt{n}|n-1\rangle$. (b) Then use (a) to show $\hat{a}^2|n\rangle = \sqrt{n(n-1)}|n-2\rangle$.
15. Use the previous result and $\hat{a}^\dagger|n\rangle = \sqrt{n+1}|n+1\rangle$ to show $\hat{a}^\dagger\hat{a}|n\rangle = n|n\rangle$.
16. Find $(\hat{a}^\dagger)^2|n\rangle$.

3.3.2 Expectation values of $\hat{x}$ and $\hat{x}^2$

We can use ϕ_n in eq. (3.2.16) to compute expectation values, but using $\hat{a}$ and $\hat{a}^\dagger$ is much simpler. We compute $\langle n|\hat{x}|n\rangle$ first. In the x-representation, $\hat{x} = x$. Recall

$$\hat{a} = \frac{1}{\sqrt{2m\hbar\omega}}(m\omega\hat{x} + i\hat{p}_x), \tag{3.3.5}$$

$$\hat{a}^\dagger = \frac{1}{\sqrt{2m\hbar\omega}}(m\omega\hat{x} - i\hat{p}_x). \tag{3.3.6}$$

We add eqs. (3.3.5) and (3.3.6) and divide by 2 to get

$$\frac{1}{2}(\hat{a} + \hat{a}^\dagger) = \frac{1}{\sqrt{2}}\sqrt{\frac{m\omega}{\hbar}}\hat{x}, \rightarrow \hat{x} = \sqrt{\frac{\hbar}{2m\omega}}(\hat{a}^\dagger + \hat{a}). \tag{3.3.7}$$

We apply $\hat{x}$ to $|n\rangle$ first and then find $\langle n|\hat{x}|n\rangle$,

$$\hat{x}|n\rangle = \sqrt{\frac{\hbar}{2m\omega}}(\hat{a}^\dagger + \hat{a})|n\rangle = \sqrt{\frac{\hbar}{2m\omega}}(\sqrt{n+1}|n+1\rangle + \sqrt{n}|n-1\rangle), \tag{3.3.8}$$

$$\langle n|\hat{x}|n\rangle = \sqrt{\frac{\hbar}{2m\omega}}(\sqrt{n+1}\langle n|n+1\rangle + \sqrt{n}\langle n|n-1\rangle). \tag{3.3.9}$$

Because all the eigenstates are orthogonal, $\langle n|n+1\rangle$ and $\langle n|n-1\rangle$ must be zero. This means in the harmonic oscillator the position's expectation value is always zero, which matches exactly the classical result (see No. 2 in Exercise 3.2.1).

Does this indicate that in QM the oscillator does not move at all? Let's compute $\langle n|\hat{x}^2|n\rangle$. We start from eq. (3.3.8), and apply $\hat{x}$ to $\hat{x}|n\rangle$ to obtain

$$\hat{x}^2|n\rangle = \frac{\hbar}{2m\omega}(\hat{a}^\dagger + \hat{a})(\hat{a}^\dagger + \hat{a})|n\rangle = \frac{\hbar}{2m\omega}[(\hat{a}^\dagger)^2 + \hat{a}^\dagger\hat{a} + \hat{a}\hat{a}^\dagger + \hat{a}^2]|n\rangle. \tag{3.3.10}$$

Using eqs. (3.3.3) and (3.3.4), we can reduce the three terms in the square bracket to

$$(\hat{a}^\dagger)^2|n\rangle = \hat{a}^\dagger\sqrt{n+1}|n+1\rangle = \sqrt{(n+1)(n+2)}|n+2\rangle,$$
$$\hat{a}^2|n\rangle = \hat{a}\sqrt{n}|n-1\rangle = \sqrt{n(n-1)}|n-2\rangle,$$
$$(\hat{a}^\dagger\hat{a} + \hat{a}\hat{a}^\dagger)|n\rangle = (2\hat{a}^\dagger\hat{a}+1)|n\rangle = (2n+1)|n\rangle.$$

Finally, employing $\langle n|m\rangle = \delta_{nm}$, we find the only nonzero term

$$\langle n|\hat{x}^2|n\rangle = \frac{\hbar}{2m\omega}\langle n|(2n+1)|n\rangle = \frac{(2n+1)\hbar}{2m\omega}, \tag{3.3.11}$$

which is slightly smaller than the classical prediction, that is $\frac{(2n+2)\hbar}{2m\omega}$. Even though $\langle\hat{x}\rangle$ is zero, $\langle\hat{x}^2\rangle$ is nonzero, so our oscillator does move, with the standard deviation being

$$\Delta x = \sqrt{\langle n|\hat{x}^2|n\rangle - \langle n|\hat{x}|n\rangle^2} = \sqrt{\frac{(2n+1)\hbar}{2m\omega}}. \tag{3.3.12}$$

This deviation has nothing to do with our measurement tool. Even if we had a perfect tool, we still would have this deviation because this originates from the wave nature of the oscillator in QM. The higher the energy is (larger n), the larger the deviation or uncertainty becomes. In the literature, this is called quantum fluctuation.

3.3.3 Expectation values of $\hat{p}_x$ and $\hat{p}_x^2$ and the uncertainty principle

Using the same method, we can find $\langle n|\hat{p}_x|n\rangle = 0$ and $\langle n|\hat{p}_x^2|n\rangle = (n+1/2)m\hbar\omega$, whose derivation is left as an exercise.

The uncertainty in momentum is (see the following exercise)

$$\Delta p_x = \sqrt{\langle n|\hat{p}_x^2|n\rangle - \langle n|\hat{p}_x|n\rangle^2} = \sqrt{(n+1/2)m\hbar\omega}.$$

If we multiply two uncertainties $\Delta x\Delta p_x$, we find the uncertainty in the position and momentum as

$$\Delta x\Delta p_x = \sqrt{\frac{(n+1/2)\hbar}{m\omega}}\sqrt{(n+1/2)m\hbar\omega} = (n+1/2)\hbar, \tag{3.3.13}$$

which satisfies the Heisenberg uncertainty principle, $\Delta x\Delta p_x \geq \frac{1}{2}\hbar$. Importantly, if we set $n = 0$ in eq. (3.3.13), we obtain an equality $\Delta x\Delta p_x = \frac{1}{2}\hbar$, which represents the lowest bound that the Heisenberg uncertainty principle allows.

Consequently, even at 0 K, the system cannot stand still and has to vibrate. So, the sum of the potential energy and kinetic energy cannot be zero, the zero-point energy. The origin of the zero-point energy is related to the wave nature of particles and is a direct consequence of the Heisenberg uncertainty principle. It is a well-known fact that helium atoms do not form solids, even if we cool them down to a very low temperature, such as 4.2 K. Research on cold atoms is a hot topic. It is possible to compute the kinetic energy from the thermal energy $k_B T/2$, where k_B is the Boltzmann constant and T is temperature. From the kinetic energy we can compute the momentum. Now, if we take this momentum as the momentum uncertainty Δp_x, we can estimate Δx, the position uncertainty of a cold atom.

Exercise 3.3.3.

17. $|n\rangle$ is an eigenstate of the harmonic oscillator. (a) Find $\langle n|\hat{p}_x|m\rangle$ and $\langle n|\hat{x}|m\rangle$ for general n and m and, if $n = m$, prove $\langle n|\hat{p}_x|n\rangle = \langle n|\hat{x}|n\rangle = 0$. (b) Find $\langle n|\hat{p}_x^2|m\rangle$; is $\langle n|\hat{p}_x^2|n\rangle$ zero?

18. The classical maximum displacement x_0 can be linked to the quantum value through $\frac{1}{2}m\omega^2 x_0^2 = (n+\frac{1}{2})\hbar\omega$. Show the average value $\bar{x^2}$ is $\frac{(2n+1)\hbar}{2m\omega}$.

19. $|n\rangle$ and $|m\rangle$ are two eigenstates of the harmonic oscillator, where $n \neq m$. Compute $\langle n|\hat{a}^2|m\rangle$ and then find the condition when $\langle n|\hat{a}^2|m\rangle \neq 0$.

20. At 4.2 K, compute the position uncertainty of a helium atom.

3.4 Time evolution in harmonic oscillator: Schrödinger picture

As clear from classical physics (see Section 3.2.1), in the harmonic oscillator both position and momentum change with time. In QM, there are several ways to describe the time dependence. We introduce the Schrödinger picture, where the state $\psi(t)$ evolves with time according to TDSE, but the operator $\hat{O}$ does not. The expectation value has time dependence from the state through $\langle\psi(t)|\hat{O}|\psi(t)\rangle$. Since the harmonic oscillator Hamiltonian is time independent, $\psi(t)$ can be found through the stationary-state evolution.[4]

3.4.1 Stationary solutions

Given an initial wavefunction $\psi(\mathbf{r}, t = 0)$ at $t = 0$, we expand it in terms of eigenstates $\phi_n(\mathbf{r})$, $\psi(\mathbf{r}, t = 0) = \sum_n f_n \phi_n(\mathbf{r})$, whose expansion coefficient f_n is $f_n = \int \phi_n^*(\mathbf{r})\psi(\mathbf{r}, 0)d\mathbf{r}$. This step is similar to the projection of a vector onto an axis to find its component. The only difference here is that we are working in a function space, Hilbert space, where eigenfunctions serve as our "coordinates" and the vector projection becomes an integration. If $\phi_n(\mathbf{r})$ is $\exp(i\mathbf{k}\cdot\mathbf{r})$, this is just a Fourier transform.

For any eigenstate $|\phi_n(\mathbf{r})\rangle$, its time-dependent state $|\phi_n(\mathbf{r}, t)\rangle$ is

$$|\phi_n(\mathbf{r}, t)\rangle = \exp(-iE_n t/\hbar)|\phi_n(\mathbf{r})\rangle,$$

which is only possible with a time-independent Hamiltonian. Then $\psi(\mathbf{r}, t)$ is

---

4 If $\hat{H}$ is time dependent, we have to solve TDSE.

$$\psi(\mathbf{r},t) = \sum_n f_n \phi_n(\mathbf{r},t) = \sum_n f_n \exp(-iE_n t/\hbar)\phi_n(\mathbf{r}).$$

Then, we compute the expectation value of a physical quantity $\hat{A}$ via

$$\langle \psi(\mathbf{r},t)|\hat{A}|\psi(\mathbf{r},t)\rangle = \int \psi^*(\mathbf{r},t)\hat{A}\psi(\mathbf{r},t)d\mathbf{r} = \sum_{nm} f_n^* f_m e^{i(E_n-E_m)t/\hbar}\langle \phi_n|\hat{A}|\phi_m\rangle.$$

3.4.2 Examples

Example 3 (Heisenberg uncertainty principle in the time domain). A harmonic oscillator is initially in a mixed state ϕ_0 and ϕ_1, $\psi(t=0) = f_0\phi_0 + f_1\phi_1$, where f_0 and f_1 are real and $f_0^2 + f_1^2 = 1$. Find (a) $\psi(t \neq 0)$, (b) $\langle \psi(t)|\hat{x}|\psi(t)\rangle$, (c) $\langle \psi(t)|\hat{x}^2|\psi(t)\rangle$, and (d) $(\Delta x \Delta p)(t)$.

(a) The initial wavefunction $\psi(\mathbf{r},0)$ is $\psi(\mathbf{r},0) = f_0|0\rangle + f_1|1\rangle$, where we use Dirac notation for ϕ_n. Its time-dependent wavefunction is

$$\psi(\mathbf{r},t) = f_0 e^{-iE_0 t/\hbar}|0\rangle + f_1 e^{-iE_1 t/\hbar}|1\rangle.$$

Since $\hat{x} = \sqrt{\frac{\hbar}{2m\omega}}(\hat{a}^\dagger + \hat{a})$, $\langle \hat{x}\rangle$ is

$$\begin{aligned}\langle \hat{x}\rangle &= \langle \psi(\mathbf{r},t)|\sqrt{\frac{\hbar}{2m\omega}}\left(\hat{a}^\dagger + \hat{a}\right)|\psi(\mathbf{r},t)\rangle \\ &= \left(f_0 e^{iE_0 t/\hbar}\langle 0| + f_1 e^{iE_1 t/\hbar}\langle 1|\right)\sqrt{\frac{\hbar}{2m\omega}}\left(\hat{a}^\dagger + \hat{a}\right)\left(f_0 e^{-iE_0 t/\hbar}|0\rangle + f_1 e^{-iE_1 t/\hbar}|1\rangle\right).\end{aligned}$$

We are going to use $\hat{a}^\dagger|n\rangle = \sqrt{n+1}|n+1\rangle$ and $a|n\rangle = \sqrt{n}|n-1\rangle$. This means that $\langle 0|\hat{a}|0\rangle = 0$, $\langle 1|\hat{a}|1\rangle = 0$, $\langle 0|\hat{a}^\dagger|0\rangle = 0$, and $\langle 1|\hat{a}^\dagger|1\rangle = 0$, so all the diagonal terms with the same states on both sides are zero. We are left with off-diagonal terms,

$$\begin{aligned}\langle \hat{x}\rangle &= \sqrt{\frac{\hbar}{2m\omega}} f_1 e^{iE_1 t/\hbar}\langle 1|\hat{a}^\dagger|0\rangle f_0 e^{-iE_0 t/\hbar} + \sqrt{\frac{\hbar}{2m\omega}} f_0 e^{iE_0 t/\hbar}\langle 0|\hat{a}|1\rangle f_1 e^{-iE_1 t/\hbar} \\ &= \sqrt{\frac{\hbar}{2m\omega}} f_0 f_1\left(e^{i(E_1-E_0)t/\hbar} + e^{-i(E_1-E_0)t/\hbar}\right) = \sqrt{\frac{2\hbar}{m\omega}} f_0 f_1 \cos\left[(E_1 - E_0)t/\hbar\right].\end{aligned}$$

One can see that $\langle \hat{x}\rangle$ oscillates with time because we are in a mixed state.

(b) Since $\hat{x} = \sqrt{\frac{\hbar}{2m\omega}}(\hat{a}^\dagger + \hat{a})$, we then have

$$\hat{x}^2 = \frac{\hbar}{2m\omega}\left(\hat{a}^\dagger + \hat{a}\right)\left(\hat{a}^\dagger + \hat{a}\right) = \frac{\hbar}{2m\omega}\left(\left(\hat{a}^\dagger\right)^2 + \hat{a}^2 + \hat{a}^\dagger\hat{a} + \hat{a}\hat{a}^\dagger\right) = \frac{\hbar}{2m\omega}\left(\left(\hat{a}^\dagger\right)^2 + \hat{a}^2 + 2\hat{a}^\dagger\hat{a} + 1\right).$$

Caution: We cannot use regular algebra, $(a+b)^2 = a^2 + 2ab + b^2$. When we have operators, this does not work if the operators do not commute. The terms $(\hat{a}^\dagger)^2$ and $\hat{a}^2$ contribute zero, and only $2\hat{a}^\dagger\hat{a} + 1$ remains:

$$\langle \hat{x}^2\rangle = \left(f_0 e^{iE_0 t/\hbar}\langle 0| + f_1 e^{iE_1 t/\hbar}\langle 1|\right)\frac{\hbar}{2m\omega}\left(2\hat{a}^\dagger\hat{a} + 1\right)\left(f_0 e^{-iE_0 t/\hbar}|0\rangle + f_1 e^{-iE_1 t/\hbar}|1\rangle\right) = \frac{\hbar}{2m\omega}\left(f_0^2 + 3f_1^2\right).$$

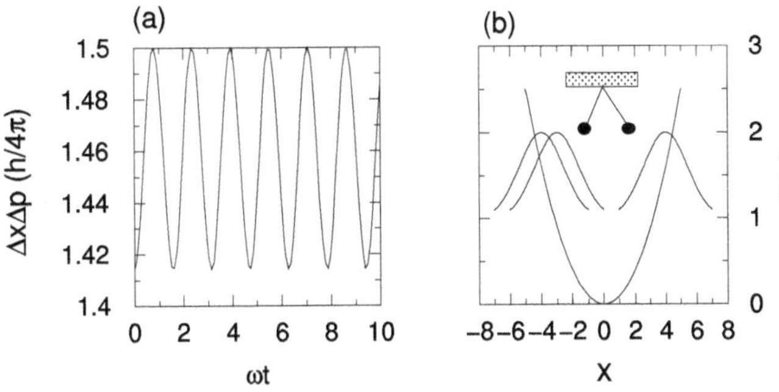

Figure 3.4: (a) Uncertainty principle in the time domain. Here, $f_0^2 = f_1^2 = 0.5$. $\omega = \frac{(E_1 - E_0)}{\hbar}$. The uncertainty is always above the minimum $\hbar/2$ or $h/(4\pi)$, due to the presence of a higher state $|1\rangle$. (b) Quantum mechanical description of a harmonic oscillator. Classically, when we move the pendulum to its maximum position x_0, in QM this corresponds to an initial Gaussian wavepack, which is a coherent state and centered at x_0. The time evolution of this pendulum is the center shifting of the initial Gaussian wavefunction.

(c) Finally, we find the uncertainty in $\hat{x}$ as

$$(\Delta x)^2 = \langle \hat{x}^2 \rangle - \langle \hat{x} \rangle^2 = \frac{\hbar}{2m\omega}\left(f_0^2 + 3f_1^2\right) - \frac{2\hbar f_1^2 f_0^2}{m\omega}\cos^2\left[(E_1 - E_0)t/\hbar\right].$$

Similarly for $\hat{p}_x$ and the uncertainty relation, we find (see the following below)

$$(\Delta p_x)^2 = \langle \hat{p}_x^2 \rangle - \langle \hat{p}_x \rangle^2 = \frac{m\hbar\omega}{2}\left(f_0^2 + f_1^2\right) - 2m\hbar\omega f_0^2 f_1^2 \sin^2\left[(E_1 - E_0)t/\hbar\right],$$

$$(\Delta x)^2 (\Delta p_x)^2 = \frac{\hbar^2}{4}\left(f_0^2 + 3f_1^2 - 4f_1^2 f_0^2 \cos^2\omega t\right)\left(f_0^2 + f_1^2 - 4f_0^2 f_1^2 \sin^2\omega t\right),$$

$$\Delta x \Delta p_x = \frac{\hbar}{2}\sqrt{\left(f_0^2 + 3f_1^2 - 4f_0^2 f_1^2 \cos^2\omega t\right)\left(f_0^2 + f_1^2 - 4f_0^2 f_1^2 \sin^2\omega t\right)},$$

where $\omega = \frac{(E_1 - E_0)}{\hbar}$ and the uncertainty is time dependent (Fig. 3.4(a)).

Example 4 (Eigenfunction of $\hat{a}$). Show $\psi(x, 0) = A_0 e^{-m\omega(x-x_0)^2/2\hbar}$ is an eigenfunction of $\hat{a}$, where A_0 is a normalization constant.

First, note that $\hat{a} = \frac{1}{\sqrt{2m\hbar\omega}}(m\omega\hat{x} + i\hat{p}_x)$. We apply it on $\psi(x, 0)$,

$$\hat{a}\psi(x, 0) = A_0 \frac{1}{\sqrt{2m\hbar\omega}}(m\omega\hat{x} + i\hat{p}_x) e^{-m\omega(x-x_0)^2/2\hbar},$$

whose second term is

$$\hat{p}_x e^{-m\omega(x-x_0)^2/2\hbar} = -i\hbar\frac{\partial}{\partial x}\left(e^{-m\omega(x-x_0)^2/2\hbar}\right) = m\omega i(x - x_0)e^{-m\omega(x-x_0)^2/2\hbar}.$$

The final expression is

$$\hat{a}\psi(x, 0) = \sqrt{\frac{m\omega}{2\hbar}} x_0 A_0 e^{-m\omega(x-x_0)^2/2\hbar} = \sqrt{\frac{m\omega}{2\hbar}} x_0 \psi(x, 0), \tag{3.4.1}$$

which concludes our proof. The eigenvalue of $\hat{a}$ is $\sqrt{\frac{m\omega}{2\hbar}}x_0$. If we repeatedly apply $\hat{a}$ to the previous equation, we have $\hat{a}^n\psi(x,0) = (\sqrt{\frac{m\omega}{2\hbar}}x_0)^n\psi(x,0)$.

Example 5 (Coherent states). In CM, we start a harmonic oscillation by moving the pendulum's bob to position x_0 (see Fig. 3.4(b)). Its quantum mechanical equivalence is to prepare an initial state at $t = 0$ as $\psi(x,0) = A_0e^{-m\omega(x-x_0)^2/2\hbar}$, where A_0 is the normalization constant. This state is an eigenfunction of the operator $\hat{a}$ and is called the coherent state. (a) Find $\psi(x,t)$ at time t. (b) Find the expectation value of $\langle\psi(x,t)|\hat{x}|\psi(x,t)\rangle$. (c) Find the expectation value $\langle\psi(x,t)|\hat{p}_x|\psi(x,t)\rangle$.

(a) According to our recipe, we must first expand $\psi(x,0)$ in terms of the eigenstates of the harmonic oscillator, $\phi_n(x)$,

$$\psi(x,0) = \sum_n c_n\phi_n(x), \tag{3.4.2}$$

where $\phi_n(x) = N_nH_n(\alpha x)e^{-\frac{1}{2}\alpha^2x^2}$, $\alpha = \sqrt{\frac{m\omega}{\hbar}}$, and $N_n = [\frac{\alpha}{\sqrt{\pi}2^nn!}]^{\frac{1}{2}}$. So the difficulty is to find c_n. Due to the orthogonality among $\phi_n(x)$, we multiply both sides of eq. (3.4.2) by $\phi_n(x)$ and integrate it to get $c_n = \int\psi(x,0)\phi_n(x)dx$.

We introduce a dimensionless quantity ξ, where $\xi = \alpha x$ and $\xi_0 = \alpha x_0$, so we can simplify c_n as

$$\begin{aligned}c_n &= A_0N_n\int e^{-(\xi-\xi_0)^2/2}e^{-\frac{1}{2}\xi^2}H_n(\xi)dx = \frac{A_0N_n}{\alpha}\int e^{-(\xi-\xi_0)^2/2-\frac{1}{2}\xi^2}H_n(\xi)d(\alpha x)\\ &= \frac{A_0N_n}{\alpha}\int e^{-(\xi^2-\xi_0\xi+\frac{1}{4}\xi^2+\frac{1}{4}\xi_0^2)}H_n(\xi)d\xi = \frac{A_0N_n}{\alpha}e^{-\frac{\xi_0^2}{4}}\int e^{-(\xi-\frac{1}{2}\xi_0)^2}H_n(\xi)d\xi.\end{aligned}$$

Using the identity relation $\int_{-\infty}^{\infty}e^{-(t-z)^2}H_n(t)dt = 2^n\sqrt{\pi}z^n$, we find

$$c_n = \frac{e^{-\frac{1}{2}(\frac{\xi_0}{\sqrt{2}})^2}}{\sqrt{n!}}\left(\frac{\xi_0}{\sqrt{2}}\right)^n.$$

So, our $\psi(x,t)$ can be written as $\psi(x,t) = \sum_n c_ne^{-iE_nt/\hbar}\phi_n(x)$, with $E_n = (n+\frac{1}{2})\hbar\omega$. Specifically,

$$\psi(x,t) = e^{-i\omega t/2}\sum_n\frac{e^{-in\omega t}e^{-\frac{1}{2}(\frac{\xi_0}{\sqrt{2}})^2}}{\sqrt{n!}}\left(\frac{\xi_0}{\sqrt{2}}\right)^n\phi_n(x) = e^{-i\omega t/2}e^{-\frac{1}{2}\eta_0^2}\sum_n\frac{1}{\sqrt{n!}}\left[b(t)\right]^n\phi_n(x),$$

where $b(t) = e^{-i\omega t}\frac{\xi_0}{\sqrt{2}} = \eta_0e^{-i\omega t}$ and $\eta_0 = \frac{\xi_0}{\sqrt{2}} = \frac{\alpha x_0}{\sqrt{2}} = \sqrt{\frac{m\omega}{2\hbar}}x_0$. The summation is the expansion of ϕ_0 but with the center moved to $x - x_0\cos\omega t$. Finally, we have

$$\psi(x,t) = \phi_0(x - x_0\cos\omega t)e^{i[\frac{m\omega x_0^2}{2\hbar}\sin^2\omega t-\frac{\omega t}{2}-\frac{m\omega x_0x}{\hbar}\sin\omega t]}.$$

This is a wavepacket that oscillates with time but keeps its original shape.

(b) To compute the expectation value of $\hat{x}$, we express it in terms of $\hat{a}$ and $\hat{a}^\dagger$ because $\psi(x,t)$ is already expressed in terms of the eigenstates of $\hat{a}$, $\hat{x} = \sqrt{\frac{\hbar}{2m\omega}}(\hat{a}^\dagger+\hat{a})$. Since $\langle\hat{a}^\dagger\rangle = b^*(t)$ and $\langle\hat{a}\rangle = b(t)$,

$$\langle\hat{x}\rangle = \sqrt{\frac{\hbar}{2m\omega}}\left[b^*(t)+b(t)\right] = \sqrt{\frac{\hbar}{2m\omega}}\sqrt{\frac{m\omega}{\hbar}}\frac{x_0}{\sqrt{2}}2[\cos\omega t] = x_0\cos\omega t,$$

where we have used $\hat{a}^\dagger|\psi(x,t)\rangle = b^*(t)|\psi(x,t)\rangle$ and $\hat{a}|\psi(x,t)\rangle = b(t)|\psi(x,t)\rangle$.

(c) We can exploit the same idea as the position operator to compute the expectation value of the momentum,

$$\langle \hat{p}_x \rangle = i\sqrt{\frac{m\omega\hbar}{2}}\left[b^*(t) - b(t)\right] = -m\omega x_0 \sin \omega t.$$

This example shows that the quantum mechanical results agree with the classical ones once we find the expectation values.

Exercise 3.4.2.
21. A harmonic oscillator is initially in a mixed state ϕ_0 and ϕ_1, $\psi(t = 0) = f_0\phi_0 + f_1\phi_1$, where f_0 and f_1 are real and $f_0^2 + f_1^2 = 1$. (a) Show $\langle \hat{p}_x \rangle(t) = -\sqrt{2m\hbar\omega} f_0 f_1 \sin[(E_1 - E_0)t/\hbar]$. (b) Sketch a diagram position-versus-momentum, $\langle \hat{x} \rangle(t)$-$\langle \hat{p}_x \rangle(t)$. $\langle \hat{x} \rangle(t)$ is from our previous example.
22. Show that an eigenfunction $|\psi\rangle$ of the lowering operator $\hat{a}$ is not an eigenfunction of the raising operator $\hat{a}^\dagger$.
23. Although $\hat{a}$ and $\hat{a}^\dagger$ do not share a common eigenfunction, it is possible to find the eigenfunction of $\hat{a}^\dagger$ from that of $\hat{a}$. Hint: Examine the result if we apply $\hat{p}_x$ to the eigenfunction of $\hat{a}$.

3.5 Heisenberg's picture and Heisenberg equation of motion

In the Schrödinger picture, states evolve with time, but operators do not. This is convenient for some problems, but not for others, in particular for the harmonic oscillator problem as already seen. The Heisenberg picture offers an alternative, where states are static, but operators evolve with time. The Heisenberg picture works with operators, so its connection to CM is closer. Table 3.1 compares these two pictures. Mathematically, these two pictures are equivalent.

3.5.1 Heisenberg equation of motion

In order to distinguish the Heisenberg picture from the Schrödinger picture, in the following, we denote wavefunctions and operators using the subscript H for the former and S for the latter. We start with the standard Schrödinger equation,

$$i\hbar\frac{\partial\psi_S}{\partial t} = \hat{H}_S\psi_S. \tag{3.5.1}$$

Table 3.1: Operators and states in the Schrödinger and Heisenberg pictures have respective subscripts S and H. t_0 is the initial time. $\hat{U}$ is a unitary operator.

Picture	State ψ	Operator $\hat{A}$	Hamiltonian
Schrödinger	$i\hbar\frac{\partial\psi_S}{\partial t} = \hat{H}_S\psi_S$	$\hat{A}_S = \hat{A}_H(t = t_0)$	$\hat{H}_S$
Heisenberg	$\psi_H = \psi_S(t = t_0)$	$\frac{d\hat{A}_H}{dt} = \frac{1}{i\hbar}[\hat{A}_H, \hat{H}_H] + \frac{\partial\hat{A}_H}{\partial t}, \hat{A}_H = \hat{U}^\dagger\hat{A}_S\hat{U}$	$\hat{H}_H = \hat{H}_S$

3.5.1.1 Unitary operator $\hat{U}$

In order to derive the form of an operator in the Heisenberg picture, we introduce the time evolution operator $\hat{U}(t, t_0)$ that transforms the initial wavefunction $\psi_S(t_0)$ at time t_0 to $\psi_S(t)$ at t as

$$\psi_S(t) = \hat{U}(t, t_0)\psi_S(t_0). \tag{3.5.2}$$

For brevity, we denote $\hat{U} = \hat{U}(t, t_0)$. To reveal what $\hat{U}$ is, we substitute $\psi_S(t)$ into both sides of eq. (3.5.1) to get

$$i\hbar\frac{d\hat{U}}{dt}\psi_S(t_0) = \hat{H}_S\hat{U}\psi_S(t_0), \tag{3.5.3}$$

where the time derivative is only on $\hat{U}$ because $\psi_S(t_0)$ does not depend on t. Here we have changed $\partial/\partial t$ to d/dt since the $\hat{U}$ operator only has time as its variable.

Since $\psi_S(t_0)$ is an arbitrary wavefunction, the coefficients of $\psi_S(t_0)$ on both sides must be the same,i. e.,

$$i\hbar\frac{d\hat{U}}{dt} = \hat{H}_S\hat{U}, \tag{3.5.4}$$

and its Hermitian conjugate is

$$-i\hbar\frac{d\hat{U}^\dagger}{dt} = \hat{U}^\dagger\hat{H}_S. \tag{3.5.5}$$

Multiplying eq. (3.5.4) by $\hat{U}^\dagger$ from the left, multiplying eq. (3.5.5) by $\hat{U}$ from the right, and then subtracting one from the other, we obtain $i\hbar(\hat{U}^\dagger\frac{d\hat{U}}{dt} + \frac{d\hat{U}^\dagger}{dt}\hat{U}) = 0$, or $\hat{U}^\dagger\hat{U}$ is constant, which can be chosen as 1, $\hat{U}^\dagger\hat{U} = 1$. $\hat{U}$ is called a unitary operator. To see what $\hat{U}^\dagger$ does to a wavefunction, we left multiply $\hat{U}^\dagger$ to eq. (3.5.2) to have

$$\hat{U}^\dagger\psi_S(t) = \hat{U}^\dagger\hat{U}\psi_S(t_0) = \psi_S(t_0) \rightarrow \hat{U}^\dagger|\psi_S(t)\rangle = |\psi_S(t_0)\rangle, \tag{3.5.6}$$

which shows that $\hat{U}^\dagger$ transforms $\psi_S(t)$ at t back to $\psi_S(t_0)$ at t_0. $\hat{U}^\dagger$ reverses the time arrow for a ket vector $|\psi_S(t)\rangle$, i. e., a column vector in the matrix mechanics. Using Dirac notation in eq. (3.5.6) is purposeful. We take the Hermitian conjugate of eq. (3.5.6) to get a row vector, or, a bra vector,

$$\langle\psi_S(t)|\hat{U} = \langle\psi_S(t_0)|, \tag{3.5.7}$$

which shows that $\hat{U}$ transforms a bra vector $\langle\psi_S(t)|$ at t to another bra vector $\langle\psi_S(t_0)|$ at t_0. $\hat{U}$ reverses the time arrow for a bra vector.

If $\hat{H}_S$ is time independent, we integrate eq. (3.5.4) to obtain

$$\hat{U}(t, t_0) = e^{-i\hat{H}_S(t-t_0)/\hbar}. \tag{3.5.8}$$

If $\hat{H}_S$ is time dependent, we relegate the discussion to Appendix 11.2. So, the time-evolution operator $\hat{U}$ essentially takes the responsibility of time integration for the wavefunction. What is more important is that it helps construct an operator.

Exercise 3.5.1.

24. Equation (3.5.7) is obtained by taking the Hermitian conjugate of eq. (3.5.6). (a) Can we take the conjugation of eq. (3.5.6) to obtain the same results? Why? (b) Show $\hat{O} = \begin{pmatrix} 0 & -i \\ i & 0 \end{pmatrix}$ is an unitary operator. (c) Suppose $|\psi\rangle = \frac{1}{\sqrt{2}}\begin{pmatrix} -i \\ i \end{pmatrix}$ is a state at time t. Use $\hat{O}$ to check your answer to (a).

25. Equation (3.5.5) has a negative sign on the left side. (a) Under what condition can we interpret it as a time reversal symmetry? (b) Prove it mathematically.

3.5.1.2 Operator in the Heisenberg picture

Suppose we want to compute the expectation value $\langle\psi_S(t)|\hat{A}_S|\psi_S(t)\rangle$ of an operator $\hat{A}_S$ in the Schrödinger picture. Employing $\hat{U}\hat{U}^\dagger = 1$, we can rewrite it as

$$\langle\psi_S(t)|\hat{A}_S|\psi_S(t)\rangle = \langle\psi_S(t)\hat{U}\hat{U}^\dagger|\hat{A}_S|\hat{U}\hat{U}^\dagger\psi_S(t)\rangle = \underbrace{\langle\psi_S(t_0)}_{\langle\psi_H}\,\underbrace{\hat{U}^\dagger|\hat{A}_S|\hat{U}}_{\hat{A}_H(t)}\,\underbrace{\psi_S(t_0)\rangle}_{\psi_H\rangle}, \tag{3.5.9}$$

where we have used eqs. (3.5.6) and (3.5.7).

According to our convention, the state in the Heisenberg's picture is same as the initial state $(\psi_S(t_0))$ of the Schrödinger picture, $\psi_H = \psi_S(t_0)$. Based on eq. (3.5.9), we define the operator in the Heisenberg's picture as

$$\hat{A}_H \equiv \hat{U}^\dagger\hat{A}_S\hat{U}, \tag{3.5.10}$$

where $\hat{U}$ is determined by eq. (3.5.4). We take the derivative of eq. (3.5.10) with respect to time to obtain

$$\frac{d\hat{A}_H}{dt} = \frac{d\hat{U}^\dagger}{dt}\hat{A}_S\hat{U} + \hat{U}^\dagger\hat{A}_S\frac{d\hat{U}}{dt} + \hat{U}^\dagger\frac{\partial\hat{A}_S}{\partial t}\hat{U}.$$

Then we substitute eqs. (3.5.4) and (3.5.5) into the previous equation, and, after a simple rearrangement, we derive

$$\frac{d\hat{A}_H(t)}{dt} = \frac{1}{i\hbar}[\hat{A}_H(t), \hat{H}] + \frac{\partial\hat{A}_H(t)}{\partial t}, \tag{3.5.11}$$

which is the Heisenberg's equation of motion, derived by Dirac first. $\frac{\partial\hat{A}_H(t)}{\partial t} \equiv \hat{U}^\dagger\frac{\partial\hat{A}_S}{\partial t}\hat{U}$.

3.5.2 Examples

To appreciate the power of the Heisenberg picture, we present some examples. To distinguish the position operator from the position variable, we put a hat over the position operator, $\hat{x}$, in all the following examples.

Example 6 (Conservation law). $\hat{A}$ is an operator and commutes with $\hat{H}$, $[\hat{A}, \hat{H}] = 0$. (a) Show $\hat{A}$ is a conserved quantity. (b) A system consists of a molecule $\hat{H}_{\text{mol}}$ and a light source $\hat{H}_{\text{light}}$, so the total Hamiltonian is $\hat{H} = \hat{H}_{\text{mol}} + \hat{H}_{\text{light}}$. Show the molecule can gain or lose energy. (c) $\hat{A}$ represents the particle number operator $\hat{N}$. If $[\hat{N}, \hat{H}] = 0$, show the number of particles of the system is conserved and does not change with time.

(a) Since $[\hat{A}, \hat{H}] = 0$, from eq. (3.5.11) we have $d\hat{A}/dt = 0$. So, $\hat{A}$ must be a conserved quantity and does not change with time.

(b) Since we are interested in the energy of the molecule, $\hat{A} = \hat{H}_{\text{mol}}$. We compute its time change $i\hbar\dot{\hat{H}}_{\text{mol}} = [\hat{H}_{\text{mol}}, \hat{H}_{\text{mol}} + \hat{H}_{\text{light}}] = [\hat{H}_{\text{mol}}, \hat{H}_{\text{light}}]$. As far as $[\hat{H}_{\text{mol}}, \hat{H}_{\text{light}}] \neq 0$, $\dot{\hat{H}}_{\text{mol}} \neq 0$. This means the molecule will gain or lose energy, which happens in photosynthesis and chemical reactions.

(c) Since $[\hat{N}, \hat{H}] = 0$ and $i\hbar d\hat{N}/dt = [\hat{N}, \hat{H}] = 0$, there is no particle number change. If we use the Schrödinger picture, we have to solve the Schrödinger equation to get $\psi(t)$, and then compute the expectation value $\langle\psi(t)|\hat{N}|\psi(t)\rangle$. In this sense, the Heisenberg picture is simpler.

Example 7 (Free particle). The Hamiltonian of a free particle is $H = \frac{\hat{p}_x^2}{2m}$. (a) Find how its momentum changes with time. (b) Find how its position changes with time.

(a) In this case, $\hat{A} = \hat{p}_x$. We have $i\hbar\dot{\hat{p}}_x = [\hat{p}_x, H]$, and $\dot{\hat{p}}_x = 0$, so $\hat{p}_x$ is a constant operator. This means that the particle moves with a constant momentum, or its momentum is conserved. This result is not surprising. But, if one wants to know how large the momentum is, then, one has to know its initial state, its wavefunction. If our initial wavefunction $\psi_H(x, t_0)$ is $e^{ikx}/\sqrt{L}$, which is normalized in a box of length L, then, $\langle\psi_H(t_0)|\hat{p}_x|\psi_H(t_0)\rangle = \hbar k$ at $t = t_0$. The momentum has the same value at any later time t.

(b) The position operator is $\hat{x}$, so its equation of motion is $i\hbar\dot{\hat{x}} = [\hat{x}, H] = [\hat{x}, \frac{\hat{p}_x^2}{2m}] = \frac{1}{2m}[\hat{x}, \hat{p}_x^2] = \frac{1}{2m}(\hat{p}_x[\hat{x}, \hat{p}_x] + [\hat{x}, \hat{p}_x]\hat{p}_x) = \frac{i\hbar\hat{p}_x}{m}$. The position changes as $\dot{\hat{x}} = \frac{\hat{p}_x}{m}$, which matches the CM. If we integrate over time, the position is $\hat{x}(t) = \hat{x}(t_0) + \frac{\hat{p}_x}{m}t$, where both $\hat{x}$ and $\hat{p}_x$ are operators. Physically, the position of the particle changes with time at a rate of $\hat{p}_x/m$ (velocity operator).

If we want to compute the expectation value of $\hat{x}(t)$, we must provide the initial wavefunction. If we adopt the same wavefunction $e^{ikx}/\sqrt{L}$, we find $\langle\psi(t_0)|\hat{x}(t)|\psi(t_0)\rangle = \langle\psi_0|\hat{x}(t_0)|\psi_0\rangle + \frac{\langle\psi(t_0)|\hat{p}_x|\psi(t_0)\rangle}{m} = \langle\hat{x}(t_0)\rangle + \frac{\hbar k}{m}t$, where the first term is the initial position. This is a constant velocity motion.

Example 8 (Time evolution of ladder operators $\hat{a}$ and $\hat{a}^\dagger$). In Heisenberg's equation of motion, any reasonable operator can be used. For instance, a translational operator of position can be used. Here, we take $\hat{a}$ and $\hat{a}^\dagger$ of the harmonic oscillator as an example. At initial time t_0, the initial operators of $\hat{a}$ and $\hat{a}^\dagger$ are $\hat{a}(t_0)$ and $\hat{a}^\dagger(t_0)$, respectively. Find their operators at time t.

The Hamiltonian, in terms of $\hat{a}$ and $\hat{a}^\dagger$, is $\hat{H} = (\hat{a}^\dagger\hat{a} + \frac{1}{2})\hbar\omega$. The equations of motion for $\hat{a}$ and $\hat{a}^\dagger$ are, respectively, $i\hbar\dot{\hat{a}} = [\hat{a}, \hat{H}] = [\hat{a}, (\hat{a}^\dagger\hat{a} + \frac{1}{2})]\hbar\omega = \hbar\omega\hat{a}$, and $i\hbar\dot{\hat{a}}^\dagger = [\hat{a}^\dagger, \hat{H}] = [\hat{a}^\dagger, (\hat{a}^\dagger\hat{a} + \frac{1}{2})]\hbar\omega = -\hbar\omega\hat{a}^\dagger$, which can be simplified as $\dot{\hat{a}} = -i\omega\hat{a}$ and $\dot{\hat{a}}^\dagger = i\omega\hat{a}^\dagger$. The solution is straightforward: $\hat{a}(t) = e^{-i\omega t}\hat{a}(t_0)$ and $\hat{a}^\dagger(t) = e^{i\omega t}\hat{a}^\dagger(t_0)$. These operators are useful if we want to compute the position and momentum operators.

For instance, at t, $\hat{x}(t) = \sqrt{\frac{\hbar}{2m\omega}}(\hat{a}(t) + \hat{a}^\dagger(t))$, $\hat{p}_x(t) = i\sqrt{\frac{m\hbar\omega}{2}}(-\hat{a}(t) + \hat{a}^\dagger(t))$. These are all operators. If we want to compute their expectation values, we need an initial wavefunction.

Exercise 3.5.2.

26. Ehrenfest's theorem. A particle moves in a potential $V(x)$. Use the Heisenberg equation of motion to derive the expectation value of position $\hat{x}$ in a state $|\psi(t_0)\rangle$ which changes as

$$m\frac{d^2}{dt^2}\langle\hat{x}\rangle = \frac{d\langle\hat{p}_x\rangle}{dt} = -\langle\nabla V(x)\rangle.$$

27. Use $\hat{a}^\dagger(t) = \hat{a}^\dagger(t_0)e^{i\omega t}$ and $\hat{a}(t) = \hat{a}(t_0)e^{-i\omega t}$ to (a) find $\langle\hat{x}(t)\rangle$ in an eigenstate $|\phi_n\rangle$ of the harmonic oscillator, (b) find $\langle\hat{x}^2(t)\rangle$, and (c) find $\sqrt{\langle\hat{x}^2(t)\rangle - (\langle\hat{x}(t)\rangle)^2}$.

28. A free particle has the Hamiltonian, $H = \frac{\hat{p}_x^2}{2m}$. Prove the position uncertainty at two different times t and t_0 fulfilling $\Delta x(t)\Delta x(t_0) \geq \frac{\hbar t}{2m}$.

3.6 Blackbody radiation

A blackbody is an ideal model object that can absorb or emit more radiation than any other objects. A common example is a cavity as illustrated in Fig. 3.5(a). Gustav Kirch-

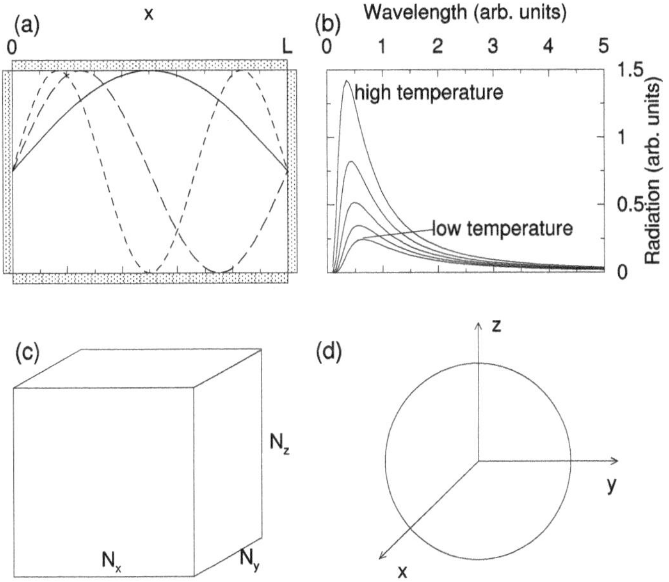

Figure 3.5: (a) Blackbody radiation in a one-dimensional cavity. Only standing waves can survive in the cavity. The number of waves N_x is inversely proportional to its wavelength λ as $N_x = 2L/\lambda$. (b) Schematic radiation as a function of wavelength and temperature. From the top to bottom, the temperature decreases. The curve is plotted using a function $\lambda^{-3}/(\exp(\beta/\lambda) - 1)$. (c) Three-dimensional system has N_x, N_y and N_z. (d) Radiation is spherical, but only 1/8th of the volume should be included since all N are positive.

hoff showed that the state of the heat radiation in a cavity is entirely independent of the properties of the substances in question. The distribution of radiation is a universal function of temperature and wavelength of radiation. A common experience with a stove is helpful here. When we heat the stove, the color of radiation is red, but as the emperature increases, it turns white, but not blue. The wavelength of the maximum radiation shifts toward to a shorter wavelength, but never to blue. Figure 3.5(b) shows that, for a fixed temperature, the radiation first increases with wavelength, reaches a maximum, and then decays to zero. This prediction is further confirmed by the cosmic microwave blackbody radiation at 2.73 K left over from the Big Bang measured by the COBE satellite. It is this blackbody radiation that constitutes a major challenge for classical thermodynamics.

3.6.1 Failure of classical physics

In the following, we show how classical physics fails to explain blackbody radiation. We adopt the linear harmonic oscillator model for emission in the cavity introduced by Heinrich Hertz. Figure 3.5(a) shows a one-dimensional example where the waves are sealed within the cavity of length L and they only can bounce back and forth.

Microscopically, energy follows the Maxwell–Boltzmann distribution, so the average energy $\bar{E}$ is

$$\bar{E} = \frac{\int_0^\infty \exp[-E/(k_B T)] E dE}{\int_0^\infty \exp[-E/(k_B T)] dE} \equiv \frac{\int_0^\infty \exp[-E/(k_B T)] E dE}{A}, \tag{3.6.1}$$

where E is the total energy of the system and A, a normalization constant, is defined through the last equation. For a one-dimensional harmonic oscillator, the total energy consists of the potential energy and the kinetic energy, i. e., $E = \frac{1}{2}mv_x^2 + \frac{1}{2}m\omega^2 x^2$ and $dE = d\tau = dxdv_x$, and the Maxwell–Boltzmann distribution is

$$f_{\mathrm{MB}}(E) = \frac{1}{A}\exp\left[-\left(\frac{1}{2}mv_x^2 + \frac{1}{2}m\omega^2 x^2\right)/(k_B T)\right], \tag{3.6.2}$$

which allows us to compute $\overline{v_x^2}$ and $\overline{x^2}$ through

$$\overline{v_x^2} = \iint_{-\infty}^{\infty} f_{\mathrm{MB}} v_x^2 dxdv_x = k_B T/m \to \frac{m\overline{v_x^2}}{2} = k_B T/2, \tag{3.6.3}$$

$$\overline{x^2} = \iint_{-\infty}^{\infty} f_{\mathrm{MB}} x^2 dxdv_x = k_B T/(m\omega^2) \to \frac{m\omega^2\overline{x^2}}{2} = k_B T/2. \tag{3.6.4}$$

This means that each degree of freedom, or each mode, either in terms of potential energy or kinetic energy, has a fixed energy of $k_B T/2$, so the total energy is $\bar{E} = k_B T$. This is the famous equipartition theorem. In order to compute the total energy, we need to find out the number of modes for our cavity.

For a one-dimensional cavity of length L, the number of modes N_x depends on the wavelength of the mode. The only mode that can survive in a cavity of length L is a standing wave, so its wavelength λ satisfies

$$L = N_x \frac{\lambda}{2}, \quad \text{or} \quad N_x = \frac{2L}{\lambda},$$

where N_x is positive and $\frac{1}{2}$ is due to the standing wave. Since the wavelength λ is related to the angular frequency ω through $\lambda = \frac{c}{\nu} = \frac{c}{\omega/2\pi} = \frac{2\pi c}{\omega}$, we can rewrite $N_x = L\omega/(\pi c)$.[5] This shows that, for the same L, the higher ω is, the larger N_x becomes.

In the three-dimensional system, the number of modes is $N_x N_y N_z$, where N_x, N_y, and N_z are the number of modes along the x-, y-, and z-axes, respectively. For a cube, $N_x N_y N_z = \frac{2L}{\lambda}\frac{2L}{\lambda}\frac{2L}{\lambda} = (\frac{2L}{\lambda})^3$. However, since radiation is spherical, not cubic, we need to compute the volume of the sphere, with radius of $(\frac{2L}{\lambda})$. We cannot use the entire sphere because only one-eighth of a sphere has positive N_x, N_y, and N_z. Therefore, our final number of modes is

$$N = \frac{1}{8}\frac{4\pi}{3}\left(\frac{2L}{\lambda}\right)^3 \times 2 = \frac{8\pi V}{3\lambda^3} = \frac{8\pi V}{3c^3}\nu^3, \tag{3.6.5}$$

where 2 accounts for two polarizations of radiation and we have set L^3 to V. One sees that the number of modes is inversely proportional to λ^3, or proportional to ν^3.

Experiments measure the radiation energy as a function of ν or λ, so we need to know how many modes dN are within an interval from ν to $\nu + d\nu$, or in terms of wavelength, λ to $\lambda + d\lambda$. From eq. (3.6.5), N is the total number of modes, but we need dN. All we need to do is to take the derivative of N with respect to ν,

$$dN = \frac{8\pi V}{c^3}\nu^2 d\nu. \tag{3.6.6}$$

Within $d\nu$, the energy per volume, E, is the number of modes dN multiplied by the average energy per mode $\bar{E} = k_B T$, then divided by volume V,

$$E = \bar{E}\frac{dN}{V} = k_B T \frac{8\pi}{c^3}\nu^2 d\nu \equiv u_\nu d\nu, \tag{3.6.7}$$

where all the terms before $d\nu$ are defined as the classical spectral energy density u_ν. The radiance S_ν is

$$S_\nu = \frac{u_\nu c}{4\pi} = \frac{2\nu^2}{c^2}k_B T \quad \text{(classical prediction)}. \tag{3.6.8}$$

5 Here, the speed of light c is used because we are considering the radiation.

This equation shows that, the higher the frequency is, the more the energy is emitted, contradicting the experimental finding as illustrated in Fig. 3.5(b). This leads to the famous ultraviolet catastrophe. Classical physics fails to explain blackbody radiation because the continuous energy leads to integration in eqs. (3.6.3) and (3.6.4).

Exercise 3.6.1.
29. The Boltzmann distribution is $f_{\rm MB}(E) = \frac{1}{A}\exp[-(\frac{1}{2}m\omega^2x^2)]$. (a) If $\int_{-\infty}^{\infty} f_{\rm MB}dx = 1$, find A. (b) Compute $\overline{x}$. (c) Compute $\overline{x^2}$.

3.6.2 Quantum theory: importance of quantization

Planck postulated that the blackbody only emits or absorbs energies in integer multiples of $h\nu$. Under thermal equilibrium, each oscillator holds only discrete energies, $E_n = 0h\nu$, $h\nu, 2h\nu, \dots$, each of which corresponds to a mode or quantum state of the harmonic oscillator. According to the Boltzmann distribution, the probabilities for the harmonics in these energy states are $\exp(-E_n/k_BT)$.[6] Using E_n, we have probabilities for each energy: $1, e^{-h\nu/k_BT}, e^{-2h\nu/k_BT}, \dots$. The probability sum is just our normalization constant A (same as A in eq. (3.6.1)),

$$A = \sum_{n=0}^{\infty} e^{-nh\nu/k_BT} = \frac{1}{1-e^{-h\nu/k_BT}}, \tag{3.6.9}$$

which converts the Boltzmann factor $\exp(-E_n/k_BT)$ into a probability [20]. The total energy of the oscillators is the sum over the product of $\exp(-E_n/k_BT)E_n$, i. e.

$$\mathcal{E}(\nu) = \sum_n nh\nu e^{-nh\nu/k_BT} = h\nu\frac{e^{-h\nu/k_BT}}{(1-e^{-h\nu/k_BT})^2}, \tag{3.6.10}$$

where $\mathcal{E}$ is written as $\mathcal{E}(\nu)$, because this average energy depends on frequency ν, a feature that CM does not have. Here we have a geometric series summation, not an integral, the mathematical reason for Planck's success, representing a key difference between CM and QM, as our first exercise in Chapter 1 is intended to explain.

To find the average energy of each oscillator, we divide $\mathcal{E}$ in eq. (3.6.10) by A in eq. (3.6.9) to find (compare with eq. (3.6.1)),[7]

$$\bar{E} = \mathcal{E}(\nu)/A = \frac{h\nu}{e^{h\nu/k_BT}-1}, \tag{3.6.11}$$

which is no longer the classical value $\bar{E} = k_BT$.

6 Quantum statistics has the same expression as classical statistics for each individual energy state.

7 $e^{-h\nu/k_BT}$ is eliminated from the numerator, so the denominator has $e^{+h\nu/k_BT}$.

Our final radiation energy per volume (in units of J/m^3) is the average energy $\bar{E}$ multiplied by dN/V (same as eq. (3.6.7))

$$E = \bar{E}\frac{dN}{V} = \underbrace{\frac{h\nu}{e^{h\nu/k_BT} - 1}\frac{8\pi}{c^3}\nu^2}_{u_\nu} d\nu \equiv u_\nu d\nu \quad \text{(quantum prediction)}, \tag{3.6.12}$$

where we have introduced the spectral energy density u_ν. The same radiation is in all directions. When the radiation exits through a small hole, the fluence S_ν (in units of J/m^2) measured experimentally is equal to $u_\nu c/(4\pi)$,[8] yielding

$$S_\nu = \frac{2h\nu^3}{c^2}\frac{1}{e^{h\nu/k_BT} - 1}, \tag{3.6.13}$$

which is Planck's law. Equation (3.6.13) is different from our classical one (eq. (3.6.8)) and has no divergence as $\nu \to \infty$ (Fig. 3.5(b)).[9] This law enables us to compute the total intensity (in units of W/m^2),

$$\int_0^\infty S_\nu d\nu = \frac{2}{15}\frac{\pi^5(k_BT)^4}{h^3c^2} \equiv \sigma T^4, \tag{3.6.14}$$

which is the Stefan–Boltzmann law. The Stefan–Boltzmann constant is $\sigma = 5.6703 \times 10^{-8}$ W/(m$^2 \cdot$ K^4).

Taking the limit $\lambda \to 0$ yields the Wien law,

$$S_\lambda d\lambda = \frac{2hc}{\lambda^5}e^{-hc/\lambda k_BT}d\lambda, \tag{3.6.15}$$

where we have converted ν to λ. In fact, the Wien law is an empirical formula with two fitted parameters in history, but here we obtain it from Planck's law. As temperature T increases, Planck's law goes back to the classical prediction (eq. (3.6.8)).

Exercise 3.6.2.

30. Show $\frac{h\nu}{e^{h\nu/k_BT}-1}$ approaches k_BT as T becomes larger.
31. Show that when $\lambda \to 0$, $S_\lambda d\lambda = \frac{2hc}{\lambda^5}e^{-hc/\lambda k_BT}d\lambda$.
32. Show that, at the limit of high T, eq. (3.6.13) can be reduced to (3.6.8).
33. Starting from eq. (3.6.13), (a) show the maximum S_ν is determined by

$$3 - \frac{h\nu}{k_BT}\frac{e^{h\nu/k_BT}}{e^{h\nu/k_BT} - 1} = 0,$$

and (b) show numerically that at the maximum S_ν, $\nu_{max} = 2.822k_BT/h$, Wien's displacement law.

---

8 4π is the solid angle over a sphere.

9 Planck proved the average radiation energy is equal to the above-average energy.

3.6.3 Insights

In classical physics, the energy is continuous, and the energy interval can be infinitely small. Mathematically, this leads to an integral in eqs. (3.6.3) and (3.6.4). The original potential energy term ($m\omega^2x^2/2$) depends on ω^2, but using the Boltzmann distribution (e^{-E/k_BT}) to compute the average potential energy, i. e., $\overline{x^2}$, produces $\overline{x^2} \propto 1/\omega^2$, which exactly cancels ω^2 in the potential energy term, so the energy for each degree of freedom is $\frac{1}{2}k_BT$, independent of frequency ν or ω. Because the number of modes N is proportional to ν^3 (eq. (3.6.5)), this leads to the ultraviolet catastrophe.

QM adopts the energy quantization, so, mathematically, the total energy is a sum, not integral. Two insights are: (i) A classical stronger oscillator with a larger amplitude corresponds to a larger quantum number n in QM; (ii) A higher frequency ν, whose effect is canceled in classical physics, affects its probability through the Boltzmann factor ($e^{-nh\nu/k_BT}$). $E_n = nh\nu$ puts the frequency of radiation at the center, where the Boltzmann factor ($e^{-nh\nu/k_BT}$) weighs the frequency ν and n from the beginning. As a result, the average energy (eq. (3.6.11)) depends on ν and T. Although both QM and CM have the same N, the exponential dependence on ν significantly suppresses the contribution from higher-energy modes and eliminates the ultraviolet catastrophe.

Exercise 3.6.3.

34. What are the mathematical reasons for the failure of CM and the success of QM?
35. What is the reason why the CM energy does not depend on ν?
36. Use a function $f(x) = \frac{x}{e^x-1}$ as an example, where x is $h\nu/k_BT$. (a) Plot $f(x)$. (b) Plot $x^2f(x)$ to explain why the ultraviolet catastrophe disappears.

3.7 Photoelectric effects

When a light photon strikes a metallic surface, electrons can be knocked out. This is the photoelectric effect. Experimentally, Heinrich Hertz showed that, when ultraviolet light illuminates metal electrodes, the light changes the voltage between these two electrodes. Classically, one would expect that the electron emission should be related to the light intensity. A stronger light field, regardless of light frequency, should be able to knock out electrons, just like a water wave where a strong enough force splashes water droplets out of water. But, this does not happen in the photoelectric effect, where experiments point out that the light frequency, not light intensity, is in control. Subsequent experiments reveal that to emit electrons, the frequency ν of the incident light must be above a particular threshold ν_{th}. If the frequency is below this threshold, no matter how strong the light is, no electron can be emitted. If ν above ν_{th}, a weaker light field can emit fewer electrons, and a stronger field emits more electrons, but each electron gains the same energy.

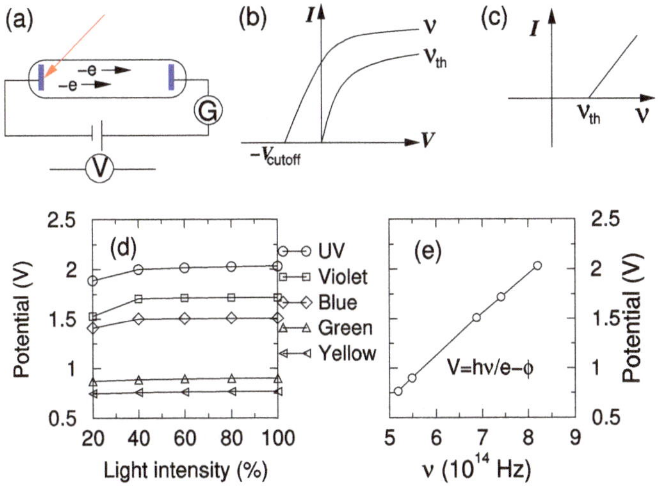

Figure 3.6: (a) Photoelectric effect. Light hits a metal plate and electrons are emitted toward the right electrode. (Bottom) A modern design replaces the battery by a voltmeter. (b) Light cannot emit electrons until its frequency ν is above ν_{th}. Current can be generated even at $V = 0\,\mathrm{V}$. A reverse bias voltage $-V_{cutoff}$ must be applied to reduce the current to zero. (c) Once ν is above the threshold frequency, ν_{th}, current increases with ν. (d) Dependence of our experimental voltage as a function of light intensity. (e) Voltage change as a function of light frequency at 100% intensity. The light source is a mercury lamp.

3.7.1 Experiment

The photoelectric effect now enters a regular advanced physics laboratory. We will use an actual experiment to demonstrate many aspects of the effect. But, first, we show a typical apparatus in Fig. 3.6(a), where two electrodes are sealed inside a vacuum tube. The incident light through a window hits one electrode. A power supply is linked to a galvanometer to measure the current. If ν is below a threshold, no matter how strong the light is, no electron is emitted, i. e., no current exists. If ν is above ν_{th}, even without a power supply, i. e., $V = 0$, it produces a current (Fig. 3.6(b)).[10] To suppress the current, one has to apply a reverse bias $-V_{cutoff}$. These observations directly contradict the classical physics prediction, where the current should be independent of light wavelength λ or frequency ν; and light with a sufficient high intensity should be able to knock out electrons. Figure 3.6(c) shows the current linearly increases with ν, once ν is above the threshold frequency.

A modern version that we used is shown in the bottom of Fig. 3.6(a), where there is no power supply and instead a voltmeter is inserted. This setup is simpler than the traditional one. The beauty of this experiment is that there is little error in the measurement, except the current leakage in the circuit. Our equipment, called h/e, is from PASCO. The

10 This is the principle of solar cells in photovoltaics.

Table 3.2: (Left) Wavelengths and frequencies of the mercury spectral lines. The wavelengths are from *Handbook of Chemistry and Physics*, 46th Ed. (1965–66). The yellow line, with a doublet of 578 and 580 nm, was determined using a 600 line/mm grating by PASCO. (Right) Work function of common materials.

Color	Wavelength (nm)	Frequency (10^{14} Hz)
Yellow	578	5.18672
Green	546.074	5.48996
Blue	435.835	6.87858
Violet	404.656	7.40858
Ultraviolet	365.483	8.20264

Element	ϕ (eV)
Na	2.28
C	4.81
Al	4.08
Ag	4.73
Ni	5.01

cathode in Fig. 3.6(d) is made of a semiconductor, Cs_3Sb, and has a high sensitivity in the visible spectrum. The anode of the photodiode is a nickel alloy.[11] Light from a mercury lamp is dispersed, through a grating of 600 lines/nm, into five spectral lines. We'll come back to the spectral lines in the next chapter. These lines have known wavelengths and frequencies, listed in Table 3.2. After illumination with light, the electrons are ejected from the cathode and reach the anode. This process continues until the electric potential in the tube becomes strong enough to completely suppress the flow of electrons. Once this equilibrium is established and the current drops to zero, a voltage reading is taken. The time to reach zero current is shorter for a stronger intensity.

By putting a filter in the light path to reduce the light intensity, we can study how the voltage depends on the light intensity. Figure 3.6(d) shows our experimental voltage as a function of light intensity for five spectrum lines. One sees that the light intensity has no effect on the voltage. This contradicts our classical prediction. The small reduction at two weaker intensities is due to current leakage in our circuit. By tuning to different colors, we measure the wavelength dependence of the voltage. Figure 3.6(e) illustrates the voltage as a function of frequency. When we change color from yellow to ultraviolet, the voltage increases linearly. This also contradicts the classical prediction.

3.7.2 Einstein's theory of the photoelectric effect

To explain the photoelectric effect, Einstein postulated the quantum concept of light, and G. N. Lewis [1] called it the photon. Light must be considered as a particle, each of which carries a fixed energy, $h\nu$. In classical EM, a stronger light intensity means a larger field amplitude. But in QM, this means the number of photons is larger, but each photon carries the same energy $h\nu$, a prediction that is verified in our experiment. We find that, for a stronger intensity, the time that it takes the system to reach equilibrium become shorter. This demonstrates that indeed more electrons are emitted with a stronger inten-

11 Technical details are provided by PASCO.

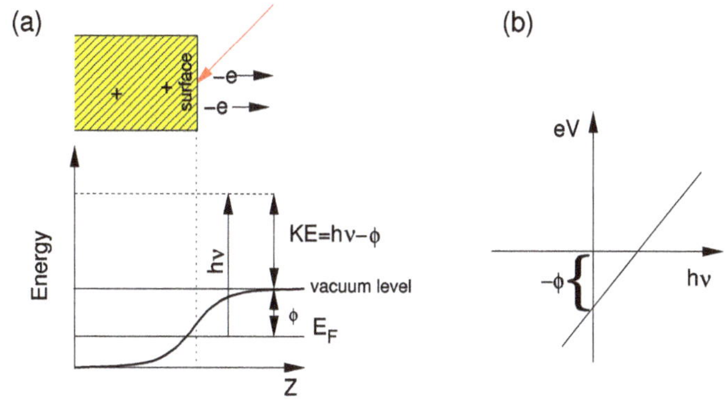

Figure 3.7: (a) (Top) Zoom-in view of the surface. (Bottom) The potential energy (thick solid curve) as a function of position z. E_F denotes the Fermi level of the electrode (the first horizontal thin line from the bottom). The second solid line from the bottom denotes the vacuum level, where the electrons are free. The difference between the vacuum level and the Fermi level is defined as the work function ϕ.

sity. But the voltage is the same at the end, regardless of light intensity (see Fig. 3.6(d)). If we used the apparatus with a battery as shown in Fig. 3.6(a), we would see a larger current with a strong radiation. Figure 3.6(e) experimentally reveals that, to knock out electrons, the light frequency must be above a threshold frequency, or its energy counterpart, ϕ, which is defined as the work function.

The physics is shown in Fig. 3.7(a). The shaded area represents a material with a surface. Once electrons leave the surface, positive charges (called holes) are left behind, so there is a net attraction between them. To liberate the electron from the surface, the electron has to overcome the potential energy difference between the vacuum level and the Fermi energy.[12] This difference is defined as the work function ϕ which is the minimum energy that the incident photon must have to free an electron. This sets the threshold frequency $\nu_{\text{th}} = \frac{\phi}{h}$ as previously discussed. If the photon energy $h\nu$ is higher than ϕ, the remaining energy goes to kinetic energy which contributes to the current in the circuit. Therefore, we have

$$KE = h\nu - \phi, \tag{3.7.1}$$

which is the Einstein photoelectric equation. In our experiment, because the potential completely suppresses the current flow, $KE = eV$. Then, our voltage V is $V = \frac{h}{e}\nu - \frac{\phi}{e}$, which explains why our V linearly depends on ν in Fig. 3.7(b). This is the reason behind the experimental data in Fig. 3.6(b). This equation allows us to measure the work function. For our sample, the average work function ϕ is 1.36 ± 0.08 eV. Table 3.2 shows the work functions for some common materials. The next exercise shows one such example.

12 The vacuum level is where electrons are free from any interaction with the surface of a material. The Fermi energy is defined as the highest energy level that electrons in a material have at 0 K.

The photoelectric effect is now developed into a major technique, photoemission, to probe surface properties of a sample. In two-photon photoemission, an electron can absorb two photons. In the angular resolved photoemission, one can measure how the electron energy distributes with angles to detect the energy dispersion in solids. Inverse photoemission uses electrons to bombard a surface to release light photons. The rich information from these techniques has made significant progress on our understanding of physics in materials of different kinds.

Exercise 3.7.2.

37. The mercury lamp is a good light source. Dispersing the light using a grating and focusing with a lens onto the Cs-Sb sample, two students, Nicole Hostetter and Tanner Latta, in 2015 collected the voltage change as a function of various colors from the h/e apparatus of PASCO. The following table summarizes their data:

Color	Ultraviolet	Violet	Blue	Green	Yellow
Voltage (V)	2.030	1.719	1.507	0.902	0.767

The first row refers to the spectral lines of the mercury lamp, whose wavelengths and frequencies are given in Table 3.2. (a) Plot a voltage–frequency graph. (b) Fit your data to a line and find the slope. Given $e = 1.602 \times 10^{-19}$ C, find Planck's constant h and compute the percentage error. (c) From the same graph, find the work function of the Cs-Sb sample.

3.8 Problems

1. (a) Prove the Cauchy–Schwarz inequality $|\langle u|v\rangle| \leq \langle u|u\rangle\langle v|v\rangle$, where u and v are wavefunctions. Hint: Start from $\langle u - v|u - v\rangle \geq 0$, and then rewrite u as u/c and v as cv. Finally, show the discriminant is not positive. (b) Prove $\sqrt{\langle u + v|u + v\rangle} \leq \sqrt{\langle u|u\rangle} + \sqrt{\langle v|v\rangle}$. Adopted from [21].
2. (a) Prove $[\hat{\mathbf{p}}, f(\mathbf{r})] = -i\hbar\nabla f(\mathbf{r})$, and then use it to prove $[e^{-i\mathbf{k}\cdot\mathbf{r}}, \hat{\mathbf{p}}] = \hbar\mathbf{k}e^{-i\mathbf{k}\cdot\mathbf{r}}$. Hint: Compute the commutation for the x-, y-, and z-components separately and then add them up. (b) Prove $[\hat{\mathbf{p}}^2, f(\mathbf{r})] = -\hbar^2\nabla^2 f(\mathbf{r}) - 2i\hbar(\nabla f)\hat{\mathbf{p}}$, and use it to prove $[e^{-i\mathbf{k}\cdot\mathbf{r}}, \hat{\mathbf{p}}^2] = e^{-i\mathbf{k}\cdot\mathbf{r}}(2\mathbf{k}\cdot\hat{\mathbf{p}} + \hbar^2 k^2)$. Adopted from [22].
3. If $\hat{H}$ is a Hermitian operator, show that its eigenvalue must be real.
4. Find the following commutations: (a) $[x, d/dx]$; (b) $[x^2, d/dx]$; and (c) $[x^n, d^2/dx^2]$.
5. Starting from $[\hat{p}_x, x] = -i\hbar$, prove $[\hat{p}_x, x^n] = -i\hbar n x^{n-1}$. Adopted from [21].
6. $\hat{A}$ and $\hat{B}$ are two operators. Prove $e^{\hat{A}}\hat{B}e^{-\hat{A}} = \hat{B} + [\hat{A}, \hat{B}] + \frac{1}{2!}[\hat{A}, [\hat{A}, \hat{B}]] + \frac{1}{3!}[\hat{A}, [\hat{A}, [\hat{A}, \hat{B}]]] + \cdots$.
7. A particle of mass m is in the infinite quantum well of length a. (a) Assume the position uncertainty $\Delta x = a$. From the Heisenberg uncertainty principle, estimate the minimum Δp_x. (b) Set the momentum square $\langle p_x^2\rangle = \Delta p_x$ and find the energy of

the particle of m, which is the minimum energy that the particle that can have. (c) Prove all the eigenenergies are higher than this minimum energy.

8. A particle is in the ground state of PIB, $\phi_1(x) = \sqrt{\frac{2}{a}}\sin(\frac{\pi x}{a})$. (a) Compute the expectation values of $\hat{x}$, $\hat{x}^2$, $\hat{p}_x$, and $\hat{p}_x^2$. (b) Find the uncertainty $\Delta x \Delta p_x$, and check whether it satisfies the Heisenberg uncertainty principle.
9. Suppose we have a wavefunction $\phi(x) = x$, where $0 < x < a$. (a) Normalize the wavefunction. (b) Show $\Delta x \Delta p_x$ does not obey the Heisenberg uncertainty principle. This shows that ϕ is not a valid wavefunction.
10. ϕ_0 is the ground-state wavefunction of the harmonic oscillator. (a) Use $\hat{a}^\dagger$ to analytically find ϕ_1. (b) Normalize ϕ_1. (c) Find the expectation values of $\hat{x}$, $\hat{p}_z$, $\hat{x}^2$, $\hat{p}_x^2$. (d) Find the uncertainty relation $\Delta x \Delta p_x$.
11. $|n\rangle$ is an eigenstate of the harmonic oscillator. Use $\hat{a}^\dagger$ and $\hat{a}$ to find the expectation values of the kinetic energy and potential energy operators, $\langle n|\hat{T}|n\rangle$ and $\langle n|V|n\rangle$.
12. Show the eigenfunction $|\psi\rangle$ of the lowering operator $\hat{a}$ can be expanded in terms of the harmonic oscillator's eigenstates $|n\rangle$,

$$|\psi\rangle = e^{-\frac{1}{2}|\alpha|^2} \sum_{n=0}^{\infty} \frac{\alpha^n}{\sqrt{n!}} |n\rangle,$$

where $\alpha = \sqrt{m\omega/\hbar}$. Hint: Expand eq. (3.4.1).

13. A particle of mass m is in the ground state of a harmonic potential $V_0(x) = \frac{1}{2}Kx^2$. (a) If, at $t = 0$, we suddenly decrease K to half $K/2$, $V_1(x) = \frac{1}{4}Kx^2$. What is the probability for the particle in the ground state of V_1? (b) If now we wait for two periods of the original period of the oscillator, find the expectation values of the position and momentum. Hint: This is a stationary problem, but in which system does the states evolve after $t = 0$?
14. The harmonic oscillator has the following Hamiltonian: $\hat{H} = \frac{\hat{p}_x^2}{2m} + \frac{1}{2}m\omega^2\hat{x}^2$. (a) Find the equation of motion for the momentum operator $\hat{p}_x$. (b) Find the equation of motion for the position operator $\hat{x}$. (c) If the initial momentum operator is $\hat{p}_x(t_0 = 0)$ and that of the position operator is $\hat{x}(t_0 = 0)$, find an analytic expression for $\hat{p}_x(t)$ and $\hat{x}(t)$. (d) Use the expression of $\hat{x}(t)$ to show that two position operators at two different times t_1 and t_2 do not commute in general, $[\hat{x}(t_1), \hat{x}(t_2)] \neq 0$. What about $[\hat{p}_x(t_1), \hat{p}_x(t_2)] \neq 0$? (e) Find the times when $[\hat{x}(t_1), \hat{x}(t_2)] = 0$. (f) Suppose that initially the expectation value of $\hat{x}$ is $\langle \hat{x}(t_0)\rangle = x_0$. Find the expectation value of $\hat{x}$ at time t.
15. Consider a particle of mass m, which is close to the surface of the earth, with the initial momentum $\langle \hat{p}_y(t_0)\rangle$ and position $\langle \hat{y}(t_0)\rangle$, and falls under gravity (acceleration is g). Ignore the air resistance. (a) Find the momentum $\langle \hat{p}_x\rangle(t)$ at time t. (b) Find its position $\langle \hat{y}\rangle(t)$ at t. Hint: You need to write down the Hamiltonian for the free fall, and choose the upward direction as the $+y$-axis. (c) Discuss what happens to your result if the particle is far above the surface of the earth with the initial momentum along the x-axis.

16. The relativistic Hamiltonian of a free particle is $\hat{H} = c\sqrt{m^2c^2 + \hat{p}_x^2 + \hat{p}_y^2 + \hat{p}_z^2}$, where m is the rest mass and c is the speed of light. (a) Using the Heisenberg equation of motion, show the velocities $\frac{d\hat{x}}{dt} = \frac{c^2\hat{p}_x}{\hat{H}}$, $\frac{d\hat{y}}{dt} = \frac{c^2\hat{p}_y}{\hat{H}}$, and $\frac{d\hat{z}}{dt} = \frac{c^2\hat{p}_z}{\hat{H}}$. (b) Show that, if $m = 0$, all the velocities are equal to c. Hint: Use **p**-representation to convert x to $i\hbar\frac{\partial}{\partial p_x}$. Adopted from [11].
17. $\hat{H}$ is a Hamiltonian and contains the kinetic energy $\hat{T} = \frac{\hat{\mathbf{p}}^2}{2m_e}$ and the potential energy $V(\mathbf{r})$. (a) Show the commutation of the position operator $\hat{\mathbf{r}}$ and $\hat{H}$, $[\hat{\mathbf{r}}, \hat{H}] = \frac{i\hbar\hat{\mathbf{p}}}{m_e}$. (b) If $|n\rangle$ and $|m\rangle$ are two eigenstates of $\hat{H}$, with eigenvalues E_n and E_m, show the matrix element of $\hat{\mathbf{r}}$ is given by

$$\langle n|\hat{\mathbf{r}}|m\rangle = \frac{i\hbar\langle n|\hat{\mathbf{p}}|m\rangle}{m_e(E_m - E_n)} = \frac{\hbar\langle n|\hat{\mathbf{p}}|m\rangle}{im_e(E_n - E_m)}.$$

This equation is used to convert the position operator to the momentum operator, or vice versa.
18. A harmonic oscillator has an initial wavefunction $\Psi(x) = N_0 e^{-m\omega(x-x_0)^2/\hbar}$. (a) Through the wavefunction normalization, determine N_0. (b) Find the expectation value $\langle x\rangle$.
19. A radio station has a frequency of 900 MHz and emits power of 10 kW. Compute the number of photons per second.
20. Infrared thermometers enable one to measure the temperature of the surface of an object, without making physical contact. This is based on the Stefan–Boltzmann law, where the intensity of radiation is σT^4. Show that $\int_0^\infty S_\nu d\nu = \frac{2}{15}\frac{\pi^5(k_BT)^4}{h^3c^2}$. Hint: This problem needs the Bose–Einstein integral, $J = \int_0^\infty \frac{x^3}{e^x-1}dx = \frac{\pi^4}{15}$.
21. A beam of light with wavelength 2,000 Å is incident on a silver surface. The work function of silver is given in Table 3.2. Find the kinetic energy of the emitted electrons.
22. For a one-dimensional harmonic oscillator, (a) prove the identity

$$\frac{1}{2m}\sum_j \langle i|\hat{p}_x|j\rangle\langle j|\hat{p}_x|i\rangle = T_i,$$

without using the completeness of the eigenstates, where T_i is the kinetic energy of eigenstate i and the summation is from $j = 0$ to ∞. (b) Prove $\frac{K}{2}\sum_j \langle i|\hat{x}|j\rangle\langle j|\hat{x}|i\rangle = V_i$, where V_i is the potential energy of eigenstate i.

4 Orbital angular momentum and hydrogen atom

The discovery of electrons by J. J. Thomson and the planetary model of nuclei formulated by Rutherford allowed Bohr to construct a semiclassical model for hydrogen. The first successful application of the Bohr model to the hydrogen atom finally connected the energy spectrum to the spectral lines observed many decades earlier than the incipient quantum idea. This paved the way for the quantum mechanical treatment of real atoms.

This chapter has four units:

- Unit 1 consists of Sections 4.1 and 4.2. It starts with the classical Bohr model of hydrogen to prepare the reader for orbital and energy quantization. Then, it discusses how two commutable operators share a common eigenfunction, a property that will be used in the next unit.
- Unit 2 includes Sections 4.3 and 4.4. It uses the above property of two commutable operators to find eigenfunctions for the orbital angular momentum operator, similar to raising and lowering operators in the harmonic oscillator. To supplement Section 4.3, Section 4.4 further explores physical and chemical meanings of orbital angular momentum. This includes the transformation between two different representations.
- Unit 3 consists of Sections 4.5, 4.6, and 4.7. The central theme is to solve the Schrödinger equation of hydrogen. Section 4.5 focuses on the concept of symmetry and the law of conservation so that Section 4.6 can separate the radial and angular parts of wavefunctions. Using the radial Hamiltonian, Section 4.7 numerically solves the radial Schrödinger equation through a code. The reader is highly encouraged to try the code.
- Unit 4 has one section, Section 4.8. It presents the experimental evidence of energy level quantization through the Franck–Hertz experiment. The experimental data are contributed by the authors.

4.1 Bohr model for the hydrogen atom: a bridge from CM to QM

The hydrogen atom consists of an electron orbiting around a proton. According to classical electrodynamics, an accelerating charge radiates energy; as the electron loses energy, it must eventually collapse toward the nucleus. But this never happens! In contrast to the prediction of classical electrodynamics, the radiation spectrum is discrete, not continuous, and it forms the so-called spectral lines as shown in Fig. 4.1(a). This apparent contradiction forces one to make radical assumptions.

Bohr postulated that the electron orbiting around the proton and its radiation must be understood as two different atomic processes. In his mind, even though the electron is orbiting around the proton, if it is in a stable orbit, i. e., a stationary state, it does not radiate. But if the electron makes a transition from one orbit to a lower orbit, it radiates photons whose energy matches the energy difference between the initial and final

https://doi.org/10.1515/9783110672152-004

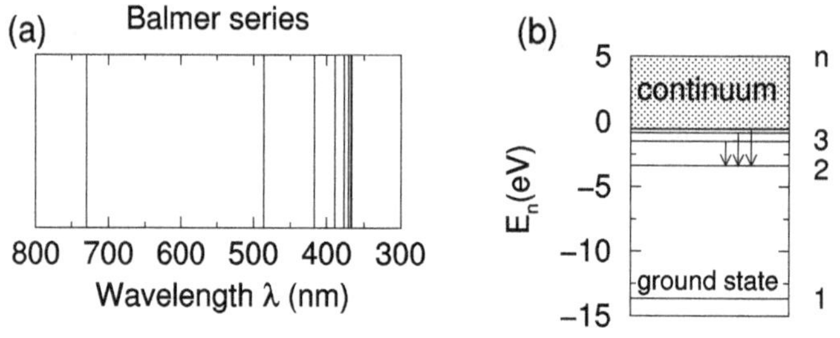

Figure 4.1: (a) Spectral lines of the Balmer series. Because they are in the visible light wavelengths, the Balmer series was the first to be discovered. Other series were found later. (b) Energy spectrum of the hydrogen atom. The electron transition from higher energy levels to level $n = 2$ produces the visible spectral lines, the Balmer series. n represents the quantum number. The ground state is at -13.6 eV. At and above 0 eV, the electron is in the continuum.

states. This is the Bohr model, where both the electron and proton are treated classically, but with one exception as explained in the next section. As the model is still classical, it is amenable to the beginner.

4.1.1 Energy

Consider an electron of mass m_e orbiting around the proton with velocity v. The proton is assumed to be at rest. The Coulomb force provides a centripetal force for the electron as $F = \frac{1}{4\pi\epsilon_0}\frac{e^2}{r^2} = m_e\frac{v^2}{r}$, where e is the elementary charge, r is the radius of the circular motion, m_e is the electron mass, and ϵ_0 is the permittivity in vacuum. Eliminating r on both sides yields

$$m_e v^2 = \frac{e^2}{4\pi\epsilon_0}\frac{1}{r}, \tag{4.1.1}$$

so the kinetic energy T is

$$T = \frac{1}{2}m_e v^2 = \frac{1}{2}\frac{e^2}{4\pi\epsilon_0 r}.$$

Since the potential energy is

$$V = \frac{-e^2}{4\pi\epsilon_0 r},$$

we have $T = -V/2$. The total energy is a sum of the kinetic energy and potential energy,

$$E_{\text{tot}} = T + V = -\frac{1}{2}V + V = \frac{1}{2}V.$$

Although this relation $E_{tot} = V/2$ is derived within CM, it is also valid in QM, which is known as the virial theorem. We will come back to this in Section 5.5.2.

4.1.2 Spatial quantization of angular momentum

To this end, the quantization really refers to the energy quantization. This is a scalar quantization. In the Bohr model, everything else is classical, except angular momentum. Bohr extended this quantum idea to vectors and quantized the angular momentum for the circular motion. He assumed that the magnitude of angular momentum is

$$|\mathbf{L}| = n\hbar, \tag{4.1.2}$$

where n is the main quantum number.[1] de Broglie showed that eq. (4.1.2) can be derived if we assume the number n of waves with wavelength λ that can be fitted into a circle is $n = 2\pi r/\lambda_{dB}$. Because $\lambda_{dB} = h/p$, one has $nh = 2\pi rp$, i. e., $|\mathbf{L}| = n\hbar$. This angular momentum is called orbital angular momentum because it is associated with the electron orbital motion.[2] By definition, $|\mathbf{L}| = |\mathbf{r} \times \mathbf{p}| = rm_e v$, so we have

$$rm_e v = n\hbar \quad \text{(Bohr's postulate)}. \tag{4.1.3}$$

This postulate completely changes our view of angular momentum: It cannot take just any arbitrary value. Instead, it takes on multiples of $\hbar$, i. e., spatial quantization. $\hbar$ has the units of $\mathrm{J \cdot s}$, which is the same as $\mathrm{kgm^2/s}$.

4.1.3 Bohr radius

Equation (4.1.3) gives the velocity $v = n\hbar/rm_e$. Using eq. (4.1.1), we find r as

$$r_n = \frac{4\pi\epsilon_0\hbar^2}{m_e e^2} n^2 \equiv n^2 a_0, \tag{4.1.4}$$

where a subscript n is added to r. If $n = 1$, r_1 is defined as the Bohr radius a_0,

$$a_0 = \frac{4\pi\epsilon_0\hbar^2}{m_e e^2} = 0.529\,\text{Å} \quad (r_1 = a_0). \tag{4.1.5}$$

a_0 is the length unit in atomic units and represents the smallest radius of the hydrogen atomic orbital. The radius r increases with n^2.

1 As will be seen later, n will be replaced by l, but, as an introduction, it is convenient to use n.

2 In QM we use the word "orbital" instead of "orbit."

4.1.4 Velocity and period

From Bohr's postulate, we can compute the electron velocity v_n as

$$v_n = \frac{n\hbar}{m_e r_n} = \frac{n\hbar}{m_e n^2 a_0} = \frac{\hbar}{n m_e a_0} = \frac{v_1}{n}. \tag{4.1.6}$$

For $n = 1$, $v_1 = \frac{\hbar}{m_e a_0} = \frac{6.626\times10^{-34}\ \mathrm{J{\cdot}s}/2\pi}{9.11\times10^{-31}\ \mathrm{kg}\times0.529\times10^{-10}\ \mathrm{m}} = 2.188 \times 10^6\ \mathrm{m/s}$. v_1 is the velocity unit in atomic units. This is the largest velocity that the electron in a hydrogen atom has, and its ratio to the speed of light is called the fine structure constant $\alpha = \frac{v_1}{c} \approx \frac{1}{137}$. The relativistic effect becomes important. In hydrogen, the effect is already observed as the hyperfine spectral lines, where the spectral lines just discussed are further split into finer lines. In heavier atoms with a large atomic number Z, v can approach the speed of light. Part of the reasons why mercury is a liquid at room temperature is due to relativity. As n increases, v_n decreases as $\frac{1}{n}$.

We can also find the period $\mathcal{T}_n$ of the circular motion as

$$\mathcal{T}_n = \frac{2\pi r_n}{v_n} = \frac{2\pi a_0}{v_1} n^3. \tag{4.1.7}$$

The period of the $n = 1$ orbital is defined as the atomic time unit.

Exercise 4.1.4.
1. Graph r_n and v_n as a function of n.
2. Taking a_0 as the position uncertainty Δr, find the momentum uncertainty Δp and compare it with the momentum p_1 which is directly computed from v_1. What can you conclude?
3. Find the period of the circular motion at $n = 1$ and convert it to femtoseconds. This period sets the atomic unit of time.

4.1.5 Eigenenergies and their connections to the spectral lines

Once we know the velocity and radius, we can compute the total energy, which is the eigenenergy of the system, with

$$E_n = T_n + V_n = \frac{1}{2} V_n = -\frac{1}{2} \frac{e^2}{4\pi\epsilon_0 r_n}. \tag{4.1.8}$$

The state with $n = 1$ is the ground state. We write it out as

$$E_1 = T_1 + V_1 = -\frac{1}{2} \frac{e^2}{4\pi\epsilon_0 r_1} = -\frac{1}{2} \frac{e^2}{4\pi\epsilon_0 a_0},$$

where

$$T_1 = \frac{1}{2}m_e v_1^2 = \frac{1}{2}m_e\left(\frac{\hbar}{m_e a_0}\right)^2 = -E_1, \quad V_1 = -\frac{e^2}{4\pi\epsilon_0 a_0} = 2E_1. \tag{4.1.9}$$

Here, we have expressed T_1 and V_1 in terms of a_0—ready for calculation for the next chapter. $E_1 = -13.6$ eV, whose absolute value is defined as the atomic energy unit, called Rydberg, 1 Ry = 13.6 eV. Another atomic energy unit is also used in the literature, called Hartree, 1 Hartree = 2 Ry. In general, since $r_n = \frac{4\pi\epsilon_0\hbar^2}{e^2 m_e}n^2$, we have

$$E_n = -\frac{1}{2}\frac{e^4 m_e}{(4\pi\epsilon_0)^2\hbar^2 n^2} = -\frac{13.6\ \text{eV}}{n^2}. \tag{4.1.10}$$

Figure 4.1(b) shows a series of eigenenergies from $n = 1$ to $n = \infty$. The first excited state, E_2, has energy of $-13.6/2^2$ eV. E_n increases (becomes less negative) as n increases. Below $n = \infty$, the electron is in a bound state, with a negative energy. Above $n = \infty$, the electron is in the continuum and has a positive energy. In the photoelectric effect discussed in Chapter 3, where electrons are liberated from the surface of a sample, they have a positive energy.

The success of the Bohr model is manifested in its explanation of the spectral lines. The spectral lines seen in Fig. 4.1(a) are the result of transitions from a higher state n_1 to a lower state n_2, with the emitted light frequency ν and wavelength λ determined by

$$h\nu = E_{n_1} - E_{n_2}, \quad hc/\lambda = E_{n_1} - E_{n_2}. \tag{4.1.11}$$

The Balmer series corresponds to transitions from $n > 2$ to $n = 2$; it was first observed because its wavelength is in the visible light regime.

Exercise 4.1.5.

4. Consider a transition $n = 3 \to n = 2$. Show the emitted light wavelength falls in the visible regime.
5. If we have a transition from $n = \infty$ to E_2, what is the wavelength of the light?

Quantization of orbital angular momentum is a crucial step in understanding the wave nature of electrons in atoms, molecules, and solids. A direct experimental measurement of orbital angular momenta is now possible with the development of X-ray magnetic circular dichroism.

This section prepares us to solve the Schrödinger equation for the hydrogen atom in the next section by taking a detour. The detour is to find eigenfunctions and eigenvalues of the orbital angular momentum operator first. Then, we employ the property between the angular momentum operator and the Hamiltonian operator: Two commutable operators share a subset of the same eigenfunctions. This partially bypasses the difficulty in solving the Schrödinger equation for the hydrogen atom directly.

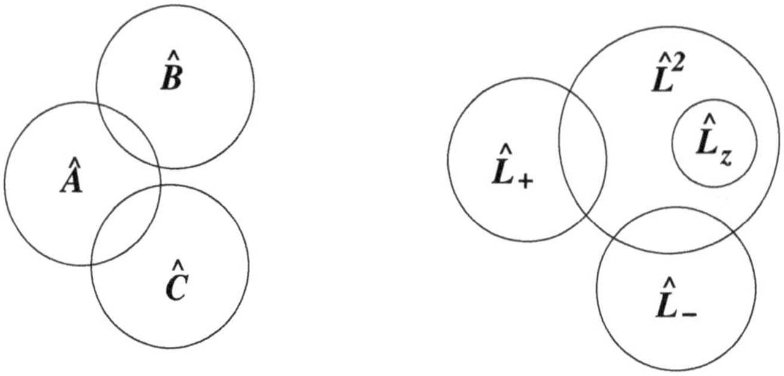

Figure 4.2: (Left) Even if two operators $\hat{A}$ and $\hat{B}$ commute, only a subset of the eigenfunctions are the same. This applies to any pair of operators. (Right) $\hat{\mathbf{L}}^2$ only commutes with $\hat{L}_z$, $\hat{L}_-$, and $\hat{L}_+$, separately, so they share a subset of eigenfunctions of $\hat{\mathbf{L}}^2$. But $\hat{L}_z$, $\hat{L}_-$, and $\hat{L}_+$ do not commute.

4.2 Commutable operators and their common eigenfunctions

First, we will prove the theorem that two commutable operators may share a subset of eigenfunctions. We solve the eigenequation of a simpler operator first and then use its eigenfunctions to solve the eigenequation of a more complicated operator. Sometimes it is necessary to solve the eigenfunctions of multiple simpler operators first, which is the case for the orbital angular momentum operator and the Hamiltonian of the hydrogen atom. Figure 4.2 (right-hand figure) schematically shows an example of the $\hat{\mathbf{L}}^2$ operator.

4.2.1 Nondegenerate eigenfunctions

We start with a case whose eigenfunctions are not degenerate, where an eigenvalue only corresponds to a single eigenfunction. To avoid confusions with the eigenstates of a system (defined by the Hamiltonian operator), we purposely use eigenvectors or eigenfunctions for a regular operator. We will prove that, if two operators, $\hat{A}$ and $\hat{B}$, commute with each other, i. e., $[\hat{A}, \hat{B}] = 0$, then $\hat{A}$ and $\hat{B}$ share a subset of eigenfunctions.[3]

Suppose $|a\rangle$ is an eigenfunction of operator $\hat{A}$, with eigenvalue a, $\hat{A}|a\rangle = a|a\rangle$. We multiply both sides by $\hat{B}$ to get

$$\hat{B}\hat{A}|a\rangle = \hat{B}a|a\rangle = a\hat{B}|a\rangle. \tag{4.2.1}$$

Since $[\hat{A}, \hat{B}] = 0, \hat{B}\hat{A} = \hat{A}\hat{B}$, we can rewrite eq. (4.2.1) as $\hat{B}\hat{A}|a\rangle = \hat{A}\hat{B}|a\rangle = a\hat{B}|a\rangle$. This means $\hat{B}|a\rangle$ is also an eigenfunction of $\hat{A}$ with the same eigenvalue a. Because we assume that

3 Suppose $\hat{A}$ only depends on x, and its eigenfunction is $\sin(x)$. Then, for any function that is not a function of x, such as $\cos(y)$, once it is multiplied by $\sin(x)$ such as $\cos(y)\sin(x)$, it is still an eigenfunction of $\hat{A}$. The subset of eigenfunctions refers to $\sin(x)$. Eigenfunctions are completely equivalent to eigenvectors in a matrix.

our eigenfunction is not degenerate, this means that $\hat{B}|a\rangle$ and $|a\rangle$ at maximum differ by a constant coefficient b, i. e.,

$$\hat{B}|a\rangle = b|a\rangle,$$

which proves that $|a\rangle$ is an eigenfunction of $\hat{B}$. $\hat{A}$ and $\hat{B}$ share the common eigenfunction. Caution must be taken here. This common eigenfunction does not mean that all the eigenfunctions of $\hat{A}$ are completely the same as those of $\hat{B}$. It only means that part of the eigenfunctions are the same. If $\hat{A}$ and $\hat{B}$ share a common eigenfunction, and $\hat{A}$ and $\hat{C}$ share another common eigenfunction, it does not mean or imply $\hat{B}$ and $\hat{C}$ would necessarily share a common eigenfunction (see Fig. 4.2).

Exercise 4.2.1.

6. It is known that $\hat{p}_x$ and $\hat{p}_y$ commute with each other. Find their common eigenfunction.
7. Suppose $\hat{A} = \begin{pmatrix} 0 & 1 \\ 1 & 0 \end{pmatrix}$. (a) Find its eigenvalues and eigenvectors. (b) If another operator $\hat{B}$ commutes with $\hat{A}$ and shares the first eigenvector, what is $\hat{B}$?

4.2.2 Degenerate eigenfunctions

Suppose two Hermitian operators $\hat{A}$ and $\hat{B}$ have N degenerate eigenfunctions: $\hat{A}|a_n\rangle = a|a_n\rangle, \hat{B}|b_n\rangle = b|b_n\rangle$, where n runs from 1 to N and a and b are their eigenvalues. One can show that, if $\hat{A}$ and $\hat{B}$ commute, their respective eigenfunctions are a linear combination of another operator's eigenfunctions. The opposite is also true.

Here we are going to prove $[\hat{A}, \hat{B}] = 0$ if eigenfunctions $|b_n\rangle$ of $\hat{B}$ can be expanded in terms of eigenfunctions $|a_j\rangle$ of $\hat{A}$, $|b_n\rangle = \sum_{j=1}^{N} c_j^n |a_j\rangle$, where c_j^n is an expansion coefficient. We multiply $\hat{A}\hat{B}$ and $\hat{B}\hat{A}$ to both sides of the equation to get

$$\hat{A}\hat{B}|b_n\rangle = b\hat{A}|b_n\rangle = \sum_{j=1}^{N} \hat{A}\hat{B}c_j^n|a_j\rangle, \quad \hat{B}\hat{A}|b_n\rangle = \sum_{j=1}^{N} \hat{B}c_j^n a|a_j\rangle.$$

We subtract these two equations from each other and then multiply $\langle b_n|$ from the left,

$$\langle b_n|(b - \hat{B})\hat{A}|b_n\rangle = \sum_{j=1}^{N} c_j^n(\langle b_n|\hat{A}\hat{B}|a_j\rangle - ab\langle b_n|a_j\rangle).$$

We use $\langle b_n|\hat{B}^\dagger = \langle b_n|\hat{B} = \langle b_n|b$ because $\hat{B}$ is assumed to be a Hermitian operator, so the left side is zero, i. e., $\sum_{j=1}^{N} c_j^n(\langle b_n|\hat{A}\hat{B}|a_j\rangle - ab\langle b_n|a_j\rangle) = 0$. In general, every c_j^n is nonzero, so the terms in the parenthesis must be zero:

$$\langle b_n|\hat{A}\hat{B}|a_j\rangle = ab\langle b_n|a_j\rangle. \tag{4.2.2}$$

Next, we prove it by contradiction. Assuming $\hat{A}\hat{B} \neq \hat{B}\hat{A}$ leads to

$$\langle b_n|\hat{A}\hat{B}|a_j\rangle \neq \langle b_n|\hat{B}\hat{A}|a_j\rangle = ab\langle b_n|a_j\rangle,$$

which contradicts eq. (4.2.2). This proves our assumption $\hat{A}\hat{B} \neq \hat{B}\hat{A}$ is invalid, so the only choice is $\hat{A}\hat{B} = \hat{B}\hat{A}$, or $[\hat{A}, \hat{B}] = 0$.

4.3 Orbital angular momentum

We are going to use the theorem that two commutable operators share common eigenfunctions. While the Bohr model treats all quantities classically, QM treats the orbital angular momentum as an operator, a cross product of the position operator and momentum operator. Because the level of difficulty to find an eigenfunction is different for three components of the orbital angular momentum operator, we will find the eigenfunction of an easier operator that commutes with the total orbital angular momentum operator. To avoid confusion, we state clearly here that we use $\hat{r}$ for the radial unit vector, $\hat{\mathbf{r}}$ for the position operator, and r for the radius. In the $\mathbf{r}$-representation, where all functions are a function of r, we have $\hat{\mathbf{r}} = \mathbf{r}$.

4.3.1 Orbital angular momentum operators: $\hat{\mathbf{L}}$, $\hat{\mathbf{L}}^2$, $\hat{L}_z$, $\hat{L}_+$, and $\hat{L}_-$

4.3.1.1 Operators in Cartesian coordinates

The orbital angular momentum operator is defined as

$$\hat{\mathbf{L}} = \hat{\mathbf{r}} \times \hat{\mathbf{p}} = \mathbf{r} \times (-i\hbar)\nabla,$$

where $\hat{\mathbf{r}}$ must be placed before $\hat{\mathbf{p}}$ because $\hat{\mathbf{p}}$ is an operator. Three components of $\hat{\mathbf{L}}$ are

$$\begin{aligned}
\hat{L}_x &= y\hat{p}_z - z\hat{p}_y = -i\hbar\left(y\frac{\partial}{\partial z} - z\frac{\partial}{\partial y}\right),\\
\hat{L}_y &= z\hat{p}_x - x\hat{p}_z = -i\hbar\left(z\frac{\partial}{\partial x} - x\frac{\partial}{\partial z}\right),\\
\hat{L}_z &= x\hat{p}_y - y\hat{p}_x = -i\hbar\left(x\frac{\partial}{\partial y} - y\frac{\partial}{\partial x}\right).
\end{aligned} \tag{4.3.1}$$

Example 1 (Commutation among orbital angular momentum operators). Show $[\hat{L}_x, \hat{L}_y] = i\hbar\hat{L}_z$.

From the definition of $\hat{L}_x$ and $\hat{L}_y$, we have $[\hat{L}_x, \hat{L}_y] = [y\hat{p}_z - z\hat{p}_y, z\hat{p}_x - x\hat{p}_z] = y\hat{p}_x[\hat{p}_z, z] + x\hat{p}_y[z, \hat{p}_z] = i\hbar(x\hat{p}_y - y\hat{p}_x) = i\hbar\hat{L}_z$.

Together we have

$$[\hat{L}_x, \hat{L}_y] = i\hbar\hat{L}_z, [\hat{L}_y, \hat{L}_z] = i\hbar\hat{L}_x, [\hat{L}_z, \hat{L}_x] = i\hbar\hat{L}_y \rightarrow \hat{\mathbf{L}} \times \hat{\mathbf{L}} = i\hbar\hat{\mathbf{L}}. \tag{4.3.2}$$

The raising and lowering operators $\hat{L}_+$ and $\hat{L}_-$ are defined as

$$\hat{L}_+ = \hat{L}_x + i\hat{L}_y; \quad \hat{L}_- = \hat{L}_x - i\hat{L}_y, \tag{4.3.3}$$

where $\hat{L}_+^\dagger = \hat{L}_-$ and $\hat{L}_-^\dagger = \hat{L}_+$.

This shows that $\hat{L}_x$, $\hat{L}_y$ and $\hat{L}_z$ do not share a common eigenfunction. Next, we check whether $\hat{\mathbf{L}}$ commutes with them, but unfortunately, it does not either. Then we move on to $\hat{\mathbf{L}}^2$ and find $[\hat{\mathbf{L}}^2, \hat{L}_x] = [\hat{\mathbf{L}}^2, \hat{L}_y] = [\hat{\mathbf{L}}^2, \hat{L}_z] = 0$, so we can use any pairs of operators. It is difficult to work with Cartesian coordinates since three coordinates are correlated to each other. Spherical coordinates are relatively simpler.

4.3.1.2 Operators in spherical coordinates

The momentum operator is $\hat{\mathbf{p}} = -i\hbar\nabla$. In spherical coordinates,

$$\nabla = \hat{r}\frac{\partial}{\partial r} + \hat{\theta}\frac{1}{r}\frac{\partial}{\partial \theta} + \hat{\phi}\frac{1}{r\sin\theta}\frac{\partial}{\partial \phi},$$

where $\hat{r}$, $\hat{\theta}$ and $\hat{\phi}$ are unit vectors, not operators, along the $\hat{r}$, $\hat{\theta}$, and $\hat{\phi}$ directions. We write the momentum operator as $\hat{\mathbf{p}} = \hat{r}\hat{p}_r + \hat{\theta}\hat{p}_\theta + \hat{\phi}\hat{p}_\phi$, and by comparison, we find $\hat{p}_r = -i\hbar\frac{\partial}{\partial r}$, $\hat{p}_\theta = -i\hbar\frac{1}{r}\frac{\partial}{\partial \theta}$ and $\hat{p}_\phi = -i\hbar\frac{1}{r\sin\theta}\frac{\partial}{\partial \phi}$.

We can now compute $\hat{\mathbf{L}} = \mathbf{r} \times \hat{\mathbf{p}}$. Because $\hat{r} \times \hat{r} = 0$, $\hat{r} \times \hat{\theta} = \hat{\phi}$, $\hat{r} \times \hat{\phi} = -\hat{\theta}$, and $\mathbf{r} = r\hat{r}$, $\hat{\mathbf{L}}$ only has two terms, with no dependence on r:

$$\hat{\mathbf{L}} = -i\hbar\left(\hat{\phi}\frac{\partial}{\partial \theta} - \hat{\theta}\frac{1}{\sin\theta}\frac{\partial}{\partial \phi}\right), \tag{4.3.4}$$

instead of three in Cartesian coordinates, greatly simplifying our calculation.

To find spherical expressions for $\hat{L}_x$, $\hat{L}_y$, and $\hat{L}_z$, the easiest way is to express $\hat{\phi}$ and $\hat{\theta}$ in Cartesian coordinates,

$$\hat{\theta} = \cos\theta\cos\phi\hat{x} + \cos\theta\sin\phi\hat{y} - \sin\theta\hat{z},$$
$$\hat{\phi} = -\sin\phi\hat{x} + \cos\phi\hat{y},$$

where $\hat{x}$, $\hat{y}$, and $\hat{z}$ are unit vectors, not an operator. We plug $\hat{\theta}$ and $\hat{\phi}$ into eq. (4.3.4) to get

$$\hat{L}_x = i\hbar\left(\sin\phi\frac{\partial}{\partial\theta} + \cot\theta\cos\phi\frac{\partial}{\partial\phi}\right),$$
$$\hat{L}_y = i\hbar\left(-\cos\phi\frac{\partial}{\partial\theta} + \cot\theta\sin\phi\frac{\partial}{\partial\phi}\right), \tag{4.3.5}$$
$$\hat{L}_z = -i\hbar\frac{\partial}{\partial\phi}.$$

Another way to derive $\hat{L}_x, \hat{L}_y$ and $\hat{L}_z$ is to start directly from eq. (4.3.1) and then transform (x, y, z) to (r, θ, ϕ). This is left as a homework assignment.

4.3.2 Eigenfunctions of $\hat{\mathbf{L}}^2$ are spherical harmonics Y_{lm}

The beginner may attempt to find the eigenfunction of $\hat{\mathbf{L}}$, but this is very difficult because essentially we have to solve three eigenequations for its three components. A better strategy is to use $\hat{\mathbf{L}}^2$. This is because $[\hat{\mathbf{L}}^2, \hat{L}_z] = 0$ and because it is much simpler to find an eigenfunction for $\hat{L}_z$. But first, to appreciate the enormous complexity of the eigenequation of $\hat{\mathbf{L}}^2$, we work out an expression for $\hat{\mathbf{L}}^2$:

$$\hat{\mathbf{L}}^2 = \hat{\mathbf{L}}\cdot\hat{\mathbf{L}} = -\hbar^2\left(\hat{\phi}\frac{\partial}{\partial\theta} - \hat{\theta}\frac{1}{\sin\theta}\frac{\partial}{\partial\phi}\right)\cdot\left(\hat{\phi}\frac{\partial}{\partial\theta} - \hat{\theta}\frac{1}{\sin\theta}\frac{\partial}{\partial\phi}\right). \tag{4.3.6}$$

This dot product harbors a hidden difficulty because $\hat{\theta}$ depends on ϕ as well, not just θ and because operators act upon functions that follow them. To demonstrate this, we take $-\hat{\theta}\frac{1}{\sin\theta}\frac{\partial}{\partial\phi}\cdot\hat{\phi}\frac{\partial}{\partial\theta}$ as an example. For a normal vector product, this must be zero because $\hat{\theta}\cdot\hat{\phi} = 0$, but here it is complicated. In the following derivation, we will temporarily switch to Cartesian coordinates since they do not have this complicated relationship:

$$\begin{aligned}
-\hat{\theta}\frac{1}{\sin\theta}\frac{\partial}{\partial\phi}\cdot\hat{\phi}\frac{\partial}{\partial\theta} &= -\hat{\theta}\frac{1}{\sin\theta}\cdot\left[\left(\frac{\partial}{\partial\phi}\hat{\phi}\right)\frac{\partial}{\partial\theta} + \hat{\phi}\frac{\partial^2}{\partial\phi\partial\theta}\right] \\
&= -\hat{\theta}\frac{1}{\sin\theta}\cdot\left[\frac{\partial}{\partial\phi}(-\sin\phi\hat{x} + \cos\phi\hat{y})\frac{\partial}{\partial\theta} + \hat{\phi}\frac{\partial^2}{\partial\phi\partial\theta}\right] \\
&= \hat{\theta}\frac{1}{\sin\theta}\cdot[\cos\phi\hat{x} + \sin\phi\hat{y}]\frac{\partial}{\partial\theta} + 0 \\
&= \frac{1}{\sin\theta}\underbrace{(\cos\theta\cos\phi\hat{x} + \cos\theta\sin\phi\hat{y} - \sin\theta\hat{z})}_{\hat{\theta}}\cdot(\cos\phi\hat{x} + \sin\phi\hat{y})\frac{\partial}{\partial\theta} \\
&= \frac{1}{\sin\theta}[\cos\theta(\cos\phi)^2 + \cos\theta(\sin\phi)^2]\frac{\partial}{\partial\theta} = \cot\theta\frac{\partial}{\partial\theta},
\end{aligned}$$

where 0 is due to $\hat{\theta}\cdot\hat{\phi} = 0$. The rest of the terms are left as an exercise. Finally, we have

$$\hat{\mathbf{L}}^2 = -\hbar^2\left[\frac{1}{\sin\theta}\frac{\partial}{\partial\theta}\left(\sin\theta\frac{\partial}{\partial\theta}\right) + \frac{1}{\sin^2\theta}\frac{\partial^2}{\partial\phi^2}\right]. \tag{4.3.7}$$

To find the eigenfunctions of $\hat{\mathbf{L}}^2$, we employ repeatedly what we have learned from the previous section: Two commutable operators share the common eigenfunction. Essentially, we first find the eigenfunctions of simpler operators and then assemble the entire solution at the end. Since $\hat{\mathbf{L}}^2$ commutes with $\hat{L}_x$, $\hat{L}_y$ and $\hat{L}_z$, we can choose anyone of them, but, in spherical coordinates, $\hat{L}_z$ is the simplest (see eq. (4.3.5)).

We denote the eigenfunction of $\hat{\mathbf{L}}^2$ as Y_{lm}, the spherical harmonics, $\hat{\mathbf{L}}^2 Y_{lm} = \lambda Y_{lm}$, where λ is its eigenvalue. Since $\hat{\mathbf{L}}^2$ and $\hat{L}_z$ commute and share part of eigenfunctions, we will use $\hat{L}_z$ to help us determine the ϕ-dependent part of Y_{lm}, $\eta(\phi)$, $\hat{L}_z\eta(\phi) = c\eta(\phi)$, where c is the eigenvalue of $\hat{L}_z$. Another part of Y_{lm} is denoted as $g(\theta)$, so we can write $Y_{lm} = g(\theta)\eta(\phi)$. g itself is independent of ϕ, so $\hat{L}_z g\eta = g\hat{L}_z\eta$, i. e., $\hat{L}_z Y_{lm} = cY_{lm}$.

Exercise 4.3.2.

8. It is easy to show $[\hat{\mathbf{L}}^2, \hat{\mathbf{L}}] = 0$. Can we say that $\hat{\mathbf{L}}^2$ and $\hat{\mathbf{L}}$ share the common eigenfunction? And why?
9. Show $[\hat{L}_x, \hat{L}_y] = i\hbar\hat{L}_z$, $[\hat{L}_y, \hat{L}_z] = i\hbar\hat{L}_x$, $[\hat{L}_z, \hat{L}_x] = i\hbar\hat{L}_y$.
10. Show $[\hat{\mathbf{L}}^2, \hat{L}_x] = [\hat{\mathbf{L}}^2, \hat{L}_y] = [\hat{\mathbf{L}}^2, \hat{L}_z] = 0$.
11. Two ladder operators are defined as $\hat{L}_\pm = \hat{L}_x \pm i\hat{L}_y$. Show $[\hat{L}_z, \hat{L}_\pm] = \pm\hbar\hat{L}_\pm$.
12. Starting from eq. (4.3.6), show that the final expression for $\hat{\mathbf{L}}^2$ is eq. (4.3.7).

4.3.3 Eigenfunction of $\hat{L}_z$

Because $\hat{L}_z$ physically represents a rotation by angle ϕ about the z-axis, it is independent of r or θ,[4] and we can change a partial differentiation equation to a full differentiation equation,

$$-i\hbar\frac{\partial}{\partial\phi}\eta = c\eta, \to -i\hbar\frac{d}{d\phi}\eta = c\eta, \tag{4.3.8}$$

where η is the eigenfunction and c is its eigenvalue. Its solution is $\eta(\phi) = \eta_0 e^{i\frac{c}{\hbar}\phi}$, where η_0 is a normalization constant. In general, c can take any value. However, if the system has a rotational symmetry, such as the hydrogen atom, where the system after 360° rotation returns to its original state, meaning $\eta(\phi + 2\pi) = \eta(\phi)$, then we have $\eta(\phi) = \eta_0 e^{ic\phi/\hbar} = \eta(\phi + 2\pi) = \eta_0 e^{ic\phi/\hbar + i2c\pi/\hbar} \to \exp(i2c\pi/\hbar) = 1$. This means the eigenvalue c only takes on some restricted values $\frac{c}{\hbar}2\pi = m2\pi$ or $c = m\hbar$, where m is an integer and is called the magnetic angular momentum quantum number.[5]

---

4 This step is not possible for $\hat{L}_x$ and $\hat{L}_y$ because they depend on both ϕ and θ. We cannot convert the partial differentiation to the full differentiation.

5 The reason we call it a magnetic angular momentum quantum number is that we assume a tiny artificial magnetic field along the z-axis to set our quantization axis. Equally, we would set the magnetic field along the x-axis, and the entire argument would remain. The only difference is that we exchange the z- with the x-axis. Note that this magnetic field is not present in reality, and it is used for our convenience.

This shows that the eigenvalues of $\hat{L}_z$ are quantized. This quantization derives from the rotational symmetry. The eigenfunction is $\eta(\phi) = \eta_0 e^{im\phi}$, where η_0 is determined by the normalization of the eigenfunction $\int_0^{2\pi} \eta^*(\phi)\eta(\phi)d\phi = \eta_0^2 2\pi = 1 \longrightarrow \eta_0 = \frac{1}{\sqrt{2\pi}}$, $\eta(\phi) = \frac{1}{\sqrt{2\pi}} e^{im\phi}$. We have the eigenequation for $\hat{L}_z$ as

$$\hat{L}_z \eta(\phi) = m\hbar\eta(\phi). \tag{4.3.9}$$

If we multiply $\eta(\phi)$ by another $g(\theta)$ that is independent of ϕ, the resultant function is still an eigenfunction of $\hat{L}_z$. This is the reason why in the last subsection our resultant function is written as $Y_{lm} = \eta g$.

Because $\hat{L}_z$ and $\hat{\mathbf{L}}^2$ share the same η, we can use $\eta(\phi)$ to get rid of $\partial^2/\partial\phi^2$ in $\hat{\mathbf{L}}^2$ (see eq. (4.3.7)).[6] Specifically, we plug $Y_{lm}(\theta,\phi)$ into $\hat{\mathbf{L}}^2 Y_{lm}(\theta,\phi) = \lambda Y_{lm}(\theta,\phi)$ to get a new eigenequation for $\hat{\mathbf{L}}^2$, $\hbar^2[\frac{1}{\sin\theta}\frac{\partial}{\partial\theta}(\sin\theta\frac{\partial}{\partial\theta}) - \frac{m^2}{\sin^2\theta}]g(\theta) = \lambda g(\theta)$, i. e.,

$$\left[\frac{1}{\sin\theta}\frac{\partial}{\partial\theta}\left(\sin\theta\frac{\partial}{\partial\theta}\right) - \frac{\lambda}{\hbar^2} - \frac{m^2}{\sin^2\theta}\right]g(\theta) = 0,$$

whose solution $g(\theta)$ is an associated Legendre function $P_l^m(\cos\theta)$. But, to find $P_l^m(\cos\theta)$ mathematically is laborious and does not add any physical insights to our solution. We prefer the ladder operators to find the expression for Y_{lm}. A crucial feature of the associated Legendre function is that, although the equation is independent of l, $P_l^m(\cos\theta)$ depends on l through m. l is called the angular momentum quantum number, and m is the magnetic angular momentum quantum number and takes only the values $m = -l, -l+1, \dots, l-1, l$. Thus, our eigenfunction Y_{lm} is completely determined once l and m are given. As we will prove in the following, $\lambda = l(l+1)\hbar^2$, where $l = 0, 1, 2, 3, \dots$. $\hat{\mathbf{L}}^2$ and $\hat{L}_z$ satisfy the following eigenequations:

$$\hat{\mathbf{L}}^2 Y_{lm} = l(l+1)\hbar^2 Y_{lm}, \quad \hat{L}_z Y_{lm} = m\hbar Y_{lm}. \tag{4.3.10}$$

4.3.4 Using the angular momentum ladder operators to find Y_{lm}

In the harmonic oscillator, the ladder operators enabled us to find a series of eigenfunctions. Here, an identical method is used. The raising and lowering operators $\hat{L}_+$ and $\hat{L}_-$ are constructed from $\hat{L}_x$ and $\hat{L}_y$ (eqs. (4.3.3) and (4.3.5)),

$$\hat{L}_\pm = \hat{L}_x \pm i\hat{L}_y = \pm\hbar e^{\pm i\phi}\left(\frac{\partial}{\partial\theta} \pm i\cot\theta\frac{\partial}{\partial\phi}\right), \tag{4.3.11}$$

6 When we say two operators share the same eigenfunction, what we really mean is that they share part of the same eigenfunction. For $\hat{L}_z$, this means the $\eta(\phi)$ part of the eigenfunction Y_{lm} of $\hat{\mathbf{L}}^2$.

where $\hat{L}_- = \hat{L}_+^\dagger$. Because $[\hat{L}_x, \hat{\mathbf{L}}^2] = 0$, $[\hat{L}_y, \hat{\mathbf{L}}^2] = 0$ and $[\hat{L}_\pm, \hat{\mathbf{L}}^2] = 0$, $\hat{L}_\pm$ and $\hat{\mathbf{L}}^2$ share the same eigenfunctions, then, if $Y_{lm}(\theta,\phi) = g(\theta)\eta(\phi)$ is an eigenfunction of $\hat{\mathbf{L}}^2$, then $\hat{L}_+ Y_{lm}(\theta,\phi)$ and $\hat{L}_- Y_{lm}(\theta,\phi)$ are also eigenfunctions of $\hat{\mathbf{L}}^2$ (see Section 4.2.1). This is the key relationship to be used to find $g(\theta)$. However, different from the harmonic oscillator, we do not know what $\hat{L}_+ Y_{lm}$ or $\hat{L}_- Y_{lm}$ produces, so we cannot find an equation like eq. (3.2.12). We need help from $\hat{L}_z$.

We apply $\hat{L}_z \hat{L}_+$ to Y_{lm} to obtain

$$\hat{L}_z \underbrace{\hat{L}_+ Y_{lm}}_{\text{focus}} = (\hat{L}_+ \hat{L}_z + \hbar \hat{L}_+) Y_{lm} = (\hat{L}_+ m\hbar + \hbar \hat{L}_+) Y_{lm} = (m+1)\hbar \underbrace{\hat{L}_+ Y_{lm}}_{\text{focus}}, \tag{4.3.12}$$

where we have used $\hat{L}_z \hat{L}_+ = \hat{L}_+ \hat{L}_z + \hbar \hat{L}_+$ and $\hat{L}_z Y_{lm} = m\hbar Y_{lm}$. $\hat{L}_+ Y_{lm}$ is a different eigenfunction of $\hat{L}_z$, but increases its eigenvalue by $+1\hbar$. This is why $\hat{L}_+$ is called the raising operator. $\hat{L}_+$ changes the eigenvalue for $\hat{L}_z$, but it keeps the eigenvalue of $\hat{\mathbf{L}}^2$. This shows a crucial difference between two sets of operators: (1) Since $[\hat{L}_+, \hat{\mathbf{L}}^2] = 0$, the application of $\hat{L}_+$ to Y_{lm} does not change the eigenvalue for $\hat{\mathbf{L}}^2$, and (2) since $[\hat{L}_+, \hat{L}_z] \neq 0$, the application of $\hat{L}_+$ to Y_{lm} does change the eigenvalue of $\hat{L}_z$.

Now we are ready to find the eigenfunction of $\hat{\mathbf{L}}^2$ in the same fashion as eq. (3.2.12) ($\hat{a}|\phi_0\rangle = 0$). Supposing that the eigenvalue of $\hat{L}_z$ already reaches its maximum value $l\hbar$, i. e., $\hat{L}_z Y_{ll} = l\hbar Y_{ll}$, a further application of $\hat{L}_+$ to Y_{ll} should produce zero,

$$\hat{L}_+ Y_{ll} = \hat{L}_+[g_l(\theta)\eta_l(\phi)] = 0, \longrightarrow \hbar e^{i\phi}\left(\frac{\partial}{\partial\theta} + i\cot\theta\frac{\partial}{\partial\phi}\right) g_l(\theta)\eta_l(\phi) = 0. \tag{4.3.13}$$

We plug $\eta_l(\phi) = \frac{1}{\sqrt{2\pi}} e^{il\phi}$ into eq. (4.3.13) to find an equation for $g_l(\theta)$,

$$\hbar e^{i\phi}\left(\frac{\partial}{\partial\theta} + i\cot\theta\frac{\partial}{\partial\phi}\right) g_l(\theta)\frac{1}{\sqrt{2\pi}} e^{il\phi} = 0,$$

which can be reduced to

$$e^{il\phi}\frac{\partial}{\partial\theta} g_l(\theta) + i\cot\theta g_l(\theta)(il)e^{il\phi} = 0 \rightarrow \frac{dg_l(\theta)}{d\theta} = l\cot\theta g_l(\theta) \tag{4.3.14}$$

because $g_l(\theta)$ is a function of θ only. We simplify eq. (4.3.14) to $\frac{dg_l(\theta)}{g_l(\theta)} = l\frac{\cos\theta d\theta}{\sin\theta} = l\frac{d\sin\theta}{\sin\theta}$ and integrate it to obtain $\ln(g_l(\theta)) = l\ln(\sin\theta) = \ln\sin^l\theta$, or $g_l(\theta) = \sin^l\theta$. This is a special solution ($m = l$) of the associated Legendre function.

Because $g_l(\theta) = \sin^l\theta$ is the θ part of the eigenfunction of $\hat{\mathbf{L}}^2$, our final eigenfunction Y_{lm} with l and $m = l$ can be written as

$$Y_{l,m=l} = g_l\eta_l = B_l \sin^l\theta e^{il\phi}, \tag{4.3.15}$$

where B_l is the normalization constant. Table 4.1 shows a few normalized spherical harmonics, where eq. (4.3.15) matches all Y_{ll} in the table. In the previous derivation, we started from $\hat{L}_+ Y_{ll} = 0$ to obtain g_l. We can also start from $\hat{L}_- Y_{l,-l} = 0$ to obtain g_{-l}.

Table 4.1: Spherical harmonics $Y_{lm}(\theta,\phi)$ with orbital characters.

(l,m)	Y_{lm}	Orbital	(l,m)	Y_{lm}	Orbital
(0,0)	$\sqrt{\frac{1}{4\pi}}$	s			
(1,0)	$\sqrt{\frac{3}{4\pi}}\cos\theta$	p_z	$(1,\pm1)$	$\mp\sqrt{\frac{3}{8\pi}}\sin\theta e^{\pm i\phi}$	p_x, p_y
(2,0)	$\sqrt{\frac{5}{16\pi}}(3\cos^2\theta-1)$	d_{z^2}	$(2,\pm1)$	$\mp\sqrt{\frac{15}{8\pi}}\cos\theta\sin\theta e^{\pm i\phi}$	d_{xz}, d_{yz}
			$(2,\pm2)$	$\sqrt{\frac{15}{32\pi}}\sin^2\theta e^{\pm 2i\phi}$	$d_{xy}, d_{x^2-y^2}$

These spherical harmonics Y_{lm} form a special series of functions. They are orthonormalized as

$$\langle Y_{lm}|Y_{l'm'}\rangle = \int_{\phi=0}^{2\pi}\int_{\theta=0}^{\pi} Y^*_{lm}(\theta,\phi)Y_{l'm'}(\theta,\phi)\sin\theta d\theta d\phi = \delta_{ll'}\delta_{mm'}. \tag{4.3.16}$$

If we apply $\hat{\mathbf{L}}^2$ (eq. (4.3.7)) to the spherical harmonics $Y_{lm=l}$ in eq. (4.3.15), we can find the eigenvalue of $\hat{\mathbf{L}}^2$ as $\lambda = l(l+1)\hbar^2$, whose detailed derivation is relegated to the ensuing exercise. Its eigenfunction is independent of m, though the eigenequation of $\hat{\mathbf{L}}^2$ depends on m. For this reason, if $m \neq l$, the θ and ϕ terms in $Y_{l,m\neq l}$ must be adjusted (see Table 4.1) in such a way that the terms with m are all canceled out. $Y_{l,m\neq l}$ can be found by repeatedly applying $\hat{L}_- Y_{ll}$, is subsequently shown.

4.3.4.1 Getting the rest of the spherical harmonics Y_{lm} from $\hat{L}_+$ and $\hat{L}_-$

To obtain the rest of the eigenfunctions, we go back to eqs. (4.3.12) and (4.3.10),

$$\hat{L}_z[\hat{L}_+Y_{lm}] = (m+1)\hbar[\hat{L}_+Y_{lm}], \quad \hat{L}_z[Y_{l,m+1}] = (m+1)\hbar[Y_{l,m+1}],$$

and compare the terms in the square brackets in the previous two equations. We conclude that these two terms must be equal to each other up to a constant C,

$$\hat{L}_+Y_{lm} = C_{lm}Y_{l,m+1}. \tag{4.3.17}$$

This shows that, if Y_{lm} is known, we can generate a new eigenfunction by applying $\hat{L}_+$ to it. But, first we need to determine C_{lm}. We take the Hermitian conjugate of eq. (4.3.17),

$$Y^*_{lm}\hat{L}_- = C^*_{lm}Y^*_{l,m+1}, \tag{4.3.18}$$

and multiply eq. (4.3.18) with eq. (4.3.17) from the left to get

$$Y^*_{lm}\hat{L}_-\hat{L}_+Y_{lm} = |C_{lm}|^2Y^*_{l,m+1}Y_{l,m+1}, \tag{4.3.19}$$

where we have used $\hat{L}^\dagger_+ = \hat{L}_-$ owing to $\hat{L}_\pm = \hat{L}_x \pm i\hat{L}_y$.

Our focus is on $\hat{L}_-\hat{L}_+Y_{lm}$. Note that $\hat{\mathbf{L}}^2 = \hat{L}_x^2 + \hat{L}_y^2 + \hat{L}_z^2 = \hat{L}_z^2 + \frac{1}{2}(\hat{L}_+\hat{L}_- + \hat{L}_-\hat{L}_+)$. Because $[\hat{L}_+, \hat{L}_-] = 2\hbar\hat{L}_z$, $\hat{L}_+\hat{L}_- - \hat{L}_-\hat{L}_+ = 2\hbar\hat{L}_z$ or $\hat{L}_+\hat{L}_- = \hat{L}_-\hat{L}_+ + 2\hbar\hat{L}_z$, we have $\hat{\mathbf{L}}^2 = \hat{L}_z^2 + \hat{L}_-\hat{L}_+ + \hbar\hat{L}_z$, i. e., $\hat{L}_-\hat{L}_+ = \hat{\mathbf{L}}^2 - \hat{L}_z^2 - \hbar\hat{L}_z$. Then $\hat{L}_-\hat{L}_+Y_{lm} = (l(l+1) - m^2 - m)\hbar^2 Y_{lm}$, where we have used eq. (4.3.10). Equation (4.3.19) can be simplified to

$$Y_{lm}^*(l(l+1) - m^2 - m)\hbar^2 Y_{lm} = |C_{lm}|^2 Y_{l,m+1}^* Y_{l,m+1}. \tag{4.3.20}$$

We integrate both sides over θ and ϕ,

$$\int Y_{lm}^*(l(l+1) - m^2 - m)\hbar^2 Y_{lm} \sin\theta d\theta d\phi = |C_{lm}|^2 \int Y_{l,m+1}^* Y_{l,m+1} \sin\theta d\theta d\phi,$$

and use the orthonormalization relation in eq. (4.3.16) to get $C_{lm} = \sqrt{l(l+1) - m(m+1)}\hbar$, where we choose a real C_{lm}. Finally, we have

$$\hat{L}_+Y_{lm} = \sqrt{l(l+1) - m(m+1)}\hbar Y_{lm+1}, \tag{4.3.21}$$
$$\hat{L}_-Y_{lm} = \sqrt{l(l+1) - m(m-1)}\hbar Y_{lm-1}, \tag{4.3.22}$$

where the derivation of the second equation is given in the following exercise. We can find the rest of Y_{lm} from these two equations.

Example 2 (Find Y_{21} from Y_{22}). Use the lowering operator to find Y_{21} from Y_{22} in Table 4.1.

We apply $\hat{L}_-$ (eq. (4.3.11)) to Y_{22} as

$$\hat{L}_-Y_{22} = -\hbar e^{-i\phi}\left(\frac{\partial}{\partial\theta} - i\cot\theta\frac{\partial}{\partial\phi}\right)\sqrt{\frac{15}{32\pi}}(\sin^2\theta)e^{2i\phi} = -4\hbar e^{i\phi}\sin\theta\cos\theta\sqrt{\frac{15}{32\pi}}.$$

From eq. (4.3.22), $\hat{L}_-Y_{22} = 2\hbar Y_{21}$. These two equations must be the same, so we can find

$$Y_{21} = -\sqrt{\frac{15}{8\pi}}\sin\theta\cos\theta e^{i\phi},$$

which matches the entry in Table 4.1.

4.3.4.2 Eigenvalues of $\hat{\mathbf{L}}^2$

Equations (4.3.21) and (4.3.22) make it possible to compute the expectation values of $\hat{L}_x$, $\hat{L}_y$, $\hat{L}_x^2$, $\hat{L}_y^2$, and $\hat{L}_x^2 + \hat{L}_y^2$. Here, we use them to find the eigenvalues of $\hat{\mathbf{L}}^2$, but there are two ways to obtain them. One is to use the ladder operators, and the other is to apply $\hat{\mathbf{L}}^2$ to Y_{ll} (see the following exercise).

We rewrite $\hat{\mathbf{L}}^2$ in terms of $\hat{L}_+$, $\hat{L}_-$ and $\hat{L}_z$,

$$\hat{\mathbf{L}}^2 = \hat{L}_x^2 + \hat{L}_y^2 + \hat{L}_z^2 = \left[\frac{1}{2}(\hat{L}_+ + \hat{L}_-)\right]^2 + \left[\frac{1}{2i}(\hat{L}_+ - \hat{L}_-)\right]^2 + \hat{L}_z^2$$

$$= \frac{1}{4}(\hat{L}_+^2 + \hat{L}_-^2 + \hat{L}_+\hat{L}_- + \hat{L}_-\hat{L}_+) - \frac{1}{4}(\hat{L}_+^2 + \hat{L}_-^2 - \hat{L}_+\hat{L}_- - \hat{L}_-\hat{L}_+) + \hat{L}_z^2$$
$$= \frac{1}{2}(\hat{L}_+\hat{L}_- + \hat{L}_-\hat{L}_+) + \hat{L}_z^2.$$

Using $[\hat{L}_+, \hat{L}_-] = 2\hbar\hat{L}_z$ to change $\hat{L}_+\hat{L}_-$ to $\hat{L}_-\hat{L}_+ + 2\hbar\hat{L}_z$, we obtain $\hat{\mathbf{L}}^2 = \frac{1}{2}(\hat{L}_-\hat{L}_+ + 2\hbar\hat{L}_z + \hat{L}_-\hat{L}_+) + \hat{L}_z^2 = \hbar\hat{L}_z + \hat{L}_-\hat{L}_+ + \hat{L}_z^2$. Because $\hat{L}_+Y_{ll} = 0$ and $\hat{L}_zY_{ll} = l\hbar$, $\hat{\mathbf{L}}^2Y_{ll} = (\hbar^2 l + l^2\hbar^2)Y_{ll} = l(l+1)\hbar^2 Y_{ll}$. So, the eigenvalue of $\hat{\mathbf{L}}^2$ is $l(l+1)\hbar^2$.

Exercise 4.3.4.
13. Starting from $\hat{L}_- g_{-l} e^{-il\phi}/\sqrt{2\pi} = 0$, find $g_{-l}(\theta)$.
14. Show $\hat{L}_- Y_{lm} = \sqrt{l(l+1) - m(m-1)}\hbar Y_{lm-1}$. Hint: Follow eq. (4.3.22).
15. Use eqs. (4.3.21) and (4.3.22) to compute $\langle Y_{lm}|\hat{L}_x|Y_{lm+1}\rangle$ and $\langle Y_{lm}|\hat{L}_y|Y_{lm+1}\rangle$.
16. Directly apply $\hat{\mathbf{L}}^2$ to Y_{ll} to find the eigenvalue of $\hat{\mathbf{L}}^2$. Hint: One has to start with the operator $\hat{\mathbf{L}}^2$ and then apply it to Y_{ll}.
17. Using $\hat{L}_-$, (a) find $Y_{1,0}$ from $Y_{1,1}$, and (b) find $Y_{1,-1}$ from $Y_{1,0}$. (c) Show $\hat{L}_- Y_{1,-1} = 0$, without using eq. (4.3.22).

4.4 Understanding orbital angular momentum

In CM, the magnitude of a vector $\mathbf{A}$ is given by $\sqrt{|\mathbf{A}|^2} = \sqrt{A_x^2 + A_y^2 + A_z^2}$, where any component A_x, A_y, or A_z must be less or equal to $|\mathbf{A}|^2$. In QM, these basic rules are different because we use the wavefunction to compute the expectation value of an operator in place of the magnitude. Vectors add another layer of difficulties. The expectation value of an operator squared $\langle\hat{\mathbf{A}}^2\rangle$ differs from the expectation value squared, $\langle\hat{\mathbf{A}}^2\rangle \neq \langle\hat{\mathbf{A}}\rangle^2$.

4.4.1 Physical meanings: magnitude of a quantum vector

The orbital angular momentum serves as an excellent example. The first difficulty is that $\hat{L}_x, \hat{L}_y, \hat{L}_z$ do not commute, where the measurement of $\hat{L}_z$ will affect those of $\hat{L}_x$ and $\hat{L}_y$, and vice versa. If the system is in the eigenfunction of $\hat{L}_z$, $\hat{L}_z$ gets a definitive value with no uncertainty, but then $\hat{L}_x$ and $\hat{L}_y$ are completely uncertain. The second difficulty is that, although we can write down at the operator level, $\hat{\mathbf{L}}^2 = \hat{L}_x^2 + \hat{L}_y^2 + \hat{L}_z^2$, we cannot use it to find the expectation value $\langle\hat{\mathbf{L}}^2\rangle$ because $\langle\hat{L}_x^2 + \hat{L}_y^2\rangle$ is unknown.

Supposing the system is in an eigenstate Y_{lm}, which is an eigenfunction of $\hat{\mathbf{L}}^2$ and $\hat{L}_z$, then we have

$$\hat{\mathbf{L}}^2 Y_{lm} = l(l+1)\hbar^2 Y_{lm}, \quad \hat{L}_z Y_{lm} = m\hbar Y_{lm}.$$

This shows that we know the expectation values of both $\hat{\mathbf{L}}^2$ and $\hat{L}_z$ immediately, before we can find $\langle\hat{L}_x^2\rangle$ and $\langle\hat{L}_y^2\rangle$. This is different from CM. Because Y_{lm} is an eigenfunction of $\hat{\mathbf{L}}^2$ and $\hat{L}_z$, there is no uncertainty in their expectation values. Since m runs from $-l$ to

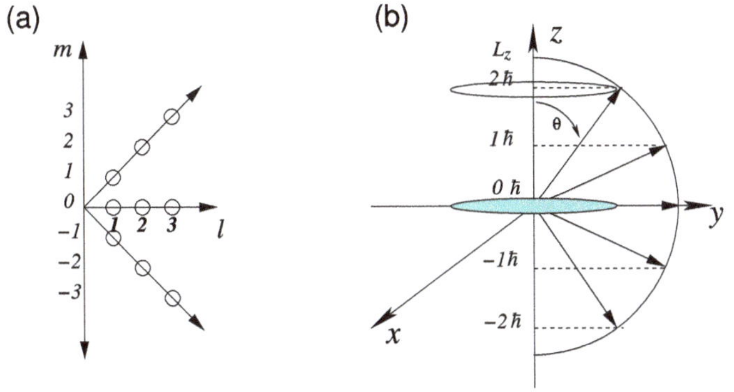

Figure 4.3: (a) m is limited between $-l$ and l. (b) Projection of angular momentum on to the z-axis. Its z-component cannot reach the full magnitude of $\hat{\mathbf{L}}^2$ because $\mathbf{L}$ has a projection on the xy-plane. The ellipse on the top denotes the case with the same $\hat{L}_z$, but with different $\hat{L}_x$ and $\hat{L}_y$.

$+l$ (see Fig. 4.3(a)), the maximum value of $\langle\hat{L}_z\rangle$ never reaches $\sqrt{l(l+1)}\hbar$, which is again different from CM. Figure 4.3(b) shows an example of $l = 2$, where $\hat{L}_z$ only takes five discrete values, the spatial quantization, all less than $\sqrt{l(l+1)}\hbar$. The difference between $l(l+1)\hbar^2$ and $m^2\hbar^2$ reflects Heisenberg's uncertainty principle.

Although the expectation values of $\hat{L}_x$ and $\hat{L}_y$ are zero, those of $\hat{L}_x^2$ and $\hat{L}_y^2$ are not. We can use $\hat{\mathbf{L}}^2 = \hat{L}_x^2+\hat{L}_y^2+\hat{L}_z^2$ to find $\hat{L}_x^2+\hat{L}_y^2 = \hat{\mathbf{L}}^2-\hat{L}_z^2$, as $\langle\hat{L}_x^2+\hat{L}_y^2\rangle = (l(l+1)-m^2)\hbar^2 \neq 0$. For our example, $l = 2$ and $m = 2$, the expectation value of $\hat{L}_x^2 + \hat{L}_y^2$ is $\langle\hat{L}_x^2\rangle + \langle\hat{L}_y^2\rangle = 2\hbar^2$. This means that the projection of $\hat{\mathbf{L}}^2$ on to the xy-plane has a length of $2\hbar^2$ which $\hat{L}_z$ misses. This explains why in Fig. 4.3(b) all the arrows are tilted away from the z-axis.

Further insights can be gained from the uncertainty principle. The uncertainty ΔA is defined as $\Delta A = \sqrt{\langle\hat{\mathbf{A}}^2\rangle - \langle\hat{\mathbf{A}}\rangle^2}$. The uncertainties ΔL_x and ΔL_y in $\hat{L}_x$ and $\hat{L}_y$ must obey

$$\Delta L_x \Delta L_y \geq \left|\frac{i}{2}\langle[\hat{L}_x, \hat{L}_y]\rangle\right| = \frac{\hbar\langle\hat{L}_z\rangle}{2} = \frac{\hbar}{2}\langle Y_{lm}|\hat{L}_z|Y_{lm}\rangle = \frac{m\hbar^2}{2}.$$

Using $m = 2$, the minimum uncertainty is $(2\hbar^2/2) = \hbar^2$, so ΔL_x and ΔL_y must not be smaller than $\hbar$. Therefore, $\langle\hat{L}_z\rangle$ cannot take the maximum length $\sqrt{6}\hbar$ of $\sqrt{\langle\hat{\mathbf{L}}^2\rangle}$.

Example 3 (Compute $\langle\hat{L}_x\rangle$ and $\langle\hat{L}_y\rangle$). Given an eigenfunction $Y_{1,-1}$, find: (a) $\langle\hat{L}_x\rangle$, $\langle\hat{L}_y\rangle$, $\langle\hat{L}_z\rangle$; (b) $\langle\hat{L}_x^2\rangle$, $\langle\hat{L}_y^2\rangle$, $\langle\hat{L}_z^2\rangle$; (c) verify $\langle\hat{L}_x^2\rangle + \langle\hat{L}_y^2\rangle + \langle\hat{L}_z^2\rangle = \langle\hat{\mathbf{L}}^2\rangle = l(l+1)\hbar^2$.

(a) A general strategy is to express $\hat{L}_x$ in terms of $\hat{L}_+$ and $\hat{L}_-$. Since $\hat{L}_\pm = \hat{L}_x \pm i\hat{L}_y$, we have $\hat{L}_x = (\hat{L}_+ + \hat{L}_-)/2$. So $\langle\hat{L}_x\rangle = (\langle\hat{L}_+\rangle + \langle\hat{L}_-\rangle)/2$. We first compute $\hat{L}_+|Y_{1,-1}\rangle = \sqrt{1(1+1)-(-1)(-1+1)}\hbar Y_{1,0} = \sqrt{2}\hbar Y_{1,0}$, where we have used the identity eq. (4.3.21). Then $\langle\hat{L}_+\rangle = \sqrt{2}\langle Y_{1,-1}|Y_{1,0}\rangle = 0$, due to the orthogonalization between spherical harmonics. Similarly, we can find $\langle Y_{1,-1}|\hat{L}_-|Y_{1,-1}\rangle = 0$. Therefore, we find $\langle\hat{L}_x\rangle = 0$. Similarly we find $\langle\hat{L}_y\rangle = 0$. $\hat{L}_z$ is in its eigenfunction, so $\langle\hat{L}_z\rangle = m\hbar = -\hbar$.

(b) We need to express the operator $\hat{L}_x^2$ in terms of $\hat{L}_+$ and $\hat{L}_-$. We note that $\hat{L}_x^2 \neq (\hat{L}_+^2+\hat{L}_-^2+2\hat{L}_+\hat{L}_-)/4$. This is because $\hat{L}_+$ and $\hat{L}_-$ do not commute. Instead, we must multiply it out as $\hat{L}_x^2 = (\hat{L}_+^2+\hat{L}_-^2+\hat{L}_+\hat{L}_-+\hat{L}_-\hat{L}_+)/4$. Because $\hat{L}_+^2(\hat{L}_-^2)$ increases (decreases) m by 2, its expectation value must be zero with $Y_{1,-1}$. In addition, $\hat{L}_-Y_{1,-1} = 0$, so we are left with the last term $\hat{L}_-\hat{L}_+Y_{1,-1} = \sqrt{2}\hbar\hat{L}_-Y_{1,0} = \sqrt{2}\sqrt{2}\hbar^2Y_{1,-1} = 2\hbar^2Y_{1,-1}$. Finally, we have $\hat{L}_x^2Y_{1,-1} = 2\hbar^2Y_{1,-1}/4$, so $\langle\hat{L}_x^2\rangle = \frac{1}{2}\hbar^2$. Similarly,we find $\langle\hat{L}_y^2\rangle = \frac{1}{2}\hbar^2$. Because $Y_{1,-1}$ is an eigenfunction of $\hat{L}_z$, $\langle\hat{L}_z^2\rangle = \hbar^2$.

(c) Because $Y_{1,-1}$ is an eigenfunction of $\hat{\mathbf{L}}^2$, $\langle\hat{\mathbf{L}}^2\rangle = l(l+1)\hbar^2 = 2\hbar^2$. On the other hand, $\langle\hat{L}_x^2\rangle+\langle\hat{L}_y^2\rangle+\langle\hat{L}_z^2\rangle = \frac{1}{2}\hbar^2 + \frac{1}{2}\hbar^2 + \hbar^2 = 2\hbar^2$. This is verified.

Exercise 4.4.1.

18. Suppose the system is in an eigenstate $Y_{2,m}$, where m takes ±2, ±1 and 0. Find (a) $\langle\hat{L}_x\rangle$, $\langle\hat{L}_y\rangle$, $\langle\hat{L}_z\rangle$; (b) $\langle\hat{L}_x^2\rangle$, $\langle\hat{L}_y^2\rangle$, $\langle\hat{L}_z^2\rangle$; (c) verify $\langle\hat{L}_x^2\rangle + \langle\hat{L}_y^2\rangle + \langle\hat{L}_z^2\rangle = \langle\hat{\mathbf{L}}\rangle = l(l+1)\hbar^2$.

19. A system is in state $Y_{3,-2}$. (a) Find the angle between $\hat{L}_z$ and the z-axis. (b) Find $\langle\hat{L}_x^2 + \hat{L}_y^2\rangle$.

4.4.2 Matrix representations and representation theory

As stated in Section 1.3.1, wave and matrix mechanics are equivalent. The angular momentum operator can be represented through matrices. This is a general property of an operator with the condition that we need to choose a basis (a coordinate) to represent it. A convenient basis for orbital angular momentum is Y_{lm}. We take $l = 1$ as an example. m has three values $\pm1, 0$, so we have three basis functions, $Y_{1,-1}$, $Y_{1,0}$, $Y_{1,1}$, just like x-, y-, and z-axes in Cartesian coordinates. Thus, we have a 3×3 matrix. But, different from the Cartesian coordinates, once we have a basis, we can represent operators in higher dimensions such as 5×5. The simplest operators are $\hat{L}_+$ and $\hat{L}_-$. Let us first start from $\hat{L}_-$ in eq. (4.3.22). If we number $Y_{1,-1}$, $Y_{1,0}$, and $Y_{1,1}$ as our basis functions 1, 2, and 3, respectively, then $\langle1|\hat{L}_-|2\rangle \equiv \langle Y_{1,-1}|\hat{L}_-|Y_{1,0}\rangle$. The rest of the matrix elements follow the same convention. These elements are found as follows. Since $\hat{L}_-|Y_{1,1}\rangle = \sqrt{2}\hbar|Y_{1,0}\rangle$, the matrix element is $\langle Y_{1,0}|\hat{L}_-|Y_{1,1}\rangle = \sqrt{2}\hbar\langle Y_{1,0}|Y_{1,0}\rangle = \sqrt{2}\hbar$, which corresponds to $\langle2|\hat{L}_-|3\rangle$. We have

$$\hat{L}_- = \begin{array}{c} \\ \langle Y_{1,-1}| \\ \langle Y_{1,0}| \\ \langle Y_{1,1}| \end{array} \begin{array}{c} \begin{array}{ccc} |Y_{1,-1}\rangle & |Y_{1,0}\rangle & |Y_{1,1}\rangle \end{array} \\ \begin{pmatrix} 0 & \sqrt{2}\hbar & 0 \\ 0 & 0 & \sqrt{2}\hbar \\ 0 & 0 & 0 \end{pmatrix} \end{array}, \tag{4.4.1}$$

which is called the matrix representation of $\hat{L}_-$. Similarly, we can find

$$\hat{L}_+ = \begin{array}{c} \\ \langle Y_{1,-1}| \\ \langle Y_{1,0}| \\ \langle Y_{1,1}| \end{array} \begin{array}{c} \begin{array}{ccc} |Y_{1,-1}\rangle & |Y_{1,0}\rangle & |Y_{1,1}\rangle \end{array} \\ \begin{pmatrix} 0 & 0 & 0 \\ \sqrt{2}\hbar & 0 & 0 \\ 0 & \sqrt{2}\hbar & 0 \end{pmatrix} \end{array}. \tag{4.4.2}$$

Since Y_{lm} is an eigenfunction of $\hat{L}_z$, the $\hat{L}_z$ matrix is diagonal, but the $\hat{L}_x$ and $\hat{L}_y$ matrices have off-diagonal elements. We can use $\hat{L}_-$ and $\hat{L}_+$ to find the matrices for $\hat{L}_x$ and $\hat{L}_y$ through $\hat{L}_x = (\hat{L}_+ + \hat{L}_-)/2$ and $\hat{L}_y = (\hat{L}_+ - \hat{L}_-)/(2i)$ as follows:

$$\hat{L}_x = \frac{\hbar}{\sqrt{2}}\begin{pmatrix} 0 & 1 & 0 \\ 1 & 0 & 1 \\ 0 & 1 & 0 \end{pmatrix}, \quad \hat{L}_y = \frac{\hbar}{\sqrt{2}i}\begin{pmatrix} 0 & -1 & 0 \\ 1 & 0 & -1 \\ 0 & 1 & 0 \end{pmatrix}, \quad \hat{L}_z = \hbar\begin{pmatrix} -1 & 0 & 0 \\ 0 & 0 & 0 \\ 0 & 0 & 1 \end{pmatrix}. \tag{4.4.3}$$

It is important to note that the order of these elements is tied to the order of the basis functions Y_{lm}. The matrix representation succinctly offers a different way to understand the underlying properties of angular momentum.

In general, any operator $\hat{O}$ can be converted, or projected, between two representations A and B that have two different bases $\{|a_i\rangle\}, \{|b_j\rangle\}$. Every element can be written as $\langle a_i|\hat{O}|a_j\rangle = \langle a_i|\sum_k |b_k\rangle\langle b_k|\hat{O}|\sum_l |b_l\rangle\langle b_l|a_j\rangle = \sum_{kl}\langle a_i|b_k\rangle\langle b_k|\hat{O}|b_l\rangle\langle b_l|a_j\rangle$, where we have used the completeness of basis functions $\sum_k |b_k\rangle\langle b_k| = 1$ and $\langle a_i|b_k\rangle$ is one element of transformation matrix R. The beauty here is that the completeness is limited to a particular l, instead of a summation over an infinite basis. In the matrix form, the operator in two representations transforms as $\hat{O}_A = R\hat{O}_B R^{-1}$. Section 4.5 extends this idea further.

Example 4 (Quantization axis along $\hat{L}_x$). $\hat{L}_x$ is given by the matrix in eq. (4.4.3). (a) Diagonalize it to find its eigenvalues and eigenvectors. (b) Within the eigenvectors of $\hat{L}_x$, compute $\langle\hat{L}_z\rangle$.

(a) To find the eigenvalues and eigenvectors of $\hat{L}_x$, we start from

$$\hat{L}_x\begin{pmatrix} a \\ b \\ c \end{pmatrix} = \frac{\hbar}{\sqrt{2}}\begin{pmatrix} 0 & 1 & 0 \\ 1 & 0 & 1 \\ 0 & 1 & 0 \end{pmatrix}\begin{pmatrix} a \\ b \\ c \end{pmatrix} = \lambda\begin{pmatrix} a \\ b \\ c \end{pmatrix}, \tag{4.4.4}$$

where λ is the eigenvalue and $(a, b, c)^T$ is the eigenvector (T is a transpose). In order to have a nontrivial root, the determinant of the coefficients must be zero,

$$\begin{vmatrix} -\lambda & \hbar/\sqrt{2} & 0 \\ \hbar/\sqrt{2} & -\lambda & \hbar/\sqrt{2} \\ 0 & \hbar/\sqrt{2} & -\lambda \end{vmatrix} = 0,$$

which is a secular equation. Solving it yields three roots, $\lambda = \hbar, 0\hbar$, and $-\hbar$. One can see that $\hat{L}_x$ in its own eigenvectors also has the same three possible values as $\hat{L}_z$. Substituting these three eigenvalues into eq. (4.4.4) gives us three eigenvectors as $|m_x = 1\rangle = (\frac{1}{2}, \frac{1}{\sqrt{2}}, \frac{1}{2})^T$, $|m_x = 0\rangle = (\frac{1}{\sqrt{2}}, 0, \frac{1}{\sqrt{2}})^T$, and $|m_x = -1\rangle = (\frac{1}{2}, -\frac{1}{\sqrt{2}}, \frac{1}{2})^T$, where we have used $|m_x\rangle$ to denote the eigenfunction of $\hat{L}_x$. If we include the basis functions, these three eigenvectors are

$$|m_x = 1\rangle = \frac{1}{2}Y_{1,-1} + \frac{1}{\sqrt{2}}Y_{1,0} + \frac{1}{2}Y_{1,1}, \tag{4.4.5}$$

$$|m_x = 0\rangle = \frac{1}{\sqrt{2}}Y_{1,-1} + \frac{1}{\sqrt{2}}Y_{1,1}, \tag{4.4.6}$$

$$|m_x = -1\rangle = \frac{1}{2}Y_{1,-1} - \frac{1}{\sqrt{2}}Y_{1,0} + \frac{1}{2}Y_{1,1}, \tag{4.4.7}$$

where we see the eigenfunction of $\hat{L}_x$ is a mixture of $Y_{1,0}$ and $Y_{1,\pm 1}$. R matrix is just the coefficient matrix from the previous three equations,

$$R = \begin{pmatrix} \frac{1}{2} & \frac{1}{\sqrt{2}} & \frac{1}{2} \\ \frac{1}{\sqrt{2}} & 0 & \frac{1}{\sqrt{2}} \\ \frac{1}{2} & -\frac{1}{\sqrt{2}} & \frac{1}{2} \end{pmatrix}.$$

(b) We have two ways to compute the expectation value of $\hat{L}_z$. We take $|m_x = 1\rangle$ as an example. The simplest one is to use the $\hat{L}_z$ matrix given in eq. (4.4.3). The expectation value of $\hat{L}_z$ is

$$\begin{pmatrix} \frac{1}{2} & \frac{1}{\sqrt{2}} & \frac{1}{2} \end{pmatrix} \hbar \begin{pmatrix} -1 & 0 & 0 \\ 0 & 0 & 0 \\ 0 & 0 & 1 \end{pmatrix} \begin{pmatrix} \frac{1}{2} \\ \frac{1}{\sqrt{2}} \\ \frac{1}{2} \end{pmatrix} = \hbar \begin{pmatrix} -\frac{1}{2} & 0 & -\frac{1}{2} \end{pmatrix} \begin{pmatrix} \frac{1}{2} \\ \frac{1}{\sqrt{2}} \\ \frac{1}{2} \end{pmatrix} = 0.$$

Another way is to compute $\hat{L}_z|m_x = 1\rangle$ from eq. (4.4.5) to get $\frac{-\hbar}{2}Y_{1,-1} + \frac{0\hbar}{\sqrt{2}}Y_{1,0} + \frac{\hbar}{2}Y_{1,1}$. Then, multiply $\langle m_x|$ from the left and integrate to get $\frac{-\hbar}{2} + \frac{\hbar}{2} = 0$.

The matrix representation is not limited to orbital angular momentum. We can start from the position operator applied to the harmonic oscillator's eigenstates (eq. (3.3.8))

$$\hat{x}|n\rangle = \sqrt{\frac{\hbar}{2m\omega}}(\hat{a}^\dagger + \hat{a})|n\rangle = \sqrt{\frac{\hbar}{2m\omega}}\left(\sqrt{n+1}|n+1\rangle + \sqrt{n}|n-1\rangle\right).$$

The position operator becomes the matrix,

$$\hat{x} = \sqrt{\frac{\hbar}{2m\omega}} \begin{pmatrix} 0 & \sqrt{1} & 0 & 0 & \cdots \\ \sqrt{1} & 0 & \sqrt{2} & 0 & \cdots \\ 0 & \sqrt{2} & 0 & \sqrt{3} & \cdots \\ 0 & 0 & \sqrt{3} & 0 & \cdots \\ \cdots & \cdots & \cdots & \cdots & \cdots \end{pmatrix}.$$

Exercise 4.4.2.

20. Find all the matrix elements of eq. (4.4.2).
21. Using the matrices from eq. (4.4.3), prove $[\hat{L}_x, \hat{L}_y] = i\hbar\hat{L}_z$. This shows that it is equivalent if we use the matrices for angular momentum operators.
22. Use R to convert $\hat{L}_z$ matrix in the $\hat{L}_z$ basis to the $\hat{L}_x$ basis.

4.4.3 Chemical meanings: atomic orbitals

We have spent a significant amount of time to introduce orbital angular momenta. It is time to get some insights into chemistry through spherical harmonics Y_{lm}.

Table 4.2: Some Legendre polynomials.

l	$P_l(x)$	$P_l(\cos\theta)$
0	1	1
1	x	$\cos\theta$
2	$\frac{1}{2}(3x^2-1)$	$\frac{1}{2}(3\cos^2\theta-1)$
3	$\frac{1}{2}(5x^3-3x)$	$\frac{1}{2}(5\cos^3\theta-3\cos\theta)$

Spherical harmonics Y_{lm} consist of the products of Legendre polynomials with polar eigenfunctions,

$$Y_{lm}(\theta,\phi)=N_{lm}P_l^m(\cos\theta)\eta_m(\phi),$$

where $\eta_m(\phi)=e^{im\phi}$. The normalization constant $N_{lm}=(-1)^m\sqrt{\frac{(2l+1)}{2\pi}\frac{(l-m)!}{(l+m)!}}$ is found by requiring

$$\int Y_{lm}^*(\theta,\phi)Y_{lm}(\theta,\phi)d\Omega=1,$$

where $d\Omega$ is the solid angle $\sin\theta d\theta d\phi$. The complicated form of $\sqrt{\frac{(2l+1)}{2\pi}\frac{(l-m)!}{(l+m)!}}$ will be explained later. Table 4.2 shows a list of Legendre polynomials from $l=0$ to 3. These mathematical polynomials take on new meanings once we replace x by $\cos\theta$: They represent the polar angular (θ) distribution of atomic orbitals in chemistry. The right-hand column is a function of $\cos\theta$, which contains the angular information of spherical harmonics Y_{lm}. However, it is not possible to display it in a two-dimensional figure because it only contains a single variable $\cos\theta$.

We find a good method by converting it to Cartesian coordinates through

$$x=|P_l(\cos\theta)|\cos\theta,$$
$$y=|P_l(\cos\theta)|\sin\theta,$$

where the "radius" is $|P_l(\cos\theta)|$; $\cos\theta$ and $\sin\theta$ convert the radius to x- and y-values. The reader can plot them directly with an `Excel` spreadsheet and is strongly encouraged to try it out. $l=0$ corresponds to a circle. $l=1$ corresponds to a dumbbell shape, and so on. Figure 4.4 shows six different orbitals. In chemistry, these are atomic orbitals, s, p, d up to m. For this reason, l is called the orbital angular momentum quantum number. The p-orbital in the figure is a p_x-orbital since its main lobes are along the x-axis. If we change $P_l(\cos\theta)$ to $P_l[\cos(\theta+\frac{\pi}{2})]$, we have a p_y-orbital. Remember the actual orbitals are three-dimensional, so we have to imagine a rotation of the orbital by 360° along its axis. The p_z-orbital can also be generated. To distinguish orbitals with the same l, m_l is introduced for $-l, -l+1, \ldots, +l$.[7] That is why we need the associated P_l^m instead of P_l.

7 We use m_l to highlight the fact that m pertains to l, though there is no confusion in this chapter.

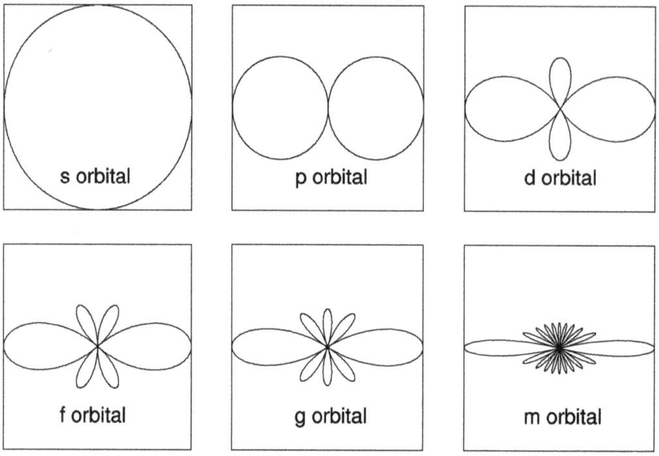

Figure 4.4: Atomic orbitals. (Top row) *s*-, *p*-, and *d*-orbitals, which correspond to $l = 0, 1$, and 2. The *s*-orbital is spherical. The *p*-orbital has a dumbbell shape. (Bottom row) *f*-, *g*-, and *m*-orbitals, with $l = 3, 4, 10$. The orbital splits as we increase the orbital angular momentum *l*. These orbitals are computed from code `legendre.f` in the Appendix listing 11.9.

The reason why we call P_l^m the associated Legendre polynomial is because P_l^m is found from P_l by taking additional derivatives.

We start from the Legendre polynomial definition $P_l(x)$ (Rodrigues' formula),

$$P_l(x) = \frac{1}{2^l l!}\frac{d^l}{dx^l}\left[(x^2-1)^l\right], \tag{4.4.8}$$

where the variable is x, but in eq. (4.3.15) the variable is $\sin\theta$. To match them, we must set $x = \cos\theta$, so $x^2 - 1 = -\sin^2\theta$, which explains why we must write $P_l(x)$ as $P_l(\cos\theta)$ in physics. The highest order in $(x^2-1)^l$ is x^{2l}. If we take l times of derivatives, we still have x^l. So we can keep on going for another l times of derivatives. This second group of derivatives is denoted by m and is defined as the associated Legendre polynomials

$$P_l^m = (-1)^m(1-x^2)^{\frac{m}{2}}\frac{d^m}{dx^m}\left[P_l(x)\right], \tag{4.4.9}$$

where the front factor is introduced to merely cancel out unwanted terms. If we substitute eq. (4.4.8) into (4.4.9), we can combine two derivatives into a single one,

$$P_l^m(x) = \frac{1}{2^l l!}(1-x^2)^{\frac{m}{2}}\frac{d^{l+m}}{dx^{l+m}}(x^2-1)^l. \tag{4.4.10}$$

Mathematically, if $m > l$, P_l^m is zero because the highest order is x^{2l}, so only $m \le l$ remains. Physically, the magnetic orbital quantum number m is bounded by the orbital quantum number l. If $m = -l$, we have no derivative, just $(1-x^2)^{-l/2}(x^2-1)^l/(2^l l!) =$

$(1-x^2)^{l/2}(-1)^l/(2^l l!) = (-1)^l \sin^l\theta/(2^l l!)$ (eq. (4.3.15)), which shows how the front factor cancels out unwanted terms. This expression matches eq. (4.3.15). Since l is an integer, $(-1)^l$ only takes two values, ±1. If $l = 1$, then P_1^0 corresponds to the p_z-orbital, P_1^1 corresponds to the p_x-orbital, and P_1^{-1} corresponds to the p_y-orbital. If there is a symmetry in the xy-plane, then p_x and p_y can take either P_1^1 or P_1^{-1}.

Exercise 4.4.3.

23. Suppose $l = 1$. Starting from eq. (4.4.10), find P_1^1, P_1^{-1}, and P_1^0.
24. Use the code `legendre.f` in the Appendix 11.9 to generate orbitals in Fig. 4.4.
25. Another way to plot atomic orbitals is to use `gnuplot`, which works on `PC/Windows` and `Linux` systems. Use the scripts that Moore [23] provided to plot the d_{yz}-orbital.

4.5 Central field potentials: symmetry and conservation law

A quantum mechanical treatment provides unprecedented insight into the electronic structures of an atom, but very few problems can be solved analytically. For this reason, one often employs the symmetry of atoms, molecules, and solids. The central field potential is a starting point. This section starts with a brief introduction of symmetry operation and its associated law of conservation and then uses the common eigenfunction properties to simplify and solve the hydrogen problem.

4.5.1 Symmetry and conservation

Symmetry in QM plays a bigger role than in CM and underlines many laws of conservation. Chapter 2 introduces the Schrödinger equation through the time-invariance symmetry. Every symmetry introduces a hidden degeneracy. An extensive discussion can be found elsewhere [24, 25, 26, 27]. Here, we introduce the rotational symmetry. The central field potential $V(\mathbf{r})$ only depends on the radius r, not the direction of $\mathbf{r}$, so we can write the potential as $V(r)$. If we rotate the system by any angle, $V(r)$ remains the same. Because the Hamiltonian is the sum of the kinetic energy and potential energy $\hat{H} = \hat{T} + V$ and $\hat{T}$ is always rotationally symmetric, the Hamiltonian is said to have rotational symmetry.

4.5.2 Rotation of coordinates

It is helpful to start with the rotation of a coordinate. Supposing we have the initial coordinate $\mathbf{r}$, a rotation, defined by $\hat{R}$, will send $\mathbf{r}$ to the new coordinate $\mathbf{r}'$,

$$\mathbf{r}' = \hat{R}\mathbf{r}, \quad r_i' = \sum_j \hat{R}_{ij} r_j \quad \text{in components,} \tag{4.5.1}$$

where the general expressions of $\hat{R}$ are given in Section 11.3.1. If one requires the length of $\mathbf{r}'$ is equal to that of $\mathbf{r}$, one can show that the inverse of $\hat{R}$, $\hat{R}^{-1}$, must satisfy $\hat{R}^{-1}\hat{R} = I$, where I is the identity matrix and $\hat{R}$ is called the unitary matrix and satisfies $\hat{R}^{-1} = \hat{R}^{\dagger}$.

Exercise 4.5.2.
26. If the length of $\mathbf{r}'$ is equal to that of $\mathbf{r}$, prove $\hat{R}^{-1}\hat{R} = I$.

4.5.3 Rotation of a wavefunction in the function space

Rotation of a wavefunction $\psi(\mathbf{r})$ means that (i) we change the variable $\mathbf{r}$ to $\mathbf{r}'$, and (ii) require the rotated wavefunction $\psi'(\mathbf{r}')$ be equal to the initial $\psi(\mathbf{r})$, i. e., $\psi(\mathbf{r}) = \psi'(\mathbf{r}')$. (ii) is purely due to the physical reason: Two wavefunctions must contain the same physics in space. Similar to eq. (4.5.1), we define the rotation of a wavefunction as

$$\psi'(\mathbf{r}) \equiv \hat{R}\psi(\mathbf{r}), \tag{4.5.2}$$

where $\mathbf{r}$ is fixed and $\psi'(\mathbf{r})$ is the rotated wavefunction to be found. We start from $\psi'(\mathbf{r}')$ and use eq. (4.5.1) to rewrite it as $\psi'(\mathbf{r}') = \psi'(\hat{R}\mathbf{r})$. Then, using (ii), we find

$$\psi'(\mathbf{r}') = \psi'(\hat{R}\mathbf{r}) = \psi(\mathbf{r}). \tag{4.5.3}$$

Next, we change $\mathbf{r} \to \hat{R}^{-1}\mathbf{r}$ in the second equation in eq. (4.5.3) to obtain

$$\psi'(\hat{R}\hat{R}^{-1}\mathbf{r}) = \psi(\hat{R}^{-1}\mathbf{r}) \to \psi'(\mathbf{r}) = \psi(\hat{R}^{-1}\mathbf{r}), \tag{4.5.4}$$

where we have used $\hat{R}^{-1}\hat{R} = I$. Now, comparing eqs. (4.5.2) and (4.5.4), we obtain the final rotated wavefunction $\psi'(\mathbf{r})$,

$$\psi'(\mathbf{r}) = \hat{R}\psi(\mathbf{r}) = \psi(\hat{R}^{-1}\mathbf{r}) \to \psi'(\mathbf{r}) = \psi(\hat{R}^{-1}\mathbf{r}), \tag{4.5.5}$$

which shows that the rotated wavefunction is the same as ψ, except the coordinate $\mathbf{r}$ rotates in the opposite direction. If we want to rotate the wavefunction by $+30°$ along the z axis, all we need to do is to rotate the coordinate $\mathbf{r}$ in $\psi(\mathbf{r})$ by $-30°$ along the z axis.

4.5.4 Rotation of an operator

Rotation of an operator is different. Consider the time-dependent Schrödinger equations for ψ and ψ', $i\hbar\frac{\partial\psi}{\partial t} = \hat{H}\psi$, $i\hbar\frac{\partial\psi'}{\partial t} = \hat{H}'\psi'$, where $\hat{H}'$ is the Hamiltonian after rotation $\hat{R}$. We plug $\psi' = \hat{R}\psi$ into the second term to get

$$i\hbar\frac{\partial(\hat{R}\psi)}{\partial t} = \hat{H}'\hat{R}\psi \longrightarrow i\hbar\hat{R}\frac{\partial\psi}{\partial t} = \hat{H}'\hat{R}\psi.$$

Multiplying both sides by $\hat{R}^{-1}$, we get

$$i\hbar\hat{R}^{-1}\hat{R}\frac{\partial\psi}{\partial t} = \hat{R}^{-1}\hat{H}'\hat{R}\psi, \longrightarrow i\hbar\frac{\partial\psi}{\partial t} = \hat{R}^{-1}\hat{H}'\hat{R}\psi.$$

If we compare it with $i\hbar\frac{\partial\psi}{\partial t} = \hat{H}\psi$, we have

$$\hat{H} = \hat{R}^{-1}\hat{H}'\hat{R} \rightarrow \hat{H}' = \hat{R}\hat{H}\hat{R}^{-1}.$$

This reveals an important relation for a general operator $\hat{O}$. If we want to rotate an operator $\hat{O}$, the operator transforms to $\hat{R}\hat{O}\hat{R}^{-1}$. This is an extension of representation theory given in Section 4.4.2, where $\hat{O}_A = R\hat{O}_B R^{-1}$. This is in contrast to a wavefunction, $\hat{R}\psi(\mathbf{r})$, where the rotation $\hat{R}$ only appears once. But, for an operator, $\hat{R}$ appears twice.

4.5.5 Conservation of angular momentum

Next, we consider what happens if $\hat{H}$ is invariant under rotation $\hat{R}$, i. e., $\hat{H}' = \hat{H}$:

$$\hat{H} = \hat{R}\hat{H}\hat{R}^{-1} \rightarrow \hat{H}\hat{R} = \hat{R}\hat{H} \rightarrow [\hat{H}, \hat{R}] = 0.$$

This means $\hat{R}$ commutes with $\hat{H}$, so $\hat{R}$ and $\hat{H}$ share some common eigenvectors.

Consider that $\hat{R}_z^\varepsilon$ represents a rotation by a small angle ε along the z-axis, so the wavefunction changes as $\psi(\phi) \rightarrow \psi(\phi - \varepsilon)$, which can be Taylor expanded to

$$\begin{aligned}
\hat{R}_z^\varepsilon\psi(\phi) = \psi(\phi-\varepsilon) &= \psi(\phi) - \varepsilon\frac{\partial}{\partial\phi}\psi(\phi) + \frac{\varepsilon^2}{2!}\frac{\partial^2}{\partial\phi^2}\psi(\phi) - \cdots \\
&= \psi(\phi) - \frac{i}{\hbar}\varepsilon\hat{L}_z\psi(\phi) + \frac{1}{2!}\left(\frac{-i\varepsilon\hat{L}_z}{\hbar}\right)^2\psi(\phi) - \cdots \\
&= \left[1 - \frac{i}{\hbar}\varepsilon\hat{L}_z + \frac{1}{2!}\left(-\frac{i\varepsilon\hat{L}_z}{\hbar}\right)^2 - \cdots\right]\psi(\phi) = e^{-i\varepsilon\hat{L}_z/\hbar}\psi(\phi),
\end{aligned}$$

where we have used $\hat{L}_z = -i\hbar\frac{\partial}{\partial\phi}$. This reveals that $\hat{R}_z^\varepsilon = e^{-\frac{i\varepsilon\hat{L}_z}{\hbar}}$ represents a rotation operator, which is realized by $\hat{L}_z$. If $[\hat{R}_z^\varepsilon, \hat{H}] = 0$, then $[\hat{L}_z, \hat{H}] = 0$. Although we have previously encountered this same commutation, the difference here is that we derive it through the rotational symmetry. The same l has $2l + 1$ different m, $m = -l, \ldots, +l$, degeneracy as previously discussed.

In summary, the rotational symmetry of a system underlines a conservation law. Mathematically, it is realized through the commutation of an operator $\hat{O}$ with $\hat{H}$. According to the Heisenberg equation of motion (Section 3.5), for $\hat{L}_z$, we have $\frac{d\hat{L}_z}{dt} = \frac{1}{i\hbar}[\hat{L}_z, \hat{H}] = 0$, which means that $\hat{L}_z$ is constant and is a conserved quantity, a key insight that underlines our solution for the hydrogen atom in the following sections.

Exercise 4.5.5.
27. Starting from eq. (4.5.2), if we require the probability density $\rho(\mathbf{r})$ to be the same for $\psi'(\mathbf{r})$ and $\psi(\mathbf{r})$, show $\hat{R}^{-1}\hat{R} = I$, where I is the identity matrix. Hint: The conjugation of eq. (4.5.2) is $\psi'^* = \psi^*(\mathbf{r})\hat{R}^\dagger$.
28. $\hat{T}$, a translation operator, transforms $f(x)$ to $f(x-a)$, i. e., $\hat{T}_a f(x) = f(x-a)$, where a is a finite distance along the x-axis. Find the expression for the operator $\hat{T}_a$.

4.6 Hydrogen atom

Our goal is to find the solution to the Schrödinger equation of the hydrogen atom, that is, the eigenfunction of the Hamiltonian operator $\hat{H}$, but this is a difficult task. The rotational symmetry of the hydrogen atom and its subsequent commutations, $[\hat{\mathbf{L}}^2, \hat{H}] = 0$ and $[\hat{L}_z, \hat{H}] = 0$, allow us to get the angular part of the eigenfunction of the hydrogen Hamiltonian, Y_{lm}, without even solving the Schrödinger equation. We will repeatedly use Y_{lm}. The central potential simplifies the equation further to a radial equation.

4.6.1 Spherical symmetry

The Hamiltonian consists of the kinetic energy and the potential energy due to the Coulombic attraction from the proton,

$$\hat{H} = \hat{T} + V = -\frac{\hbar^2\nabla^2}{2m_e} + \frac{-e^2}{4\pi\epsilon_0 r}.$$

Since the analytical solution in Cartesian coordinates is difficult to find, we convert it to spherical coordinates.

First, we change the Laplacian to spherical coordinates,

$$\begin{aligned}\nabla^2 &= \frac{1}{r^2}\frac{\partial}{\partial r}\left(r^2\frac{\partial}{\partial r}\right) + \frac{1}{r^2\sin\theta}\frac{\partial}{\partial\theta}\left(\sin\theta\frac{\partial}{\partial\theta}\right) + \frac{1}{r^2\sin^2\theta}\frac{\partial^2}{\partial\phi^2} \\ &= \frac{1}{r^2}\frac{\partial}{\partial r}\left(r^2\frac{\partial}{\partial r}\right) + \frac{1}{r^2}\left[\frac{1}{\sin\theta}\frac{1}{\partial\theta}\left(\sin\theta\frac{\partial}{\partial\theta}\right) + \frac{1}{\sin^2\theta}\frac{\partial^2}{\partial\phi^2}\right].\end{aligned} \tag{4.6.1}$$

Because

$$\hat{\mathbf{L}}^2 = -\hbar^2\left[\frac{1}{\sin\theta}\frac{\partial}{\partial\theta}\left(\sin\theta\frac{\partial}{\partial\theta}\right) + \frac{1}{\sin^2\theta}\frac{\partial^2}{\partial\phi^2}\right],$$

the second term in the brackets in eq. (4.6.1) is just $-\hat{\mathbf{L}}^2/\hbar^2$, and the Laplacian operator is

$$\nabla^2 = \frac{1}{r^2}\frac{\partial}{\partial r}\left(r^2\frac{\partial}{\partial r}\right) + \frac{-\hat{\mathbf{L}}^2/\hbar^2}{r^2} \to \hat{T} = -\frac{\hbar^2}{2m_e}\left[\frac{1}{r^2}\frac{\partial}{\partial r}\left(r^2\frac{\partial}{\partial r}\right) - \frac{\hat{\mathbf{L}}^2}{\hbar^2 r^2}\right]. \tag{4.6.2}$$

Since $\hat{\mathbf{L}}^2$ and $\hat{L}_z$ only act upon the angular part, not on the radial part of the wavefunction, the following commutation relationships are valid: $[\nabla^2, \hat{\mathbf{L}}^2] = 0$, $[\nabla^2, \hat{L}_z] = 0$, $[\hat{T}, \hat{\mathbf{L}}^2] = 0$, and $[\hat{T}, \hat{L}_z] = 0$. Since the potential does not depend on the angular variables θ and ϕ, $[V, \hat{\mathbf{L}}^2] = 0$, $[V, \hat{L}_z] = 0$. Therefore, we have $[\hat{H}, \hat{\mathbf{L}}^2] = 0$ and $[\hat{H}, \hat{L}_z] = 0$, so the eigenfunction Y_{lm} of $\hat{\mathbf{L}}^2$ and $\hat{L}_z$ is also the eigenfunction of $\hat{H}$.

4.6.2 Separation of variables and the radial Schrödinger equation

Because our potential $V(r)$ does not depend on either θ or ϕ, this allows us to separate the radial part R_{nl} from the angular part Y_{lm} of the wavefunction, i. e., $\psi_{nlm}(r, \theta, \phi) = R_{nl}(r)Y_{lm}(\theta, \phi)$, where ψ has three quantum numbers, n, l, m, the main quantum number, orbital, and magnetic orbital angular momentum quantum numbers, respectively. Because we already know the spherical harmonics $Y_{lm}(\theta, \phi)$, our job is to find R_{nl}. We substitute ψ into the Schrödinger equation $\hat{H}\psi = E\psi$ and obtain $(\hat{T} + V)R_{nl}(r)Y_{lm}(\theta, \phi) = ER_{nl}(r)Y_{lm}(\theta, \phi)$. According to eq. (4.6.2), we have

$$\begin{aligned}\hat{T}R_{nl}(r)Y_{lm}(\theta,\phi) &= -\frac{\hbar^2}{2m_e}\left[\frac{1}{r^2}\frac{\partial}{\partial r}\left(r^2\frac{\partial}{\partial r}\right)R_{nl}Y_{lm} - \frac{l(l+1)\hbar^2}{\hbar^2 r^2}R_{nl}Y_{lm}\right] \\ &= -\frac{\hbar^2 Y_{lm}}{2m_e}\left[\frac{1}{r^2}\frac{\partial}{\partial r}\left(r^2\frac{\partial}{\partial r}\right)R_{nl} - \frac{l(l+1)}{r^2}R_{nl}\right],\end{aligned}$$

where we have used the fact that the first term in the kinetic energy operator does not depend on angles θ and ϕ. We can cancel Y_{lm} on both sides to get

$$\left\{-\frac{\hbar^2}{2m_e}\left[\frac{1}{r^2}\frac{\partial}{\partial r}\left(r^2\frac{\partial}{\partial r}\right) - \frac{l(l+1)}{r^2}\right] - \frac{e^2}{4\pi\epsilon_0}\frac{1}{r}\right\}R_{nl}(r) = E_{nl}R_{nl}(r). \tag{4.6.3}$$

In the following, we suppress the subscript nl in R. The convoluted derivative in the first term can be simplified further if we let $R = \frac{u}{r}$, where u is a function of r. Taking the derivative of R with respect to r leads to $\frac{dR}{dr} = -\frac{u}{r^2} + \frac{1}{r}\frac{du}{dr}$. We multiply r^2 on both sides to get $r^2\frac{dR}{dr} = r\frac{du}{dr} - u$ and take another derivative with respect to r to obtain

$$\frac{d}{dr}\left(r^2\frac{dR}{dr}\right) = \frac{du}{dr} + r\frac{d^2u}{dr^2} - \frac{du}{dr} = r\frac{d^2u}{dr^2}.$$

Substituting the above expression into eq. (4.6.3), we find

$$-\frac{\hbar^2}{2m_e}\left[\frac{1}{r}\frac{d^2u}{dr^2} - \frac{l(l+1)R}{r^2}\right] + VR = ER.$$

Replacing the remaining R in the above equation by u yields

Table 4.3: Radial wavefunctions $R_{nl}(r)$ of the hydrogen atom. Only the radial wavefunctions can distinguish different n. All have a factor $a_0^{-\frac{3}{2}}$. The key variable is $r/(na_0)$, where a_0 is the Bohr radius. As n increases, the wavefunction decays more slowly with r.

$R_{10}(r) = 2a_0^{-\frac{3}{2}} e^{-\frac{r}{a_0}}$	(1s)
$R_{20}(r) = \frac{1}{\sqrt{2}} a_0^{-\frac{3}{2}} (1 - \frac{r}{2a_0}) e^{-\frac{r}{2a_0}}$	(2s)
$R_{21}(r) = \frac{1}{\sqrt{6}} a_0^{-\frac{3}{2}} \frac{r}{2a_0} e^{-\frac{r}{2a_0}}$	(2p)
$R_{30}(r) = \frac{2}{3\sqrt{3}} a_0^{-\frac{3}{2}} (1 - 2\frac{r}{3a_0} + \frac{2}{3}(\frac{r}{3a_0})^2) e^{-\frac{r}{3a_0}}$	(3s)
$R_{31}(r) = \frac{8}{9\sqrt{6}} a_0^{-\frac{3}{2}} \frac{r}{3a_0} (1 - \frac{1}{2}\frac{r}{3a_0}) e^{-\frac{r}{3a_0}}$	(3p)
$R_{32}(r) = \frac{4}{9\sqrt{30}} a_0^{-\frac{3}{2}} (\frac{r}{3a_0})^2 e^{-\frac{r}{3a_0}}$	(3d)

$$-\frac{\hbar^2}{2m_e}\left[\frac{d^2u}{dr^2} - \frac{l(l+1)u}{r^2}\right] + Vu = Eu,$$
$$-\frac{\hbar^2}{2m_e}\frac{d^2u}{dr^2} + \left[\frac{l(l+1)\hbar^2}{2m_e r^2} - \frac{e^2}{4\pi\epsilon_0 r}\right]u = Eu, \quad (4.6.4)$$

where in the last step we have replaced V by its Coulomb potential energy $-e^2/4\pi\epsilon_0 r$. Equation (4.6.4) is an eigenequation with a boundary condition $u \to 0$ as $r \to \infty$.

For the same l, there are infinite solutions, so we assign two quantum numbers n and l to u and E as u_{nl} and E_{nl}, where n runs from 1 to ∞ and l from 0 to ∞. R_{nl} is given as

$$R_{nl}(r) = N_{nl} e^{-\frac{r}{na_0}} \left(\frac{2r}{na_0}\right)^l L_{n-l-1}^{2l+1}\left(\frac{2r}{na_0}\right), \quad (4.6.5)$$

where N_{nl} is a normalization constant, L_{n-l-1}^{2L+1} is an associated Laguerre polynomial, and a_0 is the Bohr radius. Table 4.3 lists a few radial wavefunctions. The 1s wavefunction is $R_{10}(r) = 2a_0^{-\frac{3}{2}} e^{-\frac{r}{a_0}}$. The entire wavefunction is

$$\psi_{nlm}(r, \theta, \phi) = R_{nl}(r) Y_{lm}(\theta, \phi), \quad (4.6.6)$$

and the eigenenergies are

$$E_n = -\frac{1}{2}\frac{e^4 m_e}{(4\pi\epsilon_0)^2\hbar^2 n^2} = -\frac{1}{2}\frac{e^2}{4\pi\epsilon_0 a_0 n^2} = -\frac{1}{2}\frac{e^2}{4\pi\epsilon_0 r_n^2} = -\frac{13.6\text{ eV}}{n^2}. \quad (4.6.7)$$

Analytically solving eq. (4.6.4) is tedious, has little value for latter sections, and also hides the true meaning of n.

Exercise 4.6.2.
29. Show the ground-state wavefunction $R_{10}(r)$ is normalized, $\int_0^\infty |R_{10}(r)|^2 r^2 dr = 1$.
30. The radial probability $p(r)$ is defined as $r^2|R_{10}(r)|^2$. (a) Show the maximum probability is at $r = a_0$, and (b) find the probability.
31. Supposing an electron is in the ground state of the hydrogen atom, find: (a) $\langle \hat{\mathbf{r}} \rangle$; (b) $\langle \hat{\mathbf{r}}^2 \rangle$; (c) $\langle \hat{x} \rangle$; and (d) $\langle \hat{x}^2 \rangle$. Here, $\hat{x}$ is the position operator, not a unit vector.

4.7 Numerical solution of the hydrogen atom: hydrogen code

This section employs a numerical method that translates a differential equation into a matrix, i. e., matrix mechanics in Section 1.3.1. After diagonalizing the Hamiltonian matrix, one has both eigenvalues and eigenstates. Our hydrogen code is attached in an appendix to this book. The results are highly accurate and also contain continuum states with positive eigenenergies. The matrix mechanics is equivalent to the wave mechanics, but it is much more powerful. The only requirement is that one has a way to diagonalize a matrix, but in the 1980s Jack Dongarra and associates developed LAPACK (linear algebra package). We highly recommend this to the beginner, so he/she can appreciate insights into eigenstates.

4.7.1 Atomic units

In CM, the SI unit is the norm, but in QM it is too small to handle numerically. For instance, 1 eV = 1.602×10^{-19} J. Therefore, nearly all numerical calculations use atomic units, or a. u.. In the literature, the energy has two atomic units as briefly mentioned in Section 4.4.1: One is the Hartree (Ha), and the other is the Rydberg (Ry). 1 Ry = 13.6 eV and 1 Ha = 2 Ry = 2 × 13.6 eV = 27.2 eV. The angular momentum's a. u. is $\hbar$. The electron mass is defined as the atomic unit of mass, i. e., 1 $m_e = 1$, and its charge squared as $e^2 = 1$. The Bohr radius $a_0 = 0.529$ Å is the length unit. In addition, we set $4\pi\epsilon_0 = 1$. Doing so converts the eigenequation from

$$-\frac{\hbar^2}{2m_e}\frac{d^2u_{nl}}{dr^2} + \left[\frac{\hbar^2 l(l+1)}{2m_e r^2} - \frac{e^2}{4\pi\epsilon_0 r}\right]u_{nl} = E_{nl}u_{nl}$$

to

$$-\frac{1}{2}\frac{d^2u_{nl}}{dr^2} + \left[\frac{l(l+1)}{2r^2} - \frac{1}{r}\right]u_{nl} = E_{nl}u_{nl}.$$

This can be simplified to

$$-\frac{d^2u_{nl}}{dr^2} + \left[\frac{l(l+1)}{r^2} - \frac{2}{r}\right]u_{nl} = 2E_{nl}u_{nl}, \tag{4.7.1}$$

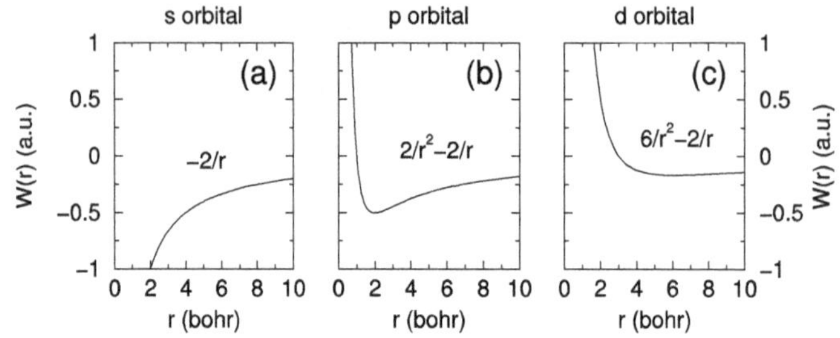

Figure 4.5: Effective potentials $W(r)$ for (a) s-, (b) p-, and (c) d-orbitals. The same type of orbitals has the same potential. For instance, all the d-orbitals have the same potential.

where the terms before u_{nl} defines the actual Hamiltonian operator in our code as

$$\hat{H} \equiv -\frac{d^2}{dr^2} + \left[\frac{l(l+1)}{r^2} - \frac{2}{r}\right] = -\frac{d^2}{dr^2} + W(r). \tag{4.7.2}$$

Here, $W(r)$ acts as an effective orbital-dependent potential[8] for our eigenequation for u_{nl}, $u_{nl}(r) = rR_{nl}(r)$. Figure 4.5(a) shows $W(r)$ for all the s states. Because $l = 0$, it only has $-2/r$, so as $r \to 0$, $W(r) \to -\infty$. This represents a strong attraction for the electron, where the electron stays closer to the proton. As we increase l to 1, this attraction becomes weaker (see Fig. 4.5(b)) because of the increase in the centrifugal potential $l(l+1)/r^2$. At $l = 2$, i. e., the d-orbital, the potential becomes much shallower (Fig. 4.5(c)), and the electron tends to become more diffusive.

4.7.2 Discretization of the eigenequation

A computer can often handle a function at discrete points, $r_1, r_2, \ldots, r_N$. These points are called mesh grid points. Our u is defined at these points, $u(r_1), u(r_2), \ldots, u(r_N)$, where N is the number of points. For a straightforward calculation, we use a uniform mesh, so the step size Δr is the same. This means that $r_1 = r_0 + \Delta r$, $r_2 = r_0 + 2\Delta r, \ldots,$ where r_0 is the first radius and can be chosen to be 0.

Now, consider $u(r + \Delta r)$ and $u(r - \Delta r)$. We can expand them around r as

$$u(r + \Delta r) = u(r) + \frac{du}{dr}\Delta r + \frac{1}{2!}\frac{d^2u}{dr^2}(\Delta r)^2 + \frac{1}{3!}\frac{d^3u}{dr^3}(\Delta r)^3 + \cdots + O((\Delta r)^n),$$
$$u(r - \Delta r) = u(r) - \frac{du}{dr}\Delta r + \frac{1}{2!}\frac{d^2u}{dr^2}(\Delta r)^2 - \frac{1}{3!}\frac{d^3u}{dr^3}(\Delta r)^3 + \cdots + O((\Delta r)^n).$$

We keep all the terms up to second order and sum both sides to get

8 The pseudopotential calculation in the actual research is similar to this potential.

$$\frac{d^2u(r)}{dr^2} = \frac{u(r+\Delta r) + u(r-\Delta r) - 2u(r)}{(\Delta r)^2}.$$

To see why $\frac{d^2u(r)}{dr^2}$ is a matrix, we introduce the mesh index k for r, $k+1$ for $r+\Delta r$, and $k-1$ for $r-\Delta r$, and the equation becomes

$$\frac{d^2u(k)}{dr^2} = \frac{u(k+1) + u(k-1) - 2u(k)}{(\Delta r)^2}. \tag{4.7.3}$$

For a fixed index k on the left, $\frac{d^2u(r)}{dr^2}$ has three nonzero elements. The coefficient of the last term $u(k)$ on the right contributes a diagonal element $-2/(\Delta r)^2$ to the Hamiltonian matrix $H(k,k)$. Two coefficients of $u(k-1)$ and $u(k+1)$, i. e., $1/(\Delta r)^2$, are just two off-diagonal elements $H(k,k-1)$ and $H(k,k+1)$. Note that $u(k)$ is an element of the eigenvector to be found and should not be considered as a matrix element.

Therefore, $-\frac{d^2}{dr^2}$ contributes three nonzero elements. $W(r)$ in eq. (4.7.2) contributes a diagonal term, $\frac{l(l+1)}{r^2} - \frac{2}{r}$.

$$H(k,k) = \frac{2}{(\Delta r)^2} + \frac{l(l+1)}{r^2} - \frac{2}{r}, \quad H(k,k\pm 1) = -\frac{1}{(\Delta r)^2}.$$

The left side of eq. (4.7.1) becomes a product of the matrix with a vector $\{u_k\}$ and the right side eq. (4.7.1) is $2E_{nl}$ multiplied by a vector,

$$\begin{pmatrix} \cdots & \cdots & \cdots & \cdots & \cdots \\ \cdots & H(k,k-1) & H(k,k) & H(k,k+1) & \cdots \\ \cdots & \cdots & \cdots & \cdots & \cdots \end{pmatrix} \begin{pmatrix} \vdots \\ u(k-1) \\ u(k) \\ u(k+1) \\ \vdots \end{pmatrix} = 2E_{nl} \begin{pmatrix} \vdots \\ u(k-1) \\ u(k) \\ u(k+1) \\ \vdots \end{pmatrix}.$$

As k runs from 1 to N, we can complete a full matrix. This matrix is programmed into our code `hydrogen.f` in the Appendix listing 11.10.

4.7.3 Hamiltonian matrix and its eigenstates on the grid mesh

The Schrödinger equation, eq. (4.7.1), has l, but has no n. How n enters our solution is not obvious.[9] Numerically, the maximum n is determined by the number of mesh grid points, N. If $N \to \infty$, then $n \to \infty$. If we choose $N = 10$, then the maximum n is 10. However, these ten eigenstates do not necessarily correspond to the same 10 lowest eigen-

9 One way to understand this is to note that the Schrödinger equation is an eigenequation with two unknowns, eigenstates and eigenenergy. For every eigenenergy, one has an eigenstate, so n is infinite.

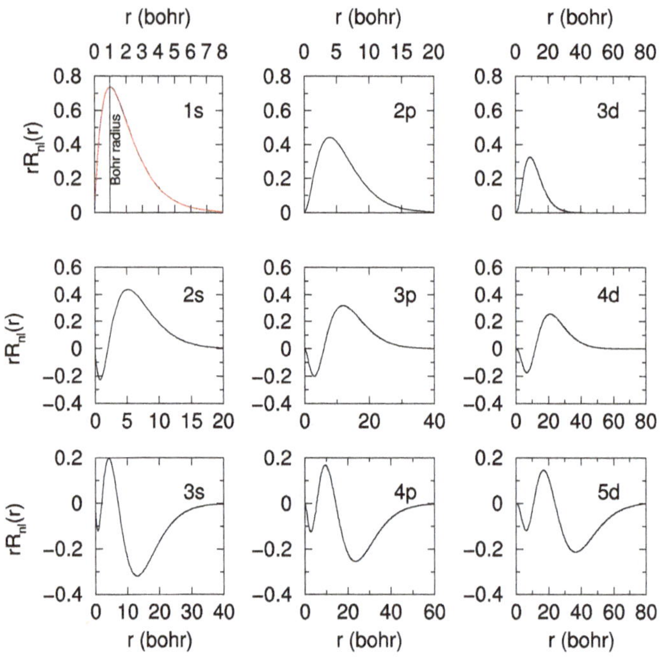

Figure 4.6: Radial wavefunctions of 1*s*, 2*s*, 3*s*, 2*p*, 3*p*, 4*p*, 3*d*, 4*d*, and 5*d*. Going from top to bottom with the same *l*, the number of nodes increases by one with the main quantum number *n*. Going from left to right, the wavefunction is similar across different *l*, but with an increase in *l*, the wavefunction becomes more diffusive or delocalized. To convert the numerical wavefunction to $rR_{nl}(r)$, one has to divide it by $\sqrt{\Delta r}$ (see the text). Caution: the horizontal axes are not the same in each subfigure.

states of hydrogen. This is because both N and Δr affect our results, as $N\Delta r$ sets the maximum spatial extension of a wavefunction.

In the code, we choose $N = 1{,}600$ and the mesh size Δr is 0.05 Bohr, so the matrix size is $1{,}600 \times 1{,}600$, $n_{\max} = 1{,}600$, and the maximum spatial extension is $N\Delta r = 80$ Bohr. After setting up the Hamiltonian matrix, we diagonalize it with the LAPACk code to find the eigenvalues and eigenvectors. The eigenvalue produced from the code is $2E_{nl}$, and because eq. (4.7.1) has $2E_{nl}$ as its eigenvalues, our actual E_{nl} is half the eigenvalues from the code.

Figure 4.6 shows our numerical solutions for the hydrogen atom using code `hydrogen.f`. To run the code, one must choose l. For instance, if $l = 0$, all the s-states are calculated. The lowest state has $n = 1$ and is a 1s-state. This is the ground state of the hydrogen atom. Our numerical wavefunction $Z(i, n)$ directly from the code differs from $rR_{nl}(r)$ by $\sqrt{\Delta r}$.[10] Figure 4.6 has a general trend of these wavefunctions. Going from

10 $Z(i,n)$ is normalized as $\sum_i |Z(i,n)|^2 = 1$, but R_{nl} is normalized as $\int_0^\infty R_{nl}^2(r)r^2 dr = 1$, so $Z^2(i,n) = R_{nl}^2(r)r^2 dr$, i. e., $rR_{nl}(r) = Z/\sqrt{\Delta r}$, where dr is written as Δr in the last step.

top to bottom with a fixed l, the number of nodes increases by one. Going from left to right, the number of nodes[11] is always the same, and the shape of the wavefunction is similar, but the wavefunction becomes more extended and delocalized as l increases. It is easy to remember that the first orbital for each l has no node, the second orbital has 1 node, and the third one has 2 nodes. In fact, it would be much simpler if we could call these first orbitals in each l, 1s, 1p, 1d, ..., but this is not used.[12] This convention is the well-known aufbau (building-up) principle.

It is interesting to focus on the 1s-orbital. Figure 4.6 (top left) shows that our numerical solution matches perfectly the analytical solution (from Table 4.3) since the two curves overlap. The wavefunction peaks at exactly one Bohr radius, $r = a_0$, as also can be seen in the prior exercise in Section 4.6.2. To be more precise, this is the square root of the probability $\sqrt{\rho(r)}$. The reader must be cautious that the radial wavefunction is normalized as

$$\int_0^\infty |R_{nl}(r)|^2 r^2 dr = \int_0^\infty \rho(r)dr, \tag{4.7.4}$$

not $\int_0^\infty |R_{nl}(r)|^2 dr$, so that the probability is $\rho(r) \equiv r^2|R_{nl}(r)|^2$.

4.7.4 Hydrogen-like ions: ionic radii and Alzheimer's disease

What we have learned so far concerns one single H atom. Many ions such as He^+, Li^{++}, Be^{+++} are similar to hydrogen. They also have a single electron in the outer shell. The main difference is that their ionic core has Z protons, so the nuclear charge increases from $+e$ to $+Ze$, which in turn reduces the atomic radius from the Bohr radius a_0 to $r_1 = a_0/Z$. Here, Z is the atomic number or the effective atomic number Z_{eff}.[13]

So the eigenenergy is changed from eq. (4.6.7) to

$$E_n = -\frac{1}{2}\frac{e^2Z^2}{4\pi\epsilon_0 a_0 n^2} = -\frac{1}{2}\frac{e^2Z^2}{4\pi\epsilon_0 r_n^2} = -\frac{Z^2 13.6\text{ eV}}{n^2}, \tag{4.7.5}$$

and in the eigenfunctions (see Table 4.3) every a_0 becomes a_0/Z. This is the reason why, in the periodic table across the same row, atomic radii decrease from left to right. Their ionic radii with the same outer-shell charges decrease in the same fashion. For instance,

11 Here, nodes refer to the wavefunction crossing the zero line.

12 Because 2s and 2p have the same eigenenergy, and 3s and 3d have the same energy, and so on, the main quantum number n for p must start from $n = 2$, and for d it must start from $n = 3$.

13 Due to the screening effect by the innermost electrons, the outermost electrons experience the charge less than the actual nuclear charge $+Ze$.

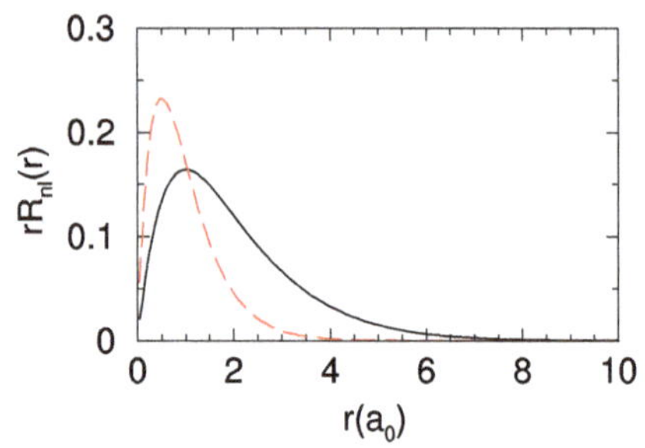

Figure 4.7: 1*s* radial wavefunctions for $Z = 1$ (solid line) and 2 (dashed line). The larger *Z* wavefunction is contracted and has a higher peak.

Na^+, Mg^{2+} and Al^{3+} have radii of 1.02, 0.72, and 0.53 Å, respectively. Different radii affect how these ions pass through blood-brain barriers and directly affect brain function. There is an ongoing investigation as to whether Al or other metals could lead to Alzheimer's disease or other types of dementia. What can be ascertained here is that the Al^{3+} ion almost has the same size as the H atom and is likely to have a higher mobility.

We can use the same hydrogen code to compute this numerically. The only change is in the function $v(r, l)$ in the code. Figure 4.7 shows two 1*s*-states for $Z = 1$ (solid line) and $Z = 2$ (dashed line). One can see a larger *Z* shifts the peak toward $r = 0$ where the ionic core situates.

Exercise 4.7.4.

32. The Schrödinger equation has no variable *n*, the main quantum number. Explain how it enters our calculation.

The following exercises are based on code `hydrogen.f` in the Appendix listing 11.10.

33. Use the existing *nm* and *dr* (a) to compute the 1*s*-, 2*p*-, and 3*d*-wavefunctions $u_{nl}(r)$, and (b) plot them and compare them with Fig. 4.6. Note that our wavefunction is really $rR_{nl}(r)$.

34. (a) Investigate how *nm* affects the wavefunctions and eigenenergies for $l = 2$ with *nm* increasing from 400 to 1,500 for $Z = 1$ and $Z = 2$. (b) For *s*-states, compare the numerical eigenenergies with the theoretical ones, as you increase `nm` from 400 to 1,200. What can you conclude by comparing these results?

35. It is possible to compute the radial transition matrix elements between 1*s* and 2*p* by properly modifying the code. Hint: One can save the wavefunctions first and then use these wavefunctions to compute the transition matrix elements.

4.8 Franck–Hertz experiment

Although spectral lines from atoms provide evidence of energy quantization, it is indirect because one relies on optics. Franck and Hertz provided a direct proof of energy quantization through electric current and voltage. They discovered that the collisions between energetic incident electrons and mercury atoms are both elastic and, more importantly, inelastic. It is the latter that directly proves the energy quantization.

However, they were unaware of Bohr's theory when they carried out the experiment. Here is an excerpt from James Franck in "On the recent past of physics" by Gerald Holton, American Journal of Physics, Vol. 29, p. 805 (1961). "It might interest you to know that when we made the experiments that we did not know Bohr's theory. We had neither read nor heard about it. We had not read it because we were negligent to read the literature well enough – and you know how that happens. ⋯ . But we made that experiment (and got the result that confirmed Bohr's theory) because we hoped that if we found out where the borderline between elastic and inelastic impact lies... only one line might appear. But we did not know whether that would be so, and we did not know whether at all an emission of an atom is of such a type that one line alone can be emitted and all the energy can be used for that purpose. The experiment gave it to us, and we were surprised about it. But we were not surprised after we read Bohr's paper later, after our publication."

4.8.1 Experiment

This experiment now can be carried out in an advanced physics lab. We carry out the experiment ourselves. Figure 4.8 schematically shows our Franck–Hertz experiment setup, which consists of a heater (H) that vaporizes the mercury liquid, a cathode (C, another heater) that ejects electrons, a grid mesh (G) which serves as an anode to accelerate the electrons, and a reverse bias (R, also called the collector) which suppresses the current noise. The accelerating voltage V_a is applied between C and G, and V_r between G and R. The ammeter A measures the current through the collector, which is converted to the voltage V_c. In the experiment, one measures V_c versus V_a.

Classically, one would expect that accelerated electrons, once having collided with the Hg atoms, lose energy and slow down. As V_a increases, one expects an increase in the collector current. Because the current is converted to the collector voltage, one would see a monotonic increase in V_c, just like a regular resistor: The larger the voltage is, the higher the current is. This process should be the same for any V_a. But, this is not what is observed. Figure 4.8(b) shows that, as we increase V_a, V_c increases smoothly, but, once it is close to 4.86 eV, V_c drops. This repeats itself for every 4.86 eV. Our experimental data are at 9.10, 13.87, 18.77, and 23.97 V, with uncertainty of 0.2 V. Beyond 27 V, V_c saturates due to our measurement limit. These peaks and troughs cannot be explained by classical physics.

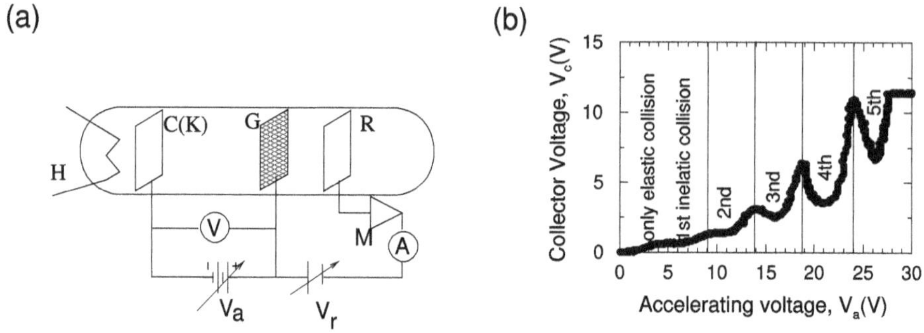

Figure 4.8: (a) Franck–Hertz scheme. A drop of mercury liquid is sealed inside the tube. A heater outside the tube heats the tube above 170°C, so the mercury liquid is vaporized. A different heater (H) heats the electrons that eject from the cathode (C). Under the influence of the acceleration voltage V_a, the electrons accelerate toward the grid mesh (G), which also serves as the anode. The collision occurs between the accelerated electrons and mercury atoms. A reverse bias (R) V_r is applied to suppress the noise current. This reverse bias electrode is also called the collector. The current, after magnification (M), is measured. (b) Collector voltage V_c versus the accelerating voltage V_a. The collector current is amplified and converted into the collector voltage V_c, which is measured by the oscilloscope. V_a is the accelerating voltage which peaks at 9.10 V, 13.87 V, 18.77 V, and 23.97 V (see the vertical lines). The separation between them is 4.77, 4.90, and 5.20 V. The average is about 4.96 V with experimental uncertainty of 0.2 V. Before 5 V, there is only pure elastic collision. The first inelastic collision occurs slightly above 5 V, but its V_a is difficult to determine. However, the second, third, fourth, and fifth collisions are very clear.

4.8.2 Verification of quantum theory

Before colliding with incident electrons, mercury atoms are in their ground state, with the two outermost electrons occupying $(6s)^2$, i. e., with $n = 6$ and $l = 0$. There is an excited state $(6s6p)$ at $\Delta E = 4.86$ eV.[14] Figure 4.9(a) illustrates the energy level.

The electron ejected from the cathode is accelerated, with maximum energy up to $-eV_a$. Figure 4.9(b) illustrates what happens inside the vacuum tube. The horizontal direction is along the tube direction. Electrons are incident from the left; and for a fixed V_a, they need to travel a distance to reach the maximum kinetic energy eV_a.[15]

If $|-eV_a| < \Delta E$, then there is no place in the tube where the electron can accumulate enough kinetic energy to match ΔE, so the electrons inside the Hg atom remain in the ground state. The collision between the incident electron and Hg atoms is elastic.

14 In fact, there are three excited states, with slightly different energies due to electron spins. Hertz's Nobel lecture includes this energy diagram. But here, we ignore this difference.

15 Collisions between the incident electron and Hg atoms occur all the time. Experiments critically depend on the temperature in the tube. If the temperature is too high, too many vaporized Hg atoms in the way shorten the mean-free path of the incident electrons, the path length that electrons can travel without collisions. On the other hand, if the temperature is too low, then there are not enough vaporized Hg atoms to collide with incident electrons. In our experiment, the temperature is set at 170°.

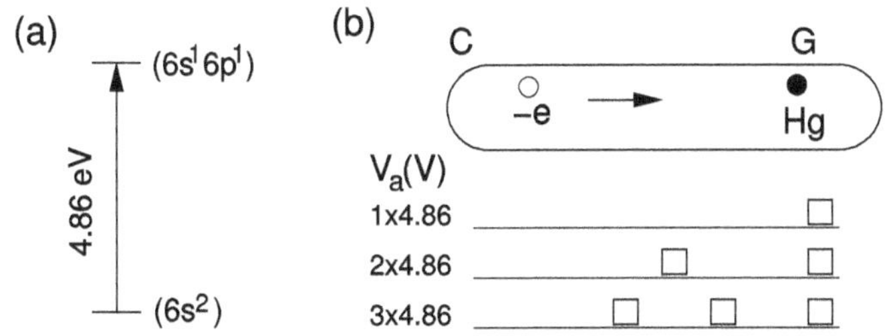

Figure 4.9: (a) Energy level scheme of Hg. The energy difference is $\Delta E = 4.86$ eV between the ground state ($6s^2$) and the excited state ($6s6p$). (b) Collision in real space. The boxes denote collision zones. At $eV_a = \Delta E$, there is one zone. The number of zones increases with eV_a.

Next, we increase V_a so eV_a matches the energy difference ΔE. Once the incident electrons reach G, they acquire the kinetic energy ΔE. When they collide with Hg atoms around G, the energy is transferred to the Hg atom's electrons. These electrons in the Hg atoms obtain ΔE to make a transition from $6s^2$ to $6s6p$. In the meantime, due to the energy conservation, the incident electrons lose almost all the kinetic energy. This is an inelastic collision. The box on the first line in Fig. 4.9(b) shows the location of collision. Very few of the incident electrons can cross the grid mesh G to contribute to the current. As a result, the current is reduced sharply, leading to the voltage trough in Fig. 4.8(b).

If we increase eV_a to $2\Delta E = 2 \times 4.86$ eV, there are two inelastic collisions. The first occurs in the middle of the line connecting the C and G electrodes (see the second line in Fig. 4.9(b)). After the collision, the incident electron can regain the energy to make a second inelastic collision (see the second box) before it hits G. Every inelastic collision slows down the incident electrons. With $eV_a = 2\Delta E$, the number of electrons that reach the collector increases, so the collector current and voltage increase, which explains a baseline increase in Fig. 4.8(b). If we increase eV_a to three or four times ΔE, we will have three or four inelastic collisions.

Importantly, every peak observed in Fig. 4.8(b) corresponds to the same two energy levels in Fig. 4.9(a). This experimental result demonstrates that, indeed, in the Hg atom the energy levels are quantized. Lastly, it might be interesting to note that collision zones in the tube can be identified by looking at the bluish disks of wavelength 4,358 Å, an emission from a high level to one of the ($6s6p$)-states.[16] These higher states are reached by the already excited electrons in the Hg atoms through successive collisions with incident electrons [28]. This is possible because not all the states can decay to the ground state at the same rate.

16 This is the transition from a spin triplet state ^3S to another spin triplet state $^3P_1$. See Chapter 6.

Exercise 4.8.2.

36. In the Franck–Hertz experiment with mercury, the first trough appears at 4.86 eV. Suppose the incident electron has zero kinetic energy. Find the minimum distance between the cathode and the grid mesh.

37. In neon, the first excitation potential to $3p$ is about 19 V. (a) Find the minimum potential to have an inelastic collision. (b) Suppose the incident electron has zero kinetic energy. What is the minimum length of the neon tube if the electron only has one inelastic collision at the minimum potential?

4.9 Problems

1. Prove the following relationships: (a) $[\hat{L}_x, x] = 0$, $[\hat{L}_y, y] = 0$, and $[\hat{L}_z, z] = 0$. (b) $[\hat{L}_x, y] = i\hbar z$, $[\hat{L}_y, z] = i\hbar x$, $[\hat{L}_z, x] = i\hbar y$. (c) $[\hat{L}_x, z] = -i\hbar y$, $[\hat{L}_y, x] = -i\hbar z$, $[\hat{L}_z, y] = -i\hbar x$.
2. Prove $[\hat{l}_\alpha, \hat{p}_\beta] = \epsilon_{\alpha\beta\gamma} i\hbar \hat{p}_\gamma$, where $\epsilon_{\alpha\beta\gamma}$ is the Levi-Civita symbol.
3. Prove (a) $[\hat{L}_\alpha, \mathbf{r}^2] = 0$ and (b) $[\hat{L}_\alpha, \mathbf{r} \cdot \hat{\mathbf{p}}] = 0$.
4. Prove $\hat{\mathbf{L}}^2 x - x\hat{\mathbf{L}}^2 = i\hbar[(\mathbf{r} \times \hat{\mathbf{L}})_x - (\hat{\mathbf{L}} \times \mathbf{r})_x]$.
5. Derive $\hat{L}_x, \hat{L}_y$, and $\hat{L}_z$ directly from eq. (4.3.1) by transforming (x, y, z) to (r, θ, ϕ). This is a hard problem. Hint: (i) To change the partial derivatives in Cartesian coordinates to the spherical ones, use $r = \sqrt{x^2 + y^2 + z^2}$, $\tan\phi = y/x$, $\cos\theta = z/\sqrt{x^2 + y^2 + z^2}$, not $\theta = \cos^{-1} z/\sqrt{x^2 + y^2 + z^2}$. (ii) For example, we compute $\partial/\partial z$ as $\frac{\partial}{\partial z} = \frac{\partial r}{\partial z}\frac{\partial}{\partial r} + \frac{\partial \cos\theta}{\partial z}\frac{\partial}{\partial \cos\theta} + \frac{\partial \tan\phi}{\partial z}\frac{\partial}{\partial \tan\phi}$. (iii) One needs to keep derivatives like $\frac{\partial}{\partial r}$, but can simplify operations like $\frac{\partial}{\partial \tan\phi}$ by taking the derivative in the denominator with respect to ϕ, and the same for $\frac{\partial}{\partial \cos\theta}$. (iv) In the end, convert all the Cartesian coordinates to spherical ones, using $x = r\sin\theta\cos\phi, y = r\sin\theta\sin\phi, z = r\cos\theta$. It is just when one converts the derivatives that one must use $r = \sqrt{x^2 + y^2 + z^2}$, $\tan\phi = y/x$, $\cos\theta = z/\sqrt{x^2 + y^2 + z^2}$.
6. Use eqs. (4.3.21) and (4.3.22) to compute $\langle Y_{lm}|\hat{L}_x|Y_{lm-1}\rangle$ and $\langle Y_{lm}|\hat{L}_y|Y_{lm-1}\rangle$.
7. The electron in hydrogen atom is in a mixed state. $|\psi\rangle = \frac{1}{\sqrt{3}}Y_{11} + \frac{2}{\sqrt{3}}Y_{10}$. (a) Find the probabilities of $\hat{L}_z$ with eigenvalues of 0 and $\pm h$. (b) Compute the expectation value of $\hat{L}_z$. (c) Find the probabilities of $\hat{L}_x$ with eigenvalues of 0 and $\pm h$. Hint: Find the eigenstates of $\hat{L}_x$ first and dot-product with $|\psi\rangle$. (d) Compute the expectation values of $\hat{L}_x$ and $\hat{L}_x^2$. (e) Find the uncertainty in $\hat{L}_x$. (f) Use the matrix form to compute (c) and (d).
8. A hydrogen atom is in a state $|\psi\rangle = c_1 Y_{11} + c_2 Y_{20}$, where c_1 and c_2 are the normalization constants. (a) Write out all the possible values of $\hat{L}_z$ and their respective probabilities. (b) Write out all the possible values of $\hat{\mathbf{L}}$ and their respective probabilities. (c) Find the expectation values of $\hat{L}_x$, $\hat{L}_y$, $\hat{L}_x^2$, and $\hat{L}_y^2$.
9. Refer to Fig. 4.3. Suppose $l = 2$. Compute all five angles of $\hat{L}_z$ with respect to the z-axis. Hint: First compute the length of $\hat{\mathbf{L}}$ and then compute the expectation value of $\hat{L}_z$.

10. Suppose a particle in state $\psi(\mathbf{r}) = C(x + y + 2z)e^{-ar}$, where $a > 0$ and C is a normalization constant. (a) Find $\langle \hat{\mathbf{L}}^2 \rangle$. (b) Find $\langle \hat{L}_z \rangle$.
11. Diagonalizing the $\hat{L}_y$ matrix (see eq. (4.4.3)), find: (a) its eigenvalues and (b) its eigenvectors $|m_y\rangle$; (c) find the expectation values of $\langle m_y|\hat{L}_z|m_y\rangle$ and $\langle m_y|\hat{L}_x|m_y\rangle$; (d) compute $\langle m_y|\hat{L}_z\hat{L}_x - \hat{L}_x\hat{L}_z|m_y\rangle$ and then check whether your results match $i\hbar\langle m_y|\hat{L}_y|m_y\rangle$.
12. Using the eigenstates $|n\rangle$ of the harmonic oscillator as the basis, write down the matrix for (a) $\hat{x}^2$ (eq. (3.3.10)), (b) $\hat{p}_x$, and (c) $\hat{p}_x^2$.
13. $|\psi\rangle = |nlm\rangle$ is an eigenstate of the hydrogen atom. Show (a) $\langle r \rangle = \frac{a_0}{2}[3n^2 - l(l+1)]$. (b) $\langle r^2 \rangle = \frac{a_0^2 n^2}{2}[5n^2 + 1 - 3l(l+1)]$. (c) $\langle \frac{1}{r} \rangle = \frac{1}{a_0 n^2}$. (d) $\langle \frac{1}{r^2} \rangle = \frac{1}{a_0^2 n^3 (l+\frac{1}{2})}$.
14. An interesting integral is the Coulomb integral,[17]

$$\int_{x=-h}^{x=+h}\int_{y=-h}^{y=+h}\int_{z=-h}^{z=+h} \frac{1}{\sqrt{x^2+y^2+z^2}}\,dxdydz. \tag{4.9.1}$$

If we integrate using Cartesian coordinates, this integral has a divergence at (0, 0, 0). A way around this is to use spherical coordinates. In Fig. 4.10 we choose a pyramid of ABCO with four surfaces—ABC, ABO, ACO, and BOC—which are described by four equations, $z = h$, $x = z$, $y = 0$, and $x = y$. These four equations are going to guide us to determine the integral limits for r, θ, and ϕ. The integration must start from dr because the volume element is $r^2 \sin\theta d\theta d\phi dr$ and r^2 cancels $1/r$ in the integrand. Since $x = r\sin\theta\cos\phi$, $y = r\sin\theta\sin\phi$, and $z = r\cos\theta$, we will use $z = h$ as our constraint for r. The limit for r is from 0 to $h/\cos\theta$. Using other relationships is difficult because x and y both change. Next, we use the second surface, $x = z$, to determine the limits for θ, while $x = 0$ and $x = y$ are used to determine the limits for ϕ. The reason is that θ starts from the z-axis and its maximum angle is bounded by the xz-plane. Equating $x = r\sin\theta\cos\phi$ to $z = r\cos\theta$, we find that $\tan\theta = 1/\cos\phi$, or $\theta = \tan^{-1}(1/\cos\phi)$. The lower limit for ϕ is determined by $x = 0$, while $x = y$ sets the upper limit, $\phi = \pi/4$. Then the equation becomes

$$\int_{\phi=0}^{\pi/4}\int_{\theta=0}^{\tan^{-1}(1/\cos\phi)}\int_{r=0}^{h/\cos\theta} \frac{1}{r} r^2 dr \sin\theta d\theta d\phi, \tag{4.9.2}$$

which can be analytically integrated.

Prove

$$\int_{x=-h}^{x=+h}\int_{y=-h}^{y=+h}\int_{z=-h}^{z=+h} \frac{1}{\sqrt{x^2+y^2+z^2}}\,dxdydz = h^2\left(-\frac{\pi}{4} + \frac{3}{2}\ln\frac{\sqrt{3}+1}{\sqrt{3}-1}\right).$$

17 Thanks to Dr. C. Zhao at Indiana State University.

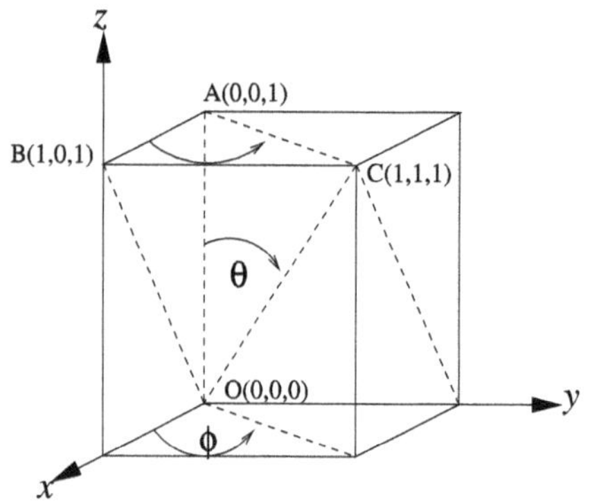

Figure 4.10: Coulomb integral. The integration is taken over the ABCO pyramid. A cube consists of 6 pyramids, each of which is defined by four planes, ABC, ABO, ACO, and BOC.

5 Time-independent approximate methods

The hydrogen or hydrogen-like atom is the only system that can be solved analytically. The rest of the systems must be solved by either numerical or approximate methods. Approximate methods are much more powerful and allow us to solve problems beyond a single-atom limit. In this chapter, we introduce two time-independent approximate methods: perturbation theory and variational principle. This opens the door to problems of practical importance.

This chapter can be grouped into three units.

- Unit 1 includes Sections 5.1 and 5.2. It first introduces nondegenerate perturbation theory. Section 5.2 provides two examples: charged harmonic oscillator and van der Waals force.
- Unit 2 includes Sections 5.3 and 5.4. Degenerate perturbation theory is more involved than the nondegenerate one as we work with a matrix. Section 5.4 shows two examples: the H_2^+ molecule and the Stark effect.
- Unit 3 includes Sections 5.5 and 5.6. Section 5.5 introduces the variational principle and provides an example. Section 5.6 takes H_2^+ as an example again to complement our degenerate treatment, so the reader can see how the variational principle works and how the theoretical results compare with the experimental ones.

5.1 Nondegenerate perturbation theory

Perturbation theory is an approximate method. It does not aim to find all the eigenstates of a system. Instead, it targets one single state. In the following, we shall assume that the target state has an index of k. This section focuses on a state that is nondegenerate, i. e., no two states having the same eigenenergy of the unperturbed system.

5.1.1 Expanding the energy and wavefunction

The goal is to find an approximate solution to a specific state ψ_k, not for all states,

$$\hat{H}\psi_k = E_k\psi_k, \tag{5.1.1}$$

where $\hat{H} = \hat{H}_0 + \lambda\hat{H}_I$ and $\hat{H}_I$ is our small perturbation. λ is a parameter to help us determine the order. If λ is zero, then we go back to the unperturbed Hamiltonian $\hat{H}_0$. If it is 1, then we have the perturbed Hamiltonian $\hat{H}$. The theory assumes that the eigenstates ϕ_i and eigenvalues ϵ_i of an unperturbed Hamiltonian $\hat{H}_0$ are known, i. e.,

$$\hat{H}_0\phi_i = \epsilon_i\phi_i. \tag{5.1.2}$$

https://doi.org/10.1515/9783110672152-005

The perturbation expands an unknown ψ_k into nth orders of corrections $\psi_k^{(n)}$,

$$\psi_k = \psi_k^{(0)} + \lambda\psi_k^{(1)} + \lambda^2\psi_k^{(2)} + \cdots, \tag{5.1.3}$$

where $\psi_k^{(0)}, \psi_k^{(1)}, \ldots$ are to be determined. The hope is that, after a few limited terms, the wavefunction converges to its exact function. The superscripts of these wavefunctions identify their orders in λ, while the subscript k denotes the state targeted. This is the only state that is computed. Similarly, the energy E_k is expanded as

$$E_k = E_k^{(0)} + \lambda E_k^{(1)} + \lambda^2 E_k^{(2)} + \cdots. \tag{5.1.4}$$

We always keep the indices of the eigenstates of $\hat{H}_0$ for a reason which will soon become clear.

We substitute eqs. (5.1.3) and (5.1.4) into $\hat{H}\psi_k = E\psi_k$. The left side of the equation must be equal to the right side of the equation,

$$\begin{aligned} &(\hat{H}_0 + \lambda\hat{H}_I)(\psi_k^{(0)} + \lambda\psi_k^{(1)} + \lambda^2\psi_k^{(2)} + \cdots) \\ &= (E_k^{(0)} + \lambda E_k^{(1)} + \lambda^2 E_k^{(2)} + \cdots)(\psi_k^{(0)} + \lambda\psi_k^{(1)} + \lambda^2\psi_k^{(2)} + \cdots). \end{aligned} \tag{5.1.5}$$

Equation (5.1.5) is the mother of perturbation theory and contains all the details. We are going to multiply them out on both sides and compare them term by term according to the order in λ.

5.1.2 Zeroth-order perturbation

We start with eq. (5.1.5) and focus on the zeroth-order terms with λ^0,

$$\lambda^0 : \hat{H}_0\psi_k^{(0)} = E_k^{(0)}\psi_k^{(0)}, \tag{5.1.6}$$

where the heading λ^0 reminds us that we are working on the zeroth order. Comparing it with eq. (5.1.2) reveals that the zeroth-order wavefunction $\psi_k^{(0)}$ is just the eigenstate ϕ_k of $\hat{H}_0$, $\psi_k^{(0)} = \phi_k$, and $E_k^{(0)}$ is just ϵ_k, $E_k^{(0)} = \epsilon_k$, This makes sense since, if there is no external perturbation, we surely get the eigenstates of $\hat{H}_0$.

5.1.3 First-order perturbation

Next, we consider λ^1 terms,

$$\lambda^1 : \hat{H}_0\psi_k^{(1)} + \hat{H}_I\psi_k^{(0)} = E_k^{(0)}\psi_k^{(1)} + E_k^{(1)}\psi_k^{(0)}. \tag{5.1.7}$$

Plugging $\psi_k^{(0)} = \phi_k$ and $E_k^{(0)} = \epsilon_k$ into this equation yields

$$\hat{H}_0\psi_k^{(1)} + \hat{H}_I\phi_k = \epsilon_k\psi_k^{(1)} + E_k^{(1)}\phi_k. \tag{5.1.8}$$

The reader can easily remember how these terms are written by noting that every term consists of two terms, where one term is in the zeroth order and the other is the first order in $\hat{H}_I$.

To get $E_k^{(1)}$ and $\psi_k^{(1)}$, we carry out two calculations: (a) multiplying eq. (5.1.8) by ϕ_k^* from the left and then integrating, and (b) doing the same as (a) but with $\phi_{m\neq k}$. This procedure is standard and works for the second-order corrections as well.

5.1.3.1 First-order energy correction $E_k^{(1)}$

Multiplying eq. (5.1.8) by ϕ_k^* from the left and integrating, we have

$$\langle\phi_k|\hat{H}_0|\psi_k^{(1)}\rangle + \langle\phi_k|\hat{H}_I|\phi_k\rangle = \epsilon_k\langle\phi_k|\psi_k^{(1)}\rangle + E_k^{(1)}\langle\phi_k|\phi_k\rangle. \tag{5.1.9}$$

Using $\langle\phi_k|\hat{H}_0 = \epsilon_k\langle\phi_k|$ and canceling terms with ϵ_k, we have the first-order energy correction,

$$E_k^{(1)} = \langle\phi_k|\hat{H}_I|\phi_k\rangle \quad \text{(First-order energy correction)}. \tag{5.1.10}$$

This shows that the first-order energy correction is the expectation value of $\hat{H}_I$ in $|\phi_k\rangle$.

5.1.3.2 First-order wavefunction correction $\psi_k^{(1)}$

Multiplying eq. (5.1.8) by $\phi_{m\neq k}^*$ from the left and then integrating, we have

$$\langle\phi_m|\hat{H}_I|\phi_k\rangle = (\epsilon_k - \epsilon_m)\langle\phi_m|\psi_k^{(1)}\rangle, \tag{5.1.11}$$

which gives

$$\langle\phi_m|\psi_k^{(1)}\rangle = \frac{\langle\phi_m|\hat{H}_I|\phi_k\rangle}{\epsilon_k - \epsilon_m} \quad (m \neq k). \tag{5.1.12}$$

This represents an overlap between $\psi_k^{(1)}$ and ϕ_m and means that the first-order perturbed wavefunction has contributions from eigenstates other than ϕ_k. We have to include all the eigenstates $\{\phi_i\}$, which is why we emphasized that we need to keep index i for ϕ. To have a correct $\psi_k^{(1)}$, we must add them up. To see this clearly, we place the completeness of eigenstates, $\sum_m |\phi_m\rangle\langle\phi_m| = 1$, before $|\psi_k^{(1)}\rangle$,

$$|\psi_k^{(1)}\rangle = \sum_m |\phi_m\rangle\langle\phi_m|\psi_k^{(1)}\rangle = \sum_{m\neq k} \frac{\langle\phi_m|\hat{H}_I|\phi_k\rangle}{\epsilon_k - \epsilon_m}|\phi_m\rangle. \tag{5.1.13}$$

Since $m \neq k$ and $\langle\phi_m|\phi_k\rangle = 0$, $\psi_k^{(1)}$ is orthogonal to ϕ_k. Equation (5.1.13) means that the wavefunction correction is a linear combination of unperturbed eigenstates ϕ_m, but it is weighted by the coefficient $\frac{\langle\phi_m|\hat{H}_I|\phi_k\rangle}{\epsilon_k-\epsilon_m}$. If a state's unperturbed eigenenergy ϵ_m is closer to our targeted state's eigenenergy ϵ_k, it has a larger contribution.

The unnormalized wavefunction up to the first-order correction is

$$|\psi_k\rangle = |\psi_k^{(0)}\rangle + |\psi_k^{(1)}\rangle = |\phi_k\rangle + \sum_{m\neq k} \frac{\langle\phi_m|\hat{H}_I|\phi_k\rangle}{\epsilon_k - \epsilon_m}|\phi_m\rangle. \tag{5.1.14}$$

To normalize it, one can compute $\langle\psi_k|\psi_k\rangle$ and rescale $|\psi_k\rangle/\sqrt{\langle\psi_k|\psi_k\rangle}$.

5.1.4 Second-order correction

We again start from eq. (5.1.5) and match terms in λ^2,

$$\begin{aligned}\lambda^2 : \hat{H}_0\psi_k^{(2)} + \hat{H}_I\psi_k^{(1)} &= E_k^{(0)}\psi_k^{(2)} + E_k^{(1)}\psi_k^{(1)} + E_k^{(2)}\psi_k^{(0)} \\ &= \epsilon_k\psi_k^{(2)} + E_k^{(1)}\psi_k^{(1)} + E_k^{(2)}\phi_k,\end{aligned} \tag{5.1.15}$$

which shows that the second-order correction depends on the first-order correction. Such a hierarchy is the hallmark of perturbation theory. The remaining procedure is exactly the same as the first-order one.

5.1.4.1 Second-order energy correction $E_k^{(2)}$

We also left multiply eq. (5.1.15) by ϕ_k^* and integrate to find

$$\langle\phi_k|\hat{H}_0|\psi_k^{(2)}\rangle + \langle\phi_k|\hat{H}_I|\psi_k^{(1)}\rangle = \epsilon_k\langle\phi_k|\psi_k^{(2)}\rangle + E_k^{(1)}\langle\phi_k|\psi_k^{(1)}\rangle + E_k^{(2)}\langle\phi_k|\phi_k\rangle. \tag{5.1.16}$$

Using $\langle\phi_k|\hat{H}_0 = \varepsilon_k\langle\phi_k|$, we can cancel the first terms on both sides of the equation. The second term on the right is zero because $\langle\phi_k|\psi_k^{(1)}\rangle = 0$ due to eq. (5.1.13). Employing the normalization condition $\langle\phi_k|\phi_k\rangle = 1$, we obtain the second-order energy correction as

$$E_k^{(2)} = \langle\phi_k|\hat{H}_I|\psi_k^{(1)}\rangle. \tag{5.1.17}$$

To appreciate the beauty of perturbation theory, we pause here to explain the structure of $E_k^{(2)}$: The entire expression is just like the "expectation value," but the wavefunction on the right is replaced by the wavefunction which is one order lower. Our first-order energy correction is $E_k^{(1)} = \langle\phi_k|\hat{H}_I|\psi_k^{(0)}\rangle$. If we compare the structure of these two equations, we find that they are similar.

This nice connection is hidden if we replace $\psi_k^{(1)}$ by ϕ_m,

$$E_k^{(2)} = \sum_{m\neq k} \frac{\langle\phi_m|\hat{H}_I|\phi_k\rangle\langle\phi_k|\hat{H}_I|\phi_m\rangle}{E_k^{(0)} - E_m^{(0)}} = \sum_{m\neq k} \frac{\langle\phi_k|\hat{H}_I|\phi_m\rangle\langle\phi_m|\hat{H}_I|\phi_k\rangle}{\epsilon_k - \epsilon_m}, \tag{5.1.18}$$

which can be rewritten as

$$E_k^{(2)} = \sum_{m \neq k} \frac{|\langle \phi_k | \hat{H}_I | \phi_m \rangle|^2}{\epsilon_k - \epsilon_m} \quad \text{(Second-order energy correction).} \tag{5.1.19}$$

5.1.4.2 Second-order wavefunction correction $\psi_k^{(2)}$

We multiply eq. (5.1.15) by a state $\phi_{m \neq k}^*$ from the left and integrate to get

$$\epsilon_m \langle \phi_m | \psi_k^{(2)} \rangle + \langle \phi_m | \hat{H}_I | \psi_k^{(1)} \rangle = \epsilon_k \langle \phi_m | \psi_k^{(2)} \rangle + E_k^{(1)} \langle \phi_m | \psi_k^{(1)} \rangle + E_k^{(2)} \langle \phi_m | \phi_k \rangle, \tag{5.1.20}$$

which yields the overlap between $\psi_k^{(2)}$ and ϕ_m,

$$\langle \phi_m | \psi_k^{(2)} \rangle = \frac{\langle \phi_m | \hat{H}_I | \psi_k^{(1)} \rangle - E_k^{(1)} \langle \phi_m | \psi_k^{(1)} \rangle}{\epsilon_k - \epsilon_m}. \tag{5.1.21}$$

Here, we have used $\langle \phi_{m \neq k} | \phi_k \rangle = 0$. Since ϵ_m is closer to ϵ_k, this overlap is larger. Similar to $\psi_k^{(1)}$, $\phi_{m \neq k}$ is not the only state that contributes to $\psi_k^{(2)}$, but all other states do. So we sum up all possible states to get our second-order wavefunction correction

$$|\psi_k^{(2)} \rangle = \sum_{m \neq k} \frac{\langle \phi_m | \hat{H}_I | \psi_k^{(1)} \rangle - E_k^{(1)} \langle \phi_m | \psi_k^{(1)} \rangle}{\epsilon_k - \epsilon_m} |\phi_m \rangle. \tag{5.1.22}$$

One sees that it depends on the first-order $\psi_k^{(1)}$ and $E_k^{(1)}$. For this reason, its expression is rather cumbersome.

5.2 Examples of nondegenerate perturbation theory

Nondegenerate perturbation theory is an important tool. We present two major examples. But first we would like to highlight the key steps:

- Separate the system Hamiltonian into two parts, $\hat{H}_0$ and $\hat{H}_I$, where $\hat{H}_0$ has a known solution and $\hat{H}_I$ is small in comparison with $\hat{H}_0$.
- Choose a targeted state k of interest.
- Construct the matrix elements of $\hat{H}_I$ among the eigenstates of $\hat{H}_0$.
- Compute the energy correction and wavefunction correction.

5.2.1 Charged harmonic oscillator in a weak electric field F

A one-dimensional harmonic oscillator carries charge q and is placed in an external electric field **F**, not to be confused with energy E. Compute (a) the energy correction up to the second order in the electric field and (b) the wavefunction to the first order.

(a) Since the electric field is weak, we will treat it as a perturbation. The total Hamiltonian of the system is split into two parts: $\hat{H} = \hat{H}_0 + \hat{H}_I$, where $\hat{H}_0$ is the unperturbed harmonic oscillator Hamiltonian and $\hat{H}_I$ is the interaction between the electric field and the system, i. e.,

$$\hat{H}_0 = -\frac{\hbar^2}{2m}\frac{d^2}{dx^2} + \frac{1}{2}m\omega x^2, \quad \hat{H}_I = -\hat{\mathbf{D}} \cdot \mathbf{F}. \tag{5.2.1}$$

Here, $\hat{\mathbf{D}}$ is the dipole $q\hat{x}$ and $\hat{H}_I = -qF_x\hat{x}$. The eigenvalues and eigenstates of the unperturbed system are $E_n^{(0)} = (n + 1/2)\hbar\omega$ and $|\phi_n\rangle$, denoted as $|n\rangle$.

Nondegenerate perturbation theory handles only one state at one time. Suppose we are interested in state $|k\rangle$, with energy up to the second order, $E_k = E_k^{(0)} + E_k^{(1)} + E_k^{(2)}$. The zeroth-order energy is $E_k^{(0)} = (k + \frac{1}{2})\hbar\omega$. Next, we compute the first-order correction, which is given by the expectation value of $\hat{H}_I$ in the zeroth-order wavefunction,

$$E_k^{(1)} = \langle\psi_k^{(0)}|\hat{H}_I|\psi_k^{(0)}\rangle = \langle k|\hat{H}_I|k\rangle = -qF_x\langle k|\hat{x}|k\rangle = 0, \tag{5.2.2}$$

where we have used $\langle k|\hat{x}|k\rangle = 0$ for the harmonic oscillator, with no contribution from the first-order perturbation. So, we need to compute the second-order energy,

$$E_k^{(2)} = \sum_{n\neq k}\frac{\langle k|\hat{H}_I|n\rangle\langle n|\hat{H}_I|k\rangle}{E_k^{(0)} - E_n^{(0)}}, \tag{5.2.3}$$

where the summation excludes $n = k$.

In order to compute $E_k^{(2)}$, we first calculate $\langle k|\hat{H}_I|n\rangle = -qF_x\langle k|\hat{x}|n\rangle$. Since $\hat{x} = \sqrt{\frac{\hbar}{2m\omega}}(\hat{a} + \hat{a}^\dagger)$, $\hat{x}|n\rangle = \sqrt{\frac{\hbar}{2m\omega}}(\sqrt{n}|n-1\rangle + \sqrt{n+1}|n+1\rangle)$, we have

$$\langle k|\hat{x}|n\rangle = \sqrt{\frac{\hbar}{2m\omega}}(\sqrt{n}\delta_{k,n-1} + \sqrt{n+1}\delta_{k,n+1}), \tag{5.2.4}$$

which yields

$$\begin{aligned} E_k^{(2)} &= \sum_{n\neq k}\frac{(qF_x)^2\langle k|\hat{x}|n\rangle\langle n|\hat{x}|k\rangle}{\hbar\omega(k-n)} \\ &= \frac{(qF_x)^2}{\hbar\omega}\left(\frac{\langle k|\hat{x}|k-1\rangle\langle k-1|\hat{x}|k\rangle}{k-(k-1)} + \frac{\langle k|\hat{x}|k+1\rangle\langle k+1|\hat{x}|k\rangle}{k-(k+1)}\right). \end{aligned}$$

Since $\langle k|\hat{x}|k-1\rangle = \sqrt{\frac{\hbar}{2m\omega}}\sqrt{k}$ and $\langle k|\hat{x}|k+1\rangle = \sqrt{\frac{\hbar}{2m\omega}}\sqrt{k+1}$, $E_k^{(2)} = -\frac{(qF_x)^2}{2m\omega^2}$. Our final energy for level k up to the second-order is

$$E_k = \left(k + \frac{1}{2}\right)\hbar\omega - \frac{(qF_x)^2}{2m\omega^2}. \tag{5.2.5}$$

This shows that the energy spectrum is downshifted by $\frac{(qF_x)^2}{2m\omega^2}$.

(b) The wavefunction up to the first order is

$$|\psi_k\rangle = |k\rangle + \sum_{n\neq k} \frac{\langle n|\hat{H}_I|k\rangle}{E_k^{(0)} - E_n^{(0)}}|n\rangle = |k\rangle + \frac{\langle k-1|\hat{H}_I|k\rangle}{E_k^{(0)} - E_{k-1}^{(0)}}|k-1\rangle + \frac{\langle k+1|\hat{H}_I|k\rangle}{E_k^{(0)} - E_{k+1}^{(0)}}|k+1\rangle, \tag{5.2.6}$$

where the second item can be written as $\frac{-qF_x\sqrt{\frac{\hbar}{2m\omega}}\sqrt{k}}{\hbar\omega}|k-1\rangle = \frac{-qF_x}{\hbar\omega}\sqrt{\frac{\hbar}{2m\omega}}\sqrt{k}|k-1\rangle$, and the third item is $\frac{qF_x\sqrt{\frac{\hbar}{2m\omega}}\sqrt{k+1}}{\hbar\omega}|k+1\rangle = \frac{qF_x}{\hbar\omega}\sqrt{\frac{\hbar}{2m\omega}}\sqrt{k+1}|k+1\rangle$. Therefore, the wavefunction up to the first order is

$$|\psi_k\rangle = |k\rangle + \frac{qF_x}{\hbar\omega}\sqrt{\frac{\hbar}{2m\omega}}(\sqrt{k+1}|k+1\rangle - \sqrt{k}|k-1\rangle), \quad k \geq 1. \tag{5.2.7}$$

5.2.2 van der Waals force

It is well known that neutral atoms and molecules experience Coulomb forces among themselves. This is collectively called the van der Waals force. In classical physics, this force is attributed to the charge fluctuation. In QM, we can clearly show this charge fluctuation is due to the wavefunction. Figure 5.1 schematically shows the diagram of two hydrogen atoms A and B separated by a distance R. Let us compute the total energy of the ground state as a function of R.

Our Hamiltonian consists of $\hat{H} = \hat{H}_0 + \hat{H}_I$. The unperturbed one, $\hat{H}_0$, represents two isolated H atoms and is the sum of two Hamiltonians at atoms A and B,

$$\hat{H}_0 = \hat{H}_A + \hat{H}_B = -\frac{\hbar^2}{2m_e}(\nabla_1^2 + \nabla_2^2) - \frac{e^2}{4\pi\epsilon_0 r_{1A}} - \frac{e^2}{4\pi\epsilon_0 r_{2B}}, \tag{5.2.8}$$

where 1(2) refers to electron 1(2). Our targeted state is the 1s-orbital. Because $\hat{H}_A$ and $\hat{H}_B$ are two isolated hydrogen atom's Hamiltonians, the zeroth-order eigenenergy is just

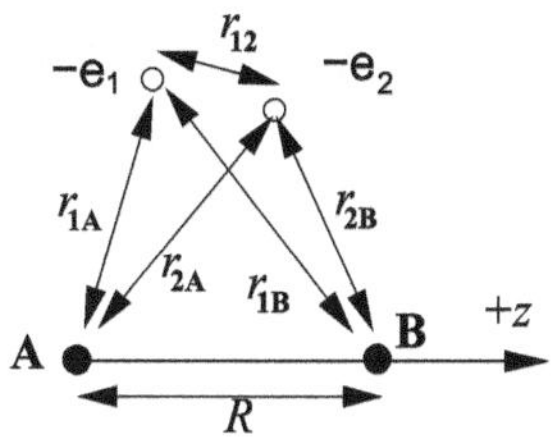

Figure 5.1: Two hydrogen atoms A and B separated by a distance R are placed along the z-axis, with A at the origin and B at $z = R$. Empty circles denote electrons $-e_1$ and $-e_2$.

$2E_{1s}$. The zeroth-order wavefunction is a product of two wavefunctions,[1]

$$\psi^{(0)} = \phi_{100}(\mathbf{r}_{1A})\phi_{100}(\mathbf{r}_{2B}). \tag{5.2.9}$$

The remaining interaction is lumped into the perturbation $\hat{H}_I$,

$$\hat{H}_I = \frac{e^2}{4\pi\epsilon_0 R} + \frac{e^2}{4\pi\epsilon_0 r_{12}} - \frac{e^2}{4\pi\epsilon_0 r_{2A}} - \frac{e^2}{4\pi\epsilon_0 r_{1B}}, \tag{5.2.10}$$

where four terms represent the nuclear repulsion between two protons A and B, the repulsion between two electrons, the attraction between A and electron 2, and the attraction between B and electron 1, respectively.

Since we are going to evaluate $\hat{H}_I$ using the wavefunction in eq. (5.2.9), we must simplify it by expanding $\hat{H}_I$ in terms of R and dropping those small terms. We choose the z-axis along the line which links A and B (see Fig. 5.1), and place atom A at the origin with coordinate $(0,0,0)$ and B at $z = R$ with coordinate $(0,0,R)$. The following table shows the details:

Electron 1	(x_1, y_1, z_1)	Electron 2	$(x_2, y_2, z_2 + R)$.
Atom A	$(0,0,0)$	Atom B	$(0,0,R)$.

With these coordinates, the distances in eq. (5.2.10) can be written out as

$$r_{12} = \sqrt{(x_1 - x_2)^2 + (y_1 - y_2)^2 + (z_1 - z_2 - R)^2}, \tag{5.2.11}$$

$$r_{2A} = \sqrt{x_2^2 + y_2^2 + (z_2 + R)^2}, \tag{5.2.12}$$

$$r_{1B} = \sqrt{x_1^2 + y_1^2 + (z_1 - R)^2}, \tag{5.2.13}$$

which must be inverted (see eq. (5.2.10)). We take r_{12} as an example, leaving r_{2A} and r_{1B} as an exercise. We introduce a shorthand notation for $x_1 - x_2 = \delta_x, y_1 - y_2 = \delta_y, z_1 - z_2 = \delta_z$, and $\delta_x^2 + \delta_y^2 + \delta_z^2 = \Delta^2$:

$$r_{12} = [\delta_x^2 + \delta_y^2 + (\delta_z - R)^2]^{\frac{1}{2}} = [R^2 + \Delta^2 - 2R\delta_z]^{\frac{1}{2}} = R\left(1 - \frac{2\delta_z}{R} + \frac{\Delta^2}{R^2}\right)^{\frac{1}{2}}, \tag{5.2.14}$$

whose inversion is $\frac{1}{r_{12}} = \frac{1}{R}(1 - \frac{2\delta_z}{R} + \frac{\Delta^2}{R^2})^{-\frac{1}{2}}$.

1 Rigorously speaking, this expression is not entirely correct, since it ignores the particle exchange in the wavefunction by assuming these two H atoms are sufficiently far apart. This will become clear in Chapter 10.

We need to expand $\frac{1}{r_{12}}$ in R. Using the relationship

$$(1 \pm x)^{-\frac{1}{2}} = 1 \mp \frac{1}{2}x + \frac{1}{2}\frac{3}{4}x^2 + \cdots, \tag{5.2.15}$$

we have

$$\frac{1}{r_{12}} \approx \frac{1}{R}\left[1 + \frac{1}{2}\left(\frac{2\delta_z}{R} - \frac{\Delta^2}{R^2}\right) + \frac{1}{2}\frac{3}{4}\left(\frac{2\delta_z}{R} - \frac{\Delta^2}{R^2}\right)^2 + \cdots\right], \tag{5.2.16}$$

where we keep the last term on the right up to δ_z^2, since the remaining higher-order terms are much smaller and can be dropped. Keeping terms up to $\frac{\delta_z^2}{R^2}$ and, after simple rearrangement, we get

$$\frac{1}{r_{12}} \approx \frac{1}{R}\left(1 + \frac{\delta_z}{R} + \frac{2\delta_z^2 - \delta_x^2 - \delta_y^2}{2R^2}\right). \tag{5.2.17}$$

Similarly, we can find

$$\frac{1}{r_{2A}} \approx \frac{1}{R}\left(1 - \frac{z_2}{R} - \frac{r_2^2}{2R^2} + \frac{3}{2}\frac{z_2^2}{R^2}\right), \tag{5.2.18}$$

$$\frac{1}{r_{1B}} \approx \frac{1}{R}\left(1 + \frac{z_1}{R} - \frac{r_1^2}{2R^2} + \frac{3}{2}\frac{z_1^2}{R^2}\right), \tag{5.2.19}$$

where $r_1^2 = x_1^2 + y_1^2 + z_1^2$ and $r_2^2 = x_2^2 + y_2^2 + z_2^2$. The detailed derivation is in the following exercise. Finally, our $\hat{H}_I$ in eq. (5.2.10) is

$$\hat{H}_I = \frac{e^2}{4\pi\epsilon_0}\left(\frac{1}{R} + \frac{1}{r_{12}} - \frac{1}{r_{2A}} - \frac{1}{r_{1B}}\right) \tag{5.2.20}$$

$$\approx \frac{e^2}{4\pi\epsilon_0 R^3}(2z_1z_2 - x_1x_2 - y_1y_2), \tag{5.2.21}$$

where all $1/R$ terms cancel out and the interaction decays much faster with R. To reveal the physics behind $e^2(2z_1z_2 - x_1x_2 - y_1y_2)$, we introduce two dipoles, $\hat{\mathbf{D}}_1 = -e\hat{\mathbf{r}}_1$ and $\hat{\mathbf{D}}_2 = -e\hat{\mathbf{r}}_2$, to rewrite $e^2(2z_1z_2 - x_1x_2 - y_1y_2) = \hat{\mathbf{D}}_1 \cdot \hat{\mathbf{D}}_2 - 3(\hat{\mathbf{D}}_1 \cdot \hat{\mathbf{R}})(\hat{\mathbf{D}}_2 \cdot \hat{\mathbf{R}})$. This reveals that this term is the dipole–dipole interaction, which is the origin of the van der Waals force. $\hat{\mathbf{R}}$ (not operator) is the unit vector along the bond direction. Physically, this means each H atom behaves like a dipole. The attraction between two H atoms occurs when the charge on one dipole attracts to the opposite charge on another dipole.

Once we have an expression for $\hat{H}_I$, we can determine its value. However, for x_1, x_2, y_1, y_2, z_1, and z_2, their expectation value is zero with the zeroth-order wavefunction (eq. (5.2.9)). This is because, due to the orthogonality in the spherical harmonics, the angular integral $\langle Y_{lm}|x(y,z)|Y_{l'm'}\rangle$ differs from zero only if $l - l' = \pm 1$, $m - m' = 0, \pm 1$. The integral for x_1x_2 is

$$\langle\psi^{(0)}|x_1x_2|\psi^{(0)}\rangle = \langle\phi_{100}(\mathbf{r}_{1A})|x_1|\phi_{100}(\mathbf{r}_{1A})\rangle\langle\phi_{100}(\mathbf{r}_{2B})|x_2|\phi_{100}(\mathbf{r}_{2B})\rangle = 0\times 0 = 0, \quad (5.2.22)$$

while other pairs are similar. This means the first-order produces no correction.

The second-order energy correction is based on eq. (5.1.19), but with one extension because we have two electrons instead of one. The zeroth-order wavefunction and energy is that of two electrons, not a single electron as intended in eq. (5.1.19). Specifically, the zeroth-order ground-state energy is $\mathcal{E}_{00} = E_{1s}(1) + E_{1s}(2) = -13.6\times 2 = -27.2\,\text{eV}$, where (1) and (2) denote electron 1 and electron 2. The zeroth-order excited-state energy is $\mathcal{E}_{n_1,n_2} = E_{n_1}(1) + E_{n_2}(2) = -13.6(\frac{1}{n_1^2} + \frac{1}{n_2^2})\,\text{eV}$, where n_1 and n_2 are the main quantum numbers of electrons 1 and 2. The intermediate state ϕ_m in eq. (5.1.19) is now a product of $\phi_{n_1,l_1,m_1}(\mathbf{r}_{1A})\phi_{n_2,l_2,m_2}(\mathbf{r}_{2B})$. We then have

$$E^{(2)} = \sum_{n1,l_1,m_1;n2,l_2,m_2;} \frac{|\langle\psi^{(0)}|\hat{H}_I|\phi_{n_1l_1m_1}(\mathbf{r}_{1A})\phi_{n_2l_2m_2}(\mathbf{r}_{2B})\rangle|^2}{\mathcal{E}_{00} - \mathcal{E}_{n_1n_2}}, \quad (5.2.23)$$

where l_1, m_1, and others are the angular momentum quantum numbers, and n_1 and n_2 both cannot be the 1s orbital. Since $\hat{H}_I$ is proportional to $1/R^3$, the numerator is proportional to $1/R^6$. The denominator is negative, $\mathcal{E}_{00} - \mathcal{E}_{n_1n_2} < 0$, so our $E^{(2)}$ is

$$E^{(2)} \approx -\frac{C}{R^6}, \quad (5.2.24)$$

where C is positive. This lowers the total energy and is responsible for the van der Waals force among neutral atoms and molecules.

Exercise 5.2.2.

1. (a) Following Example 1, compute the expectation value $\langle\psi_k|\hat{x}|\psi_k\rangle$ of $\hat{x}$ with ψ_k up to the first order. (b) This problem can be solved analytically by directly factoring $\hat{H}_I$ into $\hat{H}_0$. Find the exact eigenenergy and the wavefunction without using perturbation theory and compare both results.
2. Suppose a system has an unperturbed Hamiltonian $\hat{H}_0 = \begin{pmatrix} E_1 & 0 \\ 0 & E_2 \end{pmatrix}$, where $E_1 < E_2$. A small perturbation Hamiltonian is $\hat{H}_I = \begin{pmatrix} a & ib \\ -ib & c \end{pmatrix}$, where a and b are real. (a) Use perturbation theory to find the first-order energy correction for the lowest energy state. (b) Find the first-order wavefunction correction. (c) Solve the same problem $\hat{H} = \hat{H}_0 + \hat{H}_I$ by diagonalizing the matrix and without using perturbation theory, and then compare the perturbed results with the exact results.
3. (a) Prove eqs. (5.2.18) and (5.2.19). (b) Prove eq. (5.2.21).
4. A hydrogen atom placed in a uniform magnetic field has the following Hamiltonian: $\hat{H} = -\frac{(\hat{\mathbf{p}}-e\mathbf{A})^2}{2m_e} + V(\mathbf{r})$. Suppose the magnetic field is along the z-axis, with the vector potential $\mathbf{A}$ being $A_x = -\frac{B}{2}y, A_y = +\frac{B}{2}x, A_z = 0$. Find the eigenvalues. Hint: Rewrite $\hat{H}$ as $\hat{H}_0 + \hat{H}_I$. Ignore B^2 terms, so $\hat{H}_I$ can be written as $\frac{eB}{2m}\hat{L}_z$.

5.3 Degenerate and nearly-degenerate perturbation theory

Degenerate perturbation theory is a natural extension to nondegenerate perturbation theory. A degenerate state means that more than one state share the same eigenenergy.

These degenerate eigenstates belong to the unperturbed Hamiltonian $\hat{H}_0$. If we have a perturbation $\hat{H}_I$, we cannot use nondegenerate perturbation theory. This is because eq. (5.1.13) has the denominator $\epsilon_k - \epsilon_m$,

$$\psi_k^{(1)} = \sum_{m \neq k} \frac{\langle \phi_m | \hat{H}_I | \phi_k \rangle}{\epsilon_k - \epsilon_m} |\phi_m\rangle.$$

Although $m \neq k$, a degenerate state has $\epsilon_m = \epsilon_k$ (a nearly degenerate state has a very small $\epsilon_m - \epsilon_k$), leading to divergence, which is clearly unphysical. This points out a path forward: Degenerate states dominate over all the other states. Although we target a single state $|k\rangle$, we must include all ν degenerate states, $|k_1\rangle, |k_2\rangle, \ldots, |k_\nu\rangle$, as the first-order correction. A nearly degenerate case needs to choose ν. This constitutes a major difference from nondegenerate perturbation theory: We form and diagonalize a matrix for the interaction Hamiltonian using the eigenstates of the unperturbed Hamiltonian, while nondegenerate perturbation theory is a summation over unperturbed states.

5.3.1 Expanding the wavefunction and energy

We start with a Hamiltonian $\hat{H}$, which consists of the unperturbed Hamiltonian $\hat{H}_0$ and the perturbation $\hat{H}_I$ as

$$\hat{H} = \hat{H}_0 + \lambda \hat{H}_I, \tag{5.3.1}$$

where as before λ is a parameter to denote the order and has no impact on our solution, since λ goes to 1 at the end of calculation. The theory assumes that the eigenstates and eigenenergies of the unperturbed $\hat{H}_0$ are known: $\hat{H}_0 \phi_i = \epsilon_i \phi_i$, where i runs from 1 to ν and ν is the degeneracy. Our goal is to find an approximate solution to $\hat{H}\psi_k = E_k \psi_k$, using ϕ_i as the basis functions, where k denotes our targeted state. The entire formalism is the matrix mechanics.

Degenerate perturbation theory expands ψ_k in terms of a group of ν-degenerate eigenstates ϕ_i[2] as $\psi_k = \sum_{i=1}^{\nu} C_i \phi_i$, where $C_i = C_i^{(0)} + \lambda C_i^{(1)} + \lambda^2 C_i^{(2)} + \cdots$. Our job is to determine $C_i^{(0)}, C_i^{(1)}, \ldots$. E_k is expanded similarly as $E_k = E_k^{(0)} + \lambda E_k^{(1)} + \lambda^2 E_k^{(2)} + \cdots$. We substitute both ψ_k and E_k into $\hat{H}\psi_k = E_k \psi_k$ to get

2 Once we decide to expand ψ_k in $\{\phi_i\}$, ψ_k is restricted to a state whose eigenenergy E_k is close to the eigenvalue ϵ_i of ϕ_i. In other words, degenerate perturbation theory, similar to nondegenerate perturbation theory, provides only approximate eigenstates within a particular energy window. To get a different energy window, one has to choose a different set of states.

$$(\hat{H}_0 + \lambda\hat{H}_I)\sum_{i=1}^{\nu}(C_i^{(0)} + \lambda C_i^{(1)} + \lambda^2 C_i^{(2)} + \cdots)\phi_i$$
$$= (E_k^{(0)} + \lambda E_k^{(1)} + \lambda^2 E_k^{(2)} + \cdots)\sum_j^{\nu}(C_j^{(0)} + \lambda C_j^{(1)} + \lambda^2 C_j^{(2)} + \cdots)\phi_j, \tag{5.3.2}$$

where we purposely choose i and j as indices for two summations for C_i and C_j because they are completely independent. Equation (5.3.2) is the mother of degenerate perturbation theory. The following step resembles nondegenerate perturbation theory. For the same λ, we match terms on both sides of the equation.

5.3.2 Zeroth-order correction

For λ^0, we have

$$\lambda^0 : \sum_{i=1}^{\nu}\hat{H}_0 C_i^{(0)}\phi_i = E_k^{(0)}\sum_{j=1}^{\nu}C_j^{(0)}\phi_j. \tag{5.3.3}$$

We multiply $\langle\phi_k|$ from the left and integrate it to obtain

$$\sum_{i=1}^{\nu}C_i^{(0)}\langle\phi_k|\hat{H}_0|\phi_i\rangle = E_k^{(0)}\sum_{j=1}^{\nu}C_j^{(0)}\langle\phi_k|\phi_j\rangle. \tag{5.3.4}$$

Since ϕ_i is an eigenstate of $\hat{H}_0$, $\hat{H}_0|\phi_i\rangle = \epsilon_i|\phi_i\rangle$, we can rewrite the previous equation as

$$\sum_{i=1}^{\nu}C_i^{(0)}\epsilon_i\langle\phi_k|\phi_i\rangle = E_k^{(0)}\sum_{j=1}^{\nu}C_j^{(0)}\langle\phi_k|\phi_j\rangle \rightarrow \epsilon_k = E_k^{(0)}, \tag{5.3.5}$$

where we have used the orthogonality $\langle\phi_k|\phi_i\rangle = \delta_{ik}$. This proves that the zeroth-order energy $E_k^{(0)}$ is just ϵ_k. $C_k^{(0)}$ can choose any values, so the zeroth-order wavefunction is the same as the unperturbed wavefunction.

5.3.3 First-order correction

Now, we consider the first-order correction λ^1 term

$$\lambda^1 : \sum_{i=1}^{\nu}[\hat{H}_0 C_i^{(1)}\phi_i + \hat{H}_I C_i^{(0)}\phi_i] = \sum_{j=1}^{\nu}[E_k^{(0)}C_j^{(1)}\phi_j + E_k^{(1)}C_j^{(0)}\phi_j]. \tag{5.3.6}$$

We multiply $\langle\phi_k|$ from the left and integrate the resultant equation to obtain

$$\sum_{i=1}^{\nu}[C_i^{(1)}\langle\phi_k|\hat{H}_0|\phi_i\rangle + C_i^{(0)}\langle\phi_k|\hat{H}_I|\phi_i\rangle] = \sum_{j=1}^{\nu}[E_k^{(0)}C_j^{(1)}\langle\phi_k|\phi_j\rangle + E_k^{(1)}C_j^{(0)}\langle\phi_k|\phi_j\rangle]. \tag{5.3.7}$$

Since $\langle\phi_k|\phi_j\rangle = \delta_{kj}$ and $\langle\phi_k|\phi_i\rangle = \delta_{ki}$, the equation can be further reduced to

$$E_k^{(0)}C_k^{(1)} + \sum_{i=1}^{\nu} C_i^{(0)}\langle\phi_k|\hat{H}_I|\phi_i\rangle = E_k^{(0)}C_k^{(1)} + E_k^{(1)}C_k^{(0)}. \tag{5.3.8}$$

Canceling the first term on both sides and rewriting $E_k^{(1)}C_k^{(0)}$ as $E_k^{(1)}\sum_i C_i^{(0)}\delta_{ik}$, we have the following secular equation:

$$\sum_{i=1}^{\nu}(\langle\phi_k|\hat{H}_I|\phi_i\rangle - E_k^{(1)}\delta_{ki})C_i^{(0)} = 0, \tag{5.3.9}$$

where $C_i^{(0)}$ is no longer arbitrary, in contrast to the zeroth order.

In order to have a nontrivial solution to the secular equation, the determinant must be zero, i. e.,

$$\begin{vmatrix} \langle 1|\hat{H}_I|1\rangle - E_k^{(1)} & \langle 1|\hat{H}_I|2\rangle & \cdots & \langle 1|\hat{H}_I|\nu\rangle \\ \langle 2|\hat{H}_I|1\rangle & \langle 2|\hat{H}_I|2\rangle - E_k^{(1)} & \cdots & \vdots \\ \langle \nu|\hat{H}_I|1\rangle & \vdots & \vdots & \langle \nu|\hat{H}_I|\nu\rangle - E_k^{(1)} \end{vmatrix} = 0, \tag{5.3.10}$$

where $|i\rangle$ is our shorthand notation for ϕ_i. $E_k^{(1)}$ has ν roots. Thus, the final eigenenergy E_k, up to the first order, is

$$E_k = E_k^{(0)} + E_k^{(1)}, \tag{5.3.11}$$

where k runs from 1 to ν.

Substituting $E_k^{(1)}$ back into eq. (5.3.9), we can find ν sets of $C_i^{(0)}$ to form our wavefunction ψ_k up to the first-order correction,

$$\psi_k = \sum_{i=1}^{\nu} C_i^{(0)}\phi_i, \tag{5.3.12}$$

where the wavefunction is expanded in the basis of the unperturbed ϕ_i. If we compare it with nondegenerate perturbation theory where its wavefunction up to the first order (eq. (5.1.14)) is $\psi_k = \phi_k + \psi_k^{(1)} = \phi_k + \sum_{m\neq k}\frac{\langle\phi_m|\hat{H}_I|\phi_k\rangle}{e_k - e_m}|\phi_m\rangle$, the degenerate theory now eliminates the divergence.

Since nearly all calculations in research focus on first-order degenerate perturbation, we end here, but higher-order perturbation can be done similarly.

5.4 Examples of degenerate perturbation theory

Many aspects of degenerate perturbation theory are similar to nondegenerate perturbation theory. We outline some similar steps:

- Separate the system Hamiltonian into two parts, $\hat{H}_0$ and $\hat{H}_I$, where $\hat{H}_0$ has a known solution and $\hat{H}_I$ is small in comparison with $\hat{H}_0$.
- Target a degenerate state k of interest.
- Construct the matrix elements of $\hat{H}_I$ among the eigenstates of $\hat{H}_0$.
- Diagonalize the matrix to find the eigenvalues and eigenstates.

5.4.1 A simplified approach to the H_2^+ molecule

The H_2^+ molecule consists of two protons, A and B, and one electron. Before two hydrogen atoms form a molecule, assume both are in their respective ground states, $|1s\rangle$. So, we have a degenerate case. The new ground state should not differ from the 1s-orbital localized on two atoms. This is because, in the limit of the infinite separation between two atoms, the wavefunction should approach $\phi_A(1s)$ at atom A and $\phi_B(1s)$ at B as

$$\phi_A(1s) = \frac{1}{\sqrt{\pi a_0^3}} e^{-\frac{r}{a_0}}, \quad \phi_B(1s) = \frac{1}{\sqrt{\pi a_0^3}} e^{-\frac{r'}{a_0}}, \tag{5.4.1}$$

where $r(r')$ is referenced to $A(B)$. So, we are going to start with these two wavefunctions as our zeroth-order wavefunction. The zeroth-order energy is $E^{(0)} = E_{1s}$. We target the new ground state of the molecule.

The electron Hamiltonian for H_2^+ is given by

$$\hat{H} = -\frac{\hbar^2}{2m_e}\nabla^2 - \frac{e^2}{4\pi\epsilon_0}\left(\frac{1}{r} + \frac{1}{r'}\right), \tag{5.4.2}$$

where the first term is the kinetic energy of the electron and the second and third terms are the Coulombic attractions between the electron and atoms A and B, respectively. We separate eq. (5.4.2) into the unperturbed $\hat{H}_0$ and the perturbed $\hat{H}_I$. We note that, for A, its unperturbed Hamiltonian is just the atomic hydrogen Hamiltonian: $\hat{H}_0 = \hat{H}_A = -\frac{\hbar^2}{2m_e}\nabla^2 - \frac{e^2}{4\pi\epsilon_0}\frac{1}{r}$. So, $\hat{H}_I = -\frac{e^2}{4\pi\epsilon_0}\frac{1}{r'}$ is its perturbation. Similarly, for B, $H_I = -\frac{e^2}{4\pi\epsilon_0}\frac{1}{r}$. We note in passing that whether we use r or r', ∇^2 stays the same.[3]

Next, we are going to form the matrix elements for $\hat{H}_I$, $\langle\phi_A|\hat{H}_I|\phi_A\rangle$, $\langle\phi_A|\hat{H}_I|\phi_B\rangle$, and $\langle\phi_B|\hat{H}_I|\phi_B\rangle$. Here we take a simplified approach by denoting $\langle\phi_A|\hat{H}_I|\phi_A\rangle = -\Delta$,

3 $\frac{\partial^2}{\partial x^2} = \frac{\partial^2}{\partial(x-x_A)^2} = \frac{\partial^2}{\partial(x-x_B)^2}$, where x_A and x_B are the x-coordinates of atoms A and B, respectively. The y- and z-coordinates are the same.

$\langle\phi_B|\hat{H}_I|\phi_B\rangle = -\Delta$, and $\langle\phi_A|\hat{H}_I|\phi_B\rangle = -t$, without explicitly computing these integrals. Both t and Δ are positive. Doing so leads to the following eigenequation:

$$\begin{pmatrix} -\Delta & -t \\ -t & -\Delta \end{pmatrix}\begin{pmatrix} C_A \\ C_B \end{pmatrix} = E^{(1)}\begin{pmatrix} C_A \\ C_B \end{pmatrix}, \tag{5.4.3}$$

which has a nontrivial solution only if its determinant is zero,

$$\begin{vmatrix} -\Delta - E^{(1)} & -t \\ -t & -\Delta - E^{(1)} \end{vmatrix} = 0. \tag{5.4.4}$$

We obtain the first-order energy correction, $E^{(1)} = -\Delta \pm t$. The eigenenergies up to the first order are $E_{1,2} = E_{1s} - \Delta \pm t$, where E_{1s} is the zeroth-order energy. H_2^+ has one root above $E_{1s} - \Delta$ by t and another below the limit by t.

Plugging $E_{1,2}$ into eq. (5.4.3), we find two eigenvectors: $C_A = C_B = \frac{1}{\sqrt{2}}$, and $C_A = -C_B = \frac{1}{\sqrt{2}}$. Our eigenfunction for $E_1^{(1)}$ in the basis of ϕ_A and ϕ_B is

$$\psi_{ss\sigma} = \frac{1}{\sqrt{2}}(\phi_A + \phi_B), \tag{5.4.5}$$

which is a $ss\sigma$-bonding state because ϕ_A and ϕ_B add up. The eigenfunction for the higher energy $E_2^{(1)}$ is an antibonding state,

$$\psi_{ss\sigma^*} = \frac{1}{\sqrt{2}}(\phi_A - \phi_B). \tag{5.4.6}$$

Antibonding orbitals are not stable, but they often participate in chemical reactions.

Exercise 5.4.1.

5. A system has the Hamiltonian $\hat{H}_0 = \begin{pmatrix} E_1 & 0 \\ 0 & E_2 \end{pmatrix}$, $H_I = \begin{pmatrix} -\Delta & ia \\ -ia & -\Delta \end{pmatrix}$, where $\hat{H}_I$ is much smaller than $\hat{H}_0$. (a) If $E_1 \ll E_2$, use the degenerate perturbation theory to find the first-order energy and second-order energy corrections to the lowest eigenstate. (b) If $E_1 \approx E_2$, use the nondegenerate perturbation theory to find the first-order energy to the lowest eigenstate. (c) Under what condition will the result in (b) approach that in (a)?

5.4.2 Stark effect in hydrogen atom

The Stark effect refers to the spectral line splitting under an external electric field. Hydrogen's $n = 2$ state is fourfold degenerate: $2s$, $2p_x$, $2p_y$, and $2p_z$. When it is placed in a uniform electric field $\mathbf{F}$ along the z-axis, this state splits. (a) Compute the split eigenenergies. (b) Compute its wavefunctions in terms of unperturbed eigenstates.

(a) First, we need to find the interaction Hamiltonian, $\hat{H}_I$. From EM, we know this is the dipole interaction $\hat{H}_I = -\hat{\mathbf{D}} \cdot \mathbf{F} = +e\hat{\mathbf{r}} \cdot \mathbf{F} = erF\cos\theta$, with θ being the angle between the dipole moment and electric field. Because we have a fourfold degenerate state, we need to form a 4×4 matrix for $\hat{H}_I$, $\langle \phi_i | \hat{H}_I | \phi_j \rangle$, where $\{\phi_i\}$ are $|2s\rangle = R_{20}Y_{00}, |2p_x\rangle = R_{21}Y_{1,+1}, |2p_y\rangle = R_{21}Y_{1,-1}, |2p_z\rangle = R_{21}Y_{1,0}$. Since the spatial part $R(r)$ and the angular part $Y_{lm}(\theta, \phi)$ are separated, the integration $\langle \phi_i | \hat{H}_I | \phi_j \rangle$ can be carried out separately. Because the radial integration over r is almost always nonzero and the angular part may be zero, we carry out the angular integration first. For instance,

$$\langle 2s | \hat{H}_I | 2p_x \rangle = \iiint R_{20} Y_{00}^* erF \cos\theta R_{21} Y_{1,1} r^2 \sin\theta d\theta d\phi dr, \tag{5.4.7}$$

whose angular part is

$$\iint Y_{00}^* \cos\theta Y_{1,1} \sin\theta d\theta d\phi = \iint Y_{00}^* \cos\theta Y_{1,1} d\Omega. \tag{5.4.8}$$

One may directly plug in Y_{00} and $Y_{1,1}$ and integrate, but it is easier to use the following relationship:

$$\cos\theta Y_{lm} = \sqrt{\frac{(l+1)^2 - m^2}{(2l+1)(2l+3)}} Y_{l+1,m} + \sqrt{\frac{l^2 - m^2}{(2l-1)(2l+1)}} Y_{l-1,m}. \tag{5.4.9}$$

This shows that $\cos\theta Y_{lm}$ can be represented by two other spherical harmonics. Therefore, $\cos\theta Y_{1,1}$ only yields $Y_{2,1}$ and $Y_{0,1}$. Because all the spherical harmonics must be orthonormalized,

$$\iint d\Omega Y_{l_1 m_1}^* Y_{l_2 m_2} = \int_0^{2\pi} \int_0^{\pi} \sin\theta d\theta d\phi Y_{l_1 m_1}^* Y_{l_2 m_2} = \delta_{l_1 l_2} \delta_{m_1 m_2}, \tag{5.4.10}$$

where $d\Omega = \sin\theta d\theta d\phi$ is the solid-angle finite element. Then, we have $\iint Y_{00}^* \cos\theta Y_{1,1} d\Omega = 0$. Therefore, the only nonzero element is $\langle 2s | \hat{H}_I | 2p_z \rangle$,

$$\langle 2s | \hat{H}_I | 2p_z \rangle = \int_0^{\infty} eFR_{20}(r) r R_{21}(r) r^2 dr \times \int_{\phi=0}^{2\pi} \int_{\theta=0}^{\pi} Y_{00}(\theta, \phi) \cos\theta Y_{10} d\Omega. \tag{5.4.11}$$

The angular integral is

$$\int_{\phi=0}^{\pi} \int_{\theta=0}^{\pi} Y_{00} \left(\sqrt{\frac{(1+1)^2 - 0^2}{(2+1)(2+3)}} Y_{2,0} + \sqrt{\frac{1^2 - 0^2}{3}} Y_{0,0} \right) d\Omega = \frac{1}{\sqrt{3}}. \tag{5.4.12}$$

The radial integral is (see the exercise)

$$\int_0^{\infty} R_{20}(r) r R_{21}(r) r^2 dr = -\frac{9a_0}{\sqrt{3}}, \tag{5.4.13}$$

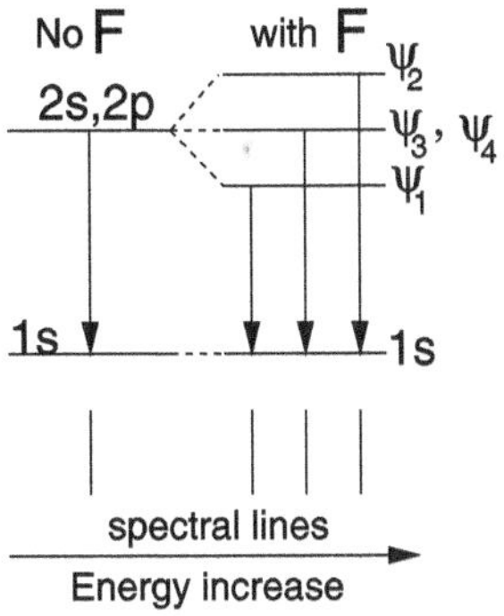

Figure 5.2: The Stark effect in a hydrogen atom. Without an electric field **F**, 2*s*, 2*p* are degenerate, but with **F**, they split, and the transition from the higher level to the lower level produces three spectral lines, instead of one without **F**.

Finally, $\langle 2s|\hat{H}_I|2p_z\rangle = eF\frac{1}{\sqrt{3}}(-\frac{9a_0}{\sqrt{3}}) = -3eFa_0$. Our final matrix $\hat{H}_I$ is

$$\begin{pmatrix} 0 & -3eFa_0 & 0 & 0 \\ -3eFa_0 & 0 & 0 & 0 \\ 0 & 0 & 0 & 0 \\ 0 & 0 & 0 & 0 \end{pmatrix}. \tag{5.4.14}$$

Diagonalizing it produces four roots, $E^{(1)} = \pm 3eFa_0, 0, 0$. The eigenenergies of $\hat{H} = \hat{H}_0 + \hat{H}_I$, to the first order, are $E^{(0)} + 3eFa_0$, $E^{(0)} - 3eFa_0$, $E^{(0)}$ and $E^{(0)}$, where $E^{(0)}$ is the external field-free 2*s*2*p* energy. The two roots, i. e., E_0, are doubly degenerate.

(b) The wavefunctions for the two *sp*-hybridized levels, with $E_1 = E^{(0)} - 3eFa_0$ and $E_2 = E^{(0)} + 3eFa_0$, are $\psi_1 = \frac{1}{\sqrt{2}}(|2s\rangle + |2p_z\rangle)$ and $\psi_2 = \frac{1}{\sqrt{2}}(|2s\rangle - |2p_z\rangle)$. In the absence of an external electric field, the transition from the 2*p*-state to 1*s*-state emits a photon with energy $\delta E = E^{(0)} - E_{1s}$. With the external field, if the electron is initially in the ψ_1-state, its transition to the $|1s\rangle$-state emits a photon with energy $\Delta E = E_1 - E_{1s} = E^{(0)} - E_{1s} - 3eFa_0$, which is $3eFa_0$ less than δE. But, if the electron initially is in ψ_2, then the photon energy is $3eFa_0$ higher than δE. The two states ψ_3 and ψ_4 that do not have energy change produce a single line. This means that spectrally one single line is split into three lines (see Fig. 5.2), the Stark effect.

Exercise 5.4.2.

6. A three-level system has double-degenerate level $E_1^{(0)}$ and one nondegenerate level $E_2^{(0)}$, with $E_2^{(0)} > E_1^{(0)}$. Its unperturbed and perturbed Hamiltonian matrices are

$$\hat{H}_0 = \begin{pmatrix} E_1^{(0)} & 0 & 0 \\ 0 & E_1^{(0)} & 0 \\ 0 & 0 & E_2^{(0)} \end{pmatrix}, \quad \hat{H}_I = \begin{pmatrix} 0 & 0 & a \\ 0 & 0 & b \\ a^* & b^* & 0 \end{pmatrix}.$$

(a) Use degenerate perturbation theory to find the energy correction to the first eigenstate up to the second order. (b) Use the exact diagonalization method to compute the first eigenstate energy and compare the results with (a).

7. Prove eq. (5.4.13), where $R_{20} = \frac{1}{\sqrt{2}a_0^{\frac{3}{2}}}(1 - \frac{r}{2a_0})e^{-\frac{r}{2a_0}}$ and $R_{21} = \frac{1}{2\sqrt{6}a_0^{\frac{3}{2}}}\frac{r}{a_0}e^{-\frac{r}{2a_0}}$. Hint: This integral uses the following relationship: $\int_0^\infty x^n e^{-ax}dx = \frac{n!}{a^{n+1}}$, $(a > 0, n > 0)$.

8. A hydrogen atom in $n = 2$ is placed in an electric field in the xy-plane which makes an angle of α with respect to the x-axis. The interaction Hamiltonian is $\hat{H}_I = -\hat{\mathbf{D}} \cdot \mathbf{F}$. (a) Compute the energy splitting as a function of α and plot the results. (b) Find the wavefunction correction as a function of α. (c) How many spectral lines are there?

5.5 Variational principle

Perturbation theory (PT) is an important approximate method. Another approximate method is the variational principle. Different from PT, we do not separate the Hamiltonian into two parts; instead, we use the system Hamiltonian directly, but rely on a trial wavefunction that contains multiple variables to be optimized. These variables can be a function or coefficients of a function. Mathematically, the variational principle is equivalent to a constrained optimization based on Lagrange multipliers. A good trial wavefunction can deliver an excellent result, but having such a good wavefunction requires good physical intuition. The variational principle is very flexible. One can optimize the wavefunction ψ directly through functionals, or additional variational variables if desired. In density-functional theory, one uses the density as a variable.

5.5.1 Functional derivatives

Functionals are a function of the function. For instance, $F[\rho]$ is a functional of ρ, where ρ itself is a function of position $\mathbf{r}$. Here, we use square brackets to denote a functional. Its functional derivative is denoted by δF or ∂F, instead of dF.

We take $F[\rho] = a\rho^2 + b$ as an example, where a and b are constants. The functional derivative of F, δF can be found through the regular Taylor expansion,

$$F[\rho + \delta\rho] = a(\rho + \delta\rho)^2 + b = a\rho^2 + 2a\rho\delta\rho + a(\delta\rho)^2 + b = F[\rho] + 2a\rho\delta\rho + a(\delta\rho)^2,$$
$$F[\rho + \delta\rho] - F[\rho] = 2a\rho\delta\rho + a(\delta\rho)^2.$$

The finite difference, $\Delta F = F[\rho + \delta\rho] - F[\rho] = 2a\rho\delta\rho + a(\delta\rho)^2$, becomes a functional derivative in the limit $\delta\rho \to 0$,

$$\frac{\delta F}{\delta\rho} = \lim_{\delta\rho\to 0}(2a\rho + a\delta\rho) = 2a\rho, \tag{5.5.1}$$

where one sees the functional derivative is similar to a regular derivative.

In QM, $F[\rho]$ may contain an integration with a gradient term $\nabla\rho$ such as

$$F[\rho] = \int f[\mathbf{r}, \rho(\mathbf{r}), \nabla\rho(\mathbf{r})]d\mathbf{r}. \tag{5.5.2}$$

We change the function ρ to $\rho + \delta\rho$ to get

$$F[\rho + \delta\rho] = \int f[\mathbf{r}, \rho + \delta\rho, \nabla\rho + \nabla\delta\rho]d\mathbf{r}. \tag{5.5.3}$$

We subtract eq. (5.5.2) from eq. (5.5.3) to generate

$$F[\rho + \delta\rho] - F[\rho] = \int (f[\mathbf{r}, \rho + \delta\rho, \nabla\rho + \nabla\delta\rho] - f[\mathbf{r}, \rho(\mathbf{r}), \nabla\rho(\mathbf{r})])d\mathbf{r} \equiv \int \frac{\delta F}{\delta\rho}\delta\rho d\mathbf{r}, \tag{5.5.4}$$

where the final equation is our definition of the functional derivative $\frac{\delta F}{\delta\rho}$ to be found.

We first carry out a similar Taylor expansion for f up to the first order since all the higher orders go to zero once we take the limit $\delta\rho \to 0$ (see eq. (5.5.1)). This leads to

$$f[\mathbf{r}, \rho + \delta\rho, \nabla\rho + \nabla\delta\rho] = f[\mathbf{r}, \rho, \nabla\rho] + \frac{\partial f}{\partial\rho}\delta\rho + \frac{\partial f}{\partial\nabla\rho} \cdot \nabla\delta\rho. \tag{5.5.5}$$

If we compare the right-most term of eq. (5.5.5) and the right-most equation of (5.5.4), we see $\nabla\delta\rho$ does not have the desired form because $\delta\rho$ is behind ∇ and does not stand alone. We employ the divergence identity for the product of a scalar ϕ with a vector $\mathbf{A}$,

$$\nabla \cdot (\mathbf{A}\phi) = \phi(\nabla \cdot \mathbf{A}) + \mathbf{A} \cdot (\nabla\phi) \to \mathbf{A} \cdot (\nabla\phi) = \nabla \cdot (\mathbf{A}\phi) - \phi(\nabla \cdot \mathbf{A}), \tag{5.5.6}$$

to rewrite (recognizing $\delta\rho$ as ϕ and $\frac{\partial f}{\partial\nabla\rho}$ as $\mathbf{A}$)

$$\frac{\partial f}{\partial\nabla\rho} \cdot \nabla\delta\rho = \nabla \cdot \left(\frac{\partial f}{\partial\nabla\rho}\delta\rho\right) - \delta\rho\left(\nabla \cdot \frac{\partial f}{\partial\nabla\rho}\right). \tag{5.5.7}$$

So, the integral becomes

$$\int \frac{\partial f}{\partial\nabla\rho} \cdot \nabla\delta\rho d\mathbf{r} = \int \nabla \cdot \left(\frac{\partial f}{\partial\nabla\rho}\delta\rho\right)d\mathbf{r} - \int \delta\rho\left(\nabla \cdot \frac{\partial f}{\partial\nabla\rho}\right)d\mathbf{r}. \tag{5.5.8}$$

The divergence theorem allows us to convert the first term to a surface integral $\oiint(\frac{\partial f}{\partial\nabla\rho}\delta\rho) \cdot d\mathbf{a}$. Because any physically meaningful quantity is zero at infinity, this surface integral must be zero. Then we have

$$F[\rho + \delta\rho] - F[\rho] = \int \left[\frac{\partial f}{\partial\rho} - \left(\nabla \cdot \frac{\partial f}{\partial\nabla\rho}\right)\right]\delta\rho d\mathbf{r}, \tag{5.5.9}$$

where both terms have our desired form, i. e., $\delta\rho$ multiplied by other terms. By the definition of the functional derivative in the last equation of eq. (5.5.4), we always can write

$$F[\rho+\delta\rho]-F[\rho]=\int\frac{\delta F}{\delta\rho}\delta\rho d\mathbf{r}. \tag{5.5.10}$$

If we compare eqs. (5.5.9) and (5.5.10), we find

$$\frac{\delta F}{\delta\rho}=\frac{\partial f}{\partial\rho}-\nabla\cdot\frac{\partial f}{\partial\nabla\rho}, \tag{5.5.11}$$

where ∂ on the right side must be understood as a functional derivative. This is a generic expression for an integral form of a functional derivative. The key message is that we can ignore the integral over $d\mathbf{r}$ if we only have one integral over $\mathbf{r}$. If we have more than one integral over $\mathbf{r}$, we keep one fewer integral. If f does not contain $\nabla\rho$, then one only has the first term $\frac{\partial f}{\partial\rho}$.

Exercise 5.5.1.

9. The Thomas–Fermi kinetic energy functional is the first functional used in the pre-density functional theory era. It connects the kinetic energy to the electron density $\rho(\mathbf{r})$ and has the form

$$T_{\mathrm{TF}}[\rho]=C_F\int\rho^{5/3}d\mathbf{r},$$

where ρ is a function of $\mathbf{r}$ and $C_F=\frac{3\hbar^2}{10m_e}(\frac{3}{\pi})^{2/3}$. Find the functional derivative $\delta T/\delta\rho$ with respect to ρ.

10. The electron–electron repulsion energy is $E[\rho]=\frac{1}{2}\frac{e^2}{4\pi\epsilon_0}\iint\frac{\rho(\mathbf{r})\rho(\mathbf{r}')}{\mathbf{r}-\mathbf{r}'}d\mathbf{r}d\mathbf{r}'$. (a) Find $\frac{\delta E}{\delta\rho(\mathbf{r})}$. (b) Find $\frac{\delta^2 E}{\delta\rho(\mathbf{r})\delta\rho(\mathbf{r}')}$.

5.5.2 Spatial scaling as a variational parameter: the virial theorem

The variational principle is very flexible. We can use the spatial scaling as a variational parameter. In the following, we take the Coulomb potential as an example for a system of N electrons. Suppose $\psi(\mathbf{r}_1,\ldots,\mathbf{r}_N)$ is a normalized eigenstate of the system,

$$\hat{H}\psi=(\hat{T}+\hat{V})\psi=\left(-\sum_i\frac{\hbar^2\nabla_i^2}{2m_e}+\frac{1}{2}\sum_{i\neq j}\frac{e^2}{4\pi\epsilon_0 r_{ij}}\right)\psi=E\psi. \tag{5.5.12}$$

If we rescale every degree of freedom from x_i to $x_i'=\lambda x_i$, i.e. setting $x_i=x_i'/\lambda$, from y_i to $y_i'=\lambda y_i$, and from z_i to $z_i'=\lambda z_i$, then the kinetic energy operator $\hat{T}$ is changed to $\lambda^2\hat{T}'$, while the potential energy operator $\hat{V}$ is changed to $\lambda\hat{V}'$. So our Hamiltonian rescales as

$$\hat{H}'=\lambda^2\hat{T}'+\lambda\hat{V}'. \tag{5.5.13}$$

The wavefunction must be rescaled as well because it must be normalized as $\int d\mathbf{r}_1\cdots d\mathbf{r}_N|\psi(\mathbf{r}_1,\ldots,\mathbf{r}_N)|^2=1$, where $d\mathbf{r}_1$ must be understood as $dx_1dy_1dz_1$ and all the rest are the same. It can be shown that

$$\psi'(\mathbf{r}_1',\ldots,\mathbf{r}_N') = \lambda^{-3N/2}\psi(\mathbf{r}_1,\ldots,\mathbf{r}_N). \tag{5.5.14}$$

The details are relegated to the following exercise.

Based on eq. (5.5.13), we find

$$E_\lambda \equiv \langle\psi'|\hat{H}'|\psi'\rangle = \lambda^2\langle\hat{T}'\rangle + \lambda\langle\hat{V}'\rangle. \tag{5.5.15}$$

Finally, we take the derivative with respect to λ to get the energy minimum,

$$\frac{\partial E_\lambda}{\partial\lambda} = 2\lambda\langle\hat{T}'\rangle + \langle\hat{V}'\rangle = 0. \tag{5.5.16}$$

This means that $\langle\hat{V}'\rangle = -2\lambda\langle\hat{T}'\rangle$. Because the true eigenenergy is at $\lambda = 1$, where $\hat{T}' = \hat{T}$ and $\hat{V}' = \hat{V}$, this leads to the virial theorem for the Coulomb potential,

$$2\langle\hat{T}\rangle + \langle\hat{V}\rangle = 0 \quad \rightarrow \quad -\frac{\langle\hat{V}\rangle}{\langle\hat{T}\rangle} = 2. \tag{5.5.17}$$

The potential energy is twice the negative kinetic energy. If the potential deviates from the Coulomb potential, the ratio $\frac{-\langle\hat{V}\rangle}{\langle\hat{T}\rangle}$ differs from 2. This is the application of the virial theorem for Coulombic potentials in QM (see Section 4.1.1).

Exercise 5.5.2.

11. Ritz method. Prove that the minimum energy of the expectation value of the Hamiltonian $\langle\psi|\hat{H}|\psi\rangle$ corresponds to the time-independent Schrödinger equation $\hat{H}|\psi\rangle = E|\psi\rangle$, where $|\psi\rangle$ is the wavefunction to be optimized and E is just λ. Hint: Choose a functional $F = \langle\psi|\hat{H}|\psi\rangle - \lambda\langle\psi|\psi\rangle$, where λ is the Lagrange multiplier.

12. (a) If we rescale from x_i to $x_i' = \lambda x_i$, y_i to $y_i' = \lambda y_i$, and z_i to $z_i' = \lambda z_i$, show that the wavefunction will be rescaled to $\psi(\mathbf{r}_1',\ldots,\mathbf{r}_N') = \lambda^{-3N/2}\psi(\mathbf{r}_1,\ldots,\mathbf{r}_N)$. (b) Following (a), prove $\langle\psi'|\hat{H}|\psi'\rangle = \langle\psi|\hat{H}|\psi\rangle$, where $|\psi'\rangle$ is the scaled wavefunction. (c) Prove, after recaling $\mathbf{r} \rightarrow \mathbf{r}' = \lambda\mathbf{r}$, $\hat{T} \rightarrow \lambda^2\hat{T}'$, while $\hat{V}$ is changed to $\lambda\hat{V}'$.

5.6 Examples of the variational principle in the H_2^+ molecule

Very few examples other than the H_2^+ molecule can demonstrate the full power of the variational principles. In the following, we carry out two variations. We start with the optimization over the coefficients of the wavefunction, which we call the first variation.

5.6.1 First variation

H_2^+ has two protons and one electron. Physically, the molecular orbital should not differ too much from its atomic orbital, so we adopt a linear combination of atomic orbitals (LCAO) as our trial wavefunction,

$$\psi = C_A\phi_A + C_B\phi_B, \tag{5.6.1}$$

where C_A and C_B are two coefficients to be optimized. The variational principle hinges on an optimization principle. Suppose we are interested in the ground-state energy. Then, we minimize the energy expectation value F,

$$\begin{aligned} F &\equiv \langle\psi|\hat{H}|\psi\rangle = \langle C_A^*\phi_A + C_B^*\phi_B|\hat{H}|C_A\phi_A + C_B\phi_B\rangle \\ &= C_A^*C_A\langle\phi_A|\hat{H}|\phi_A\rangle + C_B^*C_B\langle\phi_B|\hat{H}|\phi_B\rangle + C_A^*C_B\langle\phi_A|\hat{H}|\phi_B\rangle + C_B^*C_A\langle\phi_B|\hat{H}|\phi_A\rangle. \end{aligned} \tag{5.6.2}$$

However, we cannot minimize it without any constraint because doing so would yield $C_A = C_B = 0$. The constraint is the wavefunction normalization, i. e.,

$$\langle\psi|\psi\rangle = |C_A|^2\langle\phi_A|\phi_A\rangle + |C_B|^2\langle\phi_B|\phi_B\rangle + C_A^*C_B\langle\phi_A|\phi_B\rangle + C_B^*C_A\langle\phi_B|\phi_A\rangle = 1. \tag{5.6.3}$$

To simplify our notation, we define $H_{AA} = \langle\phi_A|\hat{H}|\phi_A\rangle$, $H_{BB} = \langle\phi_B|\hat{H}|\phi_B\rangle$, $H_{BA} = \langle\phi_B|\hat{H}|\phi_A\rangle$, and $H_{AB} = \langle\phi_A|\hat{H}|\phi_B\rangle$. We denote $S = \langle\phi_A|\phi_B\rangle$ and $S^* = \langle\phi_B|\phi_A\rangle$ and assume $\langle\phi_A|\phi_A\rangle = \langle\phi_B|\phi_B\rangle = 1$.

The function to be optimized is $G = F - E(\langle\psi|\psi\rangle - 1) = C_A^*C_AH_{AA} + C_B^*C_BH_{BB} + C_A^*C_BH_{AB} + C_B^*C_AH_{BA} - E(C_A^*C_A + C_B^*C_B + C_A^*C_BS + C_B^*C_AS^* - 1)$, where E is the Lagrange multiplier and gives the eigenenergy. Next, we decide what variables to be optimized. In order to make connections with Subsection 5.4.1, we keep ϕ_A and ϕ_B fixed and optimize C_A, C_B. The *variational principle* is realized by taking the derivative of G with respect to C_A^* and C_B^* and setting them to zero

$$\frac{\partial G}{\partial C_A^*} = C_AH_{AA} + C_BH_{AB} - E(C_A + C_BS) = 0, \tag{5.6.4}$$

$$\frac{\partial G}{\partial C_B^*} = C_BH_{BB} + C_AH_{BA} - E(C_B + C_AS^*) = 0, \tag{5.6.5}$$

which can be cast into a matrix form as

$$\begin{pmatrix} H_{AA} & H_{AB} \\ H_{BA} & H_{BB} \end{pmatrix}\begin{pmatrix} C_A \\ C_B \end{pmatrix} = E\begin{pmatrix} 1 & S \\ S^* & 1 \end{pmatrix}\begin{pmatrix} C_A \\ C_B \end{pmatrix}. \tag{5.6.6}$$

This is a generic matrix eigenvalue problem, matrix mechanics.

Before we proceed, we pause to make connections to the degenerate perturbation theory. If we set $H_{AA} = H_{BB} = E_{1s} - \Delta$, $H_{AB} = H_{BA} = -t$, and $S = 0$, then we have the eigenequation

$$\begin{pmatrix} E_{1s} - \Delta & -t \\ -t & E_{1s} - \Delta \end{pmatrix}\begin{pmatrix} C_A \\ C_B \end{pmatrix} = E\begin{pmatrix} C_A \\ C_B \end{pmatrix}, \tag{5.6.7}$$

which is exactly the same as eq. (5.4.3), except for E_{1s}.

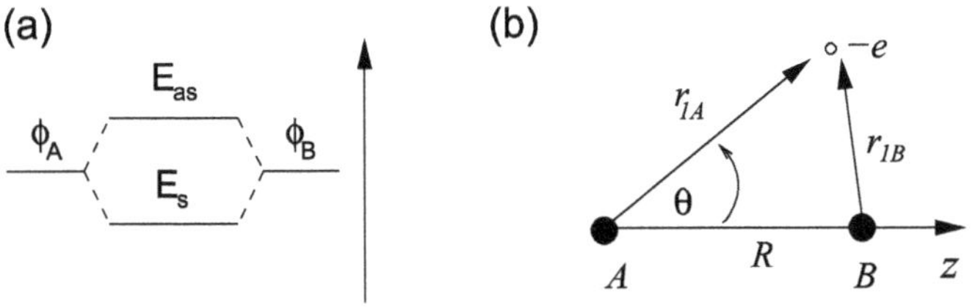

Figure 5.3: (a) Energy diagram. The symmetric bonding orbital energy E_s is lowered, while the antibonding and asymmetric orbital energy E_{as} is increased (eq. (5.6.8)). (b) H_2^+ model. We align the z-axis along the bond that connects two H atoms A and B (two filled circles) which are separated by distance R. The electron (empty circle) is r_{1A} and r_{1B} away from these two atoms.

Here, we can solve eq. (5.6.6) directly with the details left to the following exercise. Two eigenenergies that which correspond to one symmetric (bonding) and one antisymmetric (antibonding) eigenstates are

$$E_s = \frac{H_{AA} + H_{AB}}{1+S}, \qquad E_{as} = \frac{H_{AA} - H_{AB}}{1-S}. \tag{5.6.8}$$

Which one has a lower energy depends on the signs of H_{AA} and H_{AB}. If ϕ_A and ϕ_B are chosen as a localized state, such as the 1s-orbital of the H atom, then H_{AA} and H_{AB} are both negative, while S is always positive and less than one. Therefore, E_s is lower than E_{as} (see Fig. 5.3(a)). This is the same reason why in the matrix we have $-t$, where t itself is positive. As discussed before, this corresponds to a bonding state, and E_{as} corresponds to an antibonding state.

Exercise 5.6.1.
13. Prove $\begin{pmatrix} 1 & S \\ S & 1 \end{pmatrix}^{-1} = \frac{1}{1-S^2}\begin{pmatrix} 1 & -S \\ -S & 1 \end{pmatrix}$.
14. Starting from eq. (5.6.6), show its eigenvalues are given by eq. (5.6.8).

5.6.2 Second variation

The second variation includes all the features of the first variation and has an additional spatial extension of the atomic wavefunction. This new feature produces results that are accurate enough to be compared with experiments.

5.6.2.1 Geometry and trial wavefunction

We place two H atoms separated by a distance R along the z-axis (see Fig. 5.3(b)). The distance between the electron and atom A is r_{1A} and that between the electron and atom B is r_{1B}. Since we are interested in the ground-state wavefunction of H_2^+, we expect that the wavefunction should not be different from its atomic wavefunction as the molecular orbital should return to the atomic orbital as $R \to \infty$.

Our variational molecular wavefunction is a linear combination of two atomic 1s-orbital wavefunctions just as eq. (5.6.1),

$$\psi = C_A \phi_A(\mathbf{r}_{1A}) + C_B \phi_B(\mathbf{r}_{1B}), \tag{5.6.9}$$

where two 1s-orbitals for atoms A and B are

$$\phi_A(\mathbf{r}_{1A}) = \frac{\lambda^{3/2}}{\sqrt{\pi a_0^3}} e^{-\lambda r_{1A}/a_0}, \quad \phi_B(\mathbf{r}_{1B}) = \frac{\lambda^{3/2}}{\sqrt{\pi a_0^3}} e^{-\lambda r_{1B}/a_0}. \tag{5.6.10}$$

Here, $\mathbf{r}_{1A} = \mathbf{r}_1 - \mathbf{r}_A$, $\mathbf{r}_1$ is the coordinate of the electron, $\mathbf{r}_A$ is the coordinate of atom A, $r_{1A} = |\mathbf{r}_{1A}|$, and a_0 is the Bohr radius. Different from eq. (5.6.1), we have a new variational parameter λ that allows us to fine-tune the spatial extension of the original atomic orbital to get a more accurate molecular orbital since, when atoms form molecules, the atomic wavefunction must always undergo some changes.

Different from eq. (5.6.6), our wavefunction has two sets of variational parameters, $\{C_i\}$ and λ. Our first variation with respect to C_A and C_B yields eq. (5.6.6),

$$\begin{pmatrix} H_{AA} & H_{AB} \\ H_{BA} & H_{BB} \end{pmatrix} \begin{pmatrix} C_A \\ C_B \end{pmatrix} = E \begin{pmatrix} 1 & S \\ S & 1 \end{pmatrix} \begin{pmatrix} C_A \\ C_B \end{pmatrix}. \tag{5.6.11}$$

Our second variation with respect to λ allows us to further optimize the atomic orbitals. This needs the explicit expressions of H_{AA}, H_{BB}, H_{AB}, H_{BA}, and S.

5.6.2.2 Overlap integral S

Recall that the overlap integral S is defined as

$$S = \langle \phi_A(\mathbf{r}) | \phi_B(\mathbf{r}) \rangle = \frac{\lambda^3}{\pi a_0^3} \iiint e^{-\lambda(r_{1A}+r_{1B})/a_0} d\mathbf{r}. \tag{5.6.12}$$

Because ϕ_A and ϕ_B are referenced with respect to atoms A and B separately, we cannot integrate directly. We must choose an appropriate coordinate to simplify the integral. We first place atom A at the origin (see Fig. 5.3(b)), $\mathbf{r}_A = 0$, so $\mathbf{r}_{1A} = \mathbf{r}_1 - \mathbf{r}_A = \mathbf{r}$, where we have removed the subscript "1" in $\mathbf{r}$. This way we can express $\mathbf{r}_{1B}$ in terms of the polar angle θ and R as $|\mathbf{r}_{1B}| = \sqrt{R^2 + r^2 - 2Rr\cos\theta}$. This allows us to rewrite S as

$$S = \frac{\lambda^3}{\pi a_0^3} \iiint e^{-\lambda(r+\sqrt{r^2+R^2-2rR\cos\theta})/a_0} r^2 \sin\theta d\theta d\phi dr. \tag{5.6.13}$$

After a lengthy but straightforward integration, we find

$$S = \left(1 + \frac{\lambda R}{a_0} + \frac{\lambda^2 R^2}{3a_0^2}\right)e^{-\lambda R/a_0}, \tag{5.6.14}$$

whose derivation is included as a homework exercise.

5.6.2.3 Hamiltonian matrix elements

The molecular Hamiltonian of H_2^+ consists of electronic and nuclear parts, $\hat{H}_e$ and $\hat{H}_p$,

$$\begin{aligned}\hat{H}_e &= -\frac{\hbar^2\nabla^2}{2m_e} + V(\mathbf{r}_{1A}) + V(\mathbf{r}_{1B}),\\ \hat{H}_p &= -\frac{\hbar^2\nabla_A^2}{2m_p} - \frac{\hbar\nabla_B^2}{2m_p} + V(\mathbf{r}_{AB}),\end{aligned} \tag{5.6.15}$$

where m_e and m_p are the electron and proton masses, respectively, and $V(\mathbf{r}_{1A})$ and $V(\mathbf{r}_{1B})$ are the attraction between the electron and proton A and that between the electron and proton B,

$$V(\mathbf{r}_{1A}) = -\frac{1}{4\pi\varepsilon_0}\frac{e^2}{r_{1A}}, \quad V(\mathbf{r}_{1B}) = -\frac{1}{4\pi\varepsilon_0}\frac{e^2}{r_{1B}}. \tag{5.6.16}$$

The proton–proton repulsion is given by $V(\mathbf{r}_{AB}) = \frac{1}{4\pi\varepsilon_0}\frac{e^2}{R}$, where R is the distance between them. The variational calculation is carried out over $\hat{H}_e$ and $V(\mathbf{r}_{AB})$, where the kinetic energy of the protons does not enter the calculation since we ignore the zero-point vibration of the nucleus.

Because all the Hamiltonian matrix elements have forms like $\langle\phi(r_{1A})|\hat{H}_e|\phi(r_{1A})\rangle$, we compute $\hat{H}_e|\phi(r_{1A})\rangle$. The potential energy terms are straightforward since they present a simple product, but the kinetic energy operator needs some caution. Since ϕ_A has no angular dependence, we can drop the angular terms in ∇^2 to get $\frac{1}{r_{1A}}\frac{d^2}{dr_{1A}^2}(r_{1A}\phi(\mathbf{r}_{1A}))$.[4]

A direct calculation yields

$$\nabla^2\phi(\mathbf{r}_{1A}) = \left(-\frac{2}{r_{1A}}\frac{\lambda}{a_0} + \frac{\lambda^2}{a_0^2}\right)e^{-\lambda r_{1A}/a_0}. \tag{5.6.17}$$

So the kinetic energy term is

$$-\frac{\hbar^2\nabla^2}{2m_e}\phi(\mathbf{r}_{1A}) = \left(-\frac{\hbar^2\lambda^2}{2m_e a_0^2} + \frac{\hbar^2\lambda}{m_e a_0 r_{1A}}\right)\phi(\mathbf{r}_{1A}) = \left(-\frac{\hbar^2\lambda^2}{2m_e a_0^2} + \frac{e^2\lambda}{4\pi\epsilon_0 r_{1A}}\right)\phi(\mathbf{r}_{1A}), \tag{5.6.18}$$

where we have used eq. (4.1.5) to replace $\hbar^2/(m_e a_0)$ by $e^2/(4\pi\epsilon_0)$. Then, we obtain

4 In ∇, the variable is r, but, since here ϕ is centered at atom A, we replace r by r_{1A}.

$$\hat{H}_e\phi(\mathbf{r}_{1A}) = \left(-\frac{\hbar^2\nabla^2}{2m_e} - \frac{1}{4\pi\varepsilon_0}\frac{e^2}{r_{1A}} - \frac{1}{4\pi\varepsilon_0}\frac{e^2}{r_{1B}}\right)\phi(\mathbf{r}_{1A})$$
$$= \left(-\frac{\hbar^2\lambda^2}{2m_e a_0^2} + \frac{e^2}{4\pi\varepsilon_0}\frac{(\lambda-1)}{r_{1A}} - \frac{e^2}{4\pi\varepsilon_0}\frac{1}{r_{1B}}\right)\phi(\mathbf{r}_{1A}). \quad (5.6.19)$$

The diagonal matrix element $H_{AA} = \langle\phi(\mathbf{r}_{1A})|\hat{H}_e|\phi(\mathbf{r}_{1A})\rangle$ is

$$H_{AA} = -\frac{\hbar^2\lambda^2}{2m_e a_0^2} + \frac{e^2(\lambda-1)}{4\pi\varepsilon_0}\langle\phi(\mathbf{r}_{1A})|\frac{1}{r_{1A}}|\phi(\mathbf{r}_{1A})\rangle - \frac{e^2}{4\pi\varepsilon_0}\langle\phi(\mathbf{r}_{1A})|\frac{1}{r_{1B}}|\phi(\mathbf{r}_{1A})\rangle \quad (5.6.20)$$

and the off-diagonal one $H_{BA} = \langle\phi(\mathbf{r}_{1B})|\hat{H}_e|\phi(\mathbf{r}_{1A})\rangle$ is

$$H_{BA} = -\frac{\hbar^2\lambda^2}{2m_e a_0^2}S + \frac{e^2(\lambda-1)}{4\pi\varepsilon_0}\langle\phi(\mathbf{r}_{1B})|\frac{1}{r_{1A}}|\phi(\mathbf{r}_{1A})\rangle - \frac{e^2}{4\pi\varepsilon_0}\langle\phi(\mathbf{r}_{1B})|\frac{1}{r_{1B}}|\phi(\mathbf{r}_{1A})\rangle, \quad (5.6.21)$$

where the last two integrals are the same and can be combined into one,

$$H_{BA} = -\frac{\hbar^2\lambda^2}{2m_e a_0^2}S + \frac{e^2(\lambda-2)}{4\pi\varepsilon_0}\langle\phi(\mathbf{r}_{1B})|\frac{1}{r_{1A}}|\phi(\mathbf{r}_{1A})\rangle. \quad (5.6.22)$$

Now, we are left with three integrals, $\langle\phi(\mathbf{r}_{1A})|\frac{1}{r_{1A}}|\phi(\mathbf{r}_{1A})\rangle$, $\langle\phi(\mathbf{r}_{1A})|\frac{1}{r_{1B}}|\phi(\mathbf{r}_{1A})\rangle$, and $\langle\phi(\mathbf{r}_{1B})|\frac{1}{r_{1B}}|\phi(\mathbf{r}_{1A})\rangle$. We introduce a shorthand notation for them,

$$Q = \langle\phi(\mathbf{r}_{1A})|\frac{a_0}{r_{1A}}|\phi(\mathbf{r}_{1A})\rangle, \quad D = \langle\phi(\mathbf{r}_{1A})|\frac{a_0}{r_{1B}}|\phi(\mathbf{r}_{1A})\rangle, \quad X = \langle\phi(\mathbf{r}_{1B})|\frac{a_0}{r_{1B}}|\phi(\mathbf{r}_{1A})\rangle, \quad (5.6.23)$$

where we have purposely multiplied the original integrals by a_0 so that Q, D, and X are dimensionless. After a lengthy integration, we find

$$Q = \lambda, \quad D = \frac{a_0}{R} - \lambda\left(1 + \frac{a_0}{\lambda R}\right)e^{-2\lambda R/a_0}, \quad X = \lambda\left(1 + \frac{\lambda R}{a_0}\right)e^{-\lambda R/a_0}. \quad (5.6.24)$$

We can directly plot S, D, and X as a function of R. Figure 5.4(a) shows that they are all positive and decrease with R, but the "exchange" integral X decreases more slowly.

Because we have multiplied all the integrals by a_0, every term with coefficient $\frac{e^2}{4\pi\varepsilon_0}$ in eqs. (5.6.20) and (5.6.22) must be divided by a_0. Next, we use eq. (4.1.9) to rewrite the coefficients of the potential and the kinetic energy terms in eqs. (5.6.20) and (5.6.22) as $\frac{e^2}{4\pi\varepsilon_0 a_0} = -2E_1$ and $-\frac{\hbar^2}{2m_e a_0^2} = E_1$, respectively. Here, $E_1 = -13.6$ eV. This simplifies the Hamiltonian matrix elements to (note $Q = \lambda$)

$$H_{AA} = E_1\lambda^2 - 2E_1(\lambda-1)Q + 2E_1 D = E_1\lambda^2 - 2E_1(\lambda-1)\lambda + 2E_1 D,$$
$$H_{BA} = E_1\lambda^2 S - 2E_1(\lambda-2)X. \quad (5.6.25)$$

Since $E_1 < 0$, $-2E_1$ is positive. If we use $-2E_1$ as our energy unit, then

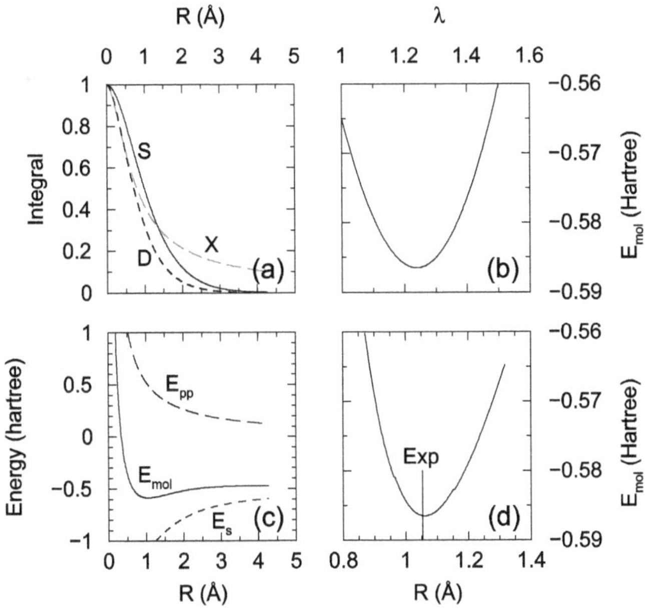

Figure 5.4: (a) Overlap integral S, direct interaction integral D, and "exchange" integral X as a function of R. (b) E_{mol} as a function of the variational parameter λ. (c) Total molecule energy E_{mol}, proton–proton repulsion E_{pp}, and electron energy E_s as a function of distance R between two protons at the optimal λ = 1.238. (d) E_{mol} is minimized with respect to both R and λ. The vertical bar is the experimental bond length. R_{min} = 1.060064 Å.

$$H_{AA} = -\frac{\lambda^2}{2} + \lambda(\lambda - 1) - D, \quad H_{BA} = -\frac{\lambda^2}{2} S + (\lambda - 2)X. \tag{5.6.26}$$

Here the off-diagonal element H_{BA} is negative since $\lambda < 2$ and both S and X are positive. This is the reason why we have consistently pointed out that the hopping integral t in eq. (5.6.7) is positive, so $-t$ is negative.

5.6.3 Variational minimization of the molecular energy

The electronic energy for the bonding state (see E_s in eq. (5.6.8)) is

$$E_s = \frac{H_{AA} + H_{BA}}{1 + S}. \tag{5.6.27}$$

Before we carry out the variational minimization of E_s with respect to both λ and R, we must further simplify H_{AA} and H_{BA}. We set $\rho = \frac{\lambda R}{a_0}$, so the ratio ρ/λ is just the distance between two protons, R/a_0. The overlap integral S is a function of ρ alone,

$$S(\rho) = \left(1 + \rho + \frac{\rho^2}{3}\right) e^{-\rho}. \tag{5.6.28}$$

We rewrite $D = \frac{a_0}{R} - \lambda(1 + \frac{a_0}{\lambda R})e^{-2\lambda R/a_0} \equiv \lambda\tilde{D}$, and $X = \lambda(1 + \lambda R/a_0)e^{-\lambda R/a_0} \equiv \lambda\tilde{X}$, where $\tilde{D}$ and $\tilde{X}$ are also a function of ρ alone,

$$\tilde{D}(\rho) = \frac{1}{\rho} - \left(1 + \frac{1}{\rho}\right)e^{-2\rho}, \quad \tilde{X}(\rho) = (1 + \rho)e^{\rho}. \tag{5.6.29}$$

This has an advantage because we can minimize the total energy with respect to two independent variables, λ and ρ.

We substitute H_{AA} and H_{BA} into E_s to get

$$E_s(\lambda, \rho) = -\frac{\lambda^2}{2} + \frac{\lambda(\lambda - 1) - \lambda\tilde{D}(\rho) + \lambda(\lambda - 2)\tilde{X}(\rho)}{1 + S(\rho)}. \tag{5.6.30}$$

The proton–proton repulsion energy is

$$E_{\rm pp} = \frac{1}{4\pi\epsilon_0}\frac{e^2}{R} = \frac{e^2}{4\pi\epsilon_0 a_0}\frac{a_0}{R} = -2E_1\frac{a_0}{R}. \tag{5.6.31}$$

We write it in the unit of $(-2E_1)$, $E_{\rm pp} = \frac{\lambda}{\rho}$, where we have used $\frac{a_0}{R} = \frac{\lambda}{\rho}$. Now, the total molecular energy is

$$E_{\rm mol} = E_s + E_{\rm pp}, \tag{5.6.32}$$

which must be minimized with respect to λ and ρ. Figure 5.4(b) plots $E_{\rm mol}$ as a function of λ for a fixed ρ. One can see a local minimum around $\lambda = 1.238$. Figure 5.4(c) shows that the proton–proton interaction $E_{\rm pp}$ is always repulsive (long dashed line) since both protons carry positive charges, but E_s provides the necessary attraction (short dashed line). The competition between them leads to an energy minimum. In this figure, we already use the optimized λ. The solid line is an extended view of the molecular energy as a function of R, with a local minimum.

We optimize both λ and ρ simultaneously. $E_{\rm mol}$ in Fig. 5.4(d) is computed using the code `flugge.f` in the Appendix listing 11.13. There is a tiny kink because we must manually choose a smaller total energy. The vertical line shows the experimentally measured bond length of H^+. One can see the theoretical value of 1.0600642 Å matches the experimental one of 1.05278 Å, within 0.69 %. This demonstrates the high accuracy of quantum theory. This small deviation is due to our limited basis (only two 1s orbitals included) and by neglecting zero-point proton fluctuation.

5.6.4 Molecular orbitals

Now, we are ready to find the molecular wavefunction with the optimal λ and R. Once all the matrix elements are known, we can find C_A and C_B by solving eq. (5.6.11) and then plugging them into eq. (5.6.9) to get the wavefunction ψ.

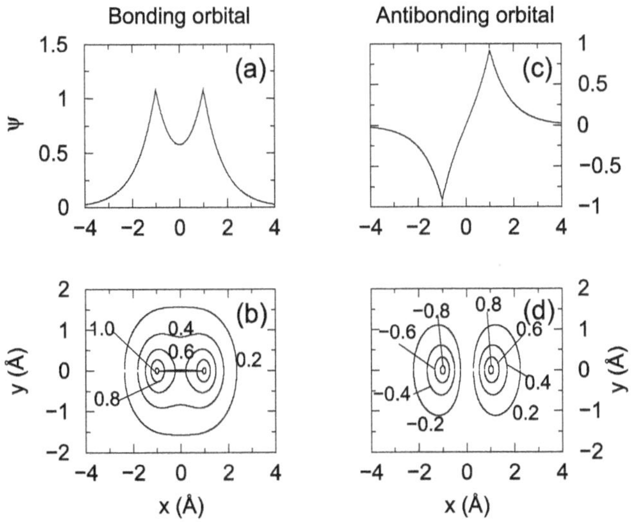

Figure 5.5: (a) One-dimensional scan of the bonding orbital in H_2^+ with nonzero electron density along the bond direction. (b) Contour of the bonding orbital on the *xy*-plane. (c) One-dimensional scan of the antibonding orbital with zero charge density in the middle of the bond. (d) Contour of the antibonding orbital on the *xy*-plane.

Figure 5.5(a) shows a line-cut of ψ through the x-axis, and its two-dimensional contour plot is shown in Fig. 5.5(b). One can see a nonzero charge density along the bond direction, typical of a σ-orbital. This reveals how the chemical bond is formed: The overlap of the atomic wavefunctions lowers the overall energy from $2E_{1s}$.

The antibonding state can be found similarly. Figure 5.5(c) shows a line-cut of ψ through the x-axis, and its two-dimensional contour plot is shown in Fig. 5.5(d). We note that, in contrast to the bonding orbital, there is no charge density between two atoms. This is the key feature of antibonding states.

Exercise 5.6.4.

15. Compute the overlap matrix element S. Hint: You need the two integrals $\int xe^{ax}dx = (\frac{x}{a} - \frac{1}{a^2})e^{ax}$ and $\int x^2e^{ax}dx = (\frac{x^2}{a} - \frac{2x}{a^2} + \frac{2}{a^3})e^{ax}$.

16. Prove $\frac{1}{r_{1A}}\frac{d^2}{dr_{1A}^2}[r_{1A}(e^{-\lambda r_{1A}/a_0})] = (-\frac{2}{r_{1A}}\frac{\lambda}{a_0} + \frac{\lambda^2}{a_0^2})e^{-\lambda r_{1A}/a_0}$.

17. (a) Show $-\frac{\hbar^2}{2m_e a_0^2} = E_1$, where E_1 is the ground-state energy of the H atom and a_0 is the Bohr radius. (b) Show $Q = \langle\phi(\mathbf{r}_{1A})|\frac{a_0}{r_{1A}}|\phi(\mathbf{r}_{1A})\rangle = \lambda$.

18. Use `flugge.f` in Appendix listing 11.13 to compute the total energy change as a function of λ.

5.7 Problems

1. A particle of mass m is in the ground state of a one-dimensional infinite quantum well with $V(x) = 0$, where $0 < x < a$, otherwise $V(x) = \infty$. It experiences a small perturbation $H_I(x) = -b$ if $0 < x < a/2$ and $H_I(x) = b$ if $a/2 < x < a$, where b is a positive constant. (a) Find the energy correction up to the second order. (b) Find the wavefunction correction up to the first order. (c) Find the probabilities between $0 < x < a/2$ and $a/2 < x < a$.
2. A particle of mass m is sealed inside an infinite quantum well of length a within $x \in (0, a)$. The particle is subject to a small perturbation, $W(x) = B\delta(x - \frac{a}{2})$, where B is a small constant in units of energy. (a) Compute the energy correction up to the first order. (b) This problem is exactly solvable without approximation. Show the exact energy E is given by k which satisfies $\tan(\frac{ka}{2}) = -\frac{\hbar^2 k}{Bm}$. (c) Take the limit $B \to 0$ to check the results from (a).
3. For the particle-in-a-box problem, suppose we add a small electric field $\mathcal{E}$ along the x-direction, $\hat{H}_I = -e\mathcal{E}x$. (a) Find the first-order ground-state energy correction. (b) Find the first-order wavefunction correction. (c) Find the second-order ground-state energy correction.
4. A particle is placed in the infinite quantum well with $x \in (0, a)$. It experiences a weak perturbation $H_I = \epsilon x\delta(x - a/2)$, where ϵ is a small positive number. (a) Find the first- and second-order energy corrections to the first-excited state $n = 2$. (b) Find the wavefunction correction to it up to the first order $|\psi^{(1)}\rangle$. (c) Find the expectation values of $\hat{x}$ and $\hat{p}_x$ with $|\psi^{(1)}\rangle$. (d) Check how the uncertainty $\Delta x \Delta p_x$ is changed.
5. A harmonic oscillator with angular frequency ω and mass m is subject to a small perturbation $\hat{H}_I = \lambda x^3$. For the ground state $\phi_{n=0}$, (a) find its first-order energy correction $E^{(1)}$; (b) compute the second-order energy correction $E^{(2)}$; (c) for a general state ϕ_n, show that the first-order wavefunction correction $\psi^{(1)}$ only contains four unperturbed wavefunctions, ϕ_{n+1}, ϕ_{n-1}, ϕ_{n-3}, and ϕ_{n+3}. Hint: Convert x^3 to $\hat{a}$ and $\hat{a}^\dagger$ and then apply them to the unperturbed eigenstates.
6. A hydrogen atom is in its ground state. We apply a uniform electric field along the z-axis, $H_I = eFz = eFr\cos\theta$, where F is the electric field. (a) Prove the ground state up to the first-order wavefunction correction is

$$\psi = \frac{1}{\sqrt{\pi a_0^3}} e^{-r/a_0}\left[1 - \frac{F\cos\theta}{e}\left(a_0 r + \frac{r^2}{2}\right)\right].$$

(b) Based on this wavefunction, show that the second-order energy correction is $-\frac{9}{4}a_0^3F^2$. (c) What if we apply the electric field along the y-axis? How do the above results change?

7. A system has an unperturbed Hamiltonian $\hat{H}_0 = \frac{\hat{\mathbf{p}}^2}{2m} + V(\mathbf{r})$ with eigenfunction $|\phi_n\rangle$ and eigenvalue ϵ_n. (a) Prove the matrix element of $\hat{\mathbf{p}}$ between two eigenstates $|m\rangle$ and $|n\rangle$ is connected with that of $\hat{\mathbf{r}}$ through

$$\langle m|\hat{\mathbf{p}}|n\rangle = \frac{im(\epsilon_m - \epsilon_n)}{\hbar}\langle m|\hat{\mathbf{r}}|n\rangle.$$

Hint: $[\hat{\mathbf{r}}, \hat{\mathbf{p}}] = i\hbar$ and $[\hat{\mathbf{r}}, \hat{H}_0] = i\hbar\hat{\mathbf{p}}/m$. (b) Suppose that the system is subject to a perturbation with the interaction Hamiltonian $\hat{H}_I = -\frac{\hat{\mathbf{p}}\cdot\mathbf{A}(\mathbf{r})}{m_e}$, where $\mathbf{A}$ is the vector potential and its spatial dependence is ignored, i. e., $\mathbf{A}(\mathbf{r} = 0)$. Show the second-order energy correction is

$$E_n^{(2)} = -\frac{A^2}{\hbar^2}\sum_k(\epsilon_k - \epsilon_n)|\hat{\mathbf{r}}_{kn}|^2.$$

8. A system is described by the unperturbed Hamiltonian matrix

$$\hat{H}_0 = \begin{pmatrix} E_1 & 0 & 0 \\ 0 & E_2 & 0 \\ 0 & 0 & E_3 \end{pmatrix}.$$

A small perturbation is given by

$$\hat{H}_I = \begin{pmatrix} 0 & \epsilon & \epsilon \\ \epsilon & 0 & \epsilon \\ \epsilon & \epsilon & 0 \end{pmatrix}.$$

(a) If $E_1 < E_2 < E_3$, find the energy and wavefunction correction for the lowest eigenstate. (b) If $E_1 = E_2 < E_3$, find the energy and wavefunction correction for the lowest eigenstate. (c) Without using the perturbation theory, find the exact solution and then find the condition that the results will reproduce (a) and (b).

9. The molecular orbital contour plot can reveal information about chemical bonding. Take H_2^+ as an example, and place two protons at $x = -R/2$ and $x = R/2$, where $R = 1.06$ Å. Suppose the wavefunction is $\psi = e^{-\sqrt{(x-R/2)^2+y^2}/a_0} + e^{-\sqrt{(x+R/2)^2+y^2}/a_0}$. (a) Given $\psi = 1, 0.8$, and 0.2, explain how you plot such an orbital in the xy-plane. (b) Plot your orbital.

10. Suppose $\phi_A = e^{-r_A^2}$ and $\phi_B = e^{-r_B^2}$, where $r_A = \sqrt{(x-R/2)^2+y^2}$ and $r_B = \sqrt{(x+R/2)^2+y^2}$. (a) Find an equation of x and y, where $\psi = \phi_A + \phi_B$ is equal to a constant c. These pairs of (x, y) are the coordinates of the contour plot. Hint: For our present ϕ_A and ϕ_B, it is possible to solve for y for a fixed x. (b) Choose $c = 0.2, 0.4, 0.8$, and plot a figure like Figure 5.5(b). (c) Do the same but for $\psi = \phi_A - \phi_B$, which mimicks an antibonding orbital.

11. Prove the Hellmann–Feynman theorem,

$$\frac{\partial}{\partial\lambda}E(\lambda) = \langle\phi(\lambda)|\frac{\partial\hat{H}(\lambda)}{\partial\lambda}|\phi(\lambda)\rangle,$$

where $E(\lambda)$ is an eigenvalue of the Hamiltonian $\hat{H}(\lambda)$, $\phi(\lambda)$ is its eigenstate, i. e., $\hat{H}(\lambda)\phi(\lambda) = E(\lambda)\phi(\lambda)$ and $\langle\phi(\lambda)|\phi(\lambda)\rangle = 1$. λ is a parameter.

12. Show $D = \langle\phi(\mathbf{r}_{1A})|\frac{a_0}{r_{1B}}|\phi(\mathbf{r}_{1A})\rangle = \frac{a_0}{R} - \lambda(1 + \frac{a_0}{\lambda R})e^{-2\lambda R/a_0}$.
13. Show $X = \langle\phi(\mathbf{r}_{1B})|\frac{a_0}{r_{1A}}|\phi(\mathbf{r}_{1A})\rangle = \lambda(1 + \frac{\lambda R}{a_0})e^{-\lambda R/a_0}$.
14. Suppose $\phi(x) = Ne^{-ax^2}$ is a variational wavefunction for the ground state of the one-dimensional harmonic oscillator with $\hat{H} = -\frac{\hbar^2}{2m}\frac{d^2}{dx^2} + \frac{1}{2}m\omega^2x^2$, where N is the normalization constant. (a) Use the variational principle to find a. (b) Find the eigenenergy. (c) Replace $\phi(x) = Ne^{-ax^2}$ by $\phi(x) = N\cos(ax)$ and find a that minimizes the total energy and find the energy.
15. $\phi_1 = x$ and $\phi_2 = x^3$ form a variational wavefunction, $\psi = c_1\phi_1 + c_2\phi_2$, for the infinite quantum well with $x \in (0, a)$. (a) Find the overlap matrix $S = \langle\phi_1|\phi_2\rangle$. (b) Set up a Hamiltonian matrix. (c) Diagonalize the matrix and find c_1 and c_2. (d) Find the energy for the ground state. (e) Sketch the wavefunction and compare it with the exact one ($\sqrt{\frac{2}{a}}\sin(\frac{\pi x}{a})$). (f) Explain why we choose odd functions for ϕ_1 and ϕ_2.
16. Consider a trial wavefunction $f(x)$ for the harmonic oscillator, $\hat{H} = \frac{\hat{p}_x^2}{2m} + \frac{1}{2}m\omega^2x^2$, where $f(x)$ is an isoceles triangle centered at $x = 0$, with a height of 1 and base length of a. (a) Use the variational principle to find a at the minimal energy. (b) Find the energy and compute the percentage error. (c) Find the overlap between $f(x)$ and the exact ground state wavefunction. (Answers: $a = 2.34$ and $E = +0.548\hbar\omega$). From [29].
17. (a) Assume a trial wavefunction for PIB, $\phi(x) = a^2 - x^2$. Normalize it, compute the expectation value of $\hat{H}$, and compare it with the exact ground-state energy. (b) Choose $\phi(x) = a^\lambda - x^\lambda$, where λ is a variational parameter. Find λ at the energy minimum and compare it with the exact ground state energy. From [30].
18. (a) For a harmonic oscillator, $\hat{H} = \frac{\hat{p}_x^2}{2m} + \frac{1}{2}m\omega^2x^2$. Find $-\frac{\langle\hat{V}\rangle}{\langle\hat{T}\rangle}$. (b) Prove that for a stationary state, a more general virial theorem is

$$2\langle\hat{T}\rangle = \langle\mathbf{r}\cdot\nabla V\rangle,$$

where $\hat{T}$ and V are the kinetic energy and potential energy operators, respectively. Hint: Starting from the Heisenberg equation of motion for the product $\mathbf{r}\cdot\hat{\mathbf{p}}$,

$$\frac{d\langle\mathbf{r}\cdot\hat{\mathbf{p}}\rangle}{dt} = \frac{1}{i\hbar}\langle[\mathbf{r}\cdot\hat{\mathbf{p}}, \hat{H}]\rangle.$$

For a stationary state, the left side must be time-averaged to zero.
19. The Weizsäcker kinetic energy functional is $E_k[\rho] = \frac{1}{8}\frac{e^2}{4\pi\epsilon_0}\int\frac{\nabla\rho(\mathbf{r})\cdot\nabla\rho(\mathbf{r})}{\rho(\mathbf{r})}d\mathbf{r}$. Find $\frac{\delta E_k}{\delta\rho(\mathbf{r})}$. (Answer: $\frac{1}{8}\frac{\nabla\rho(\mathbf{r})\cdot\nabla\rho(\mathbf{r})}{\rho^2(\mathbf{r})} - \frac{1}{4}\frac{\nabla^2\rho}{\rho}$.)

6 Electron spin

Few observables are as exceptional as spin. It is present only in QM and has no classical analogue. It is a separate degree of freedom that a quantum object possesses and runs in parallel with the spatial degree of freedom. Conceptually, it is incorrect to say that we locate spin at a particular spatial point. Mathematically, the spatial derivative (either gradient, divergence, or curl) of spin is zero. Spin is the reason why we have magnetism, just as charge is the reason for electricity, but the situation is much more complicated. Electrons, protons, and neutrons have half-integer spins, while photons and phonons have integer spins.

This chapter is grouped into four basic units.

- Unit 1 consists of the first two sections. It introduces the Stern–Gerlach experiment and four key concepts: Pauli matrices, spin operators, their commutation relations, and the total angular momentum.
- Unit 2 includes Sections 6.3 and 6.4. Section 6.3 explains how the spin interacts with the magnetic field, while Section 6.4 explains how one can explain the Stern–Gerlach experiment. The purpose is to close the loop that we started in Section 6.1.
- Unit 3 consists of Sections 6.5 and 6.6. Section 6.5 introduces the Zeeman effect that extends Section 6.4 to spectrum. While the Zeeman effect uses one static magnetic field, the electron spin resonance (ESR) in Section 6.6 uses two fields, one static and one dynamic. This dynamic field connects us to the time-dependence, which lays the ground work for the time-dependent perturbation theory in the next chapter.
- Unit 4 consists of Section 6.7 which extends all the basic knowledge previously gained to nuclear spins; this enables spatial resolution to be achieved in MRI.

6.1 Discovery of electron spin and representation of spin states

6.1.1 Stern–Gerlach experiment: spin is another degree of freedom

Between 1921 and 1922, in Germany, Walter Gerlach and Otto Stern [31] sent a beam of Ag atoms through a gradient magnetic field. If only classical physics were at work, they would have expected the atoms to form a single distribution on the screen with the maximum in the middle. But, this did not happen; instead, they found the atoms split into two groups, and so discovered the existence of spin in electrons.

Figure 6.1 schematically reproduces their experimental setup. Ag atoms pass through a collimator before entering a gradient field **B**. If we choose the vertical axis as the z-axis, $\frac{\partial B_z}{\partial z} \neq 0$. According to the Maxwell equation, the divergence of a magnetic field is zero, i. e., $\nabla \cdot \mathbf{B} = 0$, so $\frac{\partial B_x}{\partial x} \neq 0$ and/or $\frac{\partial B_y}{\partial y} \neq 0$, but their values are smaller. The Ag atom is electrically neutral because there are 47 electrons and protons. All the shells are closed ($1s^2 2s^2 2p^6 3s^2 3p^6 4s^2 3d^{10} 4p^6 4d^{10} 5s^1$), except the 47th $5s^1$ electron. On the right side, a photographic plate captures the Ag atoms. Without **B**, they observed only a single

https://doi.org/10.1515/9783110672152-006

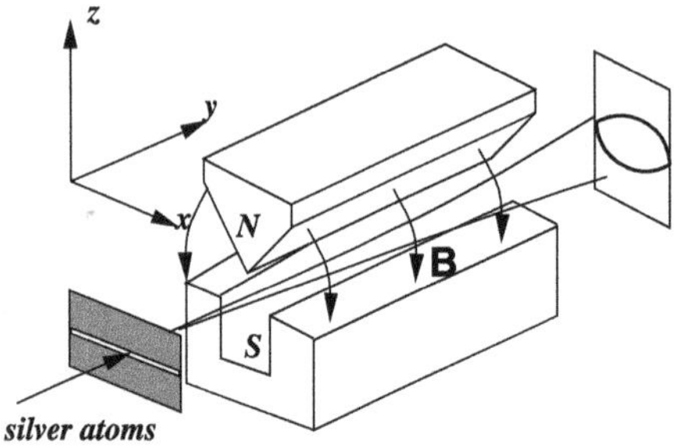

Figure 6.1: Stern–Gerlach experiment setup of silver atoms. The length of their magnetic field is about 10 cm. According to Stern's Nobel Lecture, the slit in the collimator cannot be made too narrow because the atom beam deflects too much to overshadow the deflection due to the magnetic field. Their original paper (W. Gerlach and O. Stern, Zeit für Physik, **9**, 353 (1922)) gave $B = 0.1\,\mathrm{T}$, the magnetic field gradient of 10 T/m and the vertical splitting of $\Delta z = 0.22\,\mathrm{mm}$.

line; with **B**, the single line splits into two. This is the first experimental demonstration of spin. The spin only take one of two values: up or down.

Does a uniform magnetic field work? In classical E&M, the interaction between a magnetic moment $\boldsymbol{\mu}_{\text{classical}}$ and a magnetic field $\mathbf{B}$ is described by the potential energy U,

$$U = -\boldsymbol{\mu}_{\text{classical}} \cdot \mathbf{B},$$

but, for the spin, we change it to

$$U = -\boldsymbol{\mu}_S \cdot \mathbf{B},$$

where $\boldsymbol{\mu}_S$ is the spin moment. The force along the z-axis due to spin is

$$F_z = -\frac{\partial U}{\partial z} = \boldsymbol{\mu}_S \cdot \frac{\partial \mathbf{B}}{\partial z} + \frac{\partial \boldsymbol{\mu}_S}{\partial z} \cdot \mathbf{B}.$$

If a uniform magnetic field had an effect, then this would prove that the spin moment must depend on the space. But the experiment shows that this is not the case. Therefore, the spin moment is completely decoupled from the space, $\frac{\partial \boldsymbol{\mu}_S}{\partial z} = 0$, and the electron spin is a degree of freedom separate from its charge counterpart. For this reason, in order to detect the spatial deviation, the magnetic field gradient is nonzero, $\frac{\partial B}{\partial z} \neq 0$. In an interesting historical note, 2022 was the 100th anniversary of the famous 1922 experiment.

6.1.2 Pauli matrices

We start with a three-dimensional Cartesian spatial space and consider three unit vectors $\mathbf{e}_1$, $\mathbf{e}_2$, $\mathbf{e}_3$, each of which has $\mathbf{e}_i \cdot \mathbf{e}_i = |\mathbf{e}_i|^2 = 1$. We do not assume any other relationships among them. We express another unit vector $\hat{\mathbf{r}} = c_1\mathbf{e}_1 + c_2\mathbf{e}_2 + c_3\mathbf{e}_3$, where c_i is the direction cosine ($\sum_i c_i^2 = 1$). Imposing $|\hat{\mathbf{r}}|^2 = 1$ and invoking the arbitrariness of c_i lead to $\mathbf{e}_1 \cdot \mathbf{e}_2 = 0$, $\mathbf{e}_1 \cdot \mathbf{e}_3 = 0$, and $\mathbf{e}_2 \cdot \mathbf{e}_3 = 0$. This shows that, if we impose the unit vector constraint on three spatial vectors, we can generate three unit vectors that are orthogonal to each other.

For the spin space, we do the same, but with a 2×2 matrix, represented by A [29],

$$A = \begin{pmatrix} a_{11} & a_{12} \\ a_{21} & a_{22} \end{pmatrix}. \tag{6.1.1}$$

Just like $|\hat{\mathbf{r}}|^2 = 1$, we impose one constraint: A^2 is an identity matrix I, i. e.,

$$A^2 = I = \begin{pmatrix} 1 & 0 \\ 0 & 1 \end{pmatrix}, \tag{6.1.2}$$

which leads to $a_{11} = -a_{22}$ and $a_{11}^2 + a_{12}a_{21} = 1$. This means that matrix elements, a_{11}, a_{12}, a_{21}, are not independent. A has two eigenvalues of ± 1 (see Problem No. 1).

If we choose a real $a_{11} = \cos\theta$, then $a_{12}a_{21} = \sin^2\theta$, where a_{12} and a_{21} must be both pure real or pure imaginary. The two simplest choices are (1) $a_{12} = a_{21} = \sin\theta$ and (2) $a_{12} = -a_{21} = -i\sin\theta$. If we set $\theta = \pi/2$, these two choices yield the Pauli matrices, $\hat{\sigma}_x$ (the x-component) and $\hat{\sigma}_y$ (the y-component) respectively,

$$\hat{\sigma}_x = \begin{pmatrix} 0 & 1 \\ 1 & 0 \end{pmatrix}, \quad \hat{\sigma}_y = \begin{pmatrix} 0 & -i \\ i & 0 \end{pmatrix}. \tag{6.1.3}$$

Setting $\theta = 0$ gives us the Pauli matrix $\hat{\sigma}_z$ (the z-component),

$$\hat{\sigma}_z = \begin{pmatrix} 1 & 0 \\ 0 & -1 \end{pmatrix}. \tag{6.1.4}$$

We see $\hat{\sigma}_x^2 = \hat{\sigma}_y^2 = \hat{\sigma}_z^2 = I$, with eigenvalues being ± 1. They are all Hermitian operators, $\hat{\sigma}_x^\dagger = \hat{\sigma}_x$ and so on. Then we have $\hat{\sigma}_i^\dagger\hat{\sigma}_i = I$, so they are also unitary matrices. Following the example with $\hat{\mathbf{r}}$, we express any 2×2 matrix $\hat{X}$ as $\hat{X} = c_1\hat{\sigma}_x + c_2\hat{\sigma}_y + c_3\hat{\sigma}_z$, where c_i is the direction cosine in the spin space. We require $\hat{X}$ to satisfy

$$\hat{X}^2 = (c_1\hat{\sigma}_x + c_2\hat{\sigma}_y + c_3\hat{\sigma}_z)(c_1\hat{\sigma}_x + c_2\hat{\sigma}_y + c_3\hat{\sigma}_z) = I,$$

whose diagonal terms are summed up to $c_1^2\hat{\sigma}_x^2 + c_2^2\hat{\sigma}_y^2 + c_3^2\hat{\sigma}_z^2 = (c_1^2 + c_2^2 + c_3^2)I = I$ and thus cancel I on the right side of the equation. Because $\{c_i\}$ is arbitrary, the off-diagonal terms, $c_1c_2(\hat{\sigma}_x\hat{\sigma}_y + \hat{\sigma}_y\hat{\sigma}_x) + c_2c_3(\hat{\sigma}_y\hat{\sigma}_z + \hat{\sigma}_z\hat{\sigma}_y) + c_1c_3(\hat{\sigma}_x\hat{\sigma}_z + \hat{\sigma}_z\hat{\sigma}_x)$, must be zero for each combination: $\hat{\sigma}_x\hat{\sigma}_y + \hat{\sigma}_y\hat{\sigma}_x = 0$, $\hat{\sigma}_y\hat{\sigma}_z + \hat{\sigma}_z\hat{\sigma}_y = 0$, $\hat{\sigma}_x\hat{\sigma}_z + \hat{\sigma}_z\hat{\sigma}_x = 0$. These are called the anticommutation relations, as opposed to a regular commutator $[\hat{A}, \hat{B}] = \hat{A}\hat{B} - \hat{B}\hat{A}$. They are the matrix equivalence of the orthogonalization, $\mathbf{e}_i \cdot \mathbf{e}_j = 0$ for $i \neq j$. One may use eqs. (6.1.3) and (6.1.4) to verify these relationships, but we derive it without using them.

These Pauli matrices remained mysterious until Dirac discovered through his relativistic equation that the most natural representation for the electron is four 4×4 matrices $\hat{\gamma}_0$, $\hat{\gamma}_1$, $\hat{\gamma}_2$ and $\hat{\gamma}_3$, called Dirac matrices. Each matrix consists of two 2×2 matrices,

$$\hat{\gamma}_0 = \begin{pmatrix} I & 0 \\ 0 & -I \end{pmatrix}, \quad \hat{\gamma}_1 = \begin{pmatrix} 0 & \hat{\sigma}_x \\ -\hat{\sigma}_x & 0 \end{pmatrix}, \quad \hat{\gamma}_2 = \begin{pmatrix} 0 & \hat{\sigma}_y \\ -\hat{\sigma}_y & 0 \end{pmatrix}, \quad \hat{\gamma}_3 = \begin{pmatrix} 0 & \hat{\sigma}_z \\ -\hat{\sigma}_z & 0 \end{pmatrix}, \tag{6.1.5}$$

where I is the identity matrix (eq. (6.1.2)). To Dirac's amazement, $\hat{\sigma}_x$, $\hat{\sigma}_y$, and $\hat{\sigma}_z$ (eqs. (6.1.3) and (6.1.4)) are the Pauli matrices proposed by Pauli earlier.

The Pauli matrix, denoted as $\hat{\boldsymbol{\sigma}}$, has three components $\hat{\sigma}_x$, $\hat{\sigma}_y$, and $\hat{\sigma}_z$ (eqs. (6.1.3) and (6.1.4)). If we compute $\sigma_x\sigma_y$ and $\sigma_y\sigma_x$ and subtract them from each other,

$$\hat{\sigma}_x\hat{\sigma}_y - \hat{\sigma}_y\hat{\sigma}_x = \begin{pmatrix} 0 & 1 \\ 1 & 0 \end{pmatrix}\begin{pmatrix} 0 & -i \\ i & 0 \end{pmatrix} - \begin{pmatrix} 0 & -i \\ i & 0 \end{pmatrix}\begin{pmatrix} 0 & 1 \\ 1 & 0 \end{pmatrix} = 2i\begin{pmatrix} 1 & 0 \\ 0 & -1 \end{pmatrix} = 2i\hat{\sigma}_z,$$

this can be written as a commutation as

$$[\hat{\sigma}_x, \hat{\sigma}_y] = 2i\hat{\sigma}_z, [\hat{\sigma}_y, \hat{\sigma}_z] = 2i\hat{\sigma}_x, [\hat{\sigma}_z, \hat{\sigma}_x] = 2i\hat{\sigma}_y \rightarrow \hat{\boldsymbol{\sigma}} \times \hat{\boldsymbol{\sigma}} = 2i\hat{\boldsymbol{\sigma}}. \tag{6.1.6}$$

Similar to the orbital angular momentum operators, we also introduce two auxiliary operators, raising and lowering operators, $\hat{\sigma}_+$ and $\hat{\sigma}_-$,

$$\hat{\sigma}_+ = \frac{1}{2}(\hat{\sigma}_x + i\hat{\sigma}_y), \quad \hat{\sigma}_- = \frac{1}{2}(\hat{\sigma}_x - i\hat{\sigma}_y).$$

Example 1 (Anticommutation). Prove the following anticommutation relationship: $\hat{\sigma}_z\hat{\sigma}_x + \hat{\sigma}_x\hat{\sigma}_z = 0$.

We first note that $[\hat{\sigma}_x, \hat{\sigma}_y] = 2i\hat{\sigma}_z$, which yields $\hat{\sigma}_x\hat{\sigma}_y - \hat{\sigma}_y\hat{\sigma}_x = 2i\hat{\sigma}_z$. We then left multiply it with $\hat{\sigma}_x$ to obtain $\hat{\sigma}_x^2\hat{\sigma}_y - \hat{\sigma}_x\hat{\sigma}_y\hat{\sigma}_x = 2i\hat{\sigma}_x\hat{\sigma}_z$. Using $\hat{\sigma}_x^2 = 1$, we further simplify it to $\hat{\sigma}_y - \hat{\sigma}_x\hat{\sigma}_y\hat{\sigma}_x = 2i\hat{\sigma}_x\hat{\sigma}_z$. Next, we right multiply $\hat{\sigma}_x\hat{\sigma}_y - \hat{\sigma}_y\hat{\sigma}_x = 2i\hat{\sigma}_z$ by $\hat{\sigma}_x$ to get $\hat{\sigma}_x\hat{\sigma}_y\hat{\sigma}_x - \hat{\sigma}_y = 2i\hat{\sigma}_z\hat{\sigma}_x$. We add these two equations and obtain $\hat{\sigma}_x\hat{\sigma}_z + \hat{\sigma}_z\hat{\sigma}_x = 0$.

Example 2 (Another identity). Prove $\hat{\sigma}_x\hat{\sigma}_y = -\hat{\sigma}_y\hat{\sigma}_x = i\hat{\sigma}_z$.

This relationship can be proven in two ways. We note first that $\hat{\sigma}_x\hat{\sigma}_y - \hat{\sigma}_y\hat{\sigma}_x = 2i\hat{\sigma}_z$. Since $\hat{\sigma}_x\hat{\sigma}_y + \hat{\sigma}_y\hat{\sigma}_x = 0$, we have $-\hat{\sigma}_y\hat{\sigma}_x = \hat{\sigma}_x\hat{\sigma}_y$. Then, we have $2\hat{\sigma}_x\hat{\sigma}_y = 2i\hat{\sigma}_z$, or $\hat{\sigma}_x\hat{\sigma}_y = i\hat{\sigma}_z$. Another way is to directly use the matrices: $\begin{pmatrix} 0 & 1 \\ 1 & 0 \end{pmatrix}\begin{pmatrix} 0 & -i \\ +i & 0 \end{pmatrix} = \begin{pmatrix} +i & 0 \\ 0 & -i \end{pmatrix} = i\begin{pmatrix} +1 & 0 \\ 0 & -1 \end{pmatrix}$.

Exercise 6.1.2.

1. $\hat{\mathbf{r}} = c_1\mathbf{e}_1 + c_2\mathbf{e}_2 + c_3\mathbf{e}_3$. If each $\mathbf{e}_i$ has $\mathbf{e}_i \cdot \mathbf{e}_i = |\mathbf{e}_i|^2 = 1$ and $|\hat{\mathbf{r}}|^2 = 1$, prove $\mathbf{e}_i \cdot \mathbf{e}_j = 0$ if $i \neq j$.
2. Prove that $A^2 = I$ in eq. (6.1.2) leads to $a_{11} = -a_{22}$ and $a_{11}^2 + a_{12}a_{21} = 1$.
3. (a) Prove $[\hat{\sigma}_y, \hat{\sigma}_z] = 2i\hat{\sigma}_x$. (b) Prove $[\hat{\sigma}_z, \hat{\sigma}_x] = 2i\hat{\sigma}_y$.
4. Show that $\hat{\sigma}_x^2 = \hat{\sigma}_y^2 = \hat{\sigma}_z^2 = I = \begin{pmatrix} 1 & 0 \\ 0 & 1 \end{pmatrix}$.

5. Prove $\hat{\sigma}_x\hat{\sigma}_y + \hat{\sigma}_y\hat{\sigma}_x = 0$, $\hat{\sigma}_y\hat{\sigma}_z + \hat{\sigma}_z\hat{\sigma}_y = 0$, $\hat{\sigma}_y\hat{\sigma}_z = -\hat{\sigma}_z\hat{\sigma}_y = i\hat{\sigma}_x$, $\hat{\sigma}_z\hat{\sigma}_x = -\hat{\sigma}_x\hat{\sigma}_z = i\hat{\sigma}_y$, and $\hat{\sigma}_x\hat{\sigma}_y\hat{\sigma}_z = I$, where I is an identity matrix.
6. Prove $(\hat{\boldsymbol{\sigma}} \cdot \mathbf{A})(\hat{\boldsymbol{\sigma}} \cdot \mathbf{B}) = \mathbf{A} \cdot \mathbf{B} + i\hat{\boldsymbol{\sigma}} \cdot (\mathbf{A} \times \mathbf{B})$.
7. Although in our derivation, we set $\theta = \pi/2$ to obtain $\hat{\sigma}_x$ and $\hat{\sigma}_y$, one can use $\hat{\sigma}_y\hat{\sigma}_z + \hat{\sigma}_z\hat{\sigma}_y = 0$ and $\hat{\sigma}_x\hat{\sigma}_z + \hat{\sigma}_z\hat{\sigma}_x = 0$ to prove $\theta = \pi/2$.

6.1.3 Spin matrices

At the nonrelativistic limit, one retains a subset of $\hat{\gamma}$, so we have the electron spin, $\hat{\mathbf{S}}$, which is defined as

$$\hat{\mathbf{S}} = \frac{\hbar}{2}\hat{\boldsymbol{\sigma}}, \rightarrow \hat{S}_x = \frac{\hbar}{2}\begin{pmatrix} 0 & 1 \\ 1 & 0 \end{pmatrix}, \hat{S}_y = \frac{\hbar}{2}\begin{pmatrix} 0 & -i \\ i & 0 \end{pmatrix}, \hat{S}_z = \frac{\hbar}{2}\begin{pmatrix} 1 & 0 \\ 0 & -1 \end{pmatrix}. \tag{6.1.7}$$

6.1.3.1 Spin eigenfunctions and spinors

Now we are ready to explain the spin vector. $\hat{\mathbf{S}}$ has three components, $\hat{S}_x$, $\hat{S}_y$, and $\hat{S}_z$, just like any spatial vector $\mathbf{A}$, which has A_x, A_y, and A_z. But different from $\mathbf{A}$, each component of $\hat{\mathbf{S}}$ is a 2×2 matrix, not a scalar number. This 2×2 matrix is intrinsic to spin. In Chapter 4, we learned that we can convert a differentiation operator to a matrix, but here the spin comes with a matrix and cannot be reduced to a differentiation operator. In other words, once we have spin, we are forced to adopt the matrix notation.

The eigenvalues of $\hat{S}_x$, $\hat{S}_y$, and $\hat{S}_z$ are $\pm\frac{\hbar}{2}$. Take $\hat{S}_x$ as an example

$$\frac{\hbar}{2}\begin{pmatrix} 0 & 1 \\ 1 & 0 \end{pmatrix}\begin{pmatrix} a \\ b \end{pmatrix} = \lambda\begin{pmatrix} a \\ b \end{pmatrix}.$$

In order to have a nontrivial solution, the determinant of the matrix must be zero,

$$\begin{vmatrix} 0-\lambda & \frac{\hbar}{2} \\ \frac{\hbar}{2} & 0-\lambda \end{vmatrix} = 0.$$

So we have $\lambda = \pm\frac{\hbar}{2}$. Its eigenvector is found by solving

$$\begin{pmatrix} 0 & \frac{\hbar}{2} \\ \frac{\hbar}{2} & 0 \end{pmatrix}\begin{pmatrix} a \\ b \end{pmatrix} = \pm\frac{\hbar}{2}\begin{pmatrix} a \\ b \end{pmatrix}$$

if $\lambda = +\frac{\hbar}{2}$, $\frac{\hbar}{2}b = \frac{\hbar}{2}a$, or $a = b$. Using the normalization condition $|a|^2 + |b|^2 = 1$, we have the eigenfunction η_{x_1} for $\lambda_1 = \frac{\hbar}{2}$ and η_{x_2} for $\lambda_2 = -\frac{\hbar}{2}$,

$$\eta_{x_1} = \frac{1}{\sqrt{2}}\begin{pmatrix} 1 \\ 1 \end{pmatrix} \quad \text{and} \quad \eta_{x_2} = \frac{1}{\sqrt{2}}\begin{pmatrix} 1 \\ -1 \end{pmatrix}. \tag{6.1.8}$$

These wavefunctions are called spinors since they have two components. For $\hat{S}_z$, its eigenfunctions are

$$\eta_{z_1} = \begin{pmatrix} 1 \\ 0 \end{pmatrix} \quad \text{for } \lambda_1 = \frac{\hbar}{2} \quad \text{and} \quad \eta_{z_2} = \begin{pmatrix} 0 \\ 1 \end{pmatrix} \quad \text{for } \lambda_2 = -\frac{\hbar}{2}.$$

We introduce a shorthand notation by denoting η_{z_1} as $|\alpha\rangle$, or α or $|\uparrow\rangle$ for spin-up and η_{z_2} as $|\beta\rangle$, or β or $|\downarrow\rangle$ for spin-down. So η_{x_1} and η_{x_2} can be written in terms of α and β.

6.1.3.2 Raising and lowering spin operators and spin operations

The spin's three components do not commute and instead follow a cyclic relationship

$$[\hat{S}_i, \hat{S}_j] = i\hbar\epsilon_{ijk}\hat{S}_k \rightarrow \hat{\mathbf{S}} \times \hat{\mathbf{S}} = i\hbar\hat{\mathbf{S}}, \tag{6.1.9}$$

where ϵ_{ijk} is the Levi-Civita symbol. $\epsilon_{ijk} = +1$ if $ijk = xyz, yzx, zxy$; $\epsilon_{ijk} = -1$ if any index does not follow the order; $\epsilon_{ijk} = 0$ if any two indices are the same. Equation (6.1.9) has the same form as our orbital angular momentum in eq. (4.3.2). For this reason, spin is considered as an intrinsic angular momentum, spin angular momentum. Caution must be taken that eq. (6.1.9) has a factor of 1, but eq. (6.1.6) has a factor of 2.

As a consequence of noncommutation, $\hat{S}_x$, $\hat{S}_y$, and $\hat{S}_z$ do not share a common eigenfunction. This means that, if the system is in an eigenfunction of $\hat{S}_z$, neither $\hat{S}_x$ nor $\hat{S}_y$ is in its eigenfunction, so their expectation values are zero. Take $|\alpha\rangle$ as an example

$$\langle\alpha|\hat{S}_x|\alpha\rangle = \frac{\hbar}{2}\begin{pmatrix} 1 & 0 \end{pmatrix}\begin{pmatrix} 0 & 1 \\ 1 & 0 \end{pmatrix}\begin{pmatrix} 1 \\ 0 \end{pmatrix} = \frac{\hbar}{2}\begin{pmatrix} 0 & 1 \end{pmatrix}\begin{pmatrix} 1 \\ 0 \end{pmatrix} = 0.$$

Experimentally, this means, if we place another magnetic field $\mathbf{B}_2$ along the x-axis after $\mathbf{B}_1$ ($||$ z-axis), we will change the eigenfunction, say $|\alpha\rangle$, to a mixed state of $\hat{S}_x$,

$$|\alpha\rangle = \begin{pmatrix} 1 \\ 0 \end{pmatrix} \xrightarrow{B_{2x}} \frac{1}{\sqrt{2}}(\eta_{x_1} + \eta_{x_2}), \tag{6.1.10}$$

with 50 % chance in η_{x_1} and another 50 % in η_{x_2}. This shows that the very measurement, that is placing $\mathbf{B}_2$ in the path of electron, changes the state. This is of no surprise because the measurement changes the Hamiltonian of the system. Specifically, we have the following relationships:

$$\begin{aligned} \hat{S}_x\alpha &= \frac{\hbar}{2}\beta, \quad \hat{S}_x\beta = \frac{\hbar}{2}\alpha, \\ \hat{S}_y\alpha &= i\frac{\hbar}{2}\beta, \quad \hat{S}_y\beta = -i\frac{\hbar}{2}\alpha, \\ \hat{S}_z\alpha &= \frac{\hbar}{2}\alpha, \quad \hat{S}_z\beta = -\frac{\hbar}{2}\beta. \end{aligned} \tag{6.1.11}$$

The relationships can be made clearer if we define the raising and lowering operators for spin as

$$\hat{S}_+ = (\hat{S}_x + i\hat{S}_y), \quad \hat{S}_- = (\hat{S}_x - i\hat{S}_y), \tag{6.1.12}$$

so $\hat{S}_x = (\hat{S}_+ + \hat{S}_-)/2$ and $\hat{S}_y = (\hat{S}_+ - \hat{S}_-)/(2i)$. $\hat{S}_+$ flips a spin from β to α, $\hat{S}_+\beta = \hbar\alpha$, while $\hat{S}_-$ does the opposite, $\hat{S}_-\alpha = \hbar\beta$.

6.1.3.3 Total spin for a single electron

To this end, we only compute the components of the spin. Now, we compute the total spin $\hat{\mathbf{S}}$. Similar to $\hat{\mathbf{L}}^2$, we compute $\hat{\mathbf{S}} \cdot \hat{\mathbf{S}} = \hat{\mathbf{S}}^2 = \hat{S}_x^2 + \hat{S}_y^2 + \hat{S}_z^2$. It is easy to show that $[\hat{\mathbf{S}}^2, \hat{S}_i] = 0$, where $i = x, y, z$, so $\hat{\mathbf{S}}^2$ shares common eigenfunctions with $\hat{S}_i$. To obtain $\hat{\mathbf{S}}^2$, we first compute $\hat{S}_x^2$ as

$$\hat{S}_x^2 = \frac{\hbar}{2}\hat{\sigma}_x\frac{\hbar}{2}\hat{\sigma}_x = \frac{\hbar^2}{4}\begin{pmatrix}0 & 1\\1 & 0\end{pmatrix}\begin{pmatrix}0 & 1\\1 & 0\end{pmatrix} = \frac{\hbar^2}{4}\begin{pmatrix}1 & 0\\0 & 1\end{pmatrix}.$$

$\hat{S}_y^2$ and $\hat{S}_z^2$ have the same results. Thus, our $\hat{\mathbf{S}}^2$ is

$$\hat{\mathbf{S}}^2 = \frac{3\hbar^2}{4}\begin{pmatrix}1 & 0\\0 & 1\end{pmatrix}. \tag{6.1.13}$$

To find its eigenvalue, we can apply $\hat{\mathbf{S}}^2$ to any eigenfunction of $\hat{S}_x$, $\hat{S}_y$, and $\hat{S}_z$ such as

$$\hat{\mathbf{S}}^2\eta_{x_1} = \frac{3\hbar^2}{4}\begin{pmatrix}1 & 0\\0 & 1\end{pmatrix}\frac{1}{\sqrt{2}}\begin{pmatrix}1\\1\end{pmatrix} = \frac{3\hbar^2}{4}\eta_{x_1} = \frac{1}{2}\left(\frac{1}{2}+1\right)\hbar^2\eta_{x_1}.$$

We see the eigenvalue of $\hat{\mathbf{S}}^2$ is $\frac{1}{2}(\frac{1}{2}+1)\hbar^2$, where we purposely rewrite the eigenvalue as a product of $\frac{1}{2}$ and $\frac{1}{2}+1$, in a similar fashion as $\hat{\mathbf{L}}^2 Y_{lm} = l(l+1)\hbar^2 Y_{lm}$. The magnetic spin quantum number m_s takes $+\frac{1}{2}$ and $-\frac{1}{2}$ for spin-up and spin-down. If we represent the common eigenfunctions of $\hat{\mathbf{S}}^2$ and $\hat{S}_z$ as $|sm_s\rangle$, we have $\hat{\mathbf{S}}^2|sm_s\rangle = s(s+1)\hbar^2|sm_s\rangle$ and $\hat{S}_z|sm_s\rangle = m_s\hbar|sm_s\rangle$.

Exercise 6.1.3.

8. Express $\hat{S}_+$ and $\hat{S}_-$ in matrix form.
9. Using eqs. (6.1.7), prove $[\hat{S}_x, \hat{S}_y] = i\hbar\hat{S}_z$, $[\hat{S}_y, \hat{S}_z] = i\hbar\hat{S}_x$, $[\hat{S}_z, \hat{S}_x] = i\hbar\hat{S}_y$.
10. Show $[\hat{\mathbf{S}}^2, \hat{S}_i] = 0$, where i represents x, y, and z, respectively.
11. (a) Find the eigenvalues and eigenvectors of $\hat{S}_y$. (b) Compute the expectation values of $\hat{S}_x$ and $\hat{S}_z$.
12. Compute the following commutations: $[\hat{S}_-, \hat{S}_+]$, $[\hat{S}_-, \hat{S}_x]$ and $[\hat{S}_+, \hat{S}_z]$.
13. Prove $\hat{S}_+\alpha = 0$, $\hat{S}_-\alpha = \hbar\beta$, $\hat{S}_+\beta = \hbar\alpha$, and $\hat{S}_-\beta = 0$.
14. Use the matrix form to prove all the relationships in eq. (6.1.11).

6.1.4 Total wavefunction of an electron

The wavefunction of a single electron is the product of spatial and spin wavefunctions,

$$\psi(\mathbf{r},\mathbf{s}) = \frac{1}{\sqrt{|a|^2+|b|^2}}\begin{pmatrix} a\phi_1(\mathbf{r}) \\ b\phi_2(\mathbf{r}) \end{pmatrix} = \frac{a\phi_1(\mathbf{r})\alpha + b\phi_2(\mathbf{r})\beta}{\sqrt{|a|^2+|b|^2}}, \tag{6.1.14}$$

where a and b are two normalization constants and ϕ_1 and ϕ_2 are two normalized spatial wavefunctions that are associated with the spin-up and spin-down channels, respectively. $\psi(\mathbf{r},\mathbf{s})$ is now a spinor. When we apply operators to $\psi(\mathbf{r},\mathbf{s})$, spin operators only act upon the spin wavefunction, while spatial operators act on the spatial wavefunction. For instance, $\hat{\mathbf{p}}|\alpha\rangle = |\alpha\rangle\hat{\mathbf{p}}$, and $\hat{\mathbf{S}}\phi_1(\mathbf{r}) = \phi_1(\mathbf{r})\hat{\mathbf{S}}$. In the same manner that we cannot use the spin wavefunction to find the position of an electron, neither can we find the spin from the spatial wavefunction.

Exercise 6.1.4.

15. In QM, can we say we put a spin at a position of space $\mathbf{r}$?

16. Based on eq. (6.1.14), suppose $\phi_1(\mathbf{r}) = Y_{lm_1}$ and $\phi_2(\mathbf{r}) = Y_{lm_2}$. Show $\psi(\mathbf{r},\mathbf{s})$ is an eigenfunction of (a) $\hat{\mathbf{L}}^2$, (b) $\hat{\mathbf{S}}^2$. (c) Find the condition where $\psi(\mathbf{r},\mathbf{s})$ is an eigenfunction of $\hat{L}_z + \hat{S}_z$.

17. An electron has the wavefunction (eq. (6.1.14)). Compute (a) $\langle\psi|\hat{\sigma}_x|\psi\rangle$, (b) $\langle\psi|\hat{\sigma}_y|\psi\rangle$, and (c) $\langle\psi|\hat{\sigma}_z|\psi\rangle$.

6.2 Total angular momentum $\hat{\mathbf{J}}$

With the presence of spin, we need to introduce the total angular momentum $\hat{\mathbf{J}}$, which is defined as $\hat{\mathbf{J}} = \hat{\mathbf{L}} + \hat{\mathbf{S}}$, a vector addition of spin $\hat{\mathbf{S}}$ and orbital $\hat{\mathbf{L}}$. Because $\hat{\mathbf{L}}$ belongs to the spatial space and $\hat{\mathbf{S}}$ pertains to the spin space, $[\hat{\mathbf{L}},\hat{\mathbf{S}}] = 0$. Similar to $\hat{\mathbf{S}}\times\hat{\mathbf{S}} = i\hbar\hat{\mathbf{S}}$ and $\hat{\mathbf{L}}\times\hat{\mathbf{L}} = i\hbar\hat{\mathbf{L}}$, $\hat{\mathbf{J}}$ has similar commutations,

$$[\hat{J}_x,\hat{J}_y] = i\hbar\hat{J}_z, [\hat{J}_y,\hat{J}_z] = i\hbar\hat{J}_x, [\hat{J}_z,\hat{J}_x] = i\hbar\hat{J}_y \to \hat{\mathbf{J}}\times\hat{\mathbf{J}} = i\hbar\hat{\mathbf{J}}. \tag{6.2.1}$$

$[\hat{\mathbf{J}}^2,\hat{\mathbf{L}}^2] = [\hat{\mathbf{J}}^2,\hat{\mathbf{S}}^2] = [\hat{J}_z,\hat{\mathbf{L}}^2] = [\hat{J}_z,\hat{\mathbf{S}}^2] = [\hat{J}_z,\hat{\mathbf{J}}^2] = 0$. The proof is left as an exercise.

6.2.1 Quantum number addition and spectroscopic notation

Given $\hat{\mathbf{L}}$ and $\hat{\mathbf{S}}$, how could we get $\hat{\mathbf{J}}$? In CM, when given two vectors $\mathbf{A}$ and $\mathbf{B}$, their sum $\mathbf{C} = \mathbf{A} + \mathbf{B}$ is determined. In QM, we cannot do this because we must specify which state the system is in, but then we lose the connection with CM. A compromise is that we assume $\hat{\mathbf{L}}$, $\hat{\mathbf{S}}$, and $\hat{\mathbf{J}}$ share a common eigenfunction. Since we know the quantum numbers for $\hat{\mathbf{L}}$ and $\hat{\mathbf{S}}$, all we need is to find the quantum number for $\hat{\mathbf{J}}$.

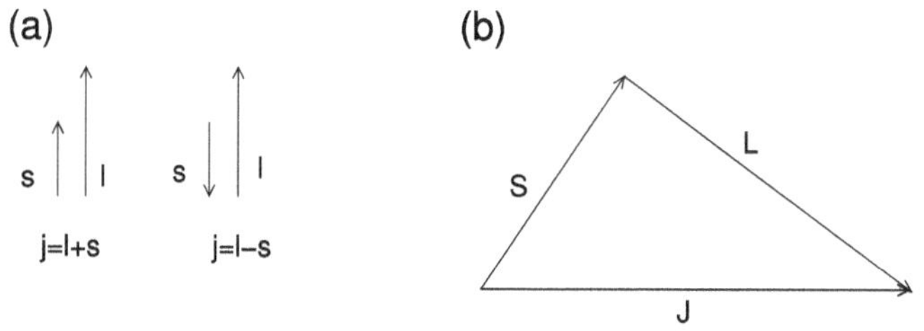

Figure 6.2: (a) Angular-momentum quantum number addition for $s = \frac{1}{2}$ and $l = 1$. (b) Addition of angular momenta. The magnitudes of $\hat{\mathbf{S}}$, $\hat{\mathbf{L}}$ and $\hat{\mathbf{J}}$ are $\sqrt{\frac{3}{4}}\hbar$, $\sqrt{2}\hbar$, $\sqrt{\frac{15}{4}}\hbar$, where $s = \frac{1}{2}$, $l = 1$ and $j = \frac{3}{2}$.

Given l and s, the total angular-momentum quantum number j is given by

$$j = l + s, l + s - 1, \dots, |l - s|,$$

where j only takes a positive number. The magnetic total angular momentum quantum number m_j is given by

$$m_j = j, j - 1, \dots, -j. \tag{6.2.2}$$

This expression is due to the projection of $\hat{\mathbf{J}}$ on to the z-axis, similar to $\hat{\mathbf{L}}$ and $\hat{\mathbf{S}}$. Let's consider a p-orbital with $l = 1$ and spin with $s = \frac{1}{2}$. Figure 6.2(a) shows that quantum number j has two possible choices: either the spin in the same direction as the orbital or in the opposite direction. For the former, j takes $\frac{3}{2}$, and for the latter $\frac{1}{2}$.

Since $\hat{\mathbf{S}}$, $\hat{\mathbf{L}}$, and $\hat{\mathbf{J}}$ are in a common eigenfunction, their magnitudes are given by $|\langle\hat{\mathbf{S}}^2\rangle| = s(s+1)\hbar^2$, $|\langle\hat{\mathbf{L}}^2\rangle| = l(l+1)\hbar^2$, and $|\langle\hat{\mathbf{J}}^2\rangle| = j(j+1)\hbar^2$. As a result, we can use the law of cosines to find angles between them. Figure 6.2(b) shows a triangle formed by three vectors.

Using three quantum numbers j, l, and s to specify a particular configuration is cumbersome. This is where the spectroscopic term notation comes in. It is used more often in many-electron configurations. To represent the electron configuration, one combines (j, l, s) into a single term

$${}^{2s+1}L_j,$$

where the superscript $2s + 1$ refers to the spin and the subscript j is the total angular momentum. L is not a quantum number. Instead, it uses a letter, S, P, D, F. and so on, for the orbital angular-momentum quantum number l being 0, 1, 2, 3 .... For instance, if $l = 1$, $j = \frac{3}{2}$, and $s = \frac{1}{2}$, we have ${}^2P_{\frac{3}{2}}$.

Example 3 (Vector addition in QM). An electron is in a p state. (a) Compute the angles between $\hat{\mathbf{J}}$ and $\hat{\mathbf{S}}$ for all possible j. (b) Compute the angles between $\hat{\mathbf{L}}$ and $\hat{\mathbf{S}}$ for all possible j.

(a) For a p-state, we have $l = 1$ and $s = \frac{1}{2}$, so $j = \frac{3}{2}, \frac{1}{2}$. The eigenvalues of $\hat{\mathbf{J}}^2$ are $j(j+1)\hbar^2 = \frac{15}{4}\hbar^2$ and $\frac{3}{4}\hbar^2$. The eigenvalue of $\hat{\mathbf{L}}^2$ is $2\hbar^2$ and that of $\hat{\mathbf{S}}^2$ is $\frac{3}{4}\hbar^2$.

To find the angle between $\hat{\mathbf{J}}$ and $\hat{\mathbf{S}}$, we start from $\hat{\mathbf{J}} - \hat{\mathbf{S}} = \hat{\mathbf{L}}$ and square both sides to get $\hat{\mathbf{J}}^2 + \hat{\mathbf{S}}^2 - 2\hat{\mathbf{J}} \cdot \hat{\mathbf{S}} = \hat{\mathbf{L}}^2$, whose expectation values lead to $j(j+1)\hbar^2 + s(s+1)\hbar^2 - 2\cos\theta_{JS}\sqrt{j(j+1)\hbar^2 s(s+1)\hbar^2} = l(l+1)\hbar^2$. We take $j = \frac{3}{2}, s = \frac{1}{2}, l = 1$ as an example to find $\cos\theta_{JS} = \frac{\sqrt{5}}{3}$ and $\theta_{JS} = 41.81°$. This case is shown in Figure 6.2(b). For $j = \frac{1}{2}$, $\cos\theta_{JS} = -\frac{1}{3}$, so $\theta_{JS} = 109.47°$.

(b) To find the angle between $\hat{\mathbf{J}}$ and $\hat{\mathbf{L}}$, we start from $\hat{\mathbf{J}} - \hat{\mathbf{L}} = \hat{\mathbf{S}}$. The rest of the steps are the same as (a). For $j = \frac{3}{2}$, $\cos\theta_{JL} = \sqrt{\frac{5}{6}}$ and $\theta_{JL} = 24.095°$. For $j = \frac{1}{2}$, $\cos\theta_{JL} = \sqrt{\frac{2}{15}}$ and $\theta_{JL} = 68.58°$. We note these angles are not arbitrary, different from CM. They are the result of quantization of $\hat{\mathbf{J}}$, $\hat{\mathbf{S}}$, and $\hat{\mathbf{L}}$.

Exercise 6.2.1.

18. Show (a) $[\hat{\mathbf{J}}^2, \hat{\mathbf{L}}^2] = 0$; (b) $[\hat{\mathbf{J}}^2, \hat{\mathbf{S}}^2] = 0$; (c) $[\hat{J}_x, \hat{J}_y] = i\hbar\hat{J}_z$, $[\hat{J}_y, \hat{J}_z] = i\hbar\hat{J}_x$, and $[\hat{J}_z, \hat{J}_x] = i\hbar\hat{J}_y$; (d) $[\hat{\mathbf{J}}^2, \hat{\mathbf{L}}^2] = [\hat{\mathbf{J}}^2, \hat{\mathbf{S}}^2] = [\hat{J}_z, \hat{\mathbf{L}}^2] = [\hat{J}_z, \hat{\mathbf{S}}^2] = [\hat{J}_z, \hat{\mathbf{J}}^2] = 0$.

19. (a) An electron is in a d state with $s = \frac{1}{2}$. What are the possible j? What are the possible m_j? (b) Following (a), write down their respective spectroscopic terms.

20. Sodium's ground state has its outer shell $3s^1$. (a) Find its ground state spectroscopic configuration. (b) Assume the $3s^1$ electron is excited to $3p$. What are the possible j? (c) Following (b), find the term notations for its excited states.

21. An electron with spin $s = 1/2$ is in a $3d$-orbital. (a) Write down all the possible total angular-momentum quantum numbers j. (b) Choose a state with the largest j and write down the spectroscopic term notation for this state. (c) Following (b), if this electron is in its eigenfunction of $\hat{\mathbf{J}}$, compute the magnitude of $\hat{\mathbf{J}}$. (d) Compute the angle between $\hat{\mathbf{L}}$ and $\hat{\mathbf{J}}$.

6.2.2 Spin-orbit coupling in a central field

In the previous subsection, we assumed that we are in a common eigenfunction of $\hat{\mathbf{J}}$, $\hat{\mathbf{L}}$, $\hat{\mathbf{S}}$, but this is rarely the case. Electrons move rapidly around the nuclei. In the reference frame of the electron, the proton is orbiting, which generates an effective magnetic field for the electron spin. The spin experiences this effective magnetic field, so the spin starts to precess around this field. This generates a coupling between the spin and the orbital motion, namely, spin-orbit coupling. Quantum mechanically, the Hamiltonian in the nonrelativistic limit of the Dirac equation is $\frac{\hat{\mathbf{p}}^2}{2m} + V - \frac{\hat{\mathbf{p}}^4}{8m^3c^2} - \frac{\hbar^2}{4m^2c^2}\frac{dV}{dr}\frac{\partial}{\partial r} + \frac{1}{2m^2c^2}\frac{1}{r}\frac{dV}{dr}\hat{\mathbf{L}} \cdot \hat{\mathbf{S}}$, where the last term is called spin-orbit coupling,

$$\hat{H}_{\text{SOC}} = \frac{1}{2m^2c^2}\frac{1}{r}\frac{dV}{dr}\hat{\mathbf{L}} \cdot \hat{\mathbf{S}} = \lambda(r)\hat{\mathbf{L}} \cdot \hat{\mathbf{S}}. \tag{6.2.3}$$

Here $\lambda(r)$ is called the spin-orbit coupling (SOC) constant. Since $\hat{\mathbf{L}}$ and $\hat{\mathbf{S}}$ are in units of $\hbar$, λ must be in units of $J/\hbar^2$. The derivation of eq. (6.2.3) is quite involved and not introduced here, but can be found in Ref. [32]. We will simply write $\lambda(r)$ as λ.

We can estimate the magnitude of SOC. Since $\hat{\mathbf{J}} = \hat{\mathbf{L}} + \hat{\mathbf{S}}$, $\hat{\mathbf{J}}^2 = \hat{\mathbf{L}}^2 + \hat{\mathbf{S}}^2 + 2\hat{\mathbf{L}} \cdot \hat{\mathbf{S}}$. We have $\hat{\mathbf{L}} \cdot \hat{\mathbf{S}} = (\hat{\mathbf{J}}^2 - \hat{\mathbf{L}}^2 - \hat{\mathbf{S}}^2)/2$. Suppose the system is in the common eigenfunction of $\hat{\mathbf{J}}, \hat{\mathbf{L}}, \hat{\mathbf{S}}$. Then the expectation value $\langle \hat{\mathbf{L}} \cdot \hat{\mathbf{S}} \rangle = (j(j+1) - l(l+1) - s(s+1))\hbar^2/2$. Now, we can compare the energy difference ΔE between ${}^2P_{1/2}$ and ${}^2P_{3/2}$, which is

$$\Delta E = E_{j=3/2} - E_{j=1/2} = \lambda\left(\frac{3}{2}\frac{5}{2} - \frac{1}{2}\frac{3}{2}\right)\hbar^2/2 = \frac{3}{2}\lambda\hbar^2. \tag{6.2.4}$$

In sodium, the magnitude of λ can be estimated from the energy difference between two D lines, $\lambda\hbar^2 = 2.136$ meV, which is smaller than the Coulomb interaction of several eV, but it is larger than the magnetic field energy of 1 T, $\mu_B B = 0.057$ meV, where μ_B is the Bohr magneton (see the next section). SOC cannot be ignored in a weak magnetic field. In heavy atoms such as Hg, SOC can be very large and the length contraction leads to a smaller radius for the orbiting electrons in the outer shell, which is largely responsible for its liquid phase.[1]

In the presence of SOC, $\hat{\mathbf{L}}$ and $\hat{\mathbf{S}}$ do not commute with the Hamiltonian, so the eigenstate of the system is no longer an eigenfunction of either $\hat{\mathbf{L}}$ or $\hat{\mathbf{S}}$. We work with $\hat{\mathbf{J}}$.

6.2.3 Eigenfunctions and eigenvalues of $\hat{\mathbf{J}}$ and $\hat{J}_z$

In the presence of SOC in a single atom, only $\hat{\mathbf{J}}$ is a conserved quantity since it commutes with the Hamiltonian. We are going to first find the eigenfunctions ξ of $\hat{\mathbf{J}}^2$.

We use the same property that two commutable operators share common eigenfunctions. For a fixed l, $\hat{\mathbf{L}}^2$'s eigenfunction is Y_{lm}. Because $[\hat{\mathbf{J}}^2, \hat{\mathbf{L}}^2] = 0$, ξ must contain Y_{lm}. But $[\hat{\mathbf{J}}^2, \hat{L}_z] \neq 0$, so ξ must contain more than one Y_{lm} such as Y_{lm_1} and Y_{lm_2}. Similarly, since $[\hat{\mathbf{J}}^2, \hat{\mathbf{S}}^2] = 0$ but $[\hat{\mathbf{J}}^2, \hat{S}_z] \neq 0$, ξ must contain both $|\alpha\rangle$ and $|\beta\rangle$. With fixed l and s, the products of the orbital and spin eigenfunctions, $Y_{lm_1}|\alpha\rangle$ and $Y_{lm_2}|\beta\rangle$, form a complete group of basis functions for $\hat{\mathbf{J}}^2$. Thus, in general, our eigenfunction is $\xi = aY_{lm_1}|\alpha\rangle + bY_{lm_2}|\beta\rangle$. Next, we need to determine the relationship between m_1 and m_2.

Recall $\hat{\mathbf{J}}^2 = (\hat{\mathbf{L}} + \hat{\mathbf{S}})^2 = \hat{\mathbf{L}}^2 + \hat{\mathbf{S}}^2 + 2\hat{\mathbf{L}} \cdot \hat{\mathbf{S}} = \hat{\mathbf{L}}^2 + \hat{\mathbf{S}}^2 + 2\hat{L}_z\hat{S}_z + 2(\hat{L}_x\hat{S}_x + \hat{L}_y\hat{S}_y)$. Since we are looking for a solution for $\hat{\mathbf{J}}^2\xi = \lambda\xi$, where λ is the eigenvalue of $\hat{\mathbf{J}}^2$, not to be confused with the spin-orbit coupling above, terms in the above equation must produce a constant multiplied by ξ. This is possible for $\hat{\mathbf{L}}^2$, and $\hat{\mathbf{S}}^2$, but not possible for the last three terms. To see this clearly, we rewrite $2(\hat{L}_x\hat{S}_x + \hat{L}_y\hat{S}_y) = \hat{L}_+\hat{S}_- + \hat{L}_-\hat{S}_+$. We apply $\hat{L}_+\hat{S}_-$ to ξ to get

$$\hat{L}_+\hat{S}_-(aY_{l,m_1}|\alpha\rangle + bY_{l,m_2}|\beta\rangle) = a\hbar^2\sqrt{l(l+1) - m_1(m_1+1)}Y_{l,m_1+1}|\beta\rangle, \tag{6.2.5}$$

and apply $\hat{L}_-\hat{S}_+$ to ξ to obtain

1 Lars J. Norrby [33] once asked, "Why is mercury liquid, or why do relativistic effects not get into chemistry textbooks?"

$$\hat{L}_-\hat{S}_+(aY_{l,m_1}|\alpha\rangle + bY_{l,m_1}|\beta\rangle) = b\hbar^2\sqrt{l(l+1) - m_2(m_2-1)}Y_{l,m_2-1}|\alpha\rangle. \tag{6.2.6}$$

If we compare eqs. (6.2.5) and (6.2.6) with $\xi = aY_{lm_1}|\alpha\rangle + bY_{lm_2}|\beta\rangle$, it is clear that in order for ξ to be the eigenfunction of $\hat{\mathbf{J}}^2$, we must require $m_2 = m_1 + 1$. Otherwise, ξ cannot be an eigenfunction $\hat{\mathbf{J}}^2$. Below, we denote m_1 by m and m_2 by $m+1$, where m is the magnetic orbital angular-momentum quantum number.[2] Next, we will use $Y_{l,m}|\alpha\rangle$ and $Y_{l,m+1}|\beta\rangle$ as the basis functions to construct a matrix for $\hat{\mathbf{J}}^2$ and then solve the eigenequation.

We denote $Y_{l,m}|\alpha\rangle$ as $|1\rangle$ and $Y_{l,m+1}|\beta\rangle$ as $|2\rangle$, and compute the matrix elements for $\hat{\mathbf{J}}^2$, $\langle i|\hat{\mathbf{J}}^2|j\rangle$, where i and j run from 1 to 2. A lengthy but straightforward calculation (see the following exercise) yields the eigenequation for $\hat{\mathbf{J}}^2$ as

$$\hbar^2\begin{pmatrix} l(l+1) + \frac{3}{4} + m & \sqrt{l(l+1) - m(m+1)} \\ \sqrt{l(l+1) - m(m+1)} & l(l+1) + \frac{3}{4} - m - 1 \end{pmatrix}\begin{pmatrix} a \\ b \end{pmatrix} = \lambda\begin{pmatrix} a \\ b \end{pmatrix}, \tag{6.2.7}$$

which has a nontrivial solution if its determinant is zero. Setting the determinant to zero yields the following equation for λ in units of $\hbar^2$:

$$\lambda^2 - \lambda\left[2l(l+1) + \frac{1}{2}\right] + l^2(l+1)^2 - \frac{1}{2}l(l+1) - \frac{3}{16} = 0. \tag{6.2.8}$$

This shows that the eigenvalue λ of $\hat{\mathbf{J}}^2$ does not depend on m. The reason is because the total angular-momentum magnitude does not depend the magnetic orbital quantum number, in the same way as $\hat{\mathbf{L}}^2$. λ has two roots, $\lambda_1 = (l-\frac{1}{2})(l+\frac{1}{2})\hbar^2$, and $\lambda_2 = (l+\frac{1}{2})(l+\frac{3}{2})\hbar^2$. Because $s = \frac{1}{2}$ and $j = l \pm s$, we rewrite λ as $j(j+1)$ to get

$$\hat{\mathbf{J}}^2\xi_j = \lambda\xi_j = j(j+1)\hbar^2\xi_j.$$

Next, we substitute λ back into eq. (6.2.7) to find two sets of a and b (the details are relegated to the following exercise):

$$\xi_{j=l+\frac{1}{2}} = \begin{pmatrix} \sqrt{\frac{l+m+1}{2l+1}}Y_{l,m} \\ \sqrt{\frac{l-m}{2l+1}}Y_{l,m+1} \end{pmatrix}; \quad \xi_{j=l-\frac{1}{2}} = \begin{pmatrix} \sqrt{\frac{l-m}{2l+1}}Y_{l,m} \\ \sqrt{\frac{l+m+1}{2l+1}}Y_{l,m+1} \end{pmatrix}. \tag{6.2.9}$$

We apply $\hat{J}_z$ to either $\xi_{j=l+\frac{1}{2}}$ or $\xi_{j=l-\frac{1}{2}}$ to find $\hat{J}_z\xi = m_j\hbar\xi$, where $m_j = m + \frac{1}{2}$. Note that m is really m_l. These eigenfunctions of $\hat{\mathbf{J}}^2$ are very peculiar. They belong to the same l and m_j, but different j. j, s, m_j, and l are good quantum numbers, but m_s and m_l are not. Except for the $l = 0$ state, the same j appears in two different l. Table 6.1 summarizes this connection, together with the X-ray edges.

2 To avoid too many subscripts, we suppress l in m_l.

Table 6.1: (Top) Relation between l, j and the spectroscopic notation. (Bottom) In X-ray spectroscopy, the spectroscopic notation carries the main quantum number n. X-ray edges start from K, L, up to P_3 edges. 1s, 2s and 3s are traditionally not written in spectroscopic terms because $l = 0$ and $j = s$.

l	0	1	2	3	4
j	1/2	(1/2, 3/2)	(3/2, 5/2)	(5/2, 7/2)	(7/2, 9/2)
Configuration	$S_{1/2}$	$(P_{1/2}, P_{3/2})$	$(D_{3/2}, D_{5/2})$	$(F_{5/2}, F_{7/2})$	$(G_{7/2}, G_{9/2})$

X-ray edge	K	L_1	L_2	L_3	M_1	M_2	M_3	M_4	M_5
Configuration	1s	2s	$2P_{1/2}$	$2P_{3/2}$	3s	$3P_{1/2}$	$3P_{3/2}$	$3D_{3/2}$	$3D_{5/2}$

Exercise 6.2.3.

22. Find the matrix elements of $\hat{\mathbf{J}}^2$ within the basis functions: $Y_{l,m}|\alpha\rangle$ denoted as $|1\rangle$ and $Y_{l,m+1}|\beta\rangle$ denoted as $|2\rangle$. And compare your results with eq. (6.2.7).

23. Starting from the determinant of eq. (6.2.7), derive and solve for λ, i. e., eq. (6.2.8).

24. Substitute two roots λ of eq. (6.2.8) into (6.2.7) to find the eigenfunctions of $\hat{\mathbf{J}}^2$, that is (a, b) in eq. (6.2.7) and then compute the eigenvalue of $\hat{J}_z$.

25. One way to appreciate why $\hat{S}_z$ and $\hat{L}_z$ are no longer conserved with the eigenfunctions of $\hat{\mathbf{J}}^2$ is to compute their expectation values. (a) Using eq. (6.2.9), compute the expectation values of $\hat{S}_z$ and $\hat{L}_z$. (b) Compute the expectation values of $\hat{S}_x$, $\hat{S}_y$, $\hat{L}_x$ and $\hat{L}_y$.

6.3 Interaction of spin and orbital moments with a magnetic field

Spin is magnetic. Only a magnetic field or another spin can affect it directly. This section focuses on a direct interaction with a magnetic field, so we can better understand the Stern–Gerlach experiment and other experiments.

6.3.1 Magnetic orbital and spin moments

A magnetic field interacts with a system through spin and orbital moments. In a classical current loop, the magnetic dipole moment is defined as $\boldsymbol{\mu}_{\text{classical}} = IA\hat{n}$, where I is the current through the loop, A is the area of the loop, and $\hat{n}$ is the surface normal defined by the right-hand rule. If a single electron is orbiting around a proton with velocity v and radius r (see Fig. 6.3(a)), its area A is πr^2, and the current is $I = \frac{-e}{T}$, where the period T is $2\pi r/v$. Then the magnitude of the orbital moment is

$$\mu_{\text{classical}} = -\frac{evr}{2} = -\frac{em_e vr}{2m_e} = -\frac{eL}{2m_e}. \tag{6.3.1}$$

In QM, we follow the same logic to define the orbital angular moment as

$$\hat{\mu}_l = -\frac{e\hat{\mathbf{L}}}{2m_e}, \tag{6.3.2}$$

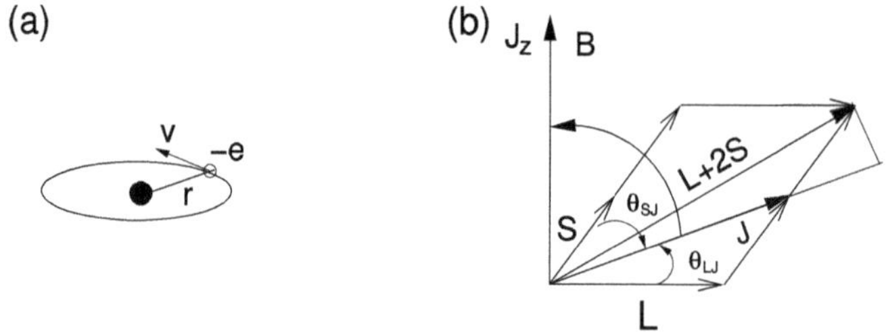

Figure 6.3: (a) Orbital angular momentum. (b) Derivation of the Lande g-factor proceeds in two steps. The first projects $\hat{\mathbf{L}} + 2\hat{\mathbf{S}}$ on to $\hat{\mathbf{J}}$, and second one projects $\hat{\mathbf{J}}$ to $\hat{J}_z$.

where the orbital moment is proportional to the orbital angular momentum. To simplify the expression, we rewrite $\hat{\boldsymbol{\mu}}_l$ as

$$\hat{\boldsymbol{\mu}}_l = -\frac{e\hbar}{2m_e}\frac{\hat{\mathbf{L}}}{\hbar} = -\mu_B\frac{\hat{\mathbf{L}}}{\hbar}, \tag{6.3.3}$$

whose coefficient of $\frac{\hat{\mathbf{L}}}{\hbar}$ is defined as the Bohr magneton

$$\mu_B = \frac{e\hbar}{2m_e} = 9.27401 \times 10^{-24}\ \mathrm{J/T}.$$

The reason why we have $\hbar$ in μ_B is to make $\hat{\mathbf{L}}/\hbar$ dimensionless. Suppose we want to compute the z-component of $\hat{\boldsymbol{\mu}}_l$ and its expectation value, we have

$$\hat{\mu}_{l_z} = -\mu_B\hat{L}_z/\hbar, \rightarrow \langle\hat{\mu}_{l_z}\rangle = -\mu_B m_l\hbar/\hbar = -m_l\mu_B, \tag{6.3.4}$$

where in the second equation we have used the eigenfunction of $\hat{L}_z$.

Although the orbital moment has a classical analogue, the spin moment does not. Instead, we define it as

$$\hat{\boldsymbol{\mu}}_s = -\frac{e}{m_e}\hat{\mathbf{S}} = -\frac{e\hbar}{2m_e}\frac{2\hat{\mathbf{S}}}{\hbar} = -\mu_B\frac{2\hat{\mathbf{S}}}{\hbar} = -\gamma\hat{\mathbf{S}} = -\mu_B\hat{\sigma}, \tag{6.3.5}$$

where γ is called the gyromagnetic ratio and the last equation uses $\hat{\mathbf{S}} = \frac{\hbar}{2}\hat{\sigma}$. For a free electron, γ is just the electron charge–mass ratio, but, in a molecule or a solid, γ may change significantly and is often used as a way to detect spin states. Equations (6.3.5) and (6.3.2) differ by a factor of 2. The expectation value of $\hat{\mu}_{s_z}$ in the eigenfunction of $\hat{S}_z$ is

$$\langle\hat{\mu}_{s_z}\rangle = -2\mu_B\frac{m_s\hbar}{\hbar} = -2m_s\mu_B. \tag{6.3.6}$$

Table 6.2: Moments used in CM, EM, and QM.

Physics	Quantity	Expression	Variables; SI units
CM	moment of inertia	$I = mr^2$	mass m, radius r. $\mathrm{kg} \cdot \mathrm{m}^2$
EM	electric dipole moment	$\mathbf{D} = q\mathbf{r}$	q: charge; $\mathbf{r}$: position. $\mathrm{C} \cdot \mathrm{m}$
EM	magnetic dipole moment	$\mathbf{M} = IA\hat{n}$	current I, area A, normal $\hat{n}$. $\mathrm{A} \cdot \mathrm{m}^2$
QM	magnetic orbital moment	$\hat{\boldsymbol{\mu}}_l = -\mu_B \hat{\mathbf{L}}/\hbar$	orbital: $\hat{\mathbf{L}}$. $\mathrm{A} \cdot \mathrm{m}^2$
QM	magnetic spin moment	$\hat{\boldsymbol{\mu}}_s = -\mu_B 2\hat{\mathbf{S}}/\hbar$	spin: $\hat{\mathbf{S}}$. $\mathrm{A} \cdot \mathrm{m}^2$

6.3.2 Interaction with the B field

It can be confusing to the beginner that different physics courses introduce multiple *moments* with entirely different meanings. Table 6.2 compares them. In electricity, the electric dipole moment describes the interaction between a charge and an electric field. In magnetism, the interaction between a magnetic field $\mathbf{B}$ and the spin and orbital is through the magnetic dipole moment, $\hat{\boldsymbol{\mu}} = \hat{\boldsymbol{\mu}}_l + \hat{\boldsymbol{\mu}}_s$. They contribute an extra term, called Zeeman energy, to our Hamiltonian as

$$\hat{U} = -\hat{\boldsymbol{\mu}} \cdot \mathbf{B} = -(\hat{\boldsymbol{\mu}}_l + \hat{\boldsymbol{\mu}}_s) \cdot \mathbf{B} = \frac{\mu_B}{\hbar}(\hat{\mathbf{L}} + 2\hat{\mathbf{S}}) \cdot \mathbf{B}. \tag{6.3.7}$$

Without spin, the potential energy is $\hat{U} = \frac{\mu_B}{\hbar}\hat{\mathbf{L}} \cdot \mathbf{B}$. If we set $\mathbf{B}$ along the z-axis, we have

$$\hat{U} = -\hat{\mu}_z B_z = \frac{\mu_B}{\hbar}\hat{L}_z B_z \to U \equiv \langle\hat{U}\rangle = m_l \mu_B B_z, \tag{6.3.8}$$

where the last step used the eigenfunction of $\hat{L}_z$. It is clear that U is directly proportional to the magnetic orbital angular-momentum quantum number m_l. If we know m_l, the interaction energy can be found immediately. Without $\hat{\mathbf{L}}$, we have only the spin term which will be used to explain the Stern–Gerlach experiment in the next section.

The most complicated case is when both spin and orbital are present. We must consider whether SOC is weaker or stronger than the external magnetic field. If the magnetic field is very strong, $\hat{\mathbf{L}}$ and $\hat{\mathbf{S}}$ are still approximately conserved and can still be represented by their respective quantum numbers, m_l and m_s, so the total moment along the z-axis is the sum of eqs. (6.3.4) and (6.3.6). The energy due to $\mathbf{B}$ is

$$U = (m_l + 2m_s)\mu_B B_z. \tag{6.3.9}$$

If the magnetic field and SOC are comparable, this does not have a simple solution, where j is no longer a good quantum number and the eigenfunctions of $\hat{\mathbf{J}}$ must be used as the basis functions for the Hamiltonian matrix of the system. If SOC is stronger than the magnetic field, we need to introduce Lande's method.

6.3.3 Derivation of Lande-*g* factor

At first glance, one would expect that the energy is now proportional to m_j. However, eq. (6.3.7) introduces a complication because it has $\hat{\mathbf{L}} + 2\hat{\mathbf{S}}$, not $\hat{\mathbf{L}} + \hat{\mathbf{S}} = \hat{\mathbf{J}}$, and there is no quantum number for $\hat{\mathbf{L}} + 2\hat{\mathbf{S}}$. Lande found a solution which consists of two steps. First, we assume that the system is in a common eigenfunction of $\hat{\mathbf{J}}$, $\hat{\mathbf{L}}$, and $\hat{\mathbf{S}}$, so $\langle\hat{\mathbf{J}}^2\rangle = j(j+1)\hbar^2$, $\langle\hat{\mathbf{L}}^2\rangle = l(l+1)\hbar^2$, and $\langle\hat{\mathbf{S}}^2\rangle = s(s+1)\hbar^2$, same as in Section 6.2.1. Since in general $\hat{\mathbf{S}}$, $\hat{\mathbf{L}}$, and $\hat{\mathbf{J}}$ point in different directions in space, we use two angles θ_{LJ} and θ_{SJ} to characterize their relative orientations with respect to $\hat{\mathbf{J}}$ (see Fig. 6.3(b)). Therefore, the projection of $\hat{\mathbf{L}} + 2\hat{\mathbf{S}}$ on $\hat{\mathbf{J}}$ has the length of

$$\sqrt{l(l+1)}\hbar\cos\theta_{LJ} + 2\sqrt{s(s+1)}\hbar\cos\theta_{SJ}. \tag{6.3.10}$$

To find $\cos\theta_{LJ}$ and $\cos\theta_{SJ}$, we square $\hat{\mathbf{S}} = \hat{\mathbf{J}} - \hat{\mathbf{L}}$ and $\hat{\mathbf{L}} = \hat{\mathbf{J}} - \hat{\mathbf{S}}$ to get

$$\hat{\mathbf{S}}^2 = \hat{\mathbf{J}}^2 + \hat{\mathbf{L}}^2 - 2\hat{\mathbf{J}}\cdot\hat{\mathbf{L}}, \quad \hat{\mathbf{L}}^2 = \hat{\mathbf{J}}^2 + \hat{\mathbf{S}}^2 - 2\hat{\mathbf{J}}\cdot\hat{\mathbf{S}}. \tag{6.3.11}$$

In an eigenfunction of $\hat{\mathbf{J}}$, its amplitude is just the square root of that of $\hat{\mathbf{J}}^2$, $\sqrt{j(j+1)}\hbar$. Similarly, $\hat{\mathbf{S}}$ and $\hat{\mathbf{L}}$ have a length of $\sqrt{s(s+1)}\hbar$ and $\sqrt{l(l+1)}\hbar$. $\langle\hat{\mathbf{J}}\cdot\hat{\mathbf{L}}\rangle$ in eq. (6.3.11) is written as $\langle\hat{\mathbf{J}}\rangle\cdot\langle\hat{\mathbf{L}}\rangle\cos\theta_{LJ}$, which is just $\sqrt{j(j+1)}\sqrt{l(l+1)}\hbar^2\cos\theta_{LJ}$. A similar expression is used for $\langle\hat{\mathbf{J}}\cdot\hat{\mathbf{S}}\rangle$. Equation (6.3.11) in their common eigenfunction yields

$$\cos\theta_{LJ} = \frac{j(j+1) + l(l+1) - s(s+1)}{2\sqrt{j(j+1)l(l+1)}}, \quad \cos\theta_{SJ} = \frac{j(j+1) + s(s+1) - l(l+1)}{2\sqrt{j(j+1)s(s+1)}},$$

which are substituted into eq. (6.3.10) to get

$$\frac{3j(j+1) + s(s+1) - l(l+1)}{2\sqrt{j(j+1)}}\hbar. \tag{6.3.12}$$

The next step is critical. In experiments, we apply a magnetic field, which breaks the spherical symmetry of the system and destroys the quantization of $\hat{\mathbf{J}}$. To salvage the situation, we let the field set a quantization z-axis, so $\hat{J}_z$, not $\hat{\mathbf{J}}$, has a good quantum number. Because we are only interested in the interaction between the magnetic field and the system, $\hat{J}_z$ is enough to save us. But, we must correct the eigenfunction assumption. To do so, we project $\hat{\mathbf{J}}$ on to $\hat{J}_z$ by rescaling the unit vector of $\hat{\mathbf{J}}$ through $\langle\hat{J}_z\rangle/\langle\hat{\mathbf{J}}\rangle = m_j/\sqrt{j(j+1)}$. We multiply eq. (6.3.12) by $m_j/\sqrt{j(j+1)}$ and $-e/2m_e$ to get the total magnetic moment along the z-axis,

$$\langle\hat{\mu}_z\rangle = -gm_j\mu_B, \tag{6.3.13}$$

where the Lande's g-factor is

$$g = 1 + \frac{j(j+1) + s(s+1) - l(l+1)}{2j(j+1)}. \tag{6.3.14}$$

Physically, g properly rescales the total magnetic moment value according to l, s and j. For example, we can compute g for a spectroscopic term $^3P_{\frac{1}{2}}$ as

$$g = 1 + \frac{\frac{1}{2}(\frac{1}{2}+1) + 1(1+1) - 1(1+1)}{2\frac{1}{2}\frac{3}{2}} = \frac{3}{2}.$$

With $\langle \hat{\mu}_z \rangle$ in eq. (6.3.13), the Zeeman energy due to the magnetic field along the z-axis is the product of $\langle \hat{\mu}_z \rangle$ and B_z,

$$U = g_j m_j \mu_B B_z \tag{6.3.15}$$

where g_j pertains to j and m_j runs from $-j, -j+1, \ldots, j$.

Exercise 6.3.3.

26. Show that the units of J/T are the same as A · m^2, where J is joules and T is teslas.
27. Why do we need to introduce the Lande-g factor?
28. (a) Find the Lande-g factor for an atom in $^2S_{\frac{1}{2}}$. (b) Find the Zeeman energies.

6.4 Explanation of Stern–Gerlach experiment and measurement

Rigorously speaking, the Stern–Gerlach experiment is a true time-dependent problem. Figure 6.4 illustrates this process. Before Ag atoms enter the magnetic field, they are described by $\hat{H}_0$, field-free. When they enter the magnetic field, the Hamiltonian is $\hat{H}_1$. After they exit the magnetic field, the Hamiltonian goes back to $\hat{H}_0$. In the following, we treat these three processes separately. The time dependence is only included through a stationary state evolution under different Hamiltonians.

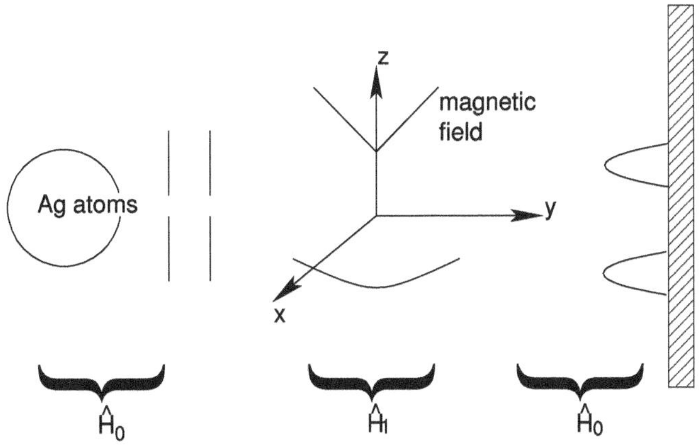

Figure 6.4: The Stern–Gerlach experiment has two different Hamiltonians, before and after the silver atoms enter the gradient magnetic field.

A beam of Ag atoms is incident along the y-axis. Because there is no magnetic field, its electron spin is described by a nonmagnetic Hamiltonian $\hat{H}_0 = \frac{\hat{\mathbf{p}}^2}{m_e}$. The electronic state must be the a spin-mixed state at $t = 0$,

$$\psi(0) = (a\alpha + b\beta)e^{iky} = e^{iky}\begin{pmatrix} a \\ b \end{pmatrix}, \tag{6.4.1}$$

where $|a|^2$ and $|b|^2$ represent the probability in the spin-up $|\alpha\rangle$ and spin-down $|\beta\rangle$ states, respectively, $|a|^2 + |b|^2 = 1$. e^{iky} is a planewave and represents the spatial part of the wavefunction of the electron that is propagating along the $+y$-axis (see eq. (6.1.14)), so the orbital angular momentum is zero.

Inside the magnetic field, the Hamiltonian is

$$\hat{H}_1 = \frac{\hat{\mathbf{p}}^2}{2m} - \hat{\boldsymbol{\mu}} \cdot \mathbf{B} = \frac{\hat{\mathbf{p}}^2}{2m} - \hat{\boldsymbol{\mu}}_s \cdot \mathbf{B} \tag{6.4.2}$$

where the total moment $\hat{\boldsymbol{\mu}}$ is just the spin moment $\hat{\boldsymbol{\mu}}_S$. Since $\mathbf{B}$ is nonuniform and must have at least two nonzero components to satisfy the zero divergence of the $\mathbf{B}$-field, $\nabla \cdot \mathbf{B} = 0$, we assume

$$B_z = B_0 + \lambda z \quad \text{and} \quad B_x = -\lambda x. \tag{6.4.3}$$

The magnetic field is mainly along the z-axis, so B_0 is much larger than the gradient term. We do not include B_y because the Ag atoms move along the y-axis and experience no force on them. λ characterizes the gradient of the B field. In the Stern–Gerlach experiment, $\lambda = 10\,\text{T/m}$. It is easy to verify that $\frac{\partial B_z}{\partial z} + \frac{\partial B_x}{\partial x} = 0$.

Since

$$\begin{aligned}
-\hat{\boldsymbol{\mu}}_S \cdot \mathbf{B} &= -(\hat{\mu}_{sx}B_x + \hat{\mu}_{sz}B_z) = -\frac{2\mu_B}{\hbar}\hat{S}_x B_x - \frac{2\mu_B}{\hbar}\hat{S}_z B_z \\
&= -\frac{2\mu_B}{\hbar}\frac{\hbar}{2}\begin{pmatrix} 0 & 1 \\ 1 & 0 \end{pmatrix} B_x - \frac{2\mu_B}{\hbar}\frac{\hbar}{2}\begin{pmatrix} 1 & 0 \\ 0 & -1 \end{pmatrix} B_z \\
&= -\mu_B \begin{pmatrix} B_z & B_x \\ B_x & -B_z \end{pmatrix} = -\mu_B B_z \hat{\sigma}_z - \mu_B B_x \hat{\sigma}_x,
\end{aligned}$$

our Hamiltonian $\hat{H}_1$ in matrix form is

$$\hat{H}_1 = \begin{pmatrix} \frac{\hat{\mathbf{p}}^2}{2m} - \mu_B B_z & -\mu_B B_x \\ -\mu_B B_x & \frac{\hat{\mathbf{p}}^2}{2m} + \mu_B B_z \end{pmatrix}.$$

6.4.1 Schrödinger picture

In the Schrödinger picture, states evolve with time, and operators are time-independent. This means that we must find the eigenstates of $\hat{H}_1$ first. The vertical motion of the electron along the z-axis is of interest, i. e., $\hat{p}_z$, the z-component of the electron momentum. Therefore, we ignore the off-diagonal elements in $\hat{H}_1$. $\frac{\hat{\mathbf{p}}^2}{2m}$ can be ignored as well since it only contributes an energy shift. Our approximate Hamiltonian is

$$\hat{H}_1 = -\mu_B \begin{pmatrix} B_z & 0 \\ 0 & -B_z \end{pmatrix}, \tag{6.4.4}$$

which has two eigenvalues: $E_{1,2} = \mp\mu_B B_z$ and eigenvectors α and β. We start with our initial state $\psi(0)$ in eq. (6.4.1), so $\psi(t)$ at time t is just

$$\psi(t) = (ae^{-iE_1 t/\hbar}|\alpha\rangle + be^{-iE_2 t/\hbar}|\beta\rangle)e^{iky}. \tag{6.4.5}$$

The expectation value of $\hat{p}_z$ is

$$\langle\psi(t)|\hat{p}_z|\psi(t)\rangle = (\langle\alpha|a^* e^{iE_1 t/\hbar} + \langle\beta|b^* e^{iE_2 t/\hbar})\hat{p}_z(ae^{-iE_1 t/\hbar}|\alpha\rangle + be^{-iE_2 t/\hbar}|\beta\rangle).$$

$\langle\hat{p}_z\rangle$ for the spin-up electron is

$$\begin{aligned}\langle\hat{p}_z\rangle &= \langle\alpha|e^{iE_1 t/\hbar}\hat{p}_z e^{-iE_1 t/\hbar}|\alpha\rangle = \langle\alpha|e^{iE_1 t/\hbar}(-i\hbar\frac{\partial E_1}{\partial z})(-it/\hbar)e^{-iE_1 t/\hbar}|\alpha\rangle \\ &= -\frac{\partial E_1}{\partial z}t = -(-1)\mu_B\frac{\partial B_z}{\partial z}t = \mu_B\lambda t.\end{aligned}$$

Since λ is positive, as t increases, the electron moves along the $+z$-axis. t is determined by the length of the nonuniform portion of the magnetic field.

For the spin-down electron, $\langle\hat{p}_z\rangle = -\mu_B\lambda t$, so it moves to the $-z$-axis. This is the origin of the Stern–Gerlach experiment. The approximation made above, i. e., ignoring the off-diagonal element, has no effect on the motion along the z-axis, which can be proved using the Heisenberg equation of motion.

6.4.2 Heisenberg picture

In the Heisenberg picture, the operator evolves with time through the Heisenberg equation of motion, $i\hbar\frac{d\hat{A}}{dt} = [\hat{A}, \hat{H}]$. Our Hamiltonian is $\hat{H}_1 = \frac{\hat{\mathbf{p}}^2}{2m} - \mu_B B_z\hat{\sigma}_z - \mu_B B_x\hat{\sigma}_x$. So the equation of motion for $\hat{p}_z$ is

$$i\hbar\dot{\hat{p}}_z = [\hat{p}_z, H_1] = \left[\hat{p}_z, \frac{\hat{\mathbf{p}}^2}{2m} - \mu_B B_z\hat{\sigma}_z - \mu_B B_x\hat{\sigma}_x\right].$$

With the help of eq. (6.4.3), the right side can be reduced to

$$-\mu_B\hat{\sigma}_z[\hat{p}_z, B_0 + \lambda z] = \mu_B\hat{\sigma}_z\lambda[z, \hat{p}_z] = \mu_B\hat{\sigma}_z\lambda i\hbar.$$

Integrating over t, we obtain $\hat{p}_z = \mu_B\hat{\sigma}_z\lambda t$. For a spin-up state, $\langle\hat{p}_z\rangle = \lambda\mu_B t$. For a spin-down state, $\langle\hat{p}_z\rangle = -\lambda\mu_B t$. One sees these results match the above results, without making any approximation. Before entering the magnetic field, the spin does not favor one particular orientation, and it is in a spin mixed state $\psi(0) = (a\alpha + b\beta)e^{iky}$. Along the x axis, $\langle\hat{p}_x\rangle$ is nonzero, i. e. a spread of electrons along the x-axis, consistent with the experiment. This proves that the electrons in the Ag atoms are not in an eigenfunction of $\hat{S}_x$. The calculation is left as a homework assignment.

Exercise 6.4.2.

29. Suppose a system is in a mixed-spin state, with $E_{1,2} = \pm\mu_B B_z$, $\psi(t = 0) = \frac{1}{\sqrt{3}}|\alpha\rangle + \sqrt{\frac{2}{3}}|\beta\rangle$. (a) Find the expectation values of $\langle\psi(t)|\hat{S}_x|\psi(t)\rangle$ and $\langle\psi(t)|\hat{S}_y|\psi(t)\rangle$. (b) Find the expectation values of $\langle\psi(t)|\hat{S}_x^2|\psi(t)\rangle$ and $\langle\psi(t)|\hat{S}_y^2|\psi(t)\rangle$.

30. In accordance with the previous question, (a) find the uncertainty relationship between $\hat{S}_x$ and $\hat{S}_y$ as a function of time. (b) Assuming $B_z = 2\,\text{T}$, plot the relationship as a function of time. (c) What can you conclude from the figure?

31. Using the Schrödinger picture: (a) for the spin-down state, show that $\langle\hat{p}_z\rangle = -\mu_B\lambda t$; (b) show the unit of $\mu_B\lambda t$ is indeed the unit of momentum; (c) Stern–Gerlach's λ is 10 T/m, and Δz is 0.22 mm. Assume that $\langle\hat{p}_z\rangle$ is the momentum of the electron. How long does it take the electron to reach Δz?

6.4.3 Quantum measurement

In Chapter 1, we stated that there are two different types of measurement. Now, prior chapters have prepared us for a comprehensive discussion of this. Specifically, Type I measurement means that the module of a state $\langle\psi|\psi\rangle$ is kept, and Type II measurement means that the module of a state $\langle\psi|\psi\rangle$ is changed. The Hamiltonians that we are about to use are all *time-independent*, leaving time-dependent Hamiltonians to Chapter 7.

We start with a state described by $|\psi(\mathbf{r}, t_0)\rangle$ at time $t_0 = 0$. Our goal is to compute the expectation value of an observable whose operator is $\hat{A}$ at several different times, with different Hamiltonians.

At t_0, the expectation value is $\langle\psi(\mathbf{r}, t_0)|\hat{A}|\psi(\mathbf{r}, t_0)\rangle$. What about at t_1? Answer: unknown. This is because how $|\psi\rangle$ evolves with time depends on the Hamiltonian $\hat{H}_1$ of a system between t_0 and t_1. For instance, in vacuum (ignoring the vacuum fluctuation), the Hamiltonian is $\hat{H}_1 = -\frac{\hat{\mathbf{p}}^2}{2m}$, and then a planewave, with the initial wavefunction of $\psi(\mathbf{r}, t_0 = 0) = e^{i(\mathbf{k}\cdot\mathbf{r})}/\sqrt[3]{V}$, becomes $\psi(\mathbf{r}, t_1) = e^{i(\mathbf{k}\cdot\mathbf{r}-\omega t_1)}/\sqrt[3]{V}$. This example uses an eigenstate, but in general $|\psi\rangle$ may not be an eigenstate. The following table shows the steps to calculate $\langle\psi(\mathbf{r}, t)|\hat{A}|\psi(\mathbf{r}, t)\rangle$, with $t_0 < t < t_1$.

Step	Procedure
1	Find the eigenstates $\{\|\phi_n\rangle\}$ and eigenenergies E_n of $\hat{H}$.
2	Project $\|\psi(\mathbf{r},t_0)\rangle$ to $\{\|\phi_n\rangle\}$ to find $c_n = \langle\psi\|\phi_n\rangle$.
3	Evolve the state in time t from t_0 to t_1, $\|\psi(\mathbf{r},t)\rangle = \sum_n c_n e^{-iE_nt/\hbar}\|\phi_n(\mathbf{r})\rangle$.
4	Find $\langle\psi(\mathbf{r},t)\|\hat{A}\|\psi(\mathbf{r},t)\rangle$.

If, after t_1, we have a new Hamiltonian $\hat{H}_2$ from t_1 to t_2. All the steps are the same, except that in step 2 we expand $|\psi(\mathbf{r},t_1)\rangle$ instead of $|\psi(\mathbf{r},t_0)\rangle$. At any time $t_0 < t < t_2$, $\langle\psi(\mathbf{r},t)|\psi(\mathbf{r},t)\rangle$ remains the same. This is the key feature of the Type I measurement.

The Type II measurement alters the state, so $\langle\psi(\mathbf{r},t)|\psi(\mathbf{r},t)\rangle$ is changed. We will take the Stern–Gerlach experiment as an example. Figure 6.5 plots a beam of electrons with mixed spin polarizations enter three different magnetic fields, two of them along the z-axis and one along the x-axis, in two time periods: (t_0,t_1) and (t_1,t_2).

At $t_0 = 0$, the electron is prepared in a state

$$\psi(\mathbf{r},t_0) = f(\mathbf{r})e^{ik_yy}(a|\alpha\rangle + b|\beta\rangle), \tag{6.4.6}$$

where $f(\mathbf{r})$ is the spatial profile of a wavepacket propagating along the y-axis, k_y is the wavevector of the wavepacket, and a and b are the coefficients of the spin up $|\alpha\rangle$ and spin down $|\beta\rangle$ states. From t_0 to t_1, the electron enters the first gradient magnetic field which is mainly along the z-axis (see Fig. 6.5(a)). We ignore a minor component along the x-axis for the reason already described. The Hamiltonian $\hat{H}_1$ is the same as eq. (6.4.4),

$$\hat{H}_1 = -\hat{\mu}_z B_z = -\mu_B B_z\hat{\sigma}_z, \tag{6.4.7}$$

and at time t the state is

$$\begin{aligned}\psi(\mathbf{r},t) &= f(\mathbf{r})e^{ik_yy}(ae^{-iE_1t/\hbar}|\alpha\rangle + be^{-iE_2t/\hbar}|\beta\rangle)\\ &\equiv f(\mathbf{r})e^{ik_yy}(ae^{i\omega_1t}|\alpha\rangle + be^{-i\omega_1t}|\beta\rangle)\\ &\equiv \underbrace{c_1e^{i\omega_1t}|\alpha\rangle}_{\psi_\uparrow \text{ to } \hat{H}_2} + \underbrace{c_2e^{-i\omega_1t}|\beta\rangle}_{\psi_\downarrow \text{ to } \hat{H}_3}\end{aligned} \tag{6.4.8}$$

where $E_1 = -\mu_B B_z$, $E_2 = +\mu_B B_z$ and $\omega_1 = \mu_B B_z/\hbar$. After t_1, the spin-up and spin-down electrons are spatially separated and are channeled into two different magnetic fields (Figs. 6.5(b) and (c)) with Hamiltonians $\hat{H}_2$ and $\hat{H}_3$. Equation (6.4.8) shows that their initial states are different for $\hat{H}_2$ and $\hat{H}_3$, i. e., $\langle\psi_\uparrow|\psi_\downarrow\rangle = 0$. This is the key feature of the Type II measurement. We purposely choose $\hat{H}_3 = \hat{H}_1$, so at any time after t_1, the spin-down channel state evolves with time as

$$\psi_\downarrow(t) = c_2e^{-i\omega_1t}|\beta\rangle. \tag{6.4.9}$$

Different from $\hat{H}_3$, $\hat{H}_2$ is chosen as $\hat{H}_2 = -\mu_B B_x\hat{\sigma}_x$. Its eigenstates are η_{x_1}, η_{x_2} (see eq. (6.1.8)), with the respective eigenvalues $E_1 = -\mu_B B_x$ and $E_2 = +\mu_B B_x$. Next, we follow

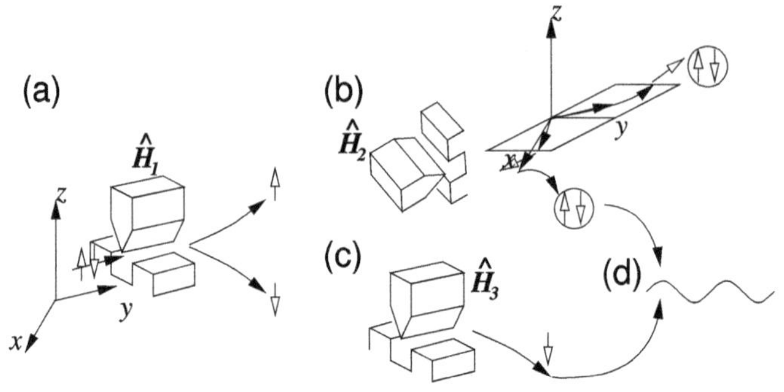

Figure 6.5: (a) A beam of Ag atoms pass through a gradient magnetic field with the field direction mainly along the z-axis and split into two groups: spin-up and spin-down. Spin-up atoms enter (b) another gradient field with the direction along the x-axis, while spin-down ones enter (c) a same magnetic field as (a). After (b) and (c), those atoms are brought together at (d) to interfere.

Step 2 in the previous table to project the initial state at time t_1 on to η_{x_1}, η_{x_2} and denote the coefficient as c_i',

$$c_1' = \langle \eta_{x_1} | \psi_\uparrow(t_1) \rangle = c_1 e^{i\omega_1 t_1}/\sqrt{2}, \quad c_2' = \langle \eta_{x_2} | \psi_\uparrow(t_1) \rangle = c_1 e^{i\omega_1 t_1}/\sqrt{2}.$$

The spin-up channel state evolves as

$$\psi_\uparrow(t) = c_1' e^{+i\omega_2 t} |\eta_{x_1}\rangle + c_2' e^{-i\omega_2 t} |\eta_{x_2}\rangle, \tag{6.4.10}$$

where $\omega_2 = \mu_B B_x/\hbar$. This is an incredible state. Although at $t = t_1$, $\psi_\uparrow(t)$ is a pure spin-up state, as far as $t > t_1$, $\psi_\uparrow(t)$ has both spin-up and spin-down contributions, which can be revealed if we express $|\eta_{x_1}\rangle$ and $|\eta_{x_2}\rangle$ in terms of α and β,

$$\psi_\uparrow(t) = c_1 e^{i\omega_1 t} [\cos(\omega_2 t)|\alpha\rangle + i \sin(\omega_2 t)]|\beta\rangle. \tag{6.4.11}$$

The second term contains the spin-down state! The reason why we still have a spin-down state is because, although we decompose the spin along x-, y- and z-axes, they are not orthogonal at all: (a) Their operators do not commute such as $[\hat{\sigma}_x, \hat{\sigma}_y] \neq 0$, and (b) their eigenstates are not orthogonal such as $\langle \eta_{x_1} | \alpha \rangle \neq 0$. Equation (6.1.10) shows that a "pure" spin-up $|\alpha\rangle$ can be expressed in terms of η_{x_1} and η_{x_2}. Despite of our best efforts to "purify" states with different magnetic fields, we cannot get a pure state. In fact, if we bring $\psi_\downarrow(t)$ and $\psi_\uparrow(t)$ together, they will interfere as shown in Fig. 6.5(d) because

$$\langle \psi_\downarrow(t) | \psi_\uparrow(t) \rangle = \frac{1}{2} c_1 c_2^* e^{i\omega_1(t+t_1)} \left(e^{i(\omega_1+\omega_2)t} - e^{i(\omega_1-\omega_2)t}\right). \tag{6.4.12}$$

The measurement $\hat{H}_2$ is in fact a perturbation and changes the states to be measured.

To this point, as in many books [30], we have never discussed the spatial part of the wavefunction, which is a shortcoming. The next exercise suggests a solution.

Exercise 6.4.3.

32. Show, for the Type I measurement, $\langle\psi(\mathbf{r},t)|\psi(\mathbf{r},t)\rangle$ remains time-independent.
33. Prove eqs. (6.4.11) and (6.4.12).
34. Suppose $f(y,t) = e^{-(y-ut)^2}$, a Gaussian wavepacket moving along the y-axis with velocity u. We further assume that this wavepacket does not change its shape as it passing through $\hat{H}_2$ and $\hat{H}_3$. (a) Compute $\langle\psi_\downarrow(t)|\psi_\uparrow(t)\rangle$ and compare it with eq. (6.4.12). (b) Explain your finding.
35. Using $\psi_\uparrow(t)$ in eq. (6.4.11), compute $\langle\psi_\uparrow(t)|\hat{S}_x|\psi_\uparrow(t)\rangle$, $\langle\psi_\uparrow(t)|\hat{S}_y|\psi_\uparrow(t)\rangle$ and $\langle\psi_\uparrow(t)|\hat{S}_z|\psi_\uparrow(t)\rangle$.

6.5 Zeeman effect

Chapter 4 discusses how atoms emit discrete spectral lines. Chapter 5 introduces how these spectral lines are further split under an electric field, i. e., Stark effect. In 1896, Pieter Zeeman discovered that, if a sodium flame is placed under a *strong* magnetic field, a single spectral line is split into several lines, on the order of 0.2 Å. This is called the normal Zeeman effect. A weaker magnetic field produces many more lines, which is now called the anomalous Zeeman effect. As will be seen in the following, the normal Zeeman effect is the making of orbital angular momentum and is not normal at all. It only occurs under zero-spin momentum. By contrast, the anomalous Zeeman effect is normal and appears more frequently. It is the joint effect of the orbital and spin angular momenta. The Zeeman effect demonstrated what the spatial quantization means optically.

Zeeman also used Cd, Fe and Zn flames. However, the Zeeman effect in the Na lamp is conceptually more difficult to explain because it does not have a zero-spin to start with, and one has to use a strong magnetic field to suppress the spin effect to get a "cleaner" normal Zeeman effect. In our experiment, we will use a cadmium lamp, instead of a Na lamp, and then present our observation and explanation of the Zeeman effect.

6.5.1 Experiment

We place our Cd lamp between two pole pieces of an electromagnet. The magnetic field $\mathbf{B}$ generated by the electromagnet goes up to 0.45 T, with a current through the coils of the electromagnet up to 9.5 A. The emitted light from the Cd lamp goes through a Fabry–Perot etalon with a resolution of approximately 4,000,000 and a telescope to form interference rings on our charge-coupled device (CCD) camera. It can resolve a wavelength change of less than 0.002 nm, and the camera is operated by a computer. Figure 6.6(a) shows that our magnetic field (the vertical hallow arrow) defines the z-axis for the spin. In the xy-plane, we draw a circular disk (shaded) with the lamp at the center, perpendicular to the z-axis. We call this the Zeeman disk.[3]

3 We are unaware of anyone else using this notation, but the idea is already in Zeeman's Nobel lecture.

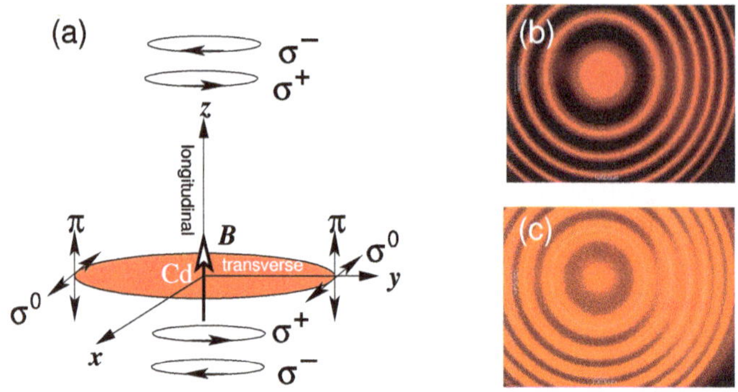

Figure 6.6: (a) Zeeman effect. The external magnetic field **B**, which is powered by an electromagnet with current up to 9.5 A, is set along the *z*-axis, and the sample, the Cd lamp, is placed at the origin. The light that propagates along the different directions has different polarizations. The circular disk, which we call the Zeeman disk, has its surface normal along **B**. (b) In the absence of a magnetic field, each order of interference has one ring. Our CCD camera is placed along the transverse direction. (c) With the presence of the magnetic field, each ring is split into three rings, where no polarizer is inserted.

Because light is a transverse wave, its propagation direction and the polarization must be perpendicular to each other. With the presence of **B**, the space is quantized. Along different directions, the light polarizes differently. If we choose to observe the light along the longitudinal direction of **B** (see Fig. 6.6(a)), either in the same or opposite direction to **B**, i. e., $\pm z$, we only see the circularly polarized light (the left σ^+ or right σ^- circularly polarized light). Here, σ denotes the circular polarization, not to be confused with the Pauli matrices above. The polarization plane is parallel to the Zeeman disk.

If we observe the light along the transversal direction of **B**, we observe two types of polarization: The linearly polarized light (π) is along **B**, and the circularly polarized light (σ^0) is perpendicular to **B**. This σ^0 light is linear! It is also in the same plane of the Zeeman disk. Figure 6.6(a) shows two examples of the transversal detection. In fact, the polarization is always tangential to the Zeeman disk if one probes the emitted light along the circumference of the Zeeman disk.

Figures 6.6(b) and 6.6(c) are our interference rings without and with a magnetic field, respectively. Our camera is placed in the transversal direction with respect to **B**. When we apply a magnetic field, each ring is split into three rings (Fig. 6.6(c)). If we place a polarizer along the **B** direction, the three rings at each order become one ring, i. e., only the center ring remains. This ring is due to the π-light. But, if we rotate the polarizer perpendicular to **B**, the three rings become two rings, i. e., only the inner and outer most rings remain. They are from the σ^0-light, with a linear polarization. This demonstrates that the magnetic field sets a spatial quantization axis and breaks the spatial symmetry.

Exercise 6.5.1.

36. (a) What is the Zeeman disk? (b) What does it tell us?

37. If we apply a magnetic field along the x-axis, show in which direction we can observe (a) the linearly polarized light, (b) the circularly polarized light, and (c) the circularly polarized light but with the linear polarization.

6.5.2 Theory and Hund's rules

What we have seen so far is the normal Zeeman effect, where the electron spin does not contribute. But how is it possible that Cd has no spin? After all, Cd has 48 electrons. The answer is that Cd has two $5s^2$ in its outermost shell. This involves the addition of spin angular momenta.

6.5.2.1 Angular momentum addition for multiple electrons

Consider two electrons, with spin s_1 and s_2. The total spin angular-momentum quantum number S and its magnetic counterpart M_S are given by, respectively,

$$S = s_1 + s_2, \ldots, |s_1 - s_2|, \quad M_S = S, S-1, \ldots -S. \tag{6.5.1}$$

Since both electrons have $s = \frac{1}{2}$, $S = 1, 0$. For $S = 0$, $M_s = 0$, called a spin singlet, with zero-spin and responsible for the normal Zeeman effect. This is the case for Cd. For $S = 1$, M_s has three values $0, \pm 1$, corresponding to three states, a spin triplet, responsible for the anomalous Zeeman effect. We will encounter them in Section 10.3.

The addition of orbital angular-momentum quantum numbers proceeds similarly:

$$L = l_1 + l_2, l_1 + l_2 - 1, \ldots, |l_1 - l_2|, \quad M_L = -L, -L+1, \ldots, L. \tag{6.5.2}$$

The eigenequations for total orbital angular momentum $\hat{\mathbf{L}}^2$ and L_z are, respectively,

$$\hat{\mathbf{L}}^2|LM_L\rangle = L(L+1)\hbar^2|LM_L\rangle, \quad L_z|LM\rangle = M_L\hbar|LM_L\rangle. \tag{6.5.3}$$

Finally, we discuss the addition of total angular momenta J. Because the total angular momentum is a sum of l_i and s_i, there are two ways to sum them up, depending on whether the spin-orbit coupling (SOC)[4] is weaker than the exchange interaction between spins and the Coulomb interaction between orbitals. If so, we add up all the l_i to get L and all the s_i to get S, and then we find J from L and S. The total angular-momentum quantum number J is given by

$$J = L + S, L + S - 1, \ldots, |L - S|, \quad M_J = J, J-1, \ldots, -J. \tag{6.5.4}$$

4 See Section 6.2.2. The exchange and Coulomb interactions will be discussed in Chapter 10.

Table 6.3: Additional rules for total angular-momentum quantum numbers L, S, and J. The right-most column lists conditions related to the state of spin-orbit coupling.

Quantum number	Sum	Total quantum numbers	Spin-orbit coupling
orbital L	$(\hat{\mathbf{L}}_1 + \hat{\mathbf{L}}_2)$	$l_1 + l_2, l_1 + l_2 - 1, \ldots, \lvert l_1 - l_2 \rvert$	absent
spin S	$(\hat{\mathbf{S}}_1 + \hat{\mathbf{S}}_2)$	$s_1 + s_2, s_1 + s_2 - 1, \ldots, \lvert s_1 - s_2 \rvert$	absent
total J	$(\hat{\mathbf{L}} + \hat{\mathbf{S}})$	$L + S, L + S - 1, \ldots, \lvert L - S \rvert$	weak
total J	$(\hat{\mathbf{J}}_1 + \hat{\mathbf{J}}_2)$	$j_1 + j_2, j_1 + j_2 - 1, \ldots, \lvert j_1 - j_2 \rvert$	strong

This is called the Russell–Saunders or LS coupling. The LS coupling is often used in light atoms, where SOC is weak.

If SOC is strong, we first add l_i and s_i to get j_i. Then, for j_1 and j_2, we add them up to get J:

$$J = j_1 + j_2, j_1 + j_2 - 1, \ldots, |j_1 - j_2|, \quad M_J = J, J - 1, \ldots, -J.$$

This is called the jj coupling, which often occurs in heavy atoms. Table 6.3 summarizes all these different scenarios. The difference between our rules and the rules on Section 6.2.1 is that those only apply to a single electron.

Example 4 (Orbital angular momentum addition). For two electrons in a p-state, find (a) all possible L and (b) the orbital angular momentum.

(a) Since two electrons are in a p-state, $l_1 = l_2 = 1$, $L = 2, 1, 0$. (b) The total orbital magnitudes, $L(L+1)\hbar^2$ are $6\hbar^2$, $2\hbar^2$ and $0\hbar^2$.

6.5.2.2 Spectroscopic notations for many-electron configurations and Hund's rules

A system consists of many interactions, from electron–electron repulsion, electron-nucleus interaction, and exchange interaction to spin-orbit coupling. Hund developed three rules to determine the ground-state electronic configurations in many-electron atoms, without a detailed calculation. The results can be compared with the experiment.

We denote the total spin and magnetic spin quantum numbers of the system by S and M_S, the total orbital and magnetic orbital quantum numbers by L and M_L, and the total angular-momentum quantum number by J. Note that the maximum positive M_S is just S, and the maximum positive M_L is just L. His rules ignore all the closed inner shells and rather focus on the unfilled shells to minimize three energies. The order of these rules matters. We must proceed as follows:

Rule 1: Minimize the exchange energy between electrons to determine M_S. Among the available electronic configurations, choose the one that gives a positive maximum $S = M_S$. Its origin is given in eq. (10.4.5).

Rule 2: Minimize the Coulomb repulsion to determine M_L. Among the configurations with the same S, choose the configuration that has a positive maximum $L = M_L$. Its origin is given in eq. (10.4.4).

Rule 3: Minimize the spin-orbit coupling energy to determine J from $J = L + S, L + S - 1, \dots, |L - S|$. If the number of electrons is less or equal to the half-filling of the shell, choose the smallest J. Otherwise, choose the largest J. The reason is due to eq. (6.2.3), where the spin and orbital directions are affected by the electron filling.

Example 5 (Use Hund's rule to find the ground-state configuration). (a) The carbon atom has the shell structure of $1s^2 2s^2 2p^2$. Find the spectroscopic term notation for its ground state. (b) Nitrogen has the shell structure of $1s^2 2s^2 2p^3$. Find its ground-state notation.

(a) Carbon has the atomic number $Z = 6$, with the shell structure of $1s^2 2s^2 2p^2$. First, we ignore two filled shells $1s^2$ and $2s^2$. So, two p electrons are going to fill six (3×2) orbitals, including spins. There are C_6^2 possible configurations, such as $\overbrace{\boxed{\uparrow}}^{1}\overbrace{\boxed{}}^{0}\overbrace{\boxed{\downarrow}}^{-1}$ and $\overbrace{\boxed{\uparrow}}^{1}\overbrace{\boxed{\uparrow}}^{0}\overbrace{\boxed{}}^{-1}$, where each box represents an orbital and accommodates 2 electrons, and the numbers over them are m_l.

Now, we apply Rule 1. Among all the possible configurations, only the configurations $\overbrace{\boxed{\uparrow}}^{1}\overbrace{\boxed{\uparrow}}^{0}\overbrace{\boxed{}}^{-1}$, $\overbrace{\boxed{}}^{1}\overbrace{\boxed{\uparrow}}^{0}\overbrace{\boxed{\uparrow}}^{-1}$, and $\overbrace{\boxed{\uparrow}}^{1}\overbrace{\boxed{}}^{0}\overbrace{\boxed{\uparrow}}^{-1}$, have the maximum $M_S = \frac{1}{2} + \frac{1}{2} = 1$, so $S = 1$.[5]

Next we apply Rule 2 to them to get a maximum M_L. So, the only first configuration, $\overbrace{\boxed{\uparrow}}^{1}\overbrace{\boxed{\uparrow}}^{0}\overbrace{\boxed{}}^{-1}$, has the maximum $M_L = 1 + 0 = 1$, or $L = 1$.

Finally, we use Rule 3. Since $S = 1$ and $L = 1$, $J = 0, 1, 2$. Then, we check the filling. Since we have 6 possible orbitals and have two electrons, the filling is less than half-filling; thus, we choose the smallest $J = 0$. The ground-state configuration for carbon is ${}^3P_0$.

(b) Nitrogen has $1s^2 2s^2 2p^3$. Rule 1 gives $M_s = m_{s_1} + m_{s_2} + m_{s_3} = \frac{1}{2} + \frac{1}{2} + \frac{1}{2} = \frac{3}{2}$. So $S = \frac{3}{2}$. Rule 2 yields $M_L = m_{l_1} + m_{l_2} + m_{l_3} = 1 + 0 - 1 = 0$. So, we have only one J because $L = 0, J = L + S = \frac{3}{2}$. The ground-state configuration is ${}^4S_{3/2}$.

6.5.2.3 Configuration for excited states

Hund's rules take care of ground-state configurations. However, the Zeeman effect involves a transition from a higher energy state to a lower energy state. So, we need to find out all possible configurations for *excited states* in spectroscopic notation. This is an extension from our single-electron notation. Here, we use L, S and J to uniquely identify a state that a system is in, ${}^{2S+1}L_J$, where L is replaced by a particular orbital symbol. For $L = 0, 1, 2$, we use S, P and D. Its corresponding g factor is computed in a similar way using eq. (6.3.14), where l, s and j are replaced by L, S, and J. We note that spectroscopic notation only works for LS coupling.

5 Since $M_S = S, S - 1, \dots, -S$, the maximum positive M_S is just S.

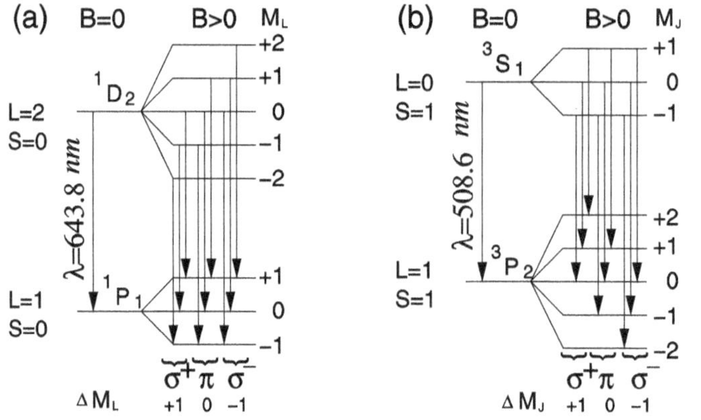

Figure 6.7: Energy diagram of Cd. (a) The transition from $^1D_2$ to $^1P_1$ leads to the normal Zeeman effect. (b) The transition from $^3S_1$ to $^3P_2$ leads to the anomalous Zeeman effect.

Example 6 (Spectroscopic notation of excited states for multielectron atoms). Cd has the ground-state configuration of $5s^2$, and three excited-state configurations of $5s6s$, $5s5p$ and $5s5d$. (a) Find all the possible J and the spectroscopic notations for $5s6s$. (b) Write down all possible spectroscopic notations for $5s5p$ and $5s5d$ states.

(a) For $5s6s$, we have $S = s_1 + s_2, s_1 + s_2 - 1, \dots, |s_1 - s_2|$. Since $s_1 = 1/2$ and $s_2 = 1/2$, $S = 0, 1$. Since $l_1 = l_2 = 0$, then $L = 0$, and J takes 0 and 1, which yields two possible notations, $^1S_0$ and $^3S_1$. (We learned this from https://physics.nist.gov/PhysRefData/Handbook/Tables/cadmiumtable5.htm).

(b) For $5s5p$, $S = 0, 1$. Since $l_1 = 0$ and $l_2 = 1$, $L = 1$. In the LS coupling, if $S = 0$ and $L = 1$, $J = 1$, we have $^1P_1$. If $S = 1$ and $L = 1$, J has three values, 0, 1 and 2, so we have $^3P_0$, $^3P_1$, and $^3P_2$. Similarly, for $5s5d$, $S = 0, 1$. Since $l_1 = 0$ and $l_2 = 2$, $L = 2$. Next, we determine J. For $S = 0$, $J = L = 2$, so our configuration is $^1D_2$. For $S = 1$, $J = 3, 2, 1$. We have three configurations, $^3D_3$, $^3D_2$, and $^3D_1$.

The normal Zeeman effect from our Cd lamp is due to a transition between two singlets, from $^1D_2$ to $^1P_1$, where the spin is zero. Figure 6.7(a) shows the energy diagram of Cd atom for this transition. These two states are not a ground state $^1S_0$. For this reason, in our experiment, we supply a voltage to light the Cd lamp to excite these states. Now, we are ready to explain the normal Zeeman effect quantitatively.

6.5.2.4 Explanation of Zeeman effects

The wavelength shift $\Delta\lambda$ observed in Zeeman effects is the single most important quantity to be calculated. We denote the photon energy by $E_{ph} = hc/\lambda$, so its derivative is $dE_{ph} = -hc\frac{d\lambda}{\lambda^2}$, and $\Delta\lambda$ in its finite difference form is

$$\Delta\lambda = -\frac{\lambda_0^2}{hc}\Delta E_{ph}, \tag{6.5.5}$$

where λ_0 is the field-free wavelength. Our job is to find ΔE_{ph}.

We consider two states i and f, with the initial magnetic-field free energy E_i and the final one by E_f, with $E_i > E_f$. With the magnetic field, we denote the energy by $\mathcal{E}$,

$$\mathcal{E}_i = E_i + U_i = E_i + g_i M_{J_i} \mu_B B_z, \quad \mathcal{E}_f = E_f + U_f = E_f + g_f M_{J_f} \mu_B B_z, \tag{6.5.6}$$

where the Zeeman energy U is from eq. (6.3.15), but with m_j replaced by M_J because we may have more than one single electron, and the same thing is true for g. The photon energy difference between two states is the difference between $\mathcal{E}_i - \mathcal{E}_f$ and $E_i - E_f$ as

$$\Delta E_{ph} = (\mathcal{E}_i - \mathcal{E}_f) - (E_i - E_f) = (g_i M_{J_i} - g_f M_{J_f}) \mu_B B_z, \tag{6.5.7}$$

so the wavelength shift becomes

$$\Delta\lambda = -\frac{\lambda_0^2}{hc} \Delta E_{ph} = (g_f M_{J_f} - g_i M_{J_i}) \mu_B B_z \frac{\lambda_0^2}{hc}. \tag{6.5.8}$$

Whether the Zeeman effect is normal or anomalous hinges on gM_J.

First, we focus on the normal Zeeman effect. In this case, $S = 0$, so $M_J = M_L$ and $g = 1$ and the wavelength shift is

$$\Delta\lambda = -\frac{\lambda_0^2}{hc} \Delta E_{ph} = (M_{L_f} - M_{L_i}) \mu_B B_z \frac{\lambda_0^2}{hc}, \tag{6.5.9}$$

where $M_{L_f} - M_{L_i}$ is subject to the dipole selection rule. The rule states that allowed transitions must have the magnetic orbital quantum number changed as $\Delta M_L = 0, \pm 1$ (see the following example and more in the next chapter).

Example 7 (Normal Zeeman effect in Cd). Cd has a transition from ${}^1D_2 \to {}^1P_1$, emitting red light at $\lambda_0 =$ 643.8 nm (Fig. 6.7(a)). If we place it under a magnetic field of 0.45 T, find the wavelength shift.

We start from the initial state ${}^1D_2$. It has $L = 2$, so $M_{L_i} = 0, \pm 1, \pm 2$. Similarly, for ${}^1P_1$, $M_{L_f} = 0, \pm 1$. Then, the allowed transitions obey $\Delta M = 0, \pm 1$. We have the energy shift $\Delta E_{ph} = \pm \mu_B B_z$, or 0, where 0 corresponds to no shift and represents the π-light, while $\pm \mu_B B_z$ represents $\sigma^{\mp}$. Using eq. (6.5.9), we find $\Delta\lambda = -0.087$ Å for a positive ΔE_{ph} or $+0.087$ Å for a negative ΔE_{ph}. Thus, we see that σ^+-light increases its wavelength by 0.087 Å, while σ^- reduces by the same amount. The spectral line is split into three lines. The following table shows the details.

Helicity	σ^+	π	σ^-
ΔM_L	+1	0	−1
$M_{L_i} g_i - M_{L_f} g_f$	−1	0	+1
$\Delta\lambda$	+0.087 Å	0 Å	−0.087 Å

Now, we explain the anomalous Zeeman effect. Since $S \neq 0$, we must use eq. (6.5.8). $\Delta\lambda$ has many more different wavelengths, so the spectral lines for the anomalous Zeeman effect are messy, or anomalous. The clean spectral line in Fig. 6.6 is obtained through a red filter to filter out the greenish color at 508.6 nm. The calculation is more complicated. We need to compute the Lande-g factors for each state. Because M_L is no longer a good quantum number, the selection rule is based on M_J, where $\Delta M_J = 0, \pm 1$.

Example 8 (Anomalous Zeeman effect in Cd). Cd has a transition from $^3S_1 \to {}^3P_2$, emitting green light at $\lambda_0 = 508.6$ nm (Fig. 6.7(b)). If we place it under a magnetic field of 0.45 T, find the wavelength shift for the transition from $M_J = -1$ to $M_J = 0$.

This transition is between two triplet states $S = 1$. First of all, we compute g_i and g_f using eq. (6.3.14). For $^3S_1$, $S = 1$, $L = 0$, and $J = 1$, $g_i = 2$; for $^3P_2$, $S = 1$, $L = 1$, and $J = 2$, $g_f = 3/2$. Next, according to the selection rule $\Delta M_J = 0, \pm 1$, there are nine possible transitions. We consider a transition from $M_{J_i} = -1$ to $M_{J_f} = 0$, and we find $(g_i M_{J_i} - g_f M_{J_f}) = 2(-1) - \frac{3}{2}(0) = -2$. Thus, the energy shift is $\Delta E_{ph} = 2\mu_B B_z$, and $\Delta\lambda = -\lambda_0^2 2\mu_B B_z/hc = -0.1086$ Å.

Exercise 6.5.2.

38. Show, if $S = 0$, then $M_J = M_L$ and $g = 1$.

39. Consider the Cd lamp placed in a magnetic field. Show how one can detect the linearly polarized light, the circularly polarized light, and the circularly polarized light with the linear polarization.

40. (a) Find the ground-state configuration for oxygen. (b) Find its Lande-g factor.

41. The nickel atom has an outer-shell electron configuration $3d^8 4s^2$. (a) Use Hund's rules to determine the ground-state configuration. (b) Express the configuration in terms of the spectroscopic term notation. (c) Compute the Lande-g factor.

42. Cd has the excited-state configuration $5s5d$. (a) Find all the possible S, L, and J. (b) Find all the possible spectroscopic notations and (c) their g.

43. Figure 6.7(b) shows the Cd emission from $^3S_1$ to $^3P_2$ with the native wavelength λ_0. Now, we place it in a magnetic field B_z. One single spectral line is split into nine lines. Find the wavelength shift analytically in units of $\lambda_0^2 \mu_B B_z/hc$ for (a) σ^+ with $\Delta M_J = +1$, (b) π with $\Delta M_J = 0$, and (c) σ^- with $\Delta M_J = -1$.

6.6 Electron spin resonance (ESR)

Zeeman used one magnetic field to split a single spectral line into several lines. In 1945, E. K. Zavoisky used two orthogonal magnetic fields,[6] where one splits the single energy level into two and the other induces the transition between these two energy levels. Our sample, 1,1-diphenyl-2-picryl-hydrazyl (DPPH), has a lone electron at N (Fig. 6.8(a)) and is placed in two orthogonal magnetic fields (Fig. 6.8(c)), where one is static and the other is alternating with a maximum frequency of 130 MHz. Two Helmholtz coils (Fig. 6.8(c)) generate a uniform static magnetic field across the sample. This magnetic field splits the energy state into the spin-up and the spin-down states (see Fig. 6.8(b)). The small coil, which holds the sample, generates a high-frequency field. If this frequency matches the

6 The two fields must be orthogonal. If these two magnetic fields are in parallel, there is no transition between split energy levels since they share the same eigenstate.

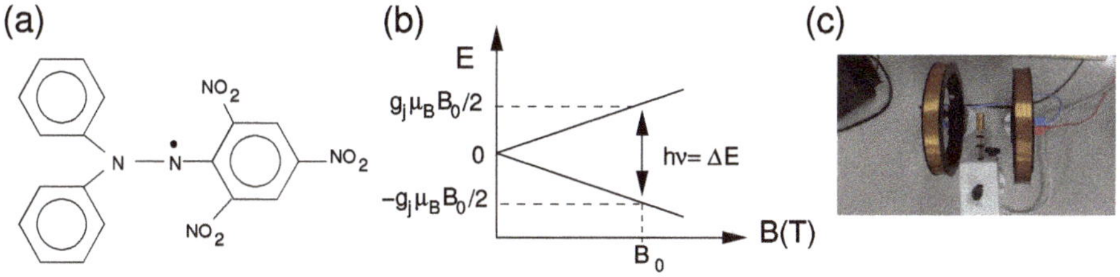

Figure 6.8: (a) 1,1-diphenyl-2-picryl-hydrazyl (DPPH) has one lone electron on the nitrogen atom. (b) The energy level splits upon the static magnetic field B_0. If $h\nu$ matches the energy splitting ΔE, the electron in the lower level absorbs the magnetic field energy from the high-frequency field and makes a transition from the lower level to the high level. (c) Two Helmholtz coils generate a uniform and static magnetic field B_0 across the sample. The small coil, which holds the sample, creates a MHz field.

frequency difference between these two levels, the sample absorbs magnetic energy from this high-frequency magnetic field and reaches resonance. This is called electron spin resonance, ESR.

6.6.1 Rotating wave approximation and Rabi Hamiltonian

We focus on the lone electron at the N site in DPPH (Fig. 6.8(a)). Without a magnetic field, the electron has an equal probability in the spin-up or spin-down states. If we apply $\mathbf{B}_0 = -B_0\hat{z}$ to the sample, the energy level splits into two (Fig. 6.8(b)). We choose a negative $\mathbf{B}_0$, so our lower energy state is a spin-up state. Experimentally, one applies the high-frequency magnetic field along either the x- or y-axis, but the field of this kind has no analytical solution. Instead, one often uses a rotating field, i. e., a high-frequency magnetic field in the xy-plane, $\mathbf{B}_1(t) = B_1(\cos\omega t\,\hat{x} - \sin\omega t\,\hat{y})$. This is called the rotating-wave approximation (RWA), under which our Hamiltonian $\hat{H}(t) = -\hat{\boldsymbol{\mu}}\cdot(\mathbf{B}_0 + \mathbf{B}_1(t))$. For a single electron without orbital moment, $\hat{\boldsymbol{\mu}} = \hat{\boldsymbol{\mu}}_s = -\mu_B\hat{\boldsymbol{\sigma}}$ (eq. (6.3.5)), so $\hat{H} = \mu_B\hat{\boldsymbol{\sigma}}\cdot(\mathbf{B}_0 + \mathbf{B}_1(t))$. Since our $\mathbf{B}_0 = -B_0\hat{z}$, we have a negative sign in the first term of our Hamiltonian,

$$\begin{aligned}\hat{H}(t) &= -\mu_B B_0\hat{\sigma}_z + \mu_B B_1(\hat{\sigma}_x\cos\omega t - \hat{\sigma}_y\sin\omega t)\\ &= -\mu_B B_0\begin{pmatrix}1 & 0\\ 0 & -1\end{pmatrix} + \mu_B B_1\begin{pmatrix}0 & \cos\omega t + i\sin\omega t\\ \cos\omega t - i\sin\omega t & 0\end{pmatrix}\\ &= \begin{pmatrix}-\mu_B B_0 & \mu_B B_1 e^{i\omega t}\\ \mu_B B_1 e^{-i\omega t} & +\mu_B B_0\end{pmatrix} = \begin{pmatrix}\hbar\omega_1 & \gamma e^{i\omega t}\\ \gamma e^{-i\omega t} & \hbar\omega_2\end{pmatrix},\end{aligned} \tag{6.6.1}$$

which is just the well-known Rabi model. Here, $\hbar\omega_1 = -\mu_B B_0$, $\hbar\omega_2 = \mu_B B_0$, $\gamma = \mu_B B_1$ (not to be confused with the gyromagnetic ratio). $|\omega_1|$ or $\omega_2 = \mu_B B_0/\hbar$ is the Larmor frequency, the precession frequency of the $\mathbf{B}_0$ field.

6.6.2 Rabi model

We start with the time-dependent Schrödinger equation,

$$i\hbar\frac{\partial\psi}{\partial t} = \hat{H}\psi. \tag{6.6.2}$$

A crucial step to solve this equation is to start with a trial wavefunction $\psi(t)$, $\psi(t) = C_1(t)e^{-i\omega_1 t}\left(\begin{smallmatrix}1\\0\end{smallmatrix}\right) + C_2(t)e^{-i\omega_2 t}\left(\begin{smallmatrix}0\\1\end{smallmatrix}\right)$, where $|C_1(t)|^2$ and $|C_2(t)|^2$ are the probability in the spin-up state $|1\rangle(|\alpha\rangle)$ and spin-down state $|2\rangle(|\beta\rangle)$, respectively. We substitute eq. (6.6.1) and $\psi(t)$ into eq. (6.6.2), whose left side is

$$\begin{aligned} i\hbar\Bigg[\dot{C}_1(t)e^{-i\omega_1 t}\begin{pmatrix}1\\0\end{pmatrix} + C_1(t)(-i\omega_1)e^{-i\omega_1 t}\begin{pmatrix}1\\0\end{pmatrix} \\ + \dot{C}_2(t)e^{-i\omega_2 t}\begin{pmatrix}0\\1\end{pmatrix} + C_2(t)(-i\omega_2)e^{-i\omega_2 t}\begin{pmatrix}0\\1\end{pmatrix}\Bigg], \end{aligned}$$

and whose right side is

$$\begin{aligned} &\begin{pmatrix}\hbar\omega_1 & \gamma e^{i\omega t}\\ \gamma e^{-i\omega t} & \hbar\omega_2\end{pmatrix}\left[C_1(t)e^{-i\omega_1 t}\begin{pmatrix}1\\0\end{pmatrix} + C_2(t)e^{-i\omega_2 t}\begin{pmatrix}0\\1\end{pmatrix}\right] \\ &= C_1(t)e^{-i\omega_1 t}\begin{pmatrix}\hbar\omega_1\\ \gamma e^{-i\omega t}\end{pmatrix} + C_2(t)e^{-i\omega_2 t}\begin{pmatrix}\gamma e^{i\omega t}\\ \hbar\omega_2\end{pmatrix}. \end{aligned}$$

We cancel terms with coefficients $\hbar\omega_1$ and $\hbar\omega_2$ from both sides to find

$$i\hbar\dot{C}_1 e^{-i\omega_1 t}\begin{pmatrix}1\\0\end{pmatrix} + i\hbar\dot{C}_2 e^{-i\omega_2 t}\begin{pmatrix}0\\1\end{pmatrix} = \gamma C_1 e^{-i(\omega_1+\omega)t}\begin{pmatrix}0\\1\end{pmatrix} + \gamma C_2 e^{i(\omega-\omega_2)t}\begin{pmatrix}1\\0\end{pmatrix}.$$

The coefficients of $\left(\begin{smallmatrix}1\\0\end{smallmatrix}\right)$ or $\left(\begin{smallmatrix}0\\1\end{smallmatrix}\right)$ must match on both sides, so the original Schrödinger equation produces two sets of equations:

$$i\hbar\dot{C}_1 e^{-i\omega_1 t} = \gamma C_2 e^{i(\omega-\omega_2)t}, \quad i\hbar\dot{C}_2 e^{-i\omega_2 t} = \gamma C_1 e^{-i(\omega+\omega_1)t}. \tag{6.6.3}$$

Eliminating C_2 yields a second-order differential equation,

$$\ddot{C}_1 + i(\omega_2 - \omega_1 - \omega)\dot{C}_1 + \frac{\gamma^2}{\hbar^2}C_1 = 0,$$

whose generic solution is $C_1 = Ae^{\lambda_1 t} + Be^{\lambda_2 t}$. A and B are two parameters to be determined by the initial condition. Suppose that at $t = 0$ the system is in the $|1\rangle$ state, $C_1(t = 0) = 1$ and $C_2(t = 0) = 0$. We find $A = -\lambda_2/(\lambda_1 - \lambda_2)$ and $B = \lambda_1/(\lambda_1 - \lambda_2)$. Two roots are $\lambda_1 = i(-\frac{\delta}{2} + \Omega)$ and $\lambda_2 = i(-\frac{\delta}{2} - \Omega)$, where the detuning is $\delta = \omega_2 - \omega_1 - \omega$ and Ω is called the Rabi frequency, $\Omega = \sqrt{\frac{\delta^2}{4} + \frac{\gamma^2}{\hbar^2}}$.

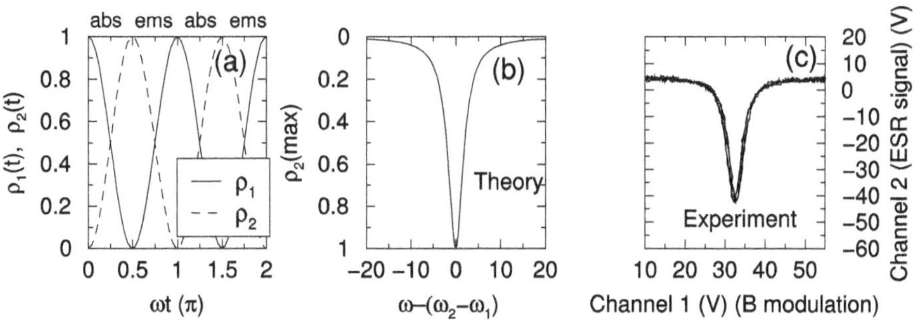

Figure 6.9: (a) Rabi oscillations in $\rho_1(t)$ and $\rho_2(t)$ in states $|1\rangle$ and $|2\rangle$. The period is determined by the Rabi frequency Ω. Every half period, the system absorbs (abs) the energy and the other half period it emits (ems) the energy. (b) Theoretical prediction of the maximum $\rho_2(t)$ as a function of the magnetic field's angular frequency ω. The plot is referenced to the frequency difference between two energy level's frequencies, ω_1 and ω_2. The y-axis is flipped. (c) Experimental ESR signal from channel 2 (providing the high-frequency magnetic field $B_1(t)$) while we sweep the voltage from channel 1 (providing the static field B_0).

6.6.3 Rabi frequency

The meaning of the Rabi frequency can be revealed by examining what happens to the population in each state. With the electron initially in $|1\rangle$, we can find $C_2 = \frac{i\hbar}{\gamma}e^{i\delta t}\frac{\lambda_1\lambda_2}{\lambda_1-\lambda_2}(-e^{\lambda_1 t}+e^{\lambda_2 t}) = -\frac{i\gamma}{\hbar\Omega}e^{i\delta t/2}\sin\Omega t$. So, the population $\rho_2(t)$ in $|2\rangle$ is

$$\rho_2(t) = C_2^* C_2 = \frac{\gamma^2}{\hbar^2\Omega^2}(\sin\Omega t)^2 = \frac{\gamma^2}{\hbar^2}\frac{(\sin\Omega t)^2}{\Omega^2}, \tag{6.6.4}$$

where we purposely rewrite the last term as $\frac{(\sin\Omega t)^2}{\Omega^2}$ because this same dependence will appear in our perturbation theory in the next chapter. Figure 6.9(a) shows that the population oscillates between these two states. In the first half of the Rabi period (π/Ω), the system absorbs the energy from the MHz magnetic field, and in the next half, it emits. The maximum population is a Lorentzian distribution (shown in Fig. 6.9(b))

$$\rho_2(\max) = \frac{\gamma^2/\hbar^2}{\delta^2/4+\gamma^2/\hbar^2}. \tag{6.6.5}$$

A key feature of the Rabi frequency Ω is that it depends on the field strength B_0 through γ and detuning $\delta = \omega_2-\omega_1-\omega$. In the photoelectric effect, the photon frequency is independent of the light intensity. What we have seen here is that the frequency of the population depends on B_0. Since γ is related to B_0 through $\gamma = \frac{1}{2}g_j\mu_B B_0$, the resonance peak is linearly proportional to the external magnetic field. Although Fig. 6.9(b) shows that one can tune the B_1's frequency to get a resonance, in the actual experiment, we do the opposite. We choose a frequency for $B_1(t)$, and then we change B_0 by sweeping the voltage to find a resonance. The experimental ESR signal is shown in Fig. 6.9(c), where the horizontal axis is our sweeping voltage and has a periodic ramp with time. The vertical axis is the voltage change due to the impedance change of the loaded circuit upon resonance. The agreement with the theory is very good, so we can find the g-factor for our sample by measuring the width of the signal.

A natural extension of the Rabi two-level system is the three-level Λ process. Experimentally, Harris and coworkers [34] demonstrated the electromagnetically induced transparency, where one control pulse can eliminate the absorption due to the probe pulse and induces coherent population trapping [35, 36].

Exercise 6.6.3.

44. Suppose that, at $t = 0$, the system is in the $|1\rangle$ state, $C_1(t = 0) = 1$ and $C_2(t = 0) = 0$. Prove $A = -\lambda_2/(\lambda_1 - \lambda_2)$ and $B = \lambda_1/(\lambda_1 - \lambda_2)$.

45. Prove (a) $-e^{\lambda_1 t} + e^{\lambda_2 t} = -2ie^{-i\delta t/2}\sin\Omega t$; (b) $\lambda_1\lambda_2/(\lambda_1 - \lambda_2) = \gamma^2/(2i\hbar^2\Omega)$.

46. If $t = 0$, the system is in the $|1\rangle$ state, $C_1(t = 0) = 1$ and $C_2(t = 0) = 0$, show (a) $A = -\frac{\lambda_2}{\lambda_1-\lambda_2} = \frac{\delta+2\Omega}{4\Omega}$, and $B = \frac{\lambda_1}{\lambda_1-\lambda_2} = \frac{-\delta+2\Omega}{4\Omega}$; (b) $C_1(t) = \frac{e^{-i\delta t/2}}{2\Omega}(i\delta\sin\Omega t + 2\Omega\cos\Omega t)$; (c) $C_2(t) = -\frac{i\gamma}{\hbar\Omega}e^{i\delta t/2}\sin\Omega t$.

47. Based on the Rabi model under the rotating-wave approximation, two roots are $\lambda_1 = i(-\frac{\delta}{2} + \Omega)$ and $\lambda_2 = i(-\frac{\delta}{2} - \Omega)$, where $\delta = \omega_2 - \omega_1 - \omega$ and $\Omega = \sqrt{\frac{\delta^2}{4} + \frac{\gamma^2}{\hbar^2}}$. γ is the coupling constant between states $|1\rangle$ and $|2\rangle$. (a) Find $|C_1(t)|^2$ and $|C_2(t)|^2$ in terms of δ, Ω, and t. (b) Prove $|C_1(t)|^2 + |C_2(t)|^2 = 1$.

6.7 Medical applications: magnetic resonance imaging

In many aspects, nuclear magnetic resonance (NMR) is similar to electron spin resonance (ESR) since they both employ the absorption of the external magnetic energy. But different from ESR, which uses the electron spin, NMR uses the nuclear spin in nuclei, such as ^{1}H and ^{13}C, where the number of nucleons is odd. Same nuclei in a different chemical environment resonate at different frequencies, contributing to a chemical shift that can pinpoint the minute difference. This makes NMR very powerful. Magnetic resonance imaging (MRI) is built upon NMR but adds a spatial resolution. In essence, MRI is the making of QM.

The nuclear spin moment is much smaller than the Bohr magneton $\mu_B = \frac{e\hbar}{2m_e}$ because the proton mass is three orders of magnitude larger than the electron mass. The nuclear magneton, the unit of nuclear spin moment, is $\mu_N = \frac{e\hbar}{2m_p}$, where m_p is the proton mass. Despite this small value, the signal from the nuclei is detectable.

6.7.1 Interaction between a nuclear spin and magnetic field

The Zeeman energy is given by[7]

$$\hat{U} = -\hat{\boldsymbol{\mu}} \cdot \mathbf{B} = -\gamma\hat{\mathbf{I}} \cdot \mathbf{B}, \tag{6.7.1}$$

7 In real samples, there is a chemical shielding from the electron density, so the equation should be multiplied by a shielding constant like $1 - \Gamma$. This contributes to the chemical shift. Γ is a tensor.

where $\hat{\mu}$ is the nuclear magnetic moment and γ is the nuclear gyromagnetic ratio (in comparison to the gyromagnetic ratio in eq. (6.3.5)). $\hat{\mathbf{I}}$ is the nuclear spin operator and follows the same commutation roles as the electron spins, $[\hat{I}_x, \hat{I}_y] = i\hbar\hat{I}_z$, or $\hat{\mathbf{I}} \times \hat{\mathbf{I}} = i\hbar\hat{\mathbf{I}}$. Assuming $\mathbf{B}$ along the z-axis, the energy splitting due to $\mathbf{B}$ is

$$\Delta U = -\gamma B_z \langle \hat{I}_z \rangle = -\omega_0 \langle \hat{I}_z \rangle, \tag{6.7.2}$$

where ω_0 is the Larmor frequency. According to the Heisenberg equation of motion,

$$\frac{d\hat{I}_x}{dt} = \frac{1}{i\hbar}[\hat{I}_x, \hat{U}], \quad \frac{d\hat{I}_y}{dt} = \frac{1}{i\hbar}[\hat{I}_y, \hat{U}], \quad \frac{d\hat{I}_z}{dt} = \frac{1}{i\hbar}[\hat{I}_z, \hat{U}], \tag{6.7.3}$$

where we divide both sides by $i\hbar$. It is easy to show that the transverse components $\hat{I}_x$ and $\hat{I}_y$ precess with $\mathbf{B}$ with ω_0. If another magnetic field oscillates at this frequency, then the energy is absorbed into the system. A 2-T magnetic field corresponds to ω_0 of 10 MHz.

A crucial feature is that resonance peaks are extremely sharp. Bloembergen et al. [37] found the peak width of 0.00005 T for a magnetic field of 0.705 T. Because the peak width is related to the relaxation time of nuclear spin, both the longitudinal relaxation time T_1 and the transverse time T_2 can be measured. T_1 is the time it takes protons to realign their spins with the external magnetic field. If we align $\mathbf{B}$ along the z-axis, the third equation for $\hat{I}_z$ in eq. (6.7.3) now must include an extra term $-\hat{I}_z/T_1$.[8] The same thing is true for $\hat{I}_x$ and $\hat{I}_y$. Recall the magnetization is the magnetic dipole moment per volume. The nuclear magnetization $M(t)$ precesses as

$$M_z(t) = M_z(0)(1 - e^{-t/T_1}). \tag{6.7.4}$$

The transverse components $M_{x,y}$ precess as

$$M_{x,y}(t) = M_{x,y}(0)e^{-t/T_2}. \tag{6.7.5}$$

T_2 characterizes how quickly M_x and M_y drop to zero. Equations (6.7.4) and (6.7.5) approximate the Bloch equation for the nuclear spins. Here, $M_x(0)$, $M_y(0)$ and $M_z(0)$ are the initial nuclear magnetizations, respectively.

Exercise 6.7.1.

48. (a) Solve three equations in eq. (6.7.3). (b) Show $\hat{I}_x$ and $\hat{I}_y$ precess with ω_0.
49. The magnetization M_z is described by $\frac{dM_z}{dt} = -(M_z - M_z(0))/T_1$. Show its solution is given by eq. (6.7.4).

8 This is because $\hat{U}$ does not include all the interactions for nuclear spins.

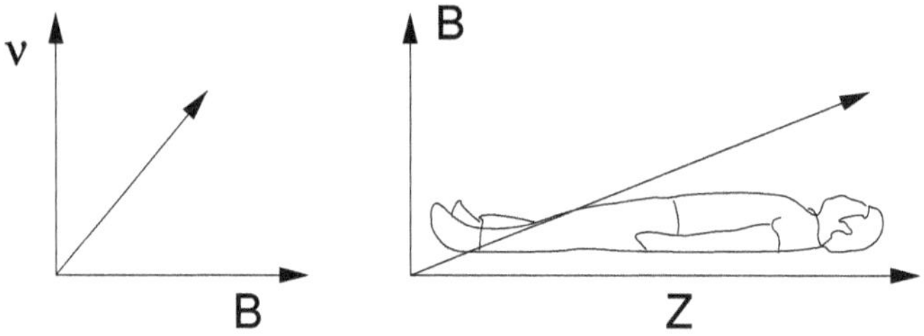

Figure 6.10: (a) The resonance frequency is linearly proportional to the magnetic field B. (b) In MRI, a gradient magnetic field is used across a patient. This allows us to correlate the frequency to the spatial location on a patient.

6.7.2 Spatial resolution

In ESR, although the signal is strong, it is difficult to correlate it to the location of the signal because electrons in many regions of a sample can contribute to the signal. In other words, it has no spatial resolution. In NMR, we use a uniform magnetic field, so we cannot pinpoint where the signal comes from either. But from eq. (6.7.2), we see the Larmor frequency is linearly proportional to the magnetic field (see Fig. 6.10(a)). A different B leads to a different ω_0. This is the most important finding.

In magnetic resonance imaging, a gradient magnetic field is used across a patient. Figure 6.10(b) shows such a design. At one end, **B** is B_{max} and at the other end B_{min}, so **B** changes along the z-axis linearly as

$$B(z) = \frac{B_{max} - B_{min}}{\Delta z} z = \frac{\partial B}{\partial z} z, \tag{6.7.6}$$

where Δz is the spatial separation between B_{max} and B_{min} (see Fig. 6.10(b)). If we substitute $B(z)$ into eq. (6.7.2), then our frequency has a spatial resolution through

$$\omega_0(z) = \gamma B(z). \tag{6.7.7}$$

This is the basic principle of MRI. Signals with different frequencies reveal the location of tissues under investigation, so a doctor can examine chemical and physical conditions of these tissues. Because the magnetic field causes no damage on human bodies, MRI is a safe alternative to X-ray and other means. By measuring T_1 and T_2, one can further enhance the sensitivity.

Exercise 6.7.2.
50. Explain how the spatial resolution is achieved in MRI.
51. The following table shows the gyromagnetic ratio for some elements.

Element	^{1}H	^{13}C	^{17}O	^{19}F
γ (MHz/T)	42.6	10.7	5.8	40.0

In hospitals, the MRI machine requires a magnetic field of 2 T. (a) Compute their resonance frequencies of MRI. (b) The gray matter in brains has $T_1 = 530$ ms at 0.25 T. How many cycles are there for each scan?

6.8 Problems

1. Show the eigenvalues of matrix A in eq. (6.1.1) are ± 1 under the constraint on Section 6.1.2.
2. On Section 6.1.2, $\hat{\sigma}_x$, $\hat{\sigma}_y$, and $\hat{\sigma}_z$ are found by setting θ to 0 and $\pi/2$. (a) Start from eq. (6.1.1) and choose a nonzero θ to form a real matrix for $\hat{\sigma}_z$. (c) Use $\hat{\sigma}_y\hat{\sigma}_z + \hat{\sigma}_z\hat{\sigma}_y = 0$ and $\hat{\sigma}_x\hat{\sigma}_z + \hat{\sigma}_z\hat{\sigma}_x = 0$ to find $\hat{\sigma}_x$ and $\hat{\sigma}_y$.
3. The Pauli matrices can be found from the square root of a two-dimensional rotation matrix. (a) Compute the eigenvalues of and eigenvectors of the matrix $R = \begin{pmatrix} \cos\theta & -\sin\theta \\ \sin\theta & \cos\theta \end{pmatrix}$. (b) Compute $\sqrt{R}$. (c) Find the relationship between $\sqrt{R}$ and the Pauli matrices.
4. (a) When Dirac developed the relativistic quantum mechanics, he realized that, if one requires the dot product $(\hat{\boldsymbol{\sigma}} \cdot \hat{\mathbf{p}})^2 = \hat{p}_x^2 + \hat{p}_y^2 + \hat{p}_z^2$, one immediately gets the permutation among the Pauli matrix. Derive them. (b) But then, if he includes the mass with the Hamiltonian $\hat{H} = c\sqrt{m^2c^2 + \hat{p}_x^2 + \hat{p}_y^2 + \hat{p}_z^2}$, Pauli matrices $\hat{\boldsymbol{\sigma}}$ are not enough. So, he extended it to Dirac matrices $\hat{\boldsymbol{\gamma}}$. Find the permutations among $\hat{\boldsymbol{\gamma}}$.
5. Suppose the spin points in the direction $\hat{n} = \sin(\theta)\sin(\phi)\hat{x} + \sin(\theta)\sin(\phi)\hat{y} + \cos(\phi)\hat{z}$. (a) Find its eigenstates in terms of θ and ϕ. (b) Supposing we want to measure the spins along the $\pm z$-axis, what are the probabilities for the spin being $\pm\hbar/2$?
6. This problem is modeled on the Stern-Gerlach experiment. A beam of electrons enteres a magnetic field along a unit vector $\mathbf{u}(\phi) = \cos\phi\hat{x} + \sin\phi\hat{z}$ in the spin $x - z$ plane, where $\hat{x}$ and $\hat{z}$ are unit vectors in the spin space, not operators. $\hat{\mathbf{S}}$ is a spin operator. (a) Find the eigenvalues and eigenvectors of the operator $\hat{\mathbf{S}} \cdot \mathbf{u}(\phi)$. (b) Suppose the system is in a state given by these two eigenvectors, find the expectation values of $\hat{S}_x$, $\hat{S}_y$ and $\hat{S}_z$. Here the external magnetic field effect is ignored, except the orientiation of the electron spin. (c) We denote the eigenvector with the lowest eigenvalue as ψ_1. Now suppose we place a new magnetic field along the direction where the electron with the smallest eigenvalue travels. This new magneic field is along a different unit vector $\mathbf{u}(\theta)$ direction, where θ is an angle in the spin space, find the expectation value of the operator $\hat{\mathbf{S}} \cdot \mathbf{u}(\theta)$. (d) In view of the new magnetic field, what are the probabilities when one finds the spin expectation values of $\hbar/2$ and $-\hbar/2$? Hints: Project ψ_1 on to the eigenvectors of the operator $\hat{\mathbf{S}} \cdot \mathbf{u}(\theta)$. (e) After the electron passes through the second magnetic field, what are the expectation values of $\hat{S}_x$, $\hat{S}_y$ and $\hat{S}_z$? Hints: There are two possible answers because there are two

eigenvectors of the operator $\hat{\mathbf{S}}\cdot\mathbf{u}(\theta)$. Adopted from [38] but we significantly modified it for clarity.

7. (a) Find the eigenvalues and eigenfunctions of $\hat{\sigma}_y$. (b) Compute the expectation value of $\langle \eta_y|\hat{\sigma}_x|\eta_y\rangle$.
8. An electron is placed in a magnetic field **B**. The Hamiltonian of the system is $\hat{H} = \frac{\hat{\mathbf{p}}^2}{2m_e} - \mu_B B_z\hat{\sigma}_z - \mu_B B_x\hat{\sigma}_x$, where $B_x = -\lambda x$ and $B_z = B_0 + \lambda z$, and λ is the magnetic field gradient. (a) Suppose that initially the electron is in a mixed-spin state, $\Psi_0(t = 0) = a\alpha + b\beta$, where a and b are the square root of the probability in the spin-up and spin-down sates. Use the Heisenberg equations of motion to compute the time-dependent $\hat{p}_x(t)$ and $\hat{p}_z(t)$. (b) Consider two cases, $a = \sqrt{1/3}$, $b = \sqrt{2/3}$ and $a = \sqrt{2/3}$, $b = \sqrt{1/3}$. Show there is no way to distinguish between these two spin states based on $\hat{p}_x$ alone.
9. An electron is in its two $\hat{S}_x$'s eigenstates. (a) If we want to measure the z-component of the spin, what are the expectation values for these two states? What about the probability? (b) At $t = 0$, the spin is in the $\hat{S}_x$'s eigenstate with eigenvalue of $-\hbar/2$. At $t_1 > 0$, we turn on the magnetic field B along the z-axis, with $\hat{H} = -\mu_B B\hat{S}_z$. Find the state at t_1 and the expectation values of $\hat{S}_x$, $\hat{S}_y$, and $\hat{S}_z$. (c) At $t_2 > t_1$, we switch the magnetic field to the x-axis, with $\hat{H} = -\mu_B B\hat{S}_x$. Find the state at $t_3 > t_2$ and the expectation values of $\hat{S}_x$, $\hat{S}_y$, and $\hat{S}_z$.
10. (a) The normal Zeeman effect in Cd is due to a transition between $^1P_1$ level at 43692.384 cm^{-1} and its $^1D_2$ level at 59219.734 cm^{-1}, where cm^{-1} is called the wavenumber, the number of waves per cm, and is the inverse of wavelength λ. A transition from $^1D_2$ to $^1P_1$ emits a single photon. Find the wavelength of the emitted photon. Hint: Compare your results with Fig. 6.7. (b) The anomalous Zeeman effect is between $^3S_1$ at 51483.980 cm^{-1} to $^3P_2$ at 31826.952 cm^{-1}. Find the wavelength of the emitted phton.
11. The sodium atom is known to have two D spectral lines, D_1 and D_2, at 5895.93 Å and 5889.96 Å, respectively. While D_1 is due to the transition from the excited state $^2P_{1/2}$ to the ground state $^2S_{1/2}$, the D_2 line in sodium is due to the transition from $^2P_{3/2}$ to $^2S_{1/2}$ (see the exercise on Section 6.2.1). Under the influence of a magnetic field of 0.2 T, D_1 splits into four lines. (a) Compute four wavelength shifts with respect to the original spectral line. (b) Compute the wavelength shifts for the D_2 line which splits into six spectral lines.
12. The Rabi Hamiltonian (eq. (6.6.1)) is very special. (a) Show that even though it is time-dependent, its instantaneous eigenvalues do not change with time. (b) Find its instantaneous eigenvectors.
13. Consider a Λ system, which has three levels, $|a\rangle$, $|b\rangle$ and $|c\rangle$. The transitions are only between $|a\rangle$ and $|b\rangle$, and $|b\rangle$ and $|c\rangle$. The Hamiltonian is

$$\hat{H} = \hbar\omega_a|a\rangle\langle a| + \hbar\omega_a|b\rangle\langle b| + \hbar\omega_a|c\rangle\langle c| - \mu_{ab}E_1(t)|a\rangle\langle b|$$
$$- \mu_{ba}E_1(t)^*|b\rangle\langle a| - \mu_{bc}E_2(t)|b\rangle\langle c| - \mu_{cb}E_2(t)^*|c\rangle\langle b|,$$

where $E_1(t)$ and $E_2(t)$ are probe and coupling laser fields, respectively, and can be either a continuous wave or pulse and with a complex envelope. $|a\rangle$ (and others) refers to the eigenstate of the system, with the eigenvalue of $\hbar\omega_a$. We assume that the electric fields of the laser are

$$E_i(t) = \frac{F_i(t)}{2}\left[e^{i\omega_i t + i\phi_i} + e^{-i\omega_i t - i\phi_i}\right].$$

The generic time-dependent wavefunction for the system takes the form

$$\Psi(t) = C_a(t)\mathrm{e}^{-i\omega_a t}|a\rangle + C_b(t)\mathrm{e}^{-i\omega_b t}|b\rangle + C_c(t)\mathrm{e}^{-i\omega_c t}|c\rangle,$$

where $C_a(t)$, $C_b(t)$ and $C_c(t)$ are probability amplitudes. Taking a rotating-wave approximation, show that $C_a(t)$, $C_b(t)$, and $C_c(t)$ obey three coupled equations,

$$\begin{aligned}
\dot{C}_a &= -\frac{\mu_{ab}}{2i\hbar}F_1(t)\mathrm{e}^{\mathrm{i}\Delta_1 t + i\phi_1}C_b;\\
\dot{C}_b &= -\frac{\mu_{ba}}{2i\hbar}F_1(t)\mathrm{e}^{-\mathrm{i}\Delta_1 t - i\phi_1}C_a - \frac{\mu_{bc}}{2i\hbar}F_2(t)\mathrm{e}^{-\mathrm{i}\Delta_2 t - i\phi_2}C_c;\\
\dot{C}_c &= -\frac{\mu_{bc}}{2i\hbar}F_2(t)\mathrm{e}^{\mathrm{i}\Delta_2 t + i\phi_2}C_b
\end{aligned}$$

where $\Delta_1 = \omega_1 - (\omega_b - \omega_a)$, and $\Delta_2 = \omega_2 - (\omega_b - \omega_c)$. (This problem is adopted from [39].)

14. The longitudinal and transverse times in MRI are derived from the Bloch equation. Suppose the Hamiltonian is $\hat{H} = -\gamma\hat{\mathbf{I}}\cdot\mathbf{B}$, where $\mathbf{B}$ is along the z-axis. (a) Derive an equation of motion for $\hat{I}_z$ and determine under what condition it approaches to eq. (6.7.4). (b) Derive an equation of motion for $\hat{I}_{x,y}$ and find under what condition it approaches to eq. (6.7.5).

7 Time-dependent perturbation theory: application to optics

Stationary state evolution is rare. Many physical, chemical and biological processes, such as molecular vibrations, spin precession, chemical reaction and photosynthesis, are truly time-dependent. Only a few simple problems, such as Rabi oscillation treated in Chapter 6, can be solved analytically; most problems demand numerical or approximate methods. Time-dependent perturbation theory (TDPT) is an approximate method and plays a role similar to time-independent perturbation theory (TIDPT) in Chapter 5. The time-dependent part of the Hamiltonian represents a weak perturbation. Different from TIDPT, TDPT does not aim to find an energy correction to the eigenenergy, instead it aims to find the probability change in the unperturbed eigenstates and describes how this change can be linked to experimental observations. The emphasis is on the transitions among the unperturbed eigenstates. While it is possible to include high-order perturbations, nearly all the textbooks exclusively focuses on the first-order because of high-level complexity encountered in high-order perturbations. In optics, the high-order perturbation theory lays the foundation for nonlinear optics, where one uses multiple pulses [40, 41, 42, 43] to detect both electron and nuclei motions [44]. Another difference is that TDPT has no well-developed degenerate theory. Some of these issues will be addressed in Chapter 10.

This chapter consists of three units:

- Unit 1 includes Section 7.1. It introduces the interaction between radiation and matter as an example of external perturbations and then discusses the Coulomb gauge. Finally, it focuses on the dipole selection rules and transition matrix elements.
- Unit 2 contains Sections 7.2 and 7.3. It formally introduces time-dependent perturbation theory. Section 7.3 provides three examples, where the perturbation is constant, periodic in time, or a pulse.
- Unit 3 includes Sections 7.4 and 7.5. It applies perturbation theory to linear optics and explains how the polarization and susceptibility are computed quantum mechanically. It introduces population inversion and stimulated emission, essential to the laser.

7.1 Interaction between radiation and matter and its selection rules

In EM, a free charge q with mass m in an electromagnetic field has the vector potential $\mathbf{A}(\mathbf{r}, t)$ and scalar potential $\varphi(\mathbf{r})$, where the electric and magnetic fields are given by $\mathbf{E} = -\nabla\varphi - \frac{\partial \mathbf{A}}{\partial t}$ and $\mathbf{B} = \nabla \times \mathbf{A}$ (in SI). Its classical Hamiltonian (kinetic energy + potential energy) is $H = \frac{1}{2m}(\mathbf{p} - q\mathbf{A})^2 + q\varphi$. The difference between the field-free Hamiltonian $\frac{\mathbf{p}^2}{2m}$ and H constitutes the interaction Hamiltonian, $H_I = -q\mathbf{p} \cdot \mathbf{A}/m + (q\mathbf{A})^2/m + q\varphi$.

https://doi.org/10.1515/9783110672152-007

7.1.1 Quantization of interaction

QM follows EM, but has some key differences. The momentum $\mathbf{p}$ becomes the momentum operator $\hat{\mathbf{p}}$. The Hamiltonian operator is

$$\hat{H} = \frac{1}{2m}(\hat{\mathbf{p}} - q\mathbf{A}(\mathbf{r}, t))^2 + q\varphi, \tag{7.1.1}$$

and the time-dependent Schrödinger equation becomes

$$i\hbar\frac{\partial\psi(\mathbf{r}, t)}{dt} = \left[\frac{1}{2m}(\hat{\mathbf{p}} - q\mathbf{A}(\mathbf{r}, t))^2 + q\varphi\right]\psi(\mathbf{r}, t). \tag{7.1.2}$$

With the presence of $\mathbf{A}(\mathbf{r}, t)$, $\hat{\mathbf{p}}$ is now called the canonical momentum because, if we calculate the change in position $\mathbf{r}$ with time through the Heisenberg equation of motion,

$$i\hbar\frac{d\mathbf{r}}{dt} = [\mathbf{r}, \hat{H}] = i\hbar\frac{\hat{\mathbf{p}} - q\mathbf{A}(\mathbf{r}, t)}{m} \to \frac{d\mathbf{r}}{dt} = \frac{\hat{\mathbf{p}} - q\mathbf{A}(\mathbf{r}, t)}{m} \equiv \frac{\hat{\mathbf{\Pi}}}{m} \equiv \hat{\mathbf{v}}, \tag{7.1.3}$$

where we see that $\hat{\mathbf{p}}$ alone does not determine the position change. Instead, $\hat{\mathbf{\Pi}} = \hat{\mathbf{p}} - q\mathbf{A}(\mathbf{r}, t)$ does, and it represents a momentum, called the mechanical momentum. In other words, when we want to compute the velocity $\hat{\mathbf{v}}$, $\frac{d\mathbf{r}}{dt}$, we must use $\hat{\mathbf{\Pi}}$, not $\hat{\mathbf{p}}$. In deriving eq. (7.1.3), we have used $[\mathbf{r}, \hat{\mathbf{p}}] = i\hbar$.

Since $\mathbf{A}(\mathbf{r}, t)$ depends on both $\mathbf{r}$ and t, $\hat{\mathbf{p}}$ does not commute with it, i. e., $[\hat{\mathbf{p}}, \mathbf{A}] \neq 0$. As a result, $(\hat{\mathbf{p}} - q\mathbf{A})^2$ must be written as $\hat{\mathbf{p}}^2 - q\hat{\mathbf{p}} \cdot \mathbf{A} - q\mathbf{A} \cdot \hat{\mathbf{p}} + (q\mathbf{A})^2$. Then, our interaction Hamiltonian for the electron ($q = -e$) is

$$\hat{H}_I = \frac{e}{2m_e}(\hat{\mathbf{p}} \cdot \mathbf{A}(\mathbf{r}, t) + \mathbf{A}(\mathbf{r}, t) \cdot \hat{\mathbf{p}}) + \frac{(e\mathbf{A}(\mathbf{r}, t))^2}{2m_e} - e\varphi, \tag{7.1.4}$$

where we have used the standard SI units for all the terms and the key term is $\hat{\mathbf{p}} \cdot \mathbf{A}(\mathbf{r}, t)$. This expression is often employed in solids and is still considered to be semiclassical because the radiation field, $\mathbf{A}(\mathbf{r}, t)$, is still classical. Such an approach is valid if we have many photons, as more photons appear to be more classical.[1]

Exercise 7.1.1.

1. Prove $[\hat{\Pi}_i, \hat{\Pi}_j] = i\hbar\epsilon_{ijk}B_k$.

2. (a) Start from eq. (7.1.3) and use the Heisenberg equation of motion to show the force on a particle of mass m and charge q is

$$m\frac{d^2\mathbf{r}}{dt^2} = q\mathbf{E} + \frac{q}{2m}(\hat{\mathbf{p}} \times \mathbf{B} - \mathbf{B} \times \hat{\mathbf{p}}) - \frac{q^2}{m}(\mathbf{A} \times \mathbf{B}),$$

---

1 In the advanced QM and quantum optics, the quantization of $\mathbf{A}(\mathbf{r}, t)$ is realized by introducing annihilation and creation operators for the radiation field.

where the electric field **E**, magnetic field **B** and vector potential **A** are functions of **r** and t. (b) Show the last term is related to the Poynting vector. Hint: Why do we have q^2? (c) In the limit, where **B** is independent of space, show

$$m\frac{d^2\mathbf{r}}{dt^2} = q\mathbf{E} + q\hat{\mathbf{u}} \times \mathbf{B},$$

where $\hat{\mathbf{u}}$ is given in eq. (7.1.3). This recovers the classical Lorentz force.

7.1.2 Coulomb gauge and dipole approximation

The simplest vector potential for an EM wave is $\mathbf{A}(\mathbf{r}, t) = \mathbf{A}_0 e^{i(\mathbf{k}\cdot\mathbf{r}-\omega t)}$. If the wavelength is very long, longer than the size of atoms and molecules, the wavevector **k** is very small, so that $\mathbf{k}\cdot\mathbf{r} \ll 1$ and $e^{i\mathbf{k}\cdot\mathbf{r}} \approx 1$. $\mathbf{A}(\mathbf{r}, t)$ can be approximated as

$$\mathbf{A}(t) = \mathbf{A}_0 e^{-i\omega t},$$

under which $\nabla \cdot \mathbf{A} = 0$. This is called the Coulomb gauge. **A** has no spatial dependence and no magnetic field. Then, $\hat{H}_I$ is very similar to the classical Hamiltonian,

$$\hat{H}_I = \frac{e}{m_e}\mathbf{A}(t)\cdot\hat{\mathbf{p}} + \frac{(e\mathbf{A}(t))^2}{2m_e} - e\varphi, \tag{7.1.5}$$

where the second term is often neglected (see the following exercise). The total Hamiltonian for the electron system is $\hat{H} = \hat{H}_0 + \hat{H}_I$, where $\hat{H}_0 = \frac{\hat{\mathbf{p}}^2}{2m_e} + V(\mathbf{r})$, as will be discussed.

Example 1 (Coulomb gauge). (a) Show that an EM wave with the vector potential $\mathbf{A}(\mathbf{r}, t)$, $\mathbf{A}(\mathbf{r}, t) = \mathbf{A}_0 e^{i(\mathbf{k}\cdot\mathbf{r}-\omega t)}$, obeys the Coulomb gauge. Here, **k** is the wavevector and denotes the light propagation direction, and $\mathbf{A}_0$ denotes the polarization direction. (b) Find **E**. (c) Find **B**. (d) Prove $|\mathbf{E}|/|\mathbf{B}| = c$.

(a) Suppose the propagation direction is along the z-axis and the polarization along the x-axis, so $\mathbf{A}(\mathbf{r}, t) = A_{0x}e^{i(k_z z-\omega t)}\hat{x}$. $\nabla \cdot \mathbf{A} = A_{0x}\partial e^{i(k_z z-\omega t)}/\partial x = 0$. (b) $\mathbf{E} = -\partial\mathbf{A}/\partial t = i\omega A_{0x}e^{i(k_z z-\omega t)}\hat{x}$. (c) $\mathbf{B} = \nabla \times \mathbf{A} = ik_z A_{0x}e^{i(k_z z-\omega t)}\hat{y}$. (d) $\frac{|\mathbf{E}|}{|\mathbf{B}|} = \frac{i\omega}{ik_z} = \lambda/T = c$.

We start from the time-dependent Schrödinger equation,

$$i\hbar\frac{\partial\psi(\mathbf{r}, t)}{\partial t} = \left[-\frac{\hbar^2}{2m_e}\left(\nabla + \frac{ie\mathbf{A}(t)}{\hbar}\right)^2 + V(\mathbf{r}) - e\varphi\right]\psi(\mathbf{r}, t). \tag{7.1.6}$$

We carry out a unitary transformation $\hat{U}$ by replacing $\psi(\mathbf{r}, t)$ in eq. (7.1.6) by $\phi(\mathbf{r}, t)$,

$$\psi(\mathbf{r}, t) = \hat{U}\phi(\mathbf{r}, t) = e^{-ie\mathbf{A}(t)\cdot\mathbf{r}/\hbar}\phi(\mathbf{r}, t),$$

to find a simplified version,

$$i\hbar\dot{\phi}(\mathbf{r},t) = [\hat{H}_0 + e\mathbf{r}\cdot\mathbf{E}(t) - e\varphi]\phi(\mathbf{r},t). \tag{7.1.7}$$

The details are relegated to the following exercise. If we drop $-e\varphi$, we can see that the new interaction Hamiltonian is just the dipole multiplied by the electric field $\mathbf{E}(t)$,

$$\hat{H}_I = -\hat{\mathbf{D}}\cdot\mathbf{E}(t) = -(-e\mathbf{r})\cdot\mathbf{E}(t) = e\mathbf{r}\cdot\mathbf{E}(t), \tag{7.1.8}$$

where $\hat{\mathbf{D}}$ is a dipole operator. This is called the dipole approximation, which has no magnetic field and thus violates the Maxwell equation (Ampere's law).

Exercise 7.1.2.

3. A laser pulse has the vector potential $|\mathbf{A}| = 0.01$ Vfs/Å. The electron's velocity is 10^6 m/s. Estimate the ratio $|e\mathbf{A}|/|\mathbf{p}|$.

The following exercises use $\hat{U} = e^{-ie\mathbf{A}\cdot\mathbf{r}/\hbar}$ and $\psi = \hat{U}\phi$, where $\mathbf{A}$ is independent of $\mathbf{r}$.

4. (a) Prove $\nabla\hat{U} = -\frac{ie\mathbf{A}}{\hbar}\hat{U}$. (b) $\nabla^2\hat{U} = (\frac{ie\mathbf{A}}{\hbar})^2\hat{U}$. (c) Show $\nabla(\hat{U}\phi(\mathbf{r},t)) = -\frac{ie\mathbf{A}}{\hbar}\hat{U}\phi(\mathbf{r},t) + \hat{U}\nabla\phi(\mathbf{r},t)$.

5. Using these relationships, show $\nabla^2(\hat{U}\phi) = (\frac{ie\mathbf{A}}{\hbar})^2\hat{U}\phi + 2\hat{U}\frac{ie\mathbf{A}}{\hbar}\cdot\nabla\phi + \hat{U}\nabla^2\phi$.

6. Show the time derivative $\dot{\psi} = \hat{U}\dot{\phi} + \frac{ie\dot{\mathbf{A}}\cdot\mathbf{r}}{\hbar}\hat{U}$.

7. Starting from $i\hbar\dot{\psi} = H\psi$, where $\hat{H} = -\frac{\hbar^2}{2m_e}(\nabla + \frac{ie\mathbf{A}}{\hbar})^2$, show the new Schrödinger equation for $\phi(\mathbf{r},t)$ is $i\hbar\dot{\phi}(\mathbf{r},t) = (-\frac{\hbar^2\nabla^2}{2m_e} + e\mathbf{E}\cdot\mathbf{r})\phi(\mathbf{r},t)$. Hint: $\mathbf{E} = -\frac{\partial\mathbf{A}}{\partial t}$.

7.1.3 Dipole selection rules

Equation (7.1.8) provides the dipole interaction between the system and the light as

$$\hat{H}_I = -\hat{\mathbf{D}}\cdot\mathbf{E}(t) = exE_x(t) + eyE_y(t) + ezE_z(t), \tag{7.1.9}$$

where E_x, E_y, and E_z are three components of the light electric field. Whether one can observe an optical transition depends on matrix elements $\mathbf{D}_{fi} = \langle f|\hat{\mathbf{D}}|i\rangle$ between the initial state $|i\rangle$ and final state $|f\rangle$. Assuming that the system has a spherical symmetry, like hydrogen atoms, then their wavefunctions ψ_{nlm} are products of the radial function $R_{nl}(r)$ and spherical harmonics $Y_{lm}(\theta,\phi)$, $\psi_{nlm} = R_{nl}(r)Y_{lm}(\theta,\phi)$.

We first convert x, y, and z to the spherical coordinates in a complex form (Euler form), $x = \frac{r}{2}\sin\theta(e^{i\phi} + e^{-i\phi})$, $y = \frac{r}{2i}\sin\theta(e^{i\phi} - e^{-i\phi})$, $z = r\cos(\theta)$, so eq. (7.1.9) is

$$\hat{H}_I = er\left[\cos\theta E_z(t) + \frac{E_x(t) - iE_y(t)}{2}\sin\theta e^{i\phi} + \frac{E_x(t) + iE_y(t)}{2}\sin\theta e^{-i\phi}\right], \tag{7.1.10}$$

which contains the information of light polarization. If both E_x and E_y are zero, then this represents the linearly polarized light along the z-axis. Other polarizations are possible. If $E_x(t) = E_0\cos(\omega t)$ and $E_y(t) = E_0\sin(\omega t)$ have the same amplitude E_0 but differ by a $\pi/2$ phase, then $E_x(t) - iE_y(t) = E_0e^{-i\omega t}$ represents right-circularly polarized light because at $t = 0$ the x-component has the largest value first. As t increases, it decreases, but the

y-component increases. Had we not made a dipole approximation, the spatial part of the light wave would be $e^{ik_z z}$, propagating along the z-axis.[2] If E_x and E_y differ in their amplitude, then we have elliptically polarized light. The third term represents the left-circularly polarized light $E_0 e^{i\omega t}$.

To compute the matrix element for a transition from $|n_i l_i m_i\rangle$ to $|n_f l_f m_f\rangle$, we first note that the matrix element $\langle n_f l_f m_f|\hat{H}_I|n_i l_i m_i\rangle$ is a product of the radial and the angular parts. The radial one is the same for all three terms in eq. (7.1.10) [29],

$$\mathcal{J} = \int_0^\infty R_{n_i l_i}(r) R_{n_f l_f}(r) r^3 dr, \tag{7.1.11}$$

but the angular part is not,

$$\int Y^*_{l_f m_f} \cos\theta Y_{l_i m_i} d\Omega, \quad \int Y^*_{l_f m_f} \sin\theta e^{i\phi} Y_{l_i m_i} d\Omega, \quad \int Y^*_{l_f m_f} \sin\theta e^{-i\phi} Y_{l_i m_i} d\Omega. \tag{7.1.12}$$

We use the recursion relationship of spherical harmonics Y_{lm} [19] with a correction,[3]

$$\sin\theta e^{i\phi} Y_{lm} = (b_{l-1,-(m+1)} Y_{l-1,m+1} - b_{l,m} Y_{l+1,m+1}) c_l, \tag{7.1.13}$$

$$\cos\theta Y_{lm} = (a_{l,m} Y_{l+1,m} + a_{l-1,m} Y_{l-1,m}) c_l, \tag{7.1.14}$$

$$\sin\theta e^{-i\phi} Y_{lm} = (-b_{l-1,m-1} Y_{l-1,m-1} + b_{l,-m} Y_{l+1,m-1}) c_l, \tag{7.1.15}$$

where a_{lm}, b_{lm}, c_l are given by

$$a_{lm} = \sqrt{(l+1)^2 - m^2}, \quad b_{lm} = \sqrt{(l+m+1)(l+m+2)}, \quad c_l = \left[(2l+1)(2l+3)\right]^{-1/2}.$$

We plug eqs. (7.1.13), (7.1.14), and (7.1.15) into eq. (7.1.12), and, then, using the orthogonality relationship for spherical harmonics

$$\int Y^*_{l'm'} Y_{lm} d\Omega = \delta_{ll'} \delta_{mm'}, \tag{7.1.16}$$

to find that the transition is not possible unless

$$\Delta l = \pm 1, \quad \Delta m_l = 0, \pm 1, \tag{7.1.17}$$

where $\Delta l = l_f - l_i$ and $\Delta m_l = m_f - m_i$. Equation (7.1.17) is called the dipole selection rule. $\Delta l = \pm 1$ means that the system (the medium) will gain from or lose to the light field by

2 If we align the thumb of our right hand along the propagation direction, the fingers are in the direction of the electric field vector.

3 a_{lm} and b_{lm} in [19] include c_l. When we compare the resultant equations with those in [29], we find c_l should stand alone and not change for the same Y_{lm}.

angular momentum $\hbar$. $\Delta m_l = 0, \pm 1$ means the system will retain, gain from, or lose to the light field by magnetic angular momentum $\hbar$. 0, +1 and −1 correspond to the light helicity, linear, right or left circular polarizations, respectively. With the selection rule, we can tell that a $1s \rightarrow 2s$ transition is forbidden, while $1s \rightarrow 2p$ is allowed.

The following table delineates the important physics behind the selection rules:

Δm_l	+1	0	−1
Polarization	σ^+	π	σ^-
Propagation direction	longitudinal	transversal	longitudinal

Example 2 (Transition matrix elements). Compute the transition matrix element $\langle n', l+1, m|z|n, l, m\rangle$.

We first note that $\langle n', l+1, m|z|n, l, m\rangle$ is

$$\langle n', l+1, m|z|n, l, m\rangle = \int_{r=0}^{\infty} R_{n',l+1}(r) r R_{n,l}(r) r^2 dr \int Y^*_{l+1,m} \cos\theta Y_{l,m} d\Omega.$$

We use eq. (7.1.14) to find $\langle n', l+1, m|z|n, l, m\rangle = \mathcal{J}\sqrt{\frac{(l+1)^2-m^2}{(2l+1)(2l+3)}}$.

Exercise 7.1.3.

8. Compute the transition matrix element due to the radial part of the wavefunctions $\mathcal{J}$ from the hydrogen 1s- to the 2p-state (see Table 4.3 on Section 4.6.2). (Answer: $\frac{192\sqrt{2}a}{243}$).
9. In the hydrogen atom, the Lyman series is due to the emission from a higher level to $n = 1(1s)$. The first two lines are from 2p to 1s and 3p to 1s, called Lyman-α and Lyman-β lines, respectively. The intensity of emission I is proportional to $\Delta E_{fi}\langle f|\mathbf{r}|i\rangle$. Find the ratio I_α/I_β. Adopted from [45]. (Answer: 3.18).
10. Compute the transition matrix element (a) between the 2s- and 3p-orbitals of the hydrogen atoms and (b) between 2s- and 4p-orbitals. These matrix elements are used in the problem at the end of the chapter.

7.2 Time-dependent perturbation theory

In many aspects, TDPT is similar to time-independent perturbation theory, though we do not have a degenerate counterpart. Consider that an unperturbed system is described by time-independent $\hat{H}_0$, whose eigenvalues E_n and eigenfunctions ϕ_n are known, i. e., $\hat{H}_0\phi_n = E_n\phi_n$. Now, we subject it to a time-dependent perturbation $\hat{H}_I(t)$. The total Hamiltonian for the entire system is

$$\hat{H} = \hat{H}_0 + \lambda\hat{H}_I(t),$$

where λ is the usual dimensionless parameter to track the order of perturbation. Our goal is to solve the time-dependent Schrödinger equation approximately,

$$i\hbar\frac{\partial\psi}{\partial t} = \hat{H}\psi. \tag{7.2.1}$$

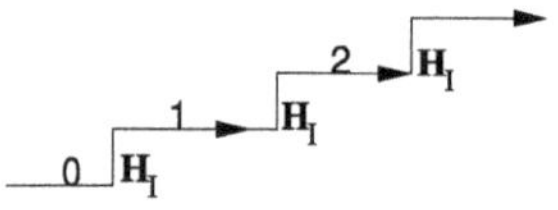

Figure 7.1: Hierarchy of time-dependent perturbation theory. The order increases by one every time the system interacts with the external field $\hat{H}_I$.

We follow the same idea of TIDPT and expand $\psi(t)$ as

$$\psi(t) = \psi^{(0)}(t) + \lambda\psi^{(1)}(t) + \lambda^2\psi^{(2)}(t) + \cdots \tag{7.2.2}$$

where $\psi^{(0)}(t), \psi^{(1)}(t), \psi^{(2)}(t), \ldots$, are the zeroth-, first-, and second-order wavefunctions, and so on. We substitute $\psi(t)$ and $\hat{H}$ into both sides of eq. (7.2.1) to get

$$i\hbar(\dot{\psi}^{(0)} + \lambda\dot{\psi}^{(1)} + \lambda^2\dot{\psi}^{(2)} + \cdots) = (\hat{H}_0 + \lambda\hat{H}_I)(\psi^{(0)} + \lambda\psi^{(1)} + \lambda^2\psi^{(2)} + \cdots).$$

Then we collect terms with the same order of λ on both sides:

$$\lambda^0 : i\hbar\dot{\psi}^{(0)} = \hat{H}_0\psi^{(0)}, \tag{7.2.3}$$
$$\lambda^1 : i\hbar\dot{\psi}^{(1)} = \hat{H}_0\psi^{(1)} + \hat{H}_I\psi^{(0)}, \tag{7.2.4}$$
$$\lambda^2 : i\hbar\dot{\psi}^{(2)} = \hat{H}_0\psi^{(2)} + \hat{H}_I\psi^{(1)}, \tag{7.2.5}$$
$$\cdots$$
$$\lambda^N : i\hbar\dot{\psi}^{(N)} = \hat{H}_0\psi^{(N)} + \hat{H}_I\psi^{(N-1)}. \tag{7.2.6}$$

Except for the zeroth order, a general pattern emerges from these equations: The current order wavefunction is connected with one lower order wavefunction through $\hat{H}_I$. Figure 7.1 shows such a hierarchy, underlying perturbative nonlinear optics [40].

7.2.1 Zeroth order

We consider the zeroth order first, i. e., $\lambda = 0$. Because $\hat{H}_0$ is time-independent, $\psi^{(0)}$ is a stationary state solution

$$\psi^{(0)}(t) = e^{-i\hat{H}_0 t/\hbar}\psi^{(0)}(t = 0).$$

If $\psi^{(0)}(t = 0)$ is an eigenstate of $\hat{H}_0$, ϕ_k, then $\psi^{(0)}(t) = e^{-iE_k t/\hbar}\phi_k$, where the exponential is called the dynamic phase factor. If $\psi^{(0)}(t = 0)$ is a mixed state, $\psi^{(0)}(t = 0) = \sum_n c_n\phi_n$, and then at time t we have $\psi^{(0)}(t) = \sum_n c_n e^{-iE_n t/\hbar}\phi_n$. Here, c_n is a constant given in the beginning and should not be confused with C_l used subsequently.

7.2.2 First order

Similar to TIDPT, we target an eigenstate ϕ_k, so the zeroth-order wavefunction is $\psi^{(0)}(t = 0) = \phi_k$, and at t, $\psi^{(0)}(t \neq 0) = e^{-iE_k t/\hbar}\phi_k$. Equation (7.2.4) becomes

$$i\hbar\dot{\psi}^{(1)} = \hat{H}_0\psi^{(1)} + \hat{H}_I e^{-iE_k t/\hbar}\phi_k. \tag{7.2.7}$$

We expand the unknown $\psi^{(1)}$ in terms of the eigenstates of the unperturbed Hamiltonian $\hat{H}_0$, called the energy representation because we use the eigenstates of $\hat{H}_0$ to represent an unknown function, $\psi^{(1)}(t = 0) = \sum_n C_{nk}^{(1)}(t = 0)\phi_n$, so, at $t \neq 0$, $\psi^{(1)}(t \neq 0) = \sum_n C_{nk}^{(1)}(t)e^{-iE_n t/\hbar}\phi_n$.[4] Doing so is tantamount to changing a partial differential equation for $\psi^{(1)}(t)$ to an ordinary differential equation for $C_{nk}^{(1)}(t)$. ψ is a function of both $\mathbf{r}$ and t, but $C_{nk}^{(1)}(t)$ is a function of t only, a key advantage in the energy representation. We substitute $\psi^{(1)}(t \neq 0)$ onto the left side of eq. (7.2.7) to find

$$\begin{aligned} i\hbar\dot{\psi}^{(1)} &= i\hbar\sum_n \dot{C}_{nk}^{(1)}(t)e^{-iE_n t/\hbar}\phi_n + i\hbar\sum_n C_{nk}^{(1)}(t)\left(\frac{-iE_n}{\hbar}\right)e^{-iE_n t/\hbar}\phi_n \\ &= i\hbar\sum_n \dot{C}_{nk}^{(1)}(t)e^{-iE_n t/\hbar}\phi_n + \sum_n C_{nk}^{(1)}(t)E_n e^{-iE_n t/\hbar}\phi_n. \end{aligned} \tag{7.2.8}$$

The first term $\hat{H}_0\psi^{(1)}$ on the right side of eq. (7.2.7) becomes

$$\hat{H}_0\psi^{(1)} = \sum_n C_{nk}^{(1)}(t)e^{-iE_n t/\hbar}\hat{H}_0\phi_n = \sum_n C_{nk}^{(1)}(t)E_n e^{-iE_n t/\hbar}\phi_n,$$

which cancels the last term on the right of eq. (7.2.8). After a straightforward simplification, eq. (7.2.7) is reduced to

$$i\hbar\sum_n \dot{C}_{nk}^{(1)}(t)e^{-iE_n t/\hbar}\phi_n = \hat{H}_I e^{-iE_k t/\hbar}\phi_k, \tag{7.2.9}$$

which is an ordinary differential equation for $C_{nk}^{(1)}(t)$, with $\hat{H}_I$ as its source. If we want to target N eigenstates of $\hat{H}_0$, then we have N differential equations.

The energy representation is convenient because it allows us to further simplify the equation. Suppose that we want to find $C_{lk}^{(1)}(t)$ for eigenstate l. We multiply eq. (7.2.9) by eigenstate $\langle\phi_l|$ from the left, integrate it, and then employ the orthonormalization condition, $\langle\phi_l|\phi_n\rangle = \delta_{ln}$, to find

$$i\hbar\dot{C}_{lk}^{(1)}(t)e^{-iE_l t/\hbar} = \langle\phi_l|\hat{H}_I|\phi_k\rangle e^{-iE_k t/\hbar} \rightarrow \dot{C}_{lk}^{(1)} = \frac{1}{i\hbar}\langle\phi_l|\hat{H}_I|\phi_k\rangle e^{i(E_l - E_k)t/\hbar},$$

4 This equation is fundamentally different from $\psi^{(0)}(t) = \sum_n c_n e^{-iE_n t/\hbar}\phi_n$, where the summation is over a list of predefined states with a time-independent c_n. Here the expansion is based on the completeness $\sum_n |\phi_n\rangle\langle\phi_n| = 1$, where the summation is over all the unperturbed eigenstates. $C_{nk}^{(1)}(t)$ is unknown and time-dependent, and may even absorb $e^{-iE_n t/\hbar}$, such as $\psi^{(1)}(t) = \sum_n C_{nk}^{(1)}(t)\phi_n$.

whose integral form is

$$C_{lk}^{(1)}(t) = \frac{1}{i\hbar}\int_{t_0}^{t} \langle\phi_l|\hat{H}_I(t')|\phi_k\rangle e^{i(E_l-E_k)t'/\hbar}dt'. \tag{7.2.10}$$

This is the first-order time-dependent wavefunction correction to ψ due to state $|\phi_l\rangle$. Including all the states, we obtain the first-order wavefunction correction as

$$\psi^{(1)}(t) = \sum_l C_{lk}^{(1)}(t)e^{-iE_l t/\hbar}\phi_l, \tag{7.2.11}$$

where eq. (7.2.10) provides $C_{lk}^{(1)}(t)$.

Physically, $|C_{lk}^{(1)}(t)|^2$ is the probability $P_{lk}(t)$ into state ϕ_l from ϕ_k under the perturbation $\hat{H}_I$,

$$P_{lk}(t) \equiv \left|C_{lk}^{(1)}(t)\right|^2 = \frac{1}{\hbar^2}\left|\int_{t_0}^{t} \langle\phi_l|\hat{H}_I(t')|\phi_k\rangle e^{i(E_l-E_k)t'/\hbar}dt'\right|^2, \tag{7.2.12}$$

and $\langle\phi_l|\hat{H}_I(t')|\phi_k\rangle$ quantifies how strong the transition from ϕ_k to ϕ_l is. Once we have $\psi^{(1)}(t)$, we can obtain the second-order correction $\psi^{(2)}(t)$ by solving eq. (7.2.5) in a similar fashion. While linear optics is based on the first-order wavefunction correction, nonlinear optics is based on the second order and higher.

7.3 Examples of time-dependent perturbation theory

TDPT is more complicated than TIDPT. Even the first-order TDPT has more nuisances. For instance, because the probability cannot be larger than 1, this requires that $\langle\phi_l|\hat{H}_I(t)|\phi_k\rangle$ cannot be too large. Here, we provide three examples for TDPT. In the next section, we will apply TDPT to linear optics.

7.3.1 $\hat{H}_I(t)$ is time independent

If $\hat{H}_I(t)$ is time independent, find $P_{lk}(t)$ and explain what it means physically.

Since $\hat{H}_I$ is constant, we can remove it from eq. (7.2.12). We introduce the angular frequencies for $E_l, \omega_l = E_l/\hbar$ and for $E_k, \omega_k = E_k/\hbar$, $\omega_{lk} = \omega_l - \omega_k$, and set $t_0 = 0$ in eq. (7.2.12), to find the probability,

$$P_{lk}(t) = \frac{1}{\hbar^2}\langle\phi_l|\hat{H}_I|\phi_k\rangle^2\left|\frac{e^{i\omega_{lk}t}-1}{\omega_{lk}}\right|^2 = \frac{1}{\hbar^2}\langle\phi_l|\hat{H}_I|\phi_k\rangle^2\frac{\sin^2(\omega_{lk}t/2)}{(\omega_{lk}/2)^2},$$

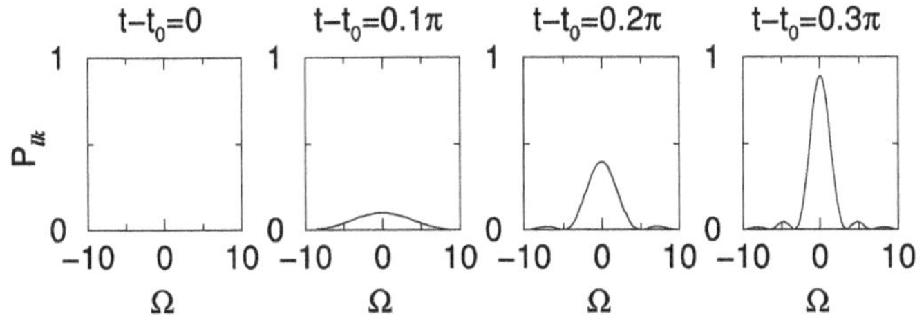

Figure 7.2: Transition probability P_{lk} as a function of energy detuning Ω at four different times $t - t_0$, from right to left. As time increases, the energy domain narrows, and only a few states can make the transition, due to the energy-time uncertainty principle. The figure is plotted using a function $\frac{\sin^2(\Omega(t-t_0)/2)}{(\Omega/2)^2}$, where $t - t_0$ is in units of π.

which reveals that $P_{lk}(t)$ oscillates with time t, a dynamical interference. Figure 7.2 shows the probability with $\Omega = \omega_{lk}$ at four different times. The probability must be smaller than 1. But as time goes by, it grows. The largest probability is at $\omega_{lk} \to 0$ and proportional to t^2. This reveals a difficulty with TDPT. To overcome this difficulty, one can diagonalize the new Hamiltonian with the presence of $\hat{H}_I$ to find new eigenstates, and then expand the initial state in terms of new eigenstates. The following exercise includes a case with $(E_l - E_k) = 0$.

7.3.2 $\hat{H}_I(t)$ is a periodic perturbation

A light field $\hat{H}_I(t) = e\mathbf{r} \cdot \mathbf{E}(t) = e\mathbf{r} \cdot \mathbf{E}_0 \cos(\omega t)$ (eq. (7.1.8)) is applied to a hydrogen atom, where ω is the light angular frequency and $\mathbf{E}_0$ is the amplitude. The atom is initially in the 1s-state. (a) Find the probability $P_{1s,2s}$ that the electron is in the 2s-state. (b) Find the probability $P_{1s,2p}$ for the electron to be in the 2p-state. (c) Determine the condition for first-order perturbation theory. (d) Find the rate of the probability of the transition.

(a) We choose that $|\phi_k\rangle$ is the 1s-state and $|\phi_l\rangle$ is the 2s-state. Then, $\langle\phi_l|\hat{H}_I(t)|\phi_k\rangle = e\langle\phi_l|\mathbf{r}|\phi_k\rangle\cdot\mathbf{E}_0\cos(\omega t)$. $\langle\phi_l|\mathbf{r}|\phi_k\rangle$ is the transition matrix element. According to Section 7.1, this element is zero, so there is no transition between them and $P_{1s,2s} = |C_{lk}^{(1)}|^2 = 0$. For this reason, the dipole selection rule is so powerful.

(b) Here, $|\phi_l\rangle$ is a 2p-state, and the transition is allowed. All the spectral lines on Section 4.1 in Chapter 4 follow this rule. The magnitude of these transitions is determined by $C_{lk}^{(1)}(t)$. We denote $\langle\phi_l|\mathbf{r}|\phi_k\rangle$ as $\mathbf{r}_{lk}$ and $\omega_{lk} = \omega_l - \omega_k$. To ease our integration, we write $\cos(\omega t)$ as $\frac{1}{2}(e^{i\omega t} + e^{-i\omega t})$. For the $e^{-i\omega t}$ term, eq. (7.2.10) yields

$$C_{lk}^{(1)}(t) = \frac{e\mathbf{r}_{lk}\cdot\mathbf{E}_0}{2i\hbar}\int_{t_0}^{t} e^{i(\omega_{lk}-\omega)t'}dt' = \frac{e\mathbf{r}_{lk}\cdot\mathbf{E}_0}{2i\hbar}\,\frac{e^{i(\omega_{lk}-\omega)t} - e^{i(\omega_{lk}-\omega)t_0}}{i(\omega_{lk}-\omega)}. \tag{7.3.1}$$

For the $e^{i\omega t}$ term, we replace every $-\omega$ by $+\omega$ in eq. (7.3.1). Because $\omega_{lk} > 0$ and $\omega > 0$, $\omega_{lk} + \omega$ is positive and the resultant term is smaller than eq. (7.3.1) and ignored in the absorption process. We set $\Omega = \omega_{lk} - \omega$ and denote $e\mathbf{r}_{lk} \cdot \mathbf{E}_0$ as A, so $C_{lk}^{(1)}(t) = -\frac{A}{2\hbar\Omega} e^{i\frac{\Omega(t+t_0)}{2}} (2i) \sin(\frac{\Omega(t-t_0)}{2})$. The probability $P_{1s,2p}(\Omega) = |C_{lk}^{(1)}(t)|^2$ is

$$P_{1s,2p}(\Omega) = \frac{A^2}{4\hbar^2} \frac{\sin^2(\frac{\Omega(t-t_0)}{2})}{(\Omega/2)^2} = \frac{e^2|\mathbf{r}_{lk} \cdot \mathbf{E}_0|^2}{4\hbar^2} \frac{\sin^2(\frac{\Omega(t-t_0)}{2})}{(\Omega/2)^2}. \tag{7.3.2}$$

Figure 7.2 shows the probability with Ω at four different times $(t - t_0) = 0, 0.1\pi, 0.2\pi$ and 0.3π. At $t - t_0 = 0$, there is no transition. If $t - t_0$ is smaller, the peak is broader, but, as $t - t_0$ increases, the peak becomes narrower. This is the manifestation of the frequency-time or energy-time uncertainty principle.

Now, if we compare it with eq. (6.6.4) in Section 6.3.3, we see they agree with each other very well, except that in eq. (6.6.4) the Rabi frequency Ω has an additional contribution from the strength of the magnetic field there. But, here our frequency is just Ω.

(c) If $\Omega \to 0$, i. e., $\omega = \omega_l - \omega_k$, we reach a resonance, and $|C_{lk}^{(1)}(t)|^2 \to \frac{A^2(t-t_0)^2}{\hbar^2}$. Because $|C_{lk}^{(1)}(t)|^2$ or P_{lk} cannot be larger than 1, this means that $|e\mathbf{r}_{lk} \cdot \mathbf{E}_0|^2(t - t_0)^2 \le \hbar^2$, which sets the limit for $\mathbf{E}_0$ in order for first-order perturbation theory to work.

(d) To compute the rate of the probability, we need to extract the time first from eq. (7.3.2). As $t \to \infty$, we can use the identify $\lim_{a\to\infty} \frac{\sin^2 ax}{\pi a x^2} = \delta(x)$ to simplify

$$\frac{\sin^2(\frac{\Omega(t-t_0)}{2})}{(\Omega/2)^2} = 2\pi(t - t_0)\delta(\Omega).$$

Next, we need to take into account the orientation of the electric field with respect to $\mathbf{r}$. Suppose the angle between $\mathbf{r}$ and $\mathbf{E}_0$ is θ, so $\mathbf{r}_{lk} \cdot \mathbf{E}_0 = |\mathbf{r}_{lk}||\mathbf{E}_0| \cos\theta$ and $|\mathbf{r}_{lk} \cdot \mathbf{E}_0|^2 = |\mathbf{r}_{lk}|^2|\mathbf{E}_0|^2 \cos^2\theta$. If the orientation is random, we can take the average of $\cos^2\theta$ as

$$\overline{\cos^2\theta} = \frac{1}{4\pi} \int_0^{2\pi} d\phi \int_0^{\pi} \cos^2\theta \sin\theta d\theta = \frac{1}{3} \to \overline{|\mathbf{r}_{lk} \cdot \mathbf{E}_0|^2} = \frac{1}{3}|\mathbf{r}_{lk}|^2|\mathbf{E}_0|^2. \tag{7.3.3}$$

For a single angular frequency ω (monochromatic light), the energy density for a single cycle of light is $u(\omega) = \epsilon_0 \int E_0^2 \cos^2(\omega t) dt = \frac{1}{2}\epsilon_0 E_0^2$, so $|\mathbf{E}_0|^2$ in eq. (7.3.3) can be replaced by $2u(\omega)/\epsilon_0$. However, for a broad spectrum of many different frequencies, $u(\omega)$ should be further replaced by the energy density per frequency $\rho(\omega)$ because this way we can compare it with experiments. Mathematically, we rewrite $u(\omega) = \rho(\omega)d\omega$, and convert eq. (7.3.2) to an integral over ω,

$$P_{1s,2p} = (t - t_0)\frac{\pi e^2|\mathbf{r}_{lk}|^2}{3\epsilon_0\hbar^2} \int \rho(\omega)\delta(\Omega)d\omega = (t - t_0)\frac{\pi e^2|\mathbf{r}_{lk}|^2}{3\epsilon_0\hbar^2}\rho(\omega_{lk}), \tag{7.3.4}$$

where the second equation uses the δ function property, that is $\int f(x)\delta(x-a)dx = f(a)$. So, we can define the transition rate $R_{lk} = \frac{dP_{lk}}{dt}$,

$$R_{lk} = \frac{\pi e^2 |\mathbf{r}_{lk}|^2}{3\epsilon_0 \hbar^2}\rho(\omega_{lk}), \tag{7.3.5}$$

which represents the rate of the probability for the transition from 1s- to 2p-states.

7.3.3 Excitation of a charged harmonic oscillator by a pulse

A charged harmonic oscillator carries charge q and is initially in the ground state $|n=0\rangle$ at $t_0 = -\infty$. We excite it with a pulsed laser and $\hat{H}_I = -qE_0\hat{x}e^{-t^2/\tau^2}e^{-i\Omega t}$, where E_0 is the field amplitude, Ω is the angular frequency, and τ is the pulse duration. (a) Find the probabilities in all the excited states at $t=\infty$. (b) Find the probability in the ground state. (c) Suppose the laser has $E_0 = 0.01\,\mathrm{V/Å}$, $\tau = 100\,\mathrm{fs}$, and $\hbar\Omega = \hbar\omega = 0.01\,\mathrm{eV}$. Take $q = e = 1.602\times10^{-19}$ C and $m = 1.66\times10^{-27}$ kg. Compute numerically $C_{10}^{(1)}(\infty)$ and discuss the issue with TDPT.

(a) According to eq. (7.2.10), we need to compute the transition matrix elements between the ground state $|k=0\rangle$ and state $|l\rangle$, $\langle l|\hat{x}|0\rangle$. We convert $\hat{x}$ to $\hat{a}$ and $\hat{a}^\dagger$, $\hat{x} = \sqrt{\frac{\hbar}{2m\omega}}(\hat{a}+\hat{a}^\dagger)$, so $\langle l|\hat{x}|0\rangle = \delta_{l,1}\sqrt{\frac{\hbar}{2m\omega}}$, where m is the mass of the harmonic oscillator and ω is its angular frequency. This means only one state can be excited. Our $C_{10}^{(1)}(t)$ is

$$C_{10}^{(1)}(t) = -\frac{qE_0}{i\hbar}\sqrt{\frac{\hbar}{2m\omega}}\int_{t_0}^{t} e^{-t'^2/\tau^2}e^{-i\Omega t'}e^{i\omega t'}dt'.$$

In the limit $t=\infty$ and $t_0 = -\infty$, we integrate over t' to get

$$C_{10}^{(1)}(\infty) = iqE_0\sqrt{\frac{\pi}{2\hbar\omega m}}\tau e^{-\tau^2(\Omega-\omega)^2/4},$$

which has a Gaussian dependence on Ω, instead of the Bessel function in the periodic perturbation. This shows that the pulse shape will alter the shape of the probability. The maximum probability is at $\Omega=\omega$, whose probability $P_{10} = |C_{10}^{(1)}|^2$ is

$$P_{10} = \left|C_{10}^{(1)}(\infty)\right|^2 = \frac{\pi q^2 E_0^2}{2\hbar\omega m}\tau^2.$$

(b) The probability in the ground state is just $1-|C_{10}^{(1)}|^2 = 1-\frac{\pi q^2E_0^2}{2\hbar\omega m}\tau^2$.

(c) We plug these parameters into the expression to find $C_{10}^{(1)}(\infty) = 1.23122i$, which is larger than 1. This is clearly incorrect since one cannot have a probability larger than 1. We note that the parameters used are quite realistic for an actual laser pulse. The failure reflects the limitation of first-order perturbation theory, where we have to keep the field amplitude low. Otherwise, we have to use higher-order perturbation theories.

Exercise 7.3.3.

11. Show that $C_{lk}^{(1)}$ in eq. (7.2.10) has no unit.

12. (a) Show that, if in eq. (7.3.2) $\Omega \to 0$, i. e., $\omega = \omega_l - \omega_k$, $|C_{lk}^{(1)}(t)|^2 \to \frac{A^2(t-t_0)^2}{\hbar^2}$. (b) Show eq. (7.3.2) indeed obeys the energy-time uncertainty principles, $\hbar\Delta\Omega\Delta t \geq \hbar/2$. Hint: Find two first zeros of $\sin^2(\frac{\Omega(t-t_0)}{2})(\Omega/2)^2$ and their difference is $\Delta\Omega$, and set Δt to $t - t_0$.

13. A hydrogen atom is in the ground state. Suddenly, we apply an electric field pulse along the z axis, $E_z = E_0\delta(t)$, where E_0 is the amplitude of the field. (a) Find the probabilities in each eigenstate under the dipole approximation. (b) Find the probability for the electron remaining in the ground state.

14. A system has two degenerate states $|a\rangle$ and $|b\rangle$, with eigenenergies $E_a = E_b$. A constant perturbation, $\hat{H}_I$, which couples them as $\langle a|\hat{H}_I|b\rangle = c$, is turned on at $t = 0$. Suppose initially the system is in $|a\rangle$. (a) Find $P_{ab}(t)$ without using TDPT. (b) Suppose now that E_b is slightly higher than E_a by δ. Use TDPT to find $P_{ab}(t)$ and find the condition when first-order TDPT is still valid.

7.4 Linear optics: polarization and susceptibility

Many predictions of time-dependent perturbation theory can be realized in optics. This section provides the connection between theory and experiment.

7.4.1 Classical optics

In classical optics, the material response is characterized by the electric polarization **P**, which can be expanded according to the orders of interaction with the external electric field **E** in SI units [42]:

$$\mathbf{P} = \mathbf{P}_0 + \epsilon_0\chi_e^{(1)} : \mathbf{E} + \epsilon_0\chi_e^{(2)} :: \mathbf{EE} + \epsilon_0\chi_e^{(3)} ::: \mathbf{EEE} + \cdots,$$

where $\chi_e^{(n)}$ is the nth-order susceptibility in SI units of $(\mathrm{m/V})^{n-1}$, $\mathbf{P}_0$ is the static polarization, and ":" denotes the tensor product. Nonlinear optics is concerned with $\chi_e^{(2)}$ and higher. When the external electric field is weak, we are in the linear optical regime.[5] At this limit, the induced polarization $\mathbf{P}_{\mathrm{ind}}^{(1)}$ (the second term in the equation) by an external electric field **E** is given by

$$\mathbf{P}_{\mathrm{ind}}^{(1)} = \epsilon_0\chi_e^{(1)}\mathbf{E}, \tag{7.4.1}$$

where $\chi_e^{(1)}$ is the linear electric susceptibility. The SI units of **P** are $\mathrm{C/m^2}$, the same as the surface charge density. $\chi_e^{(1)}$ is dimensionless [42] and is a 3×3 matrix. For simplicity, we

---

5 A strong field, such as one produced by a laser, can induce a strong nonlinear optical effect [40, 41, 42].

suppress its indices. Since $1+\chi_e^{(1)} = \epsilon_r$, ϵ_r can be found once $\chi_e^{(1)}$ is known. In general, $\chi_e^{(1)}$ and ϵ_r are functions of the incident field angular frequency ω, which will be discussed later.

According to the Maxwell equation, the speed of light in a vacuum is $c = \frac{1}{\sqrt{\mu_0\epsilon_0}}$. In a medium, the speed of light is reduced to $v = \frac{1}{\sqrt{\mu\epsilon}}$, where μ and ϵ are the permeability and permittivity in the medium, respectively. The index of refraction n is defined as

$$n = \frac{c}{v} = \frac{1/\sqrt{\mu_0\epsilon_0}}{1/\sqrt{\mu\epsilon}} = \sqrt{\epsilon/\epsilon_0}\sqrt{\frac{\mu}{\mu_0}}.$$

Since $\mu \approx \mu_0$, $n = \sqrt{\epsilon/\epsilon_0} = \sqrt{\epsilon_r}$, where ϵ_r is the relative permittivity and is also called the dielectric constant. ϵ_r is dimensionless as is n. If we find $\chi_e^{(1)}$, both n and ϵ_r can be obtained.

7.4.2 Quantum optics

In QM, similarly we start with the polarization operator $\hat{\mathbf{P}}$ defined through the dipole moment per volume Ω. The dipole moment is $\hat{\mathbf{D}} = -e\hat{\mathbf{r}}$, in units of Cm, so $\hat{\mathbf{P}}$ is

$$\hat{\mathbf{P}} = \frac{\hat{\mathbf{D}}}{\Omega} = \frac{-e\hat{\mathbf{r}}}{\Omega},$$

whose expectation value in a state $|\psi(t)\rangle$ is $\langle\psi(t)|\hat{\mathbf{P}}|\psi(t)\rangle$, a real quantity. We utilize eq. (7.2.2) to expand $|\psi(t)\rangle$, so $\langle\psi(t)|\hat{\mathbf{P}}|\psi(t)\rangle$ becomes

$$\begin{aligned}\langle\psi(t)|\hat{\mathbf{P}}|\psi(t)\rangle &= \langle\psi^{(0)}(t) + \psi^{(1)}(t) + \cdots|\hat{\mathbf{P}}|\psi^{(0)}(t) + \psi^{(1)}(t) + \cdots\rangle \\ &= \langle\psi^{(0)}(t)|\hat{\mathbf{P}}|\psi^{(0)}(t)\rangle + \langle\psi^{(1)}(t)|\hat{\mathbf{P}}|\psi^{(0)}(t)\rangle + \langle\psi^{(0)}(t)|\hat{\mathbf{P}}|\psi^{(1)}(t)\rangle + \cdots,\end{aligned}$$

where the order of interaction in each is just the sum of the superscripts on two wavefunctions. The first term is the zeroth order, a static polarization, present in polar molecules, semiconductors, and insulators. The second and third are the first order, induced by $\mathbf{E}$, and constitute the first-order induced polarization,

$$\mathbf{P}^{(1)}(t) = \langle\psi^{(1)}(t)|\hat{\mathbf{P}}|\psi^{(0)}(t)\rangle + \langle\psi^{(0)}(t)|\hat{\mathbf{P}}|\psi^{(1)}(t)\rangle. \quad (7.4.2)$$

We first note that $\mathbf{P}^{(1)}(t)$ is time-dependent and is always real because two terms form a pair of complex conjugation.[6] Only one term is unique, and we choose the second term.

6 This is because the second term represents the light absorption and the first term the light emission. However, as will be seen, if we only compute the absorption or emission, we must choose one term, and the $\mathbf{P}^{(1)}(t)$ becomes a complex number.

7.4.3 Linear susceptibility: specialized to state $|k\rangle$

We suppose that the unperturbed system is already solved with eigenstates $\{\phi_i\}$ and eigenvalues $\{E_i\}$, $\hat{H}_0\phi_i = E_i\phi_i$. A cw electromagnetic radiation perturbs the system through the interaction Hamiltonian,

$$\hat{H}_I = -\hat{\mathbf{D}} \cdot \mathbf{E}(t), \tag{7.4.3}$$

where $\hat{\mathbf{D}}$ is the dipole moment and $\mathbf{E}(t)$ is the light field. We choose a complex form for the field[7] $\mathbf{E}(t) = \mathbf{E}_0 e^{-i\omega t}$ because we only include the second term in eq. (7.4.2).

Equation (7.4.2) is quite generic. We specialize it to an initial state $|k\rangle$. This means that our electron is initially in $|k\rangle$ or ϕ_k, so its zeroth-order wavefunction at time t is $\psi^{(0)}(t) = \phi_k e^{-iE_k t/\hbar}$. We expand the first-order $\psi^{(1)}(t)$ in terms of eigenstates (see eq. (7.2.11)), so the first-order polarization becomes

$$\begin{aligned}\langle\psi^{(0)}(t)|\hat{\mathbf{P}}|\psi^{(1)}(t)\rangle &= \langle\phi_k e^{iE_k t/\hbar}|\hat{\mathbf{P}}|\sum_n C_{nk}^{(1)}(t)e^{-iE_n t/\hbar}|\phi_n\rangle \\ &= \sum_n C_{nk}^{(1)}(t)e^{i(E_k-E_n)t/\hbar}\langle\phi_k|\hat{\mathbf{P}}|\phi_n\rangle = \frac{1}{\Omega}\sum_n C_{nk}^{(1)}(t)e^{i(E_k-E_n)t/\hbar}\langle\phi_k|\hat{\mathbf{D}}|\phi_n\rangle,\end{aligned}$$

where in the last equation we have replaced $\hat{\mathbf{P}}$ by $\hat{\mathbf{D}}$. To simplify our notation, we denote $\frac{E_n-E_k}{\hbar} = \omega_n - \omega_k = \omega_{nk}$, and $\langle\phi_k|\hat{\mathbf{D}}|\phi_n\rangle = \mathbf{D}_{kn}$, so

$$\langle\psi^{(0)}(t)|\hat{\mathbf{P}}|\psi^{(1)}(t)\rangle = \frac{1}{\Omega}\sum_n C_{nk}^{(1)}(t)e^{-i\omega_{nk}t}\mathbf{D}_{kn}. \tag{7.4.4}$$

The number of times that $\hat{\mathbf{D}}$ enters polarization is always equal to the order plus one. For the linear process, $\hat{\mathbf{D}}$ enters the polarization, $1+1=2$, twice, once through the definition of polarization and the other through the interaction Hamiltonian (eq. (7.4.3)).

7.4.3.1 Lifetime and initial condition

Because excited states have limited lifetimes, electrons in those states will decay to lower-energy states. To properly take this into account, for eigenstate $|n\rangle$, we multiply $e^{-i\omega_n t}$ by $e^{-\Gamma_n t}$ to get $e^{-i\omega_n t-\Gamma_n t}$, where Γ_n is the inverse of the lifetime of state ϕ_n and is positive. Similarly, the term $e^{-i\omega_{nk}t}$ in eq. (7.4.4) becomes $e^{-i\omega_{nk}t-\Gamma_{nk}t}$.

7 In principle, this also has a spatial component like $e^{i\mathbf{k}\cdot\mathbf{r}}$, but, once we take the dipole approximation, the spatial dependence is dropped. This approach is valid as long as the wavelength of the light is much larger than the system size. Otherwise, one must keep it.

We find $C_{nk}^{(1)}$ from eq. (7.2.10), with $\hat{H}_I$ replaced by eq. (7.4.3),

$$C_{nk}^{(1)}(t) = \frac{1}{i\hbar}\int_{t_0}^{t} -\mathbf{D}_{nk}\cdot\mathbf{E}(t')e^{i\omega_{nk}t'+\Gamma_{nk}t'}dt'. \tag{7.4.5}$$

We want to alert the reader about a subtlety in $\mathbf{D}$: in eq. (7.4.5), $\mathbf{D}$ is $\mathbf{D}_{nk}$, but in (7.4.4) it is $\mathbf{D}_{kn}$. They are conjugate to each other, called a parametric process in optics. We substitute $\mathbf{E}(t') = \mathbf{E}_0 e^{-i\omega t'}$ into eq. (7.4.5) and then integrate over t' from t_0 to t to get

$$C_{nk}^{(1)}(t) = -\frac{1}{i\hbar}\mathbf{D}_{nk}\cdot\mathbf{E}_0\frac{e^{i\omega_{nk}t'+\Gamma_{nk}t'-i\omega t'}\Big|_{t_0}^{t}}{i\omega_{nk}+\Gamma_{nk}-i\omega},$$

where the upper limit gives us $C_{nk}^{(1)}(t)$ at t, but the lower limit at t_0 must be treated carefully. In the beginning, the external field $\mathbf{E}(t)$ is off, so $C_{nk}^{(1)}(t_0)$ must be zero. Mathematically, we set t_0 to $-\infty$ and assume $\Gamma_{nk} > 0$, so $C_{nk}^{(1)}(t_0 = -\infty) = 0$. Our final $C_{nk}^{(1)}(t)$ is

$$C_{nk}^{(1)}(t) = \frac{\mathbf{D}_{nk}\cdot\mathbf{E}_0 e^{i\omega_{nk}t+\Gamma_{nk}t-i\omega t}}{\hbar(\omega_{nk}-i\Gamma_{nk}-\omega)}.$$

7.4.3.2 Susceptibility

We substitute $C_{nk}^{(1)}(t)$ into eq. (7.4.4) and cancel two factors, $e^{i\omega_{nk}t}$ and $e^{-i\omega_{nk}t}$, to obtain

$$\langle\psi^{(0)}|\hat{\mathbf{P}}|\psi^{(1)}\rangle = \frac{1}{\hbar\Omega}\sum_n\frac{\mathbf{D}_{kn}\mathbf{D}_{nk}\cdot\mathbf{E}_0e^{-i\omega t}}{\omega_{nk}-\omega-i\Gamma_{nk}} = \frac{1}{\hbar\Omega}\sum_n\frac{\mathbf{D}_{kn}\mathbf{D}_{nk}\cdot\mathbf{E}}{\omega_{nk}-\omega-i\Gamma_{nk}}. \tag{7.4.6}$$

Now, we compare it with our definition (eq. (7.4.1)) to identify $\chi_{\alpha,\beta}^{(1)}(\omega)$ in its component format as

$$\chi_{\alpha,\beta}^{(1)}(\omega) = \frac{1}{\hbar\epsilon_0\Omega}\sum_n\frac{D_{kn}^{\alpha}D_{nk}^{\beta}}{\omega_{nk}-\omega-i\Gamma_{nk}}, \tag{7.4.7}$$

where α and β denote the Cartesian coordinates x, y, and z. It is important to emphasize that the order of $\mathbf{D}_{kn}$ and $\mathbf{D}_{nk}$ matters because the second $\mathbf{D}$ is dot-producted with the field, so its direction β is really the direction of the incident external field. On the other hand, α is the direction of the outgoing field. $\chi_{\alpha,\beta}^{(1)}(\omega)$ has both real and imaginary parts. If $\omega = \omega_{nk}$, we have a resonance, with the peak broadening determined by Γ. Figure 7.3(a) shows an example, where we choose the energy difference of 1.5 eV. The dashed line is $\chi_I^{(1)}(\omega)$ and represents an absorption of light and has a peak at $\hbar\omega = 1.5$ eV. The solid line is $\chi_R^{(1)}(\omega)$, and represents the reflection. $\chi_I^{(1)}(\omega)$ and $\chi_R^{(1)}(\omega)$ are linked through the Kramer–Kronig relationship (see the problem).

We can work out a similar formula for the emission (see the exercise) and add up the absorption and emission to obtain

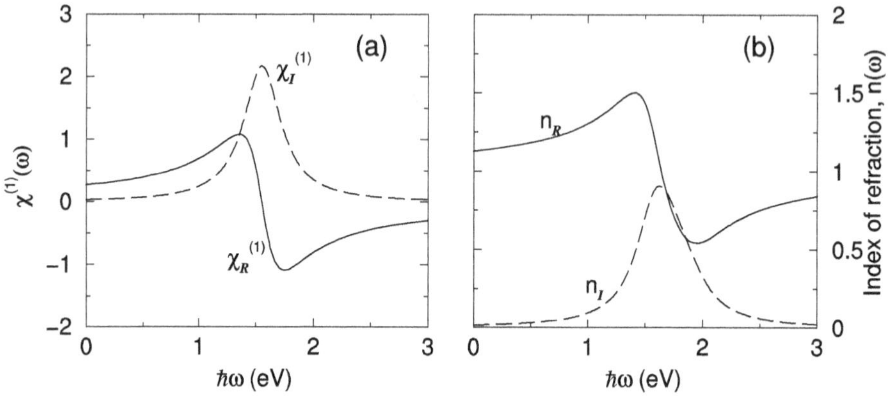

Figure 7.3: (a) Real and imaginary parts of $\chi^{(1)}$. The imaginary part is positive, with a peak at the same location as the kink in the real part. (b) Real and imaginary parts of index of refraction n. Both are positive.

$$\mathbf{P}^{(1)} = \frac{1}{\hbar\Omega}\sum_n\left(\frac{\mathbf{D}_{kn}\mathbf{D}_{nk}}{\omega_{nk}-\omega-i\Gamma_{nk}}+\frac{\mathbf{D}_{nk}\mathbf{D}_{kn}}{\omega_{nk}+\omega+i\Gamma_{nk}}\right)\cdot\underbrace{\mathbf{E}_0e^{-i\omega t}}_{\mathbf{E}}, \tag{7.4.8}$$

which yields the susceptibility with both absorption and emission as

$$\chi_e^{(1)}(\omega) = \frac{1}{\epsilon_0\hbar\Omega}\sum_n\left(\frac{\mathbf{D}_{kn}\mathbf{D}_{nk}}{\omega_{nk}-\omega-i\Gamma_{nk}}+\frac{\mathbf{D}_{nk}\mathbf{D}_{kn}}{\omega_{nk}+\omega+i\Gamma_{nk}}\right). \tag{7.4.9}$$

If $\omega_{nk} > 0$, the first term refers to the absorption and the second one the emission. The reader can that check $\chi_e^{(1)}$ is dimensionless.

Exercise 7.4.3.

15. Starting from eq. (7.4.9), show $\boldsymbol{\chi}_e^{(1)}$ is dimensionless.
16. Prove the first-order polarization $\mathbf{P}^{(1)}$ is real. Hint: Start with $\mathbf{P}^{(1)} = \langle\Psi^{(0)}|\hat{\mathbf{P}}|\Psi^{(1)}\rangle + \langle\Psi^{(1)}|\hat{\mathbf{P}}|\Psi^{(0)}\rangle$.
17. A system is under a periodic perturbation of $\mathbf{E}(t) = \mathbf{E}_0e^{-i\omega t}$. The polarization expectation value is $\langle\Psi^{(0)}|\hat{\mathbf{P}}|\Psi^{(1)}\rangle = \frac{1}{\Omega}\sum_n C_n^{(1)}e^{i\omega_{kn}t}\mathbf{D}_{kn}$. (a) It is known that $C_n^{(1)}(t) = \frac{1}{i\hbar}\int_{t_0}^{t}(-1)\mathbf{D}_{nk}\cdot\mathbf{E}(t')e^{i\omega_{nk}t'}dt'$. Integrate it to find $C_n^{(1)}(t)$. (b) In (a), the integral does not converge. We need to introduce the lifetime, or damping term for the excited states. Which one, $e^{-\Gamma t'}$ or $e^{+\Gamma t'}$, can converge the integral?

7.4.4 Experimental connection: penetration depth and Beer–Lambert law

To connect with the experimental observation, consider that a beam of light impinges on a sample. Before the light hits the sample, the light intensity I_0 (SI units of $\frac{\mathrm{W}}{\mathrm{m}^2}$) is[8]

---

8 The intensity is defined as the power per unit area. The fluence is defined as the energy per unit area. To compute the fluence, we need to integrate the intensity over time. Here, 1/2 is from the time integration.

$$I_0 = \frac{1}{2} c\epsilon_0 n_0(\omega)|\mathbf{E}_0|^2 \quad \text{(in vacuum)},$$

where $\mathbf{E}_0$ is the light electric field, with the index of refraction $n_0(\omega) = 1$ in a vacuum. Suppose that the light is propagating along the z-axis with the polarization along the x-axis, so $E_0(z,t) = A_x e^{i(k_z^{(0)} z - \omega t)}$, where $k_z^{(0)}$ is the wavevector in a vacuum, ω is the angular frequency, and $|\mathbf{E}_0|^2 = A_x^2$. After the light impinges the sample, some of the light is absorbed, or reflected, or transmitted through the sample, so the light intensity is reduced from I_0 to I. Inside the medium, the intensity is given by

$$I = \frac{1}{2} c\epsilon_0 n(\omega)|\mathbf{E}|^2 \quad \text{(in medium)},$$

where $\mathbf{E}$ is the field in the medium. To simplify our derivation, we further assume that $\mathbf{E}$ polarizes in the same direction as $\mathbf{E}_0$, so its $E(z,t) = A_x e^{i(k_z z - \omega t)}$, where k_z is changed from $k_z^{(0)}$, but ω remains unchanged as discussed in Chapter 1. To find k_z, recall that the wavevector is connected to the wavelength and the index of refraction $n(\omega)$ through

$$k(\omega) = \frac{2\pi}{\lambda} = \frac{2\pi\nu}{v} = \underbrace{\frac{2\pi\nu}{c/n(\omega)}}_{\text{Definition of } n(\omega)} = \frac{\omega}{c} n(\omega).$$

Since $n = \sqrt{\epsilon_r}$ and $\epsilon_r = 1 + \chi_e^{(1)}(\omega)$, the index of refraction is $n(\omega) = \sqrt{1 + \chi_e^{(1)}(\omega)}$ whose real and imaginary parts, n_R, and n_I, are both positive as can been in Fig. 7.3(b). Because $n(\omega) = n_R(\omega) + i n_I(\omega)$, $k(\omega)$ also has both real and imaginary parts.

We find $E(z,t) = A_x e^{i(k_z^R z + i k_z^I z - \omega t)} = A_x e^{i(k_z^R z - \omega t)} e^{-k_z^I z}$, so $|E(z,t)|^2 = A_x^2 e^{-2k_z^I z}$. The intensity in the medium is reduced from I_0 to

$$I(z) = \frac{1}{2} c\epsilon_0 n(\omega)|\mathbf{E}_0|^2 e^{-2k_z^I z},$$

which is now a function of z. Here, we have used $|A_x^2| = |\mathbf{E}_0|^2$. As the light intensity drops to $\frac{1}{e}$ of the initial intensity, this z is defined as the light penetration depth λ_{pen},

$$e^{-2k_z^I \lambda_{\text{pen}}} = e^{-1} \rightarrow \lambda_{\text{pen}} = \frac{1}{2k_z^I} = \frac{c}{2 \times \omega n_I(\omega)} = \frac{c}{4\pi\nu n_I(\omega)} = \frac{\lambda_0}{4\pi n_I(\omega)},$$

where we have used $k_z^I = \frac{\omega}{c} n_I(\omega)$ and λ_0 is the wavelength in vacuum. λ_{pen} is a measurable quantity.

If we rewrite I in terms of I_0, we have the Beer–Lambert law,

$$I(z) = n(\omega) I_0 e^{-az}.$$

a (in units of 1/m) is called the absorption coefficient,

$$a(\omega) = \frac{4\pi n_i(\omega)}{\lambda_0} = \frac{4\pi}{\lambda_0} \text{Im}(\sqrt{1 + \chi_e^{(1)}(\omega)}).$$

Since χ_I is positive, $\alpha(\omega)$ is always positive. It is clear that the absorption $\alpha(\omega)$ is related to the imaginary part of $\chi_e^{(1)}(\omega)$. Since $\chi_e^{(1)}(\omega)$ is inversely proportional to $(\omega_{nk} - \omega + i\Gamma_{nk})$ (see eq. (7.4.9)), when ω approaches ω_{nk}, $\chi_e^{(1)}(\omega)$ becomes very large and so does $\alpha(\omega)$. In this case, the material strongly absorbs light, and $I(z)$ decays very quickly. On the other hand, if ω is away from the frequency difference between two energy states, $\alpha(\omega)$ becomes very small. This happens in a transparent material. Therefore, we reveal a crucial physical insight. The absorption significantly depends on whether the incident photon energy matches the energy difference between the eigenstates.

Exercise 7.4.4.

18. Use $n = \sqrt{1 + \chi_e^{(1)}(\omega)}$ to prove n_r and n_i are both positive.

19. Pure gold has the index of refraction $n = 0.18 + 4.91i$ at 780 nm [46]. (a) Find its $\chi_e^{(1)}(\omega)$. (b) Find the penetration depth at this wavelength. (c) Find the absorption coefficient. (d) Use the Beer–Lambert law to plot the light intensity change as a function of z.

20. In a Heusler compound Mn_2RuGa, there are two eigenstates $|a\rangle$ and $|b\rangle$, with eigenenergies of $E_a = 0.5578$ Ry and $E_b = 0.672$ Ry, respectively. Note that 1 Ry is 13.6 eV. The momentum matrix element is $\langle a|\hat{p}_x|b\rangle = 0.03493 - 0.01616i$ in the unit of $\hbar/a_0$, where $\hbar$ is the reduced Planck constant, $h/2\pi$, and a_0 is the Bohr radius. (a) Use the relationship $\langle a|\hat{x}|b\rangle = \frac{i\hbar}{m_e}\frac{\langle a|\hat{p}_x|b\rangle}{E_b - E_a}$ to compute $\langle a|\hat{x}|b\rangle$ and then the dipole matrix element $\langle a|\hat{D}_x|b\rangle$ in units of Cm. (b) The volume of Mn_2RuGa is 358.97158 a_0^3. Just use these two eigenstates to compute $\chi^{(1)}(\omega)$ as a function of the photon energy $\hbar\omega$ in units of eV. (c) Plot and compare your figure with Fig. 7.3.

7.5 Laser

Laser stands for Light Amplification of Stimulated Emission of Radiation. In 1960, Theodore Maiman was the first to develop a solid-state ruby laser [47, 48]. The ruby rod is 10-cm long and has diameter of 0.8 cm and is surrounded by a helical xenon flashtube. Ruby is made of Al_2O_3 with chromium impurities. When he irradiated the ruby rod, he discovered a pulse of red light of wavelength 694.3 nm emitting through the partially silvered mirror. This discovery changed the world forever. Lexicologically, lase is a verb. A lasing material such as ruby, is also called the gain medium. This section aims to explain what the light amplification is and why one needs a stimulated emission.

7.5.1 Absorption and emission

In electronics, transistors are used to amplify electric signals. In photonics, a laser is used to amplify light signals. Both transistors and lasers are quantum mechanical, but work differently. In order to understand the working principles of the laser, first we need to explain how the light interacts with a medium. Consider a beam of light incident on a material. The light can be reflected, absorbed, or simply transmitted through the medium. Figure 7.4(a) shows the reflection and transmission from the medium.

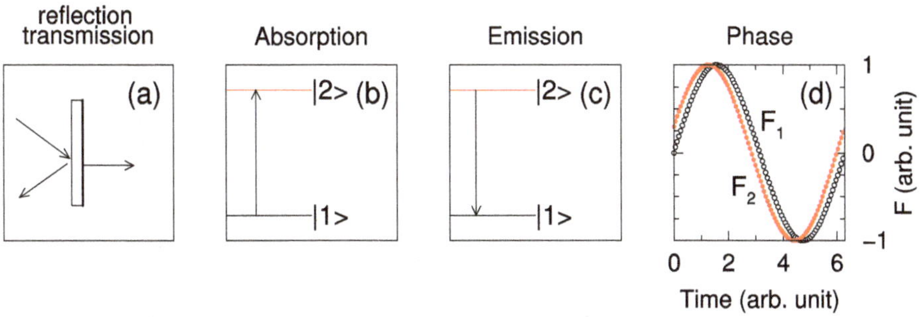

Figure 7.4: (a) Reflection and transmission of light through a medium. (b) During absorption an electron absorbs a light photon and makes a transition from a lower level $|1\rangle$ to a higher level $|2\rangle$. (c) Emission is reversed where the electron is initially in the excited state $|2\rangle$ and transitions back to $|1\rangle$, with a photon emitted. (d) The emitted light intensity depends on the phase between two light fields emitted from the two separate atoms.

However, central to the laser is the absorption (Fig. 7.4(b)) and emission (Fig. 7.4(c)). During absorption, if the incident photon energy $h\nu$ matches the gap $\Delta E = E_2 - E_1$, an electron in state $|1\rangle$ absorbs a photon and transitions to state $|2\rangle$. In a laser, the initial excitation to a high level may not be optically driven. In semiconductor lasers, it is electric bias. The emission is opposite, where the electron jumps to a lower state $|1\rangle$ by emitting a photon (Fig. 7.4(c)). We denote the electric field of the emitted light from atom 1 by $\mathbf{E}_1$. If we have two identical atoms, the total electric field is $\mathbf{E}_1 + \mathbf{E}_2$, where its time-dependence is omitted for simplicity. According to the classical EM, the light intensity is defined as

$$I = \frac{1}{2}c\epsilon_0|\mathbf{E}_{\text{tot}}|^2 = \frac{1}{2}c\epsilon_0(|\mathbf{E}_1|^2 + |\mathbf{E}_2|^2 + 2\mathbf{E}_1 \cdot \mathbf{E}_2). \tag{7.5.1}$$

If the emission is spontaneous among atoms, the phases of the fields from different atoms are random (see Fig. 7.4(d)), so $\mathbf{E}_1 \cdot \mathbf{E}_2$ is time-averaged to zero. This shows $I = I_1 + I_2 = 2I_1$. If we have N atoms, $I = NI_1$. This does not represent a gain.

If the emitted light fields all have the same phase, $\langle \mathbf{E}_1 \cdot \mathbf{E}_2 \rangle = |\mathbf{E}_1|^2 = |\mathbf{E}_2|^2$ and $I_{\text{tot}} = I_1 + I_2 + 2I_1 = 4I_1$. For N atoms, $I = N^2 I_1$, a huge gain. But, how can one synchronize these fields?

7.5.2 Population inversion and stimulated emission

A typical laser consists of a cavity and a lasing medium. At both ends of the cavity are two reflecting mirrors, one completely and one partially silvered (Fig. 7.5(a)). Surviving light waves inside the cavity are standing waves, whose wavelength is given by

$$\lambda_n = \frac{L}{2n}, \tag{7.5.2}$$

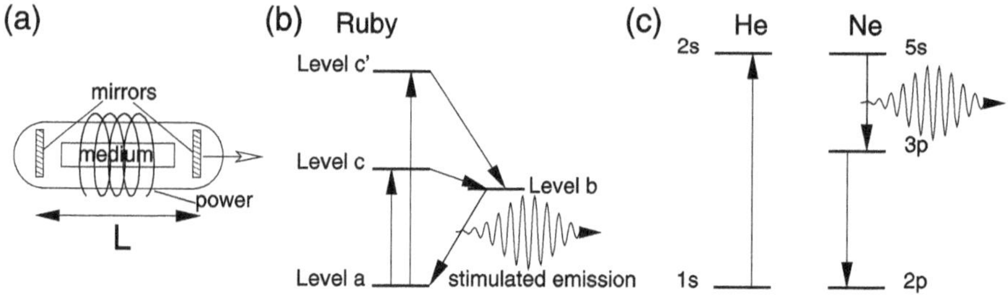

Figure 7.5: (a) Laser cavity consists of a lasing medium, an external power, and two reflecting mirrors, one of which is only partially reflecting so the laser light can emit. The cavity length is L. (b) Three-level systems. Here, ruby is used as the example, where the levels are a, b, and c. (c) Four-level system of the He–Ne laser.

where n takes a positive integer and L is the cavity length. Since L is usually much larger than the light wavelength λ_n, n can be a big number. As will be seen, the light amplification is limited to those wavelengths.

The lasing medium determines the type of a system. A two-level system cannot lead to lasing because statistically there is a higher population in a lower state than in a higher state, and there is no way to maintain the electron in a higher state. A minimum system must have three levels. Figure 7.5(b) shows a three-level system in ruby. All the atoms have exactly the same level scheme. First, an external field, such as a flash lamp, excites electrons from level a to c or c' with the very short lifetime of 10 ns. Then, the electrons transition down to level b (b is a metastable state), with a lifetime of 5 ms,[9] where electrons can stay for a while, without transiting back to a immediately. It is much like a waiting room. During this waiting period, unexcited atoms get excited so their electrons proceed to level b. This creates a situation, where more atoms have electrons on the high level b than those in the lower level a. This is called population inversion.

Now, suppose that atom 1 undergoes spontaneous emission from b to a. The emitted light starts the clock and is going to stimulate emission in other atoms. But there is no guarantee that it will stimulate others to radiate. This is where two reflecting mirrors come in. The light from atom 1 is reflected back and forth from the mirrors and passes through other atoms many times since the cavity length is very short, until the rest of the atoms are stimulated to irradiate almost simultaneously. Because almost all the atoms radiate in phase, this generates a coherent (in phase) light emission, lasing.

The emitted photon energy is $\Delta E = E_b - E_a$, so its wavelength is

$$\lambda = \frac{hc}{\Delta E} = \frac{hc}{E_b - E_a}. \tag{7.5.3}$$

However, as already stated, not all the wavelengths can survive in the cavity. λ must match one of $\lambda_n = \frac{L}{2n}$. Therefore, one type of medium corresponds to one type of cavity

9 If such a level does not exist, then we do not have lasing.

length. In ruby, only the light of wavelength 694.3 nm is amplified, the rest is suppressed. This is quite different from the electric amplification in transistors, where all signals, regardless of their frequencies, are amplified.

If we want to amplify light of different wavelengths, we must find a different lasing medium. There are other types of systems with four levels. Figure 7.5(c) is a 4-level system, representing the He-Ne laser, where the pumping prompts electrons from the He 1*s*-state to its 2*s*-state. Through the collision between He and Ne atoms, the electron occupies Ne's 5*s*-state. The lasing occurs between Ne's 5*s*- and 3*p*-states. Because there is no electron in the 3*p*-state, the population inversion is readily achieved. The He–Ne laser radiates 632.8 nm.

Exercise 7.5.2.

21. The ruby laser emits red light of 694.3 nm. Find the energy difference between levels *a* and *b* in Fig. 7.5.
22. Suppose a ruby rod is 10-cm long. Find the number of the possible waves for the red light of 694.3 nm.

7.6 Problems

1. Starting from eqs. (7.1.13), (7.1.14), (7.1.15), (a) prove the following equations: [29]

$$\langle n', l+1, m+1|x+iy|n,l,m\rangle = -\mathcal{J}\sqrt{\frac{(l+m+2)(l+1+m)}{(2l+1)(2l+3)}},$$

$$\langle n', l+1, m+1|x-iy|n,l,m\rangle = 0,$$

$$\langle n', l+1, m+1|z|n,l,m\rangle = \mathcal{J}\sqrt{\frac{(l+1)^2-m^2}{(2l+1)(2l+3)}},$$

$$\langle n', l+1, m-1|x+iy|n,l,m\rangle = 0,$$

$$\langle n', l+1, m-1|x-iy|n,l,m\rangle = \mathcal{J}\sqrt{\frac{(l+1-m)(l+2-m)}{(2l+1)(2l+3)}}.$$

(b) Prove $\frac{1}{2}|\langle n', l+1, m+1|(x+iy)|n,l,m\rangle|^2 + \frac{1}{2}|\langle n', l+1, m-1|(x-iy)|n,l,m\rangle|^2 + |\langle n', l+1, m+1|z|n,l,m\rangle|^2 = \mathcal{J}^2\frac{l+1}{2l+1}$, independent of m.

2. Upon being radiated with light, an electron in the hydrogen atom is knocked out from the 1*s*-orbital to the vacuum. The electron in the vacuum has a continuous wavefunction $|k'\rangle = e^{-i\mathbf{k}'\cdot\mathbf{r}}/\sqrt{\Omega}$, where $\mathbf{k}'$ is the wavevector of the outgoing electron and Ω is the volume. Without making the dipole approximation of the light field, the transition matrix element is $\langle k'|e^{i\mathbf{k}\cdot\mathbf{r}}|1s\rangle$, where $|1s\rangle$ is the electron 1*s*-wavefunction and $\mathbf{k}$ is the incident photon wavevector. Show the matrix element is $\psi_0(q)/\sqrt{\Omega}$, where $\psi_0(q) = 8\sqrt{\pi}a_0^{3/2}/(1+q^2a_0^2)^2$ and $\mathbf{q} = \mathbf{k} - \mathbf{k}'$. This problem is adopted from Gottfried and Yang [49], with some modifications.

3. An electron of m_e and $-e$ is in the ground state of the one-dimensional infinite square well of length a, which is centered at $x = 0$. An electric field is suddenly

applied along the x axis, $E(t) = \mathcal{E}_0\delta(t)$. (a) Based on the parity and within the dipole approximation, derive the selection rule for the electric field. (b) Find the probability in the first and second odd states.

4. An electric field pulse is applied to a ground-state hydrogen atom. The pulse starts at $t = 0$ and decays as $\mathbf{E}(t) = \mathcal{E}_0 e^{-t\tau}\hat{r}$, where τ is a decay constant and $\hat{r}$ is the unit vector along the radial direction. (a) Take the dipole approximation and write down the interaction Hamiltonian, $\hat{H}_I(t)$. (b) Use perturbation theory to find the probability that the electron is in $|2s\rangle$, $|2p_x\rangle$, $|2p_y\rangle$ and $|2p_z\rangle$ at time $t \to \infty$.
5. A periodic electric field $\mathbf{E}(t) = \mathcal{E}_0 e^{i\omega t}$ is applied to the one-dimensional infinite well of length a, which is centered at $x = 0$. Initially, the electron is in the ground state. Compute the susceptibility $\chi^{(1)}(\omega)$.
6. A charged (q) one-dimensional harmonic oscillator is in the ground state. A small electric field, $E(t) = \frac{E_0}{\sqrt{\pi}\tau} e^{-t^2/\tau^2}$, is applied to the oscillator. (a) Use first-order perturbation theory to find the probability in all the states. (b) Change $E(t)$ to $E(t) = E_0 e^{i\omega t}$ and find the susceptibility $\chi^{(1)}(\omega)$.
7. A hydrogen atom is in the 2s state. Suddenly, we apply an electric field pulse along the x axis, $E_x = E_0\delta(t)$, where E_0 is the amplitude of the field. (a) Find the probabilities in each eigenstate. (b) Find the probability for the electron remaining in the ground state.
8. An electron is initially in a state $\phi_i = e^{ikx}/\sqrt{L}$ with $k > 0$, normalized in a line segment, $x \in (0, L)$, and is perturbed by a potential $V(x,t) = V_0\delta(t)$ for $x \in (L/2, L)$, where V_0 is positive. (a) Find the probability for the electron in $\phi_f = e^{iqx}/\sqrt{L}$. (b) What is the probability for the electron bouncing back? Hint: Choose $q < 0$.
9. (a) Use the expression of $\langle\psi^{(0)}|\hat{\mathbf{P}}|\psi^{(1)}\rangle$ to find $\langle\psi^{(1)}|\hat{\mathbf{P}}|\psi^{(0)}\rangle$ which represents an emission process. (b) In the emission process, the excited state is occupied in the beginning. This is equivalent to the time-reversal, or the frequency changes sign from $+\omega$ to $-\omega$. Use this fact to prove the susceptibility for the emission is

$$\chi_e^{(1)}(\omega) = \frac{1}{\epsilon_0\hbar\Omega}\sum_n \frac{\mathbf{D}_{nk}\mathbf{D}_{kn}}{\omega_{nk} + \omega + i\Gamma_{nk}}.$$

10. (a) Prove the Kramers–Kronig relationship for susceptibility χ,

$$P.V. \int_{-\infty}^{\infty} \frac{\chi(\omega')}{\omega' - \omega} d\omega' = i\pi\chi(\omega),$$

where the integral is over the principle value only. (b) Prove the real and imaginary parts of χ are related to each other through

$$\chi_I(\omega) = -\frac{2\omega}{\pi} P \int_0^{\infty} \frac{\chi_R(\omega)}{(\omega')^2 - \omega^2} d\omega', \quad \chi_R(\omega) = \frac{2\omega}{\pi} P \int_0^{\infty} \frac{\chi_I(\omega)}{(\omega')^2 - \omega^2} d\omega'.$$

11. An electron is in the ground state (1s) of the hydrogen atom. (a) Find $\chi^{(1)}(\omega)$ with the highest states up to 2s and 2p. (b) Plot $\chi^{(1)}(\omega)$ as a function of $\hbar\omega$, where $\hbar\omega$ starts from 0 to 20 eV.
12. The Balmer series covers the visible wavelength, which is the reason why it was discovered first. Here, the emission is from a high level to $n = 2$. (a) Find $\chi^{(1)}(\omega)$ at the lowest energy $\hbar\omega$ for the first line. (b) Find $\chi^{(1)}(\omega)$ at the lowest energy $\hbar\omega$ for the second line. (c) Compare their magnitudes. Hint: You need to compute the transition matrix elements first for several different light polarization $(\sigma^{\pm}, \pi)$.
13. An electron is confined in the one-dimensional infinite quantum well $x \in (0, a)$. (a) Show their matrix elements $\langle\phi_n|\hat{x}|\phi_m\rangle$ and $\langle\phi_n|\hat{p}_x|\phi_m\rangle$ (see Section 2.8) obey the Hellmann–Feynman theorem $(E_m - E_n)\langle\phi_n|\hat{x}|\phi_m\rangle = \frac{i\hbar}{m_e}\langle\phi_n|\hat{p}_x|\phi_m\rangle$. (b) Assume the electron initially is in the ground state $n = 1$. Find the expression for $\chi^{(1)}(\omega)$. (c) Compute the real and imaginary parts of the index of refraction.
14. $\phi_n(x)$ and $\phi_m(x)$ are two eigenstates of the one-dimensional harmonic oscillator. (a) Compute the transition matrix elements $\langle n|\hat{x}|m\rangle$ and $\langle n|\hat{p}_x|m\rangle$ (see Section 3.3.3). (b) Find $\chi^{(1)}(\omega)$.

8 Electrons in molecules and one-dimensional solids

The progress in applying using quantum mechanics to atoms and small molecules has encouraged us to advance toward larger and more complicated systems. However, solids have 10^{23} atoms and many more electrons. DNA double helices have a radius of about 10 Å, and as long as 34 Å per full helical turn. Each complete turn consists of about 10 base pairs. Each pair consists of either adenine (A), thymine (T), or cytosine (C)–guanine (G), each of which consists of dozens of atoms attached to the sugar-phosphate backbone. The length of human DNA is close to 1 m. Charge transfer in the photosynthesis reaction center is a real-time dependent process. These structures do not have a symmetry to be exploited. A quantum-mechanical description of this size is difficult and still an active research field. For these reasons, we will employ smaller molecules and crystalline solids with orderly structures. Even within these systems, we encounter many new interactions. While the electrons still interact with other electrons and protons in the same atom, they also interact with those from other atoms. A direct treatment of these interactions requires a self-consistent approach that falls beyond the scope of this chapter.

This chapter can be grouped into three units:

- Unit 1 consists of Sections 8.1 and 8.2, which introduce molecular orbitals and hybridization.
- Unit 2 consists of Sections 8.3 and 8.4 on the two molecules: allyl and polyacetylene, where both molecular orbital wavefunctions and molecular eigenvalues are introduced.
- Unit 3 consists of Section 8.5, an introduction to the 1D solid, where we provide a significant amount of detail of all the derivations, going beyond that in other books. If there is not sufficient time for a course to cover Chapter 9, this section serves as a good overview of solids.

8.1 Molecular orbitals

Molecules are made of atoms. When atoms move closer, their mutual interaction lowers the total energy, below the energy sum of isolated atoms. As a result, forming bonds between atoms becomes energetically favorable, with atomic orbitals (AOs) forming a molecular orbital (MO). Chemical intuition helps to guide us to develop the machinery. This section introduces the basic framework for the quantum-mechanical treatment of molecules. We consider the σ-orbitals first, then π-orbitals, and finally orbital hybridization. Different from atoms, the quantization axis is no longer arbitrary and not simple. For orbitals between two atoms, the quantization axis can be chosen along the line connecting these two atoms. If more than two atoms contribute to bonding, a unique quantization axis is not possible, so the concept of σ- and π-orbitals becomes ill-defined because of orbital hybridization.

https://doi.org/10.1515/9783110672152-008

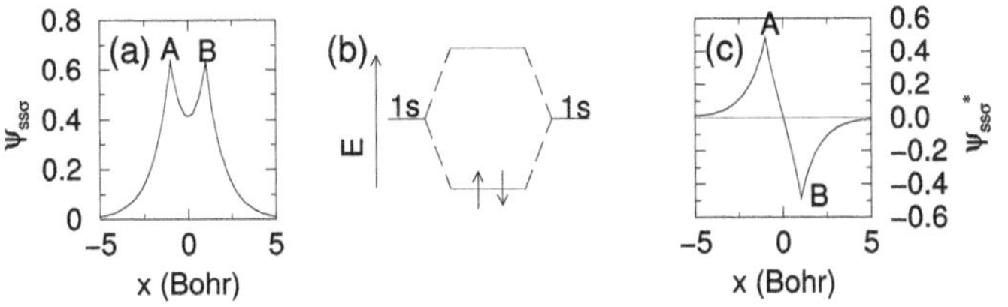

Figure 8.1: (a) σ bonding orbital in H_2, where the charge density appears along the line linking two atoms. (b) How two atomic orbitals form a molecular orbital. (c) σ^* anti-bonding orbital in H_2.

8.1.1 σ-orbitals

σ-orbitals are formed when two atomic orbitals approach each other along the line that links two atoms. In H_2, the molecular σ-wavefunction is formed from two s-orbitals as,

$$\psi_{ss\sigma} = \frac{1}{\sqrt{2}}(\phi_s^{\mathrm{at}}(A) + \phi_s^{\mathrm{at}}(B)), \tag{8.1.1}$$

where $\phi_s^{\mathrm{at}}(A)$ is the s-orbital on A and $\phi_s^{\mathrm{at}}(B)$ is the s-orbital on B. $\frac{1}{\sqrt{2}}$ is the normalization constant. In Dirac notation, we write it as $|\psi_{ss\sigma}\rangle = (|A\rangle + |B\rangle)/\sqrt{2}$. Figure 8.1(a) shows that the σ-orbital of H_2 has charge density between two H atoms. Energetically, two electrons occupy the lower MO (see Fig. 8.1(b)). The higher MO, the antibonding orbital, is shifted upward energetically, with the wavefunction

$$\psi_{ss\sigma^*} = \frac{1}{\sqrt{2}}(\phi_s^{\mathrm{at}}(A) - \phi_s^{\mathrm{at}}(B)), \tag{8.1.2}$$

which has a node (zero charge density) between two atoms crossing the zero line (Fig. 8.1(c)).

Figure 8.2(a) shows that spatially s-orbitals at A and B atoms overlap, so the charge density is present along the AB line. This is called the $ss\sigma$-orbital and the bond is called the σ-bond. The Greek letter σ has two meanings: First, it is an atomic equivalent of the s-orbital; second, it is symmetric with respect to the bond direction. If we choose this bond direction as a quantization axis for the molecule,[1] the molecular orbital momentum quantum number is $M = 0$, just as $m = 0$ for atoms. Therefore, all σ-orbitals have a zero magnetic quantum number.

Different from s-orbitals, p-orbitals are not spherically symmetric. If we have an s-orbital and a p-orbital, there are many possible arrangements between them. One example is shown in Fig. 8.2(b). Here the overlap of two orbitals is along the AB line. Although the p-orbital at atom B is along the horizontal direction, this orbital is still called

1 This is the z-axis for the molecule, though the AB line is along the x-axis in the figure. This demonstrates that, once we have a multiple-atom system, the quantization axis must be chosen along a high-symmetry direction, in order to simplify our analysis.

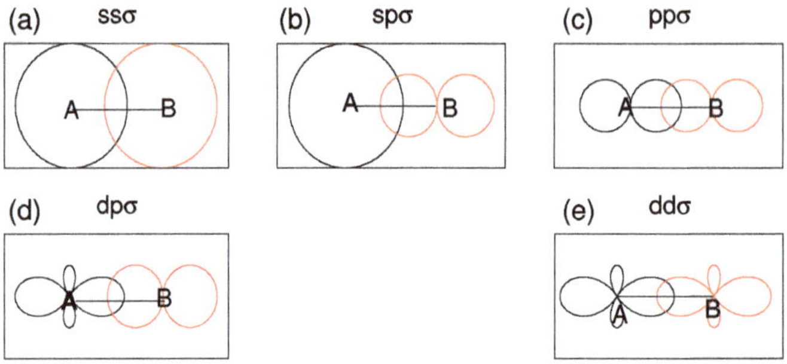

Figure 8.2: Molecular orbitals. Atoms A and B are linked with a line which is defined as the bond direction. σ-bonds have charge density along the line, but π-bonds have the charge density above and below the line. (a) σ-orbital is formed by two s-orbitals. (b) σ-orbital is formed by s- and p-orbitals. (c) σ-orbital is formed by p- and p-orbitals. (d) σ-orbital is formed by d- and p-orbitals. (e) σ-orbital is formed by d- and d-orbitals.

p_z, not p_x, because the horizontal direction is our quantization axis, or our z-axis. The molecular wavefunction for this case is

$$\psi_{sp\sigma} = \frac{1}{\sqrt{2}}(\phi_s^{\text{at}}(A) + \phi_{p_z}^{\text{at}}(B)). \tag{8.1.3}$$

Since the magnetic orbital angular-momentum quantum number for s- and p_z-orbitals is $m_1 = m_2 = 0$, the total orbital quantum number of the molecule is $M = m_1 + m_2 = 0$, corresponding to a σ-orbital, denoted as $sp\sigma$, to differentiate it from the previous $ss\sigma$-orbital. We can use the same method to form many different σ-orbitals, such as $pp\sigma$-orbital in Fig. 8.2(c), $dp\sigma$-orbital in Fig. 8.2(d), and $dd\sigma$-orbital in Fig. 8.2(e).

8.1.2 π-orbitals

If we put two p-orbitals side by side, then we have a π-orbital (see Fig. 8.3(a)). The π-orbital has no charge density along the AB line, but has density above and below the line (Fig. 8.3(b)), and in 3D is a torus along the AB line. Its molecular wavefunction is

$$\psi_{pp\pi} = \frac{1}{\sqrt{2}}(\phi_{p_z}^{\text{at}}(A) + \phi_{p_z}^{\text{at}}(B)). \tag{8.1.4}$$

Different from σ orbitals, π MOs can be in different planes. Figure 8.3(a) is in the yz plane, while that in Fig. 8.3(b) is in the xy plane.

8.1.3 Orbital hybridization and linear combination of atomic orbitals

σ- and π-orbitals are a result of atomic orbital hybridization. In molecules with more than an atom, the molecular wavefunction is often a linear combination of atomic or-

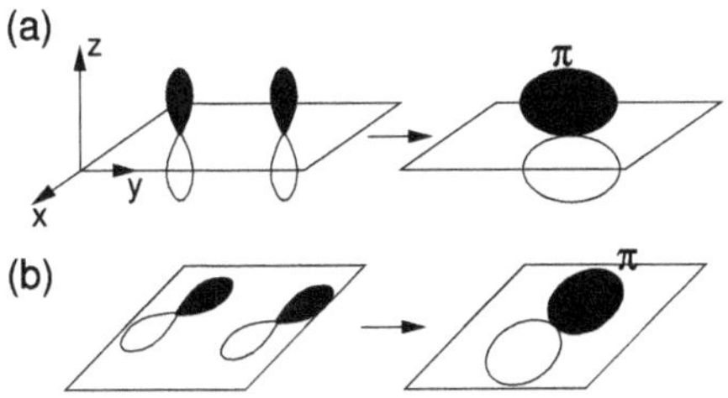

Figure 8.3: (a) π MO is in the yz plane. (b) π MO is in the xy plane.

bitals (LCAOs). For instance, in benzene, the molecular orbital at minimum consists of six AOs. These orbitals are highly delocalized. In chemistry, these orbitals are called conjugated molecular orbitals. The level of delocalization in conjugated orbitals is in general less extended than the plane wave encountered in metals, except for some synthetic metals such as polyacetylene [50].

LCAO expresses the MO as

$$\psi_{\text{mol}} = \sum_{i=1}^{N} \sum_{\alpha=1}^{M} c_{i\alpha} \phi_{i\alpha}^{\text{at}}, \tag{8.1.5}$$

where i is the atomic index, α is the orbital index, and $c_{i\alpha}$ is the coefficient to be found below. We show one example how hybridization enters our wavefunction.

Consider a sp^2-hybridization for two atoms. Its wavefunction is

$$\psi_{sp^2} = c_1 \phi_s^{\text{at}}(A) + c_2 \phi_{p_x}^{\text{at}}(B) + c_3 \phi_{p_y}^{\text{at}}(B), \tag{8.1.6}$$

which includes s-, p_x-, and p_y-orbitals. They are called sp^2-hybridization, but the level of hybridization is determined by c_i. If we set $c_3 = 0$ and change p_x to p_z, we have a standard sp-hybridization and $sp\sigma$-orbital. Conceptually, LCAO is a simple method to describe MOs, and is very powerful. While we choose AOs as our basis functions for MOs to describe complex molecular orbitals, we can include extra orbitals. This is because what we are doing is to use the variational principle to convert a difficult molecular Schrödinger equation, a complicated Schrödinger equation in three-dimensional space, to a matrix that is formed by AOs, or more precisely the basis functions, matrix mechanics. In the following, we explain how to form such a molecular Hamiltonian matrix.

Exercise 8.1.3.
1. For a diatomic molecule, write down the MO wavefunction for the sp^3-hybridization.
2. Explain why we have to change p_x to p_z when we want to describe the sp-hybridization and $sp\sigma$-orbital.
3. What are the main differences between the σ- and π-orbitals?

8.2 Molecular Hamiltonian, overlap and hopping integrals

The molecular Hamiltonian for $N_{\rm at}$ atoms and M orbitals is given by

$$\hat{H}_{\rm mol} = \sum_{i=1}^{N_{\rm at}} \hat{H}_i^{\rm at} + \frac{1}{2}\sum_{i=1}^{N_{\rm at}}\sum_{j=1}^{N_{\rm at}} U_{ij}, \tag{8.2.1}$$

where $\hat{H}_i^{\rm at}$ is the native atomic Hamiltonian of atom i and U_{ij} is the mutual interaction between atoms i and j that provides the attraction to form molecules. $\frac{1}{2}$ is added to eliminate double counting. We start from eq. (8.1.5) with our molecular wavefunction assumed to be $\psi_{\rm mol} = \sum_{i=1}^{N_{\rm at}}\sum_{\alpha=1}^{M} c_{i\alpha}\phi_{i\alpha}^{\rm at}$.[2] Because i runs from 1 to $N_{\rm at}$ and α runs from 1 to M, we combine them into a single index I running from 1 to $N_{\rm at}M$. We employ the variational principle to minimize the total energy with respect to $c_{i\alpha}$ through

$$F[\psi] = \langle\psi_{\rm mol}|\hat{H}_{\rm mol}|\psi_{\rm mol}\rangle - \lambda(\langle\psi_{\rm mol}|\psi_{\rm mol}\rangle - 1)$$

to get an eigenequation spanned by those AOs,

$$\begin{pmatrix} H_{11} & H_{12} & \cdots & H_{1,N_{\rm at}M} \\ H_{12} & H_{22} & \cdots & H_{2,N_{\rm at}M} \\ \cdots & \cdots & \cdots & \cdots \\ H_{N_{\rm at}M,1} & H_{N_{\rm at}M,2} & \cdots & H_{N_{\rm at}M,N_{\rm at}M} \end{pmatrix}\begin{pmatrix} c_1 \\ c_2 \\ \vdots \\ c_{N_{\rm at}M}\end{pmatrix} = \lambda\begin{pmatrix} S_{11} & S_{12} & \cdots & S_{1,N_{\rm at}M} \\ S_{12} & S_{22} & \cdots & S_{2,N_{\rm at}M} \\ \cdots & \cdots & \cdots & \cdots \\ S_{N_{\rm at}M,1} & S_{N_{\rm at}M,2} & \cdots & S_{N_{\rm at}M,N_{\rm at}M} \end{pmatrix}\begin{pmatrix} c_1 \\ c_2 \\ \vdots \\ c_{N_{\rm at}M}\end{pmatrix}, \tag{8.2.2}$$

where $H_{IJ} = \langle\phi_{i\alpha}^{\rm at}|\hat{H}_{\rm mol}|\phi_{j\beta}^{\rm at}\rangle$ and $S_{IJ} = \langle\phi_{i\alpha}^{\rm at}|\phi_{j\beta}^{\rm at}\rangle$. The procedure is exactly the same as Subsection 5.6.1.

Whether and how atoms can form chemical bonds depends on the molecular Hamiltonian and overlap matrix elements in eq. (8.2.2). This is where AOs come in. AOs have different shapes and orientations, and, when they form MOs, their traits matter. Only s-orbitals are spherical, and the rest are not. Consider two atoms A and B separated by a distance R, with an s-orbital on A and p_x-orbital on B. They can have many different relative orientations. Figure 8.4(a) shows that both A and B are along the x-axis, and

2 The variational principle always starts from prior knowledge. This wavefunction is based on our prior discussion on the molecular orbital, but it may not be the best one. Here, $\phi_{i\alpha}^{\rm at}$ refers to the atomic wavefunction of atom i and orbital α. For instance, for the carbon atom, if we only include four $2s$ and $2p$ orbitals, then α runs from 1 to 4.

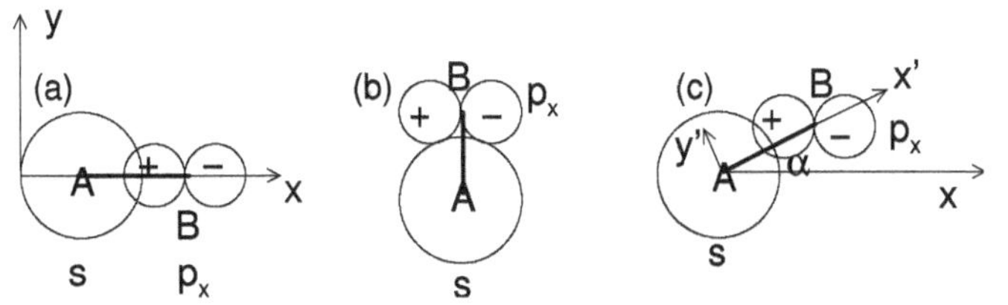

Figure 8.4: (a) s on atom A and p_x on B are placed along the x-axis. Atomic wavefunctions have a nonzero overlap integral and are able to form bonds. (b) Atom B is placed along the y-axis. The overlap integral is zero due to the cancellation of two lobes of the p_x-orbital. (c) Atom B is shifted upward. The overlap is reduced by a factor of $\cos\alpha$.

the axis of the p_x-orbital is through the center of the s-orbital. The overlap integral S between them is given by

$$S_{s,p_x} = \int \phi_s^{\mathrm{at}*}(r_A)\phi_{p_x}^{\mathrm{at}}(r_B)d\mathbf{r}.$$

On the other hand, if we move atom B along the y-axis, we have another configuration. Here p_x is on the top of A (see Fig. 8.4(b)), where the p_x-axis does not go through the center of the s-orbital. The negative and positive lobes of the p_x-orbital inside the previous integrals cancel out. Therefore, Fig. 8.4(b) has zero overlap, and A and B do not form a bond. We can show how this is realized mathematically. Suppose the p_x-orbital is shifted sideways of atom A (Fig. 8.4(b)). We must carry out a projection of p_x on the line AB. Assume AB makes an angle of α with respect to the x-axis (see Fig. 8.4(c)). The original $\phi_{p_x}^{\mathrm{at}}$ must be projected to the line AB using the new x'- and y'-axes as

$$\phi_{p_x}^{\mathrm{at}} \to \phi_{p_x}^{\mathrm{at}}\cos\alpha + \phi_{p_y}^{\mathrm{at}}\sin\alpha.$$

The reason behind this projection is that the "true" p_x- and p_y-orbitals are only defined with respect to the bond, line AB. Then, both S_{IJ} and H_{IJ} pick up a factor $\cos\alpha$, $S_{IJ} \to S_{IJ}\cos(\alpha)$, and $H_{IJ} \to H_{IJ}\cos(\alpha)$. If $\alpha = 0$, we have a standard sp-hybridization. If $\alpha = 90^\circ$, we have no bonding, $S_{IJ} = H_{IJ} = 0$.

We can use the same method for two p-orbitals

$$\begin{aligned}\phi_{p_1}^{\mathrm{at}}(A) &= \cos\alpha_1\phi_{p_z}^{\mathrm{at}}(A) + \sin\alpha_1\phi_{p_x}^{\mathrm{at}}(A),\\ \phi_{p_2}^{\mathrm{at}}(B) &= \cos\alpha_2\phi_{p_z}^{\mathrm{at}}(B) + \sin\alpha_2\phi_{p_x}^{\mathrm{at}}(B),\end{aligned} \tag{8.2.3}$$

which represent hybridization of two p-orbitals.

The Hamiltonian matrix can be formulated similarly. For instance, the element between ϕ_s^{at} at atom A and $\phi_{p_x}^{\mathrm{at}}$ at B is given by

$$H_{s,p_x} = \int \phi_s^{\mathrm{at}*}(r_A)\hat{H}_{\mathrm{mol}}\phi_{p_x}^{\mathrm{at}}(r_B)d\mathbf{r},$$

which can be calculated once a specific Hamiltonian is known. In a qualitative calculation, the Hamiltonian matrix element is denoted by $-t$, often called the hopping integral. For some well-known materials, these matrix elements are already tabulated [51]; depending on the atomic orbitals and orientation of the orbitals, their values vary considerably.

Exercise 8.2.0.

4. Starting from $F[\psi] = \langle\psi_{\mathrm{mol}}|\hat{H}_{\mathrm{mol}}|\psi_{\mathrm{mol}}\rangle - \lambda(\langle\psi_{\mathrm{mol}}|\psi_{\mathrm{mol}}\rangle - 1)$, prove eq. (8.2.2).
5. Consider two p-orbitals given in eq. (8.2.3). Find the overlap matrix S.

8.3 Application to small molecules: the allyl radical

We consider the allyl radical, C_3H_5, a planar molecule which has eight atoms, but the major backbone comprises three carbon atoms that form a chain [52]. Each carbon has four valence electrons in four orbitals, $2s$, $2p_x$, $2p_y$ and $2p_z$. We limit ourselves to three $2p_z$-orbitals which stick out of the plane of the molecule (see Fig. 8.5), ignoring the rest of the orbitals. Whether such an approach is adequate must be checked by experiment or a higher-level theory.

We first construct the Hamiltonian for allyl. As in many molecules, in allyl each carbon atom has two contributions. One is its original atomic energy and constitutes the main part of the diagonal elements for the Hamiltonian. For atom C_1, we have

$$H_{p_z,p_z}(C_1,C_1) = \int \phi_{p_z}^{\mathrm{at}*}(\mathbf{r}_{C_1})\hat{H}_{\mathrm{mol}}\phi_{p_z}^{\mathrm{at}}(\mathbf{r}_{C_1})d\mathbf{r} \equiv E_0, \tag{8.3.1}$$

where $\phi_{p_z}^{\mathrm{at}}(\mathbf{r}_{C_1})$ is referenced to atom C_1. We assume that the diagonal element for the rest of atoms is the same $H_{ii} = E_0$. This energy is commonly called the site energy because it is on the atom itself. The energy contributed by neighboring atoms constitutes the off-diagonal elements. For C_1 and C_2, the element is given by

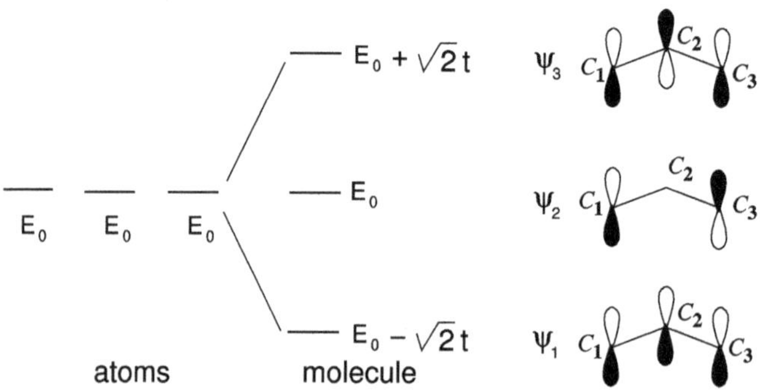

Figure 8.5: (left) A single atomic energy level is split into three levels. (right) The molecular orbitals of the allyl radical, where only three carbon atoms, C_1, C_2 and C_3, are shown.

$$H_{p_z,p_z}(C_1,C_2) = \int \phi^{at*}_{p_z}(\mathbf{r}_{C_1})\hat{H}_{mol}\phi^{at}_{p_z}(\mathbf{r}_{C_2})d\mathbf{r} \equiv -t, \tag{8.3.2}$$

which describes the electron hopping from one atom to another and are mainly contributed by the kinetic energy term. t is called the hopping integral. We keep the off-diagonal elements to the nearest neighbor atoms only, i. e., $H_{12} = -t$, $H_{23} = -t$, while setting the hopping integral between the next neighbors, atoms 1 and 3, to zero, $H_{13} = 0$. This approach is called the tight-binding approximation. Now, our Hamiltonian is a matrix and its eigenequation is

$$\begin{pmatrix} E_0 & -t & 0 \\ -t & E_0 & -t \\ 0 & -t & E_0 \end{pmatrix} \begin{pmatrix} c_1 \\ c_2 \\ c_3 \end{pmatrix} = E \begin{pmatrix} c_1 \\ c_2 \\ c_3 \end{pmatrix}, \tag{8.3.3}$$

which can be solved analytically by setting the determinant to zero. We find $E_1 = E_0 - \sqrt{2}t$, $E_2 = E_0$ and $E_3 = E_0 + \sqrt{2}t$. Figure 8.5 (left) shows that the original single atomic energy is split into three (see Fig.8.5 (middle)).

Their respective eigenstates are

$$\psi_1 = \frac{1}{2}|C_1\rangle + \frac{\sqrt{2}}{2}|C_2\rangle + \frac{1}{2}|C_3\rangle, \tag{8.3.4}$$

$$\psi_2 = \frac{1}{\sqrt{2}}|C_1\rangle - \frac{1}{\sqrt{2}}|C_3\rangle, \tag{8.3.5}$$

and

$$\psi_3 = \frac{1}{2}|C_1\rangle - \frac{\sqrt{2}}{2}|C_2\rangle + \frac{1}{2}|C_3\rangle, \tag{8.3.6}$$

where $|C_i\rangle$ is a shorthand notation for $\phi^{at}_{p_z}(\mathbf{r}_{C_i})$. We plot them in Fig. 8.5 (right). ψ_1 is a bonding state and spreads over three atoms. This is a π-state since the $2p_z$-orbitals overlap above and below the bond between two carbon atoms. The second state ψ_2 is a nonbonding state, isolated by C_2. ψ_3 is an antibonding state because the positive and negative lobes are not on the same side of the molecule.

Finally, we should point out the limitations of our model. First, we only include one orbital for each carbon atom, where whether three carbon atoms are in line or not has no effect on our Hamiltonian. So, our MO is approximate. But in reality, molecular structures often twist. Besides $2p_z$-orbitals, we may add $2p_x$, $2p_y$ and, in some cases d- or f-orbitals, even though the orbital in question is primarily a p-orbital.

8.3.1 Occupying the molecular orbitals and many-body wavefunctions

The neutral allyl radical has three π-electrons, denoted as a, b, c. The Pauli exclusion principle (see Chapter 10) dictates that each molecular orbital can take two electrons at maximum: one for spin-up and the other for spin-down. Three electrons need to occupy at minimum two MOs, ψ_1 and ψ_2. The highest occupied molecular orbital, HOMO, is ψ_2. The lowest unoccupied molecular orbital, LUMO, is ψ_3. The total wavefunction is a many-body wavefunction, denoted as Ψ, which is a Slater determinant,[3]

$$\Psi(a,b,c) = \sqrt{\frac{1}{3!}} \begin{vmatrix} \psi_1(a)\alpha(a) & \psi_1(b)\alpha(b) & \psi_1(c)\alpha(c) \\ \psi_1(a)\beta(a) & \psi_1(b)\beta(b) & \psi_1(c)\beta(c) \\ \psi_2(a)\alpha(a) & \psi_2(b)\alpha(b) & \psi_2(c)\alpha(c) \end{vmatrix}, \tag{8.3.7}$$

where $\psi_1(a)\alpha(a)$ means that the electron a is in ψ_1 with spin-up and the remaining terms have a similar meaning. This wavefunction is rather complicated. We take the diagonal element as an approximation because doing so does not affect the chemistry here: $\Psi(a,b,c) \approx \psi_1(a)\alpha(a)\psi_1(b)\beta(b)\psi_2(c)\alpha(c)$. We further drop the spin wavefunction, so $\Psi(a,b,c) \approx \psi_1(a)\psi_1(b)\psi_2(c)$. This simplified many-body wavefunction already contains some new information that we have not encountered before.

First, $\Psi(a,b,c)$ is a product of single-electron wavefunctions which is a many-body wavefunction, different from a sum of single-electron wavefunctions which is the superposition of single-electron wavefunctions and remains a single-electron wavefunction.

Second, when we compute the electron probability density from $\Psi(a,b,c)$ for a particular electron, we must integrate coordinates over other electrons. For instance, if we want to find out the density of electron a, we must integrate over the coordinates of electrons b and c as

$$\iint |\Psi(a,b,c)|^2 d\mathbf{r}_b d\mathbf{r}_c = |\psi_1(a)|^2 \int |\psi_1(b)|^2 d\mathbf{r}_b \int |\psi_2(c)|^2 d\mathbf{r}_c = |\psi_1(a)|^2, \tag{8.3.8}$$

which is significantly more complicated than a single-electron wavefunction. We need to introduce a new operator.

8.3.2 Charge density

We use the Dirac notation to define the charge density operator $\hat{\rho}_a = |a\rangle\langle a|$. $\hat{\rho}_a$ is better considered as a projector, which projects out the single-particle wavefunction pertaining to electron a, without integrating over its coordinates. For instance, eq. (8.3.8) is exactly equivalent to

$$\langle\Psi|\hat{\rho}_a|\Psi\rangle = \langle\psi_1(a)\psi_1(b)\psi_2(c)|\hat{\rho}_a|\psi_1(a)\psi_1(b)\psi_2(c)\rangle = \langle\psi_1(a)|a\rangle\langle a|\psi_1(a)\rangle, \tag{8.3.9}$$

where the integration is only over electrons b and c and leaves electron a intact. Importantly, $\langle\psi_1(a)|a\rangle = \psi_1^*(a)$ and $\langle a|\psi_1(a)\rangle = \psi_1(a)$ are not integral, which extends our Dirac

3 In Chapter 10, we will present more details on this, but for now we only need the determinant itself.

notation. This notation simplifies our procedure. If we want to compute the total charge density, our total density operator $\hat{\rho}$ is the sum of all three electrons,

$$\hat{\rho} = \hat{\rho}_a + \hat{\rho}_b + \hat{\rho}_c,$$

which allows us to compute the total charge density (see the ensuing exercise for details),

$$\langle\Psi|\hat{\rho}|\Psi\rangle = \underbrace{2|\psi_1|^2}_{\text{electrons } a,b} + \underbrace{|\psi_2|^2}_{\text{electron } c} = 2|\psi_1\rangle\langle\psi_1| + |\psi_2\rangle\langle\psi_2|. \tag{8.3.10}$$

If we integrate it over the coordinates of three electrons, we find

$$\iiint \langle\Psi|\hat{\rho}|\Psi\rangle d\mathbf{r}_a d\mathbf{r}_b d\mathbf{r}_c = 2\int |\psi_1\rangle\langle\psi_1| d\mathbf{r}_a + 1\int |\psi_2\rangle\langle\psi_2| d\mathbf{r}_c = 2 + 1 = 3, \tag{8.3.11}$$

which shows the normalization of a many-body wavefunction is equal to the number of electrons that we have, not 1 anymore (see eq. (1.4.9)). These new features are crucial to understanding the chemistry of each atom.

Next, we can find out how many electrons each atom has. We start with eq. (8.3.10) and substitute each term with their eigenstates (eqs. (8.3.4), (8.3.5) and (8.3.6)),

$$2|\psi_1\rangle\langle\psi_1| = 2\left(\frac{1}{2}|C_1\rangle + \frac{\sqrt{2}}{2}|C_2\rangle + \frac{1}{2}|C_3\rangle\right)\left(\frac{1}{2}\langle C_1| + \frac{\sqrt{2}}{2}\langle C_2| + \frac{1}{2}\langle C_3|\right). \tag{8.3.12}$$

To find the number of electrons at carbon atom C_1, all we need is to find the coefficient of $|C_1\rangle\langle C_1|$, where $|C_1\rangle\langle C_1|$ identifies atom C_1. In this case, we find $2 \times \frac{1}{2} \times \frac{1}{2} = \frac{1}{2}$. We can work this out for $|\psi_2\rangle\langle\psi_2|$ as well.

If a molecule loses or gains electrons, due to oxidation or reduction, the numbers of electrons at C_1, C_2, and C_3 change. This is how a chemical reaction is understood on the molecular level.

Exercise 8.3.2.

6. Cyclopropenyl is a triangular system, where three carbons are all linked to each other. The Hamiltonian matrix is

$$\begin{pmatrix} E_0 & -t & -t \\ -t & E_0 & -t \\ -t & -t & E_0 \end{pmatrix}.$$

(a) Compute the eigenvalues. (b) Find the eigenvectors. Hint: This problem has a degeneracy. Any linear combination is a good wavefunction.

7. Prove eqs. (8.3.10) and (8.3.11).

8. (a) The neutral allyl has the ground state as $\Psi = \psi_1(a)\psi_1(b)\psi_2(c)$. Find the number of electrons at the carbon atoms C_1, C_2, and C_3. (b) If we remove one electron from the system, we have the ground state as $\Psi = \psi_1(a)\psi_1(b)$. Find the number of electrons at C_1, C_2, and C_3. (c) If we add one electron to the system, we have $\Psi = \psi_1(a)\psi_1(b)\psi_2(c)\psi_2(d)$. Find the number of electrons at C_1, C_2, and C_3.

8.3.3 Dirac notation

To ease our presentation with the Hamiltonian matrix, we introduce the Dirac notation to represent a matrix without using a matrix.[4] We take a 2×2 matrix as an example. The matrix form is

$$\hat{H} = \begin{pmatrix} H_{11} & H_{12} \\ H_{21} & H_{22} \end{pmatrix}.$$

If we apply it to a column vector $|\psi\rangle = (c_1, c_2)^T$, we have

$$\hat{H}\begin{pmatrix} c_1 \\ c_2 \end{pmatrix} = \begin{pmatrix} H_{11} & H_{12} \\ H_{21} & H_{22} \end{pmatrix}\begin{pmatrix} c_1 \\ c_2 \end{pmatrix} = \begin{pmatrix} H_{11}c_1 + H_{12}c_2 \\ H_{21}c_1 + H_{22}c_2 \end{pmatrix}. \tag{8.3.13}$$

In the Dirac notation, $\hat{H}$ becomes

$$\hat{H} = H_{11}|1\rangle\langle 1| + H_{12}|1\rangle\langle 2| + H_{21}|2\rangle\langle 1| + H_{22}|2\rangle\langle 2| = \sum_{ij} H_{ij}|i\rangle\langle j|, \tag{8.3.14}$$

where $|i\rangle\langle j|$ holds row i and column j indices for an entry H_{ij}. More importantly, $|i\rangle\langle j|$ represents a projector that projects state $|j\rangle$ to a state $|i\rangle$. For instance, if $|1\rangle\langle 2|$ applies to $|2\rangle$, we obtain $|1\rangle\langle 2|2\rangle = |1\rangle$; if $|1\rangle\langle 2|$ applies to $|1\rangle$, we get $|1\rangle\langle 2|1\rangle = 0$, where we have used $\langle 2|2\rangle = 1$ and $\langle 2|1\rangle = 0$. This explains what the projector means.

In the Dirac notation, the column vector $\left(\begin{smallmatrix} c_1 \\ c_2 \end{smallmatrix}\right)$ is now written as $\psi = c_1|1\rangle + c_2|2\rangle$. The same product $\hat{H}\psi$ is

$$\begin{aligned}\hat{H}\psi &= \sum_{ij}(H_{ij}|i\rangle\langle j|)(c_1|1\rangle + c_2|2\rangle) = \sum_{ij} c_1 H_{ij}|i\rangle\langle j|1\rangle + \sum_{ij} c_2 H_{ij}|i\rangle\langle j|2\rangle \\ &= \sum_i (c_1 H_{i1}|i\rangle + c_2 H_{i2}|i\rangle) = (c_1H_{11} + c_2H_{12})|1\rangle + (c_1H_{21} + c_2H_{22})|2\rangle,\end{aligned}$$

where the coefficient of $|1\rangle$ is the first row of eq. (8.3.13) and that of $|2\rangle$ is the second row of eq. (8.3.13). The Dirac notation allows us to carry out matrix operations without the complexity of a matrix.

Exercise 8.3.3.

9. Suppose we have the following eigenequation:

$$\begin{pmatrix} E_0 & -t \\ -t & E_0 \end{pmatrix}\begin{pmatrix} c_1 \\ c_2 \end{pmatrix} = E\begin{pmatrix} c_1 \\ c_2 \end{pmatrix}.$$

Rewrite it in terms of the Dirac notation.

---

4 In advanced quantum mechanics, this is equivalent to the second quantization operator.

8.4 Applications to conjugated polymers: polyacetylene

In physical chemistry, conjugation means that π-electrons from multiple atoms of a molecule strongly overlap across the entire molecule. Polyacetylene (PA) is a conjugated polymer. It has extraordinary electronic and optical properties and opened the era of synthetic metals. Three pioneers, Hideki Shirakawa [53], Alan G. MacDiarmid [54], and Alan J. Heeger [55], shared the Nobel Prize in Chemistry in 2000. PA consists of C_2H_2 repeating units, forming a quasi-one-dimensional chain. It can adopt two conformations: cis and trans. Figure 8.6 shows trans-polyacetylene, with similar ligands on the opposite side of the chain, whereas the cis structure has similar ligands on the same side of the chain. The bond lengths for single and double bonds are 1.36 and 1.44 Å, [50, 53]. Essential to PA are the p_z-orbitals of carbon atoms [50], which appear either above or below the chain. These orbitals are our focus.

We consider a simple model chain of N units of (CH), instead of (C_2H_2). This is tantamount to ignoring the difference between the single and double bonds in Fig. 8.6. Each unit has a single $2p_z$-orbital (due to the carbon atom) denoted as $|i\rangle$. The minimum distance between two neighboring units is denoted by a constant a, called lattice constant. So, the total length of our system is $L = Na$. We adopt the tight-binding idea and only consider the interaction between the nearest-neighbor units. Although there is a difference in the hopping integral between the single and double bonds, we ignore this difference as already stated, so the hopping integral is $-t$. We further assume that orbitals are orthonormalized, i. e., $\langle i|j\rangle = \delta_{ij}$. With these approximations in mind, our Hamiltonian in the Dirac notation is

$$\hat{H} = \sum_i [E_s|i\rangle\langle i| - t(|i\rangle\langle i+1| + |i+1\rangle\langle i|)], \tag{8.4.1}$$

where $|i\rangle$ is the atomic orbital at atomic site i and serves as a basis function. The first term in eq. (8.4.1) is the site energy and is diagonal. The second term has two parts: The first is the hopping from site $i+1$ to i, corresponding to the upper triangle of the matrix, while the second is from site i to $i+1$, corresponding to the lower triangle.

Next, we find the eigenvalues and eigenvectors of $\hat{H}$. For a chain of N sites, $\hat{H}$ corresponds to a $N \times N$ matrix with N eigenstates, which is difficult to diagonalize. So, we must examine the physics first. The eigenstate ψ_n must be a linear superposition of the AOs as $\psi_n = \sum_j C_j^n|j\rangle$, where j runs from 1 to N, and n is the eigenstate index from 1 to N. C_j^n is unknown. Since our chain length is $L = Na$, the wavefunction at $x = 0$ and $x_N = Na$ must be zero to match the boundary conditions, similar to PIB in Chapter 2. In addition, in order for the wavefunction to fit inside the chain, the wavelength λ of the wavefunction must take a limited number of values $\lambda = \frac{2L}{n}$, where $n = 1, 2 \dots N$. To satisfy both conditions, our wavefunction, or more precisely the coefficient C_j^n at site j, must be a sine function of position x_j as $C_j^n = C\sin(\beta_n x_j)$, where $\beta_n = 2\pi/\lambda = \pi n/L$ and $x_j = ja$. Therefore, our unnormalized trial function is

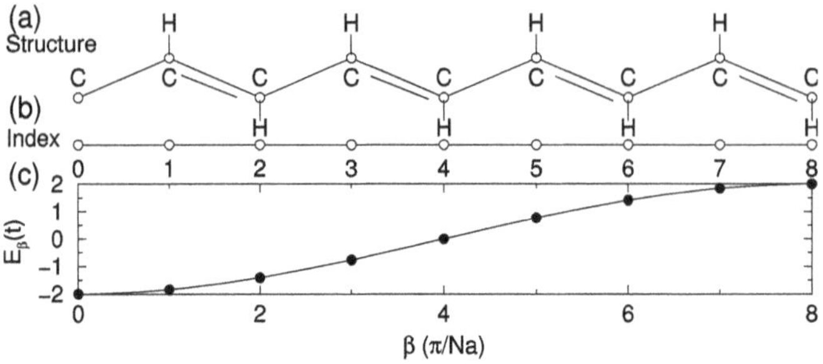

Figure 8.6: (a) Short segment of *trans*-polyacetylene. Single and double bonds alternate along the chain, but our model ignores this difference. (b) Site index starts from 1 to N. 0 is ignored. (c) Energy dispersion as a function of β.

$$|\psi_n\rangle = \sum_j c_j^n|j\rangle = C\sum_j \sin(\beta_n x_j)|j\rangle,$$

where $x_j = ja$. We can find C by requiring $\langle\psi_n|\psi_n\rangle = 1$ (see the exercise for details) to get $C = \sqrt{\frac{2}{N}}$. The normalized trial wavefunction is

$$|\psi_n\rangle = \sqrt{\frac{2}{N}}\sum_j \sin(\beta_n x_j)|j\rangle.$$

Now, we prove $|\psi_n\rangle$ is indeed an eigenstate of $\hat{H}$, $\hat{H}|\psi_n\rangle = E_n|\psi_n\rangle$. We write

$$\begin{aligned}\hat{H}|\psi_n\rangle &= \hat{H}\sqrt{\frac{2}{N}}\sum_j \sin(\beta_n x_j)|j\rangle \\ &= \sqrt{\frac{2}{N}}\Big(\sum_i [E_s|i\rangle\langle i| - t(|i\rangle\langle i+1| + |i+1\rangle\langle i|)]\Big)\sum_j \sin(\beta_n x_j)|j\rangle,\end{aligned}$$

which has three terms $\langle i|j\rangle$, $\langle i+1|j\rangle$, and $\langle i|j\rangle$.

Because we assume the orthogonality among the atomic orbitals, $\langle i|j\rangle = \delta_{ij}$, we can eliminate $\sum_i$ from the equation to obtain

$$\begin{aligned}\hat{H}|\psi_n\rangle &= \sqrt{\frac{2}{N}}\Big(\sum_j E_s \sin(\beta_n x_j)|j\rangle - t\sum_j \sin(\beta_n x_j)|j-1\rangle - t\sum_j \sin(\beta_n x_j)|j+1\rangle\Big) \\ &= E_s\psi_n - t\sqrt{\frac{2}{N}}\sum_j |j\rangle\sin(\beta_n x_{j+1}) - t\sqrt{\frac{2}{N}}\sum_j |j\rangle\sin(\beta_n x_{j-1}),\end{aligned} \tag{8.4.2}$$

where in the second equation, we have rewritten the second and third terms by increasing and decreasing j by 1, respectively. This shift has no effect on the summation (see the exercise). Now we can combine $\sin(\beta_n x_{j+1}) + \sin(\beta_n x_{j-1})$. Since $x_{j+1} = x_j + a$ and $x_{j-1} = x_j - a$, $\sin(\beta_n x_{j+1}) + \sin(\beta_n x_{j-1}) = 2\sin(\beta_n x_j)\cos(\beta_n a)$. Equation (8.4.2) is reduced to $\hat{H}\psi_n = E_s\psi_n - 2t\sqrt{\frac{2}{N}}\cos\beta_n a \sum_j \sin(\beta_n x_j)|j\rangle$, i. e.,

$$\hat{H}\psi_n = [E_s - 2t\cos(\beta_n a)]\psi_n \equiv E_n|\psi_n\rangle. \tag{8.4.3}$$

This shows that $|\psi_n\rangle$ is indeed an eigenstate of $\hat{H}$ and E_n is the eigenenergy.

Each n, i. e., β_n, has an eigenenergy E_n. Figure 8.6 shows the energy dispersion of E_n versus β_n, with $N = 8$. There are only 8 eigenstates. As N gets larger, discrete eigenenergies form a band of energy, known as energy bands. Here we only consider p_z-orbitals for each carbon atom. If we include M orbitals for each atom, we will have M bands. The number of bands can also be changed if we change the unit size. For instance, if we distinguish double and single bonds, i. e., each unit containing units of $(CH)_2$, instead of (CH), for the same length $L = 8a$ only four units of $(CH)_2$ are possible, so there are four β_n. But, at each β_n, it now has two eigenstates. These two eigenstates form two distinctive bands. This leads us to the formal presentation of solids.

Exercise 8.4.0.

10. Prove $\sum_j \sin(\beta_l x_j)|j-1\rangle = \sum_j \sin(\beta_l x_{j+1})|j\rangle$.
11. Prove $\sum_j \sin(\beta_l x_j)|j+1\rangle = \sum_j \sin(\beta_l x_{j-1})|j\rangle$.
12. Assume the lattice constant $a = 1.42$ Å for PA, and $N = 16$. (a) Find all the possible β_n. (b) Set $E_s = 0$ and $t = 1.0$ eV, and plot E_n as a function of $\beta_n a$.

8.5 One-dimensional solid: Kronig–Penney model

Nearly all the models for crystalline solids cannot be solved analytically. Kronig and Penney [56] proposed a model that bears their names and can be solved analytically up to a convoluted equation. The rest must be solved numerically. It has all the features that a one-dimensional periodic system possesses. We introduce it here.

The model adopts a square-well potential (Chapter 2), but repeats itself periodically every $a + b$, where $a + b$ is called the lattice constant denoted as $L = a + b$. Figure 8.7 shows that, between $x = -b$ and $x = 0$, the potential $V(x)$ is V_0 and between $x = 0$ and $x = a$, $V(x)$ is 0, where V_0 is positive. We consider three regions.

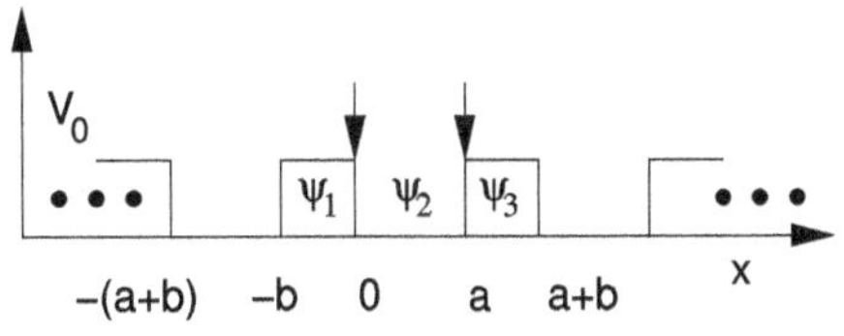

Figure 8.7: Kronig–Penny model. The barrier height is set by V_0. Two boundary conditions are highlighted by two vertical arrows. Three wavefunctions in three regions are denoted by ψ_1, ψ_2, and ψ_3. ψ_3 is obtained by the Bloch theorem from ψ_1.

8.5.1 Region 1: $-b < x < 0$

Between $x = -b$ and $x = 0$, the potential is $V(x) = V_0$, so our Schrödinger equation is

$$\frac{-\hbar^2}{2m}\frac{d^2\psi_1}{dx^2} + V_0\psi_1 = E\psi_1 \longrightarrow \frac{d^2\psi_1}{dx^2} = \frac{2m(V_0 - E)}{\hbar^2}\psi_1 \equiv \beta^2\psi_1, \tag{8.5.1}$$

where $\beta^2 = \frac{2m(V_0-E)}{\hbar^2}$, $\beta = \frac{\sqrt{2m(V_0-E)}}{\hbar}$. If $E < V_0$, this gives a bound state. If $E > V_0$, then our solution is a continuous state, and we simply change β to $i\beta$, as seen in the next equation. For either case, ψ_1 can be written as

$$\psi_1 = Ce^{\beta x} + De^{-\beta x}. \tag{8.5.2}$$

8.5.2 Region 2: $0 < x < a$

Between $x = 0$ and $x = a$, $V(x) = 0$, the Schrödinger equation is

$$-\frac{\hbar^2}{2m}\frac{d^2\psi_2}{dx^2} = E\psi_2 \rightarrow \frac{d^2\psi_2}{dx^2} = -\frac{2mE}{\hbar^2}\psi_2 \equiv -\alpha^2\psi_2, \tag{8.5.3}$$

where $\alpha^2 = \frac{2mE}{\hbar^2}$, i. e., $\alpha = \frac{\sqrt{2mE}}{\hbar}$. Because E is positive, α is real. Its solution is

$$\psi_2 = Ae^{i\alpha x} + Be^{-i\alpha x}, \tag{8.5.4}$$

where α is the wavevector in the zero-potential.

Boundary conditions require that, at $x = 0$, ψ_1 and ψ_2 must have the same values as well as its derivatives, $\psi_1(x = 0) = \psi_2(x = 0)$, leading to $A + B = C + D$, while its derivative, $\psi_1'(x = 0) = \psi_2'(x = 0)$, yielding $i\alpha(A - B) = \beta(C - D)$. However, these two equations are not enough to determine four unknown coefficients: A, B, C, and D.

8.5.3 Bloch theorem

Crystalline solids have the special property that the potential repeats periodically after a distance L. It can be proven that, if we shift our wavefunction by L, it only differs from our original wavefunction by a phase,

$$\psi(x + L) = e^{ikL}\psi(x), \tag{8.5.5}$$

which is called the Bloch theorem. Its formal approval is left to the next chapter. k is a wavevector for our wavefunction. $\hbar k$ is called the crystal momentum, but sometimes one also calls k the crystal momentum, for brevity.

8.5.4 Region 3: $a < x \le a + b$

We denote the wavefunction as ψ_3 for $a < x \le a + b$. Then, ψ_1 and ψ_3 just separate by L (see Fig. 8.7), so we can use the Bloch theorem (eq. (8.5.5)) to get $\psi_3(x+L) = e^{ikL}\psi_1(x)$ and $\psi_3'(x+L) = e^{ikL}\psi_1'(x)$. Caution must be taken as how to find the expression for ψ_3 at $x = a$ from ψ_1. Because ψ_3's variable is $x+L$, we must set $x+L = a$. Then, $x = a-L = a-a-b = -b$, which enters $\psi_1(x = -b)$. So, $\psi_3(x + L) = e^{ikL}\psi_1(x)$ yields $\psi_3(a) = e^{ikL}\psi_1(-b)$, while $\psi_3'(x+L) = e^{ikL}\psi_1'(x)$ yields $\psi_3'(a) = e^{ikL}\psi_1'(-b)$.

Since ψ_2 and ψ_3 are connected at $x = a$, this gives a new boundary condition. Specifically, at $x = a$, we have

$$\psi_3(a) = e^{ikL}\psi_1(-b) = e^{ikL}(Ce^{-\beta b} + De^{\beta b}) = \psi_2(a) = Ae^{i\alpha a} + Be^{-i\alpha a},$$

which can be simplified to $Ae^{i\alpha a} + Be^{-i\alpha a} = e^{ik(a+b)}(Ce^{-\beta b} + De^{\beta b})$. The derivative $\psi_3'(a) = \psi_2'(a)$ leads to $i\alpha(Ae^{i\alpha a} - Be^{-i\alpha a}) = \beta e^{ik(a+b)}(Ce^{-\beta b} - De^{\beta b})$.

In total, we have four equations,

$$A + B = C + D, \tag{8.5.6}$$

$$i\alpha(A - B) = \beta(C - D), \tag{8.5.7}$$

$$Ae^{i\alpha a} + Be^{-i\alpha a} = (Ce^{-\beta b} + De^{\beta b})e^{ik(a+b)}, \tag{8.5.8}$$

$$i\alpha(Ae^{i\alpha a} - Be^{-i\alpha a}) = \beta(Ce^{-\beta b} - De^{\beta b})e^{ik(a+b)}. \tag{8.5.9}$$

Eliminating all the unknowns leads to

$$\frac{\beta^2 - \alpha^2}{2\alpha\beta}\sin(\alpha a)\sinh(\beta b) + \cos(\alpha a)\cosh(\beta b) = \cos[k(a + b)], \tag{8.5.10}$$

whose solution is the eigenenergy E for every k. Before we solve it numerically, we show next how to find this complicated equation.

8.5.5 Derivation of the secular equation

A common strategy is to compute the 4×4 determinant and then set it to zero. However, this is quite complicated. There has been no detailed derivation that we can find, except a short note by McQuarrie [57].

In the following, we introduce a relatively simpler method. We start from eqs. (8.5.6) and (8.5.7) to get

$$2A = \left(1 + \frac{\beta}{i\alpha}\right)C + \left(1 - \frac{\beta}{i\alpha}\right)D, \tag{8.5.11}$$

$$2B = \left(1 - \frac{\beta}{i\alpha}\right)C + \left(1 + \frac{\beta}{i\alpha}\right)D, \tag{8.5.12}$$

and from eqs. (8.5.8) and (8.5.9) to get

$$2Ae^{i\alpha a} = \left(1 + \frac{\beta}{i\alpha}\right)e^{ik(a+b)-\beta b}C + \left(1 - \frac{\beta}{i\alpha}\right)e^{ik(a+b)+\beta b}D, \tag{8.5.13}$$

$$2Be^{-i\alpha a} = \left(1 - \frac{\beta}{i\alpha}\right)e^{ik(a+b)-\beta b}C + \left(1 + \frac{\beta}{i\alpha}\right)e^{ik(a+b)+\beta b}D. \tag{8.5.14}$$

The next step is most crucial. We substitute eq. (8.5.11) into (8.5.13) and eq. (8.5.12) into (8.5.14) to replace both $2A$ and $2B$. Two resultant equations contain C and D terms only. We then combine the C and D terms and put them on two sides of the equations,

$$\left(1 + \frac{\beta}{i\alpha}\right)[e^{i\alpha a} - e^{ik(a+b)-\beta b}]C = \left(1 - \frac{\beta}{i\alpha}\right)[-e^{i\alpha a} + e^{ik(a+b)+\beta b}]D, \tag{8.5.15}$$

$$\left(1 - \frac{\beta}{i\alpha}\right)[e^{-i\alpha a} - e^{ik(a+b)-\beta b}]C = \left(1 + \frac{\beta}{i\alpha}\right)[-e^{-i\alpha a} + e^{ik(a+b)+\beta b}]D. \tag{8.5.16}$$

We divide eq. (8.5.15) by (8.5.16) to eliminate C and D on both sides,

$$\frac{(1 + \frac{\beta}{i\alpha})[e^{i\alpha a} - e^{ik(a+b)-\beta b}]}{(1 - \frac{\beta}{i\alpha})[e^{-i\alpha a} - e^{ik(a+b)-\beta b}]} = \frac{(1 - \frac{\beta}{i\alpha})[-e^{i\alpha a} + e^{ik(a+b)+\beta b}]}{(1 + \frac{\beta}{i\alpha})[-e^{-i\alpha a} + e^{ik(a+b)+\beta b}]}.$$

We cross-multiply it out to get a single equation, whose left side is

$$\left(1 + \frac{\beta}{i\alpha}\right)^2 [-1 - e^{2ik(a+b)} + e^{i\alpha a+ik(a+b)+\beta b} + e^{-i\alpha a+ik(a+b)-\beta b}],$$

and whose right side is

$$\left(1 - \frac{\beta}{i\alpha}\right)^2 [-1 - e^{2ik(a+b)} + e^{-i\alpha a+ik(a+b)+\beta b} + e^{i\alpha a+ik(a+b)-\beta b}].$$

We now multiply out both $(1 + \frac{\beta}{i\alpha})^2$ and $(1 - \frac{\beta}{i\alpha})^2$ to get ready for a combination of various terms. We combine the coefficients of the $-1 - e^{2ik(a+b)}$ terms and then those of the third and fourth terms on both sides to obtain three terms, respectively,

$$\frac{4\beta}{i\alpha}(-1 - e^{2ik(a+b)}), \tag{8.5.17}$$

$$\left(1 - \frac{\beta^2}{\alpha^2}\right)2i\sin(\alpha a)e^{ik(a+b)+\beta b} + \frac{2\beta}{i\alpha}2\cos(\alpha a)e^{ik(a+b)+\beta b}, \tag{8.5.18}$$

$$\left(1 - \frac{\beta^2}{\alpha^2}\right)(-2i)\sin(\alpha a)e^{ik(a+b)-\beta b} + \frac{2\beta}{i\alpha}2\cos(\alpha a)e^{ik(a+b)-\beta b}, \tag{8.5.19}$$

the sum of which is zero.

We add the terms (8.5.18) and (8.5.19) and then multiply all the terms by $\frac{i\alpha}{4\beta}e^{-ik(a+b)}$, so that the term (8.5.17) is reduced to $-2\cos[k(a+b)]$. The resultant equation is

$$-2\cos k(a+b) - \frac{\alpha}{\beta}\left(1 - \frac{\beta^2}{\alpha^2}\right)\sin(\alpha a)\sinh(\beta b) + 2\cos(\alpha a)\cosh(\beta b) = 0, \quad (8.5.20)$$

which can be further simplified to our final equation,

$$\frac{\beta^2 - \alpha^2}{2\alpha\beta}\sin(\alpha a)\sinh(\beta b) + \cos(\alpha a)\cosh(\beta b) = \cos[k(a+b)]. \quad (8.5.21)$$

Exercise 8.5.5.
13. For the Kronig–Penny model, using the Bloch theorem, (a) write down the wavefunction $\psi_4(x)$ for $a+b < x < 2a+b$ in terms of $\psi_2(x)$. (b) Show at $x = a+b$ that the boundary conditions for ψ_3 and ψ_4 generate two equations of the coefficients A, B, C, and D, exactly the same as $x = 0$.

8.5.6 Band structure

Now, we follow McQuarrie [57] to introduce shorthand notations for our variables, so eq. (8.5.21) only depends on three parameters: $\varepsilon = \frac{E}{V_0}$, $r = \frac{b}{a}$, and $\mathcal{A} = a\sqrt{\frac{2mV_0}{\hbar^2}}$, or $\frac{2mV_0}{\hbar^2} = \frac{\mathcal{A}^2}{a^2}$. The right side of eq. (8.5.21) becomes $\cos[k(a+b)] = \cos[ka(1+r)]$.

Next, we treat αa and βb. From eq. (8.5.3), we have $\alpha^2 = \frac{2mV_0}{\hbar^2}\frac{E}{V_0} = \frac{2mV_0}{\hbar^2}\varepsilon = \frac{\mathcal{A}^2}{a^2}\varepsilon$, so $\alpha a = \mathcal{A}\sqrt{\varepsilon}$. Equation (8.5.1) leads to $\beta^2 = \frac{2m(V_0-E)}{\hbar^2} = \frac{2mV_0}{\hbar^2}(1-\varepsilon) = \frac{\mathcal{A}^2}{a^2}(1-\varepsilon)$, so $\beta = \frac{\mathcal{A}}{a}\sqrt{1-\epsilon}$, and $\beta b = r\mathcal{A}\sqrt{1-\varepsilon}$. With these substitutions, eq. (8.5.21) becomes

$$\begin{aligned}&\frac{1-2\varepsilon}{2\sqrt{(1-\varepsilon)\varepsilon}}\sin(\mathcal{A}\sqrt{\varepsilon})\sinh(r\mathcal{A}\sqrt{1-\varepsilon})\\&\quad+\cos(\mathcal{A}\sqrt{\varepsilon})\cosh(r\mathcal{A}\sqrt{1-\varepsilon}) = \cos[ka(1+r)].\end{aligned} \quad (8.5.22)$$

Because this equation depends on ε, we consider three cases. If $0 < \varepsilon < 1$, this corresponds to a bound state, and eq. (8.5.22) can be used directly.

If $\varepsilon = 1$, the first term must be treated with care since both the denominator and numerator go to zero. We expand it to

$$\begin{aligned}\sinh(r\mathcal{A}\sqrt{1-\varepsilon})/\sqrt{1-\varepsilon} &= \frac{1}{2}(e^{r\mathcal{A}\sqrt{1-\varepsilon}} - e^{-r\mathcal{A}\sqrt{1-\varepsilon}})/\sqrt{1-\varepsilon}\\&= \frac{\frac{1}{2}[1 + r\mathcal{A}\sqrt{1-\varepsilon} + \cdots - (1 - r\mathcal{A}\sqrt{1-\varepsilon} + \cdots)]}{\sqrt{1-\varepsilon}} = r\mathcal{A}.\end{aligned}$$

Thus, the first term in eq. (8.5.22) is $-\frac{1}{2}r\mathcal{A}\sin\mathcal{A}$, and eq. (8.5.22) becomes

$$-\frac{1}{2}r\mathcal{A}\sin\mathcal{A} + \cos\mathcal{A} = \cos[ka(1+r)]. \quad (8.5.23)$$

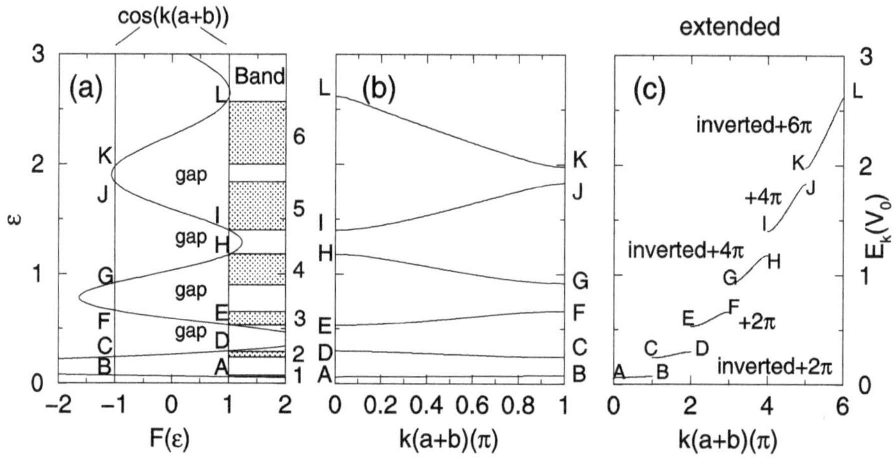

Figure 8.8: (a) $F(\varepsilon)$ as a function of ε. F is given in the text. Allowed bands fall between two vertical lines at −1 and 1 that are given by $\cos[k(a+b)]$. There are six bands labeled 1 to 6 from bottom to top, between which are the band gaps. (b) Band dispersion (in units of V_0) in the first reduced Brillouin zone. (c) Band energy dispersion E_k (in units of V_0) in the extended Brillouin zone by shifting and inverting those bands in (b). This figure's data is from code KP.f.

If $\varepsilon > 1$, this corresponds to an unbound state since the electron energy is higher than V_0, and β becomes imaginary. This requires us to change β to $i\beta$. So, all $\sinh\beta b$ and $\cosh\beta b$ become $i\sin\beta b$, and $\cos\beta b$, $\frac{1}{\sqrt{1-\varepsilon}}$ becomes $\frac{1}{i\sqrt{\varepsilon-1}}$:

$$\frac{1-2\varepsilon}{2\sqrt{\varepsilon(\varepsilon-1)}}\sin(\mathcal{A}\sqrt{\varepsilon})\sin(r\mathcal{A}\sqrt{1-\varepsilon})+\cos(\mathcal{A}\sqrt{\varepsilon})\cos(r\mathcal{A}\sqrt{\varepsilon-1})=\cos[ka(1+r)]. \quad (8.5.24)$$

We denote eqs. (8.5.22), (8.5.23), and (8.5.24) as $F(\varepsilon)=\cos[ka(1+r)]$. F depends on $\mathcal{A}$ and r. As an example, in Fig. 8.8(a), we choose $\mathcal{A}=10$ and $r=0.2$, which uniquely determine $F(\varepsilon)$. We plot $F(\varepsilon)$ as a function of ε up to $\varepsilon=3$. One can see $F(\varepsilon)$ swings strongly with a large amplitude for a small ε, very similar to the first-order Bessel function $\sin(x)/x$, but then it slows down as ε becomes larger. Because $\varepsilon=E/V_0$, ε reflects how the eigenenergy changes. When ε is small, the Kronig–Penney model reduces to the infinite well problem, where only a discrete energy is allowed. One thing is new. Because $\cos[ka(1+r)]$ is bound within $[-1,1]$, not all ε are allowed for a set of $(r,\mathcal{A})$. Two vertical lines are drawn at −1 and 1. Allowed energies are only between these two lines. Because k is a wavevector and changes continuously, allowed regions form a band of energy, or an energy band, denoted as E_k in units of V_0, while those forbidden regions represent the band gaps. We also call E_k the band structure or band dispersion. We always plot E_k versus k. If one plots E_k versus $\cos[ka(1+r)]$, then the band disperses as just as shown in Fig. 8.8(a). The bending of the band at $k(a+b)=0$ and π in Fig. 8.8(b) is due mainly to the cosine function.

Now, we show how to obtain E_k as a function of k for a fixed set of $\mathcal{A}$ and r. Because k enters through $\cos[ka(1+r)]$, i. e., $\cos[k(a+b)]$, k and $-k$ produce the same $\cos[k(a+b)]$. This shows there is an intrinsic inversion symmetry, i. e., $E_k = E_{-k}$. In the band dispersion, we only need to compute results for a positive k.

We choose a value, say 0.5, for $\cos[ka(1+r)]$, draw a vertical line at 0.5 in Fig. 8.8(a), and then find ε at all the possible intercepts between this vertical line and $F(\varepsilon)$. The first intercept, counting from the bottom, is our first band, $E_{n=1,k}$, which falls on the segment AB in Fig. 8.8(a). Point A is at $\cos[k(a+b)] = 1$, so its phase is $k(a+b) = 0$, while B has $k(a+b) = \pi$. What about C? Since C and B are on the same segment just like F and G, C also has phase $k(a+b) = \pi$. But D is different from A. Because C already has π and $k(a+b)$ only can increase, D must take $k(a+b) = 2\pi$ higher than A. Similar to B and C, E has the same phase as D. In summary, phases are A(0), B(π), C(π), D(2π), E(2π), F(3π), G(3π), H(4π), I(4π), J(5π), K(5π), and L(6π). To distinguish multiple roots for the same phase, we introduce an index n, the band index, besides k. Band 2 is on the CD segment, band 3 is on the EF segment, and so on. In principle, we have an infinite number of bands, and Fig. 8.8(a) only shows six bands. We read off ε from those intercepts.[5] Then, we choose a different value for $\cos[ka(1+r)]$, which gives us a different set of ε. Because $\varepsilon = E/V_0$, we can find every E_k for every k.

Figure 8.8(b) shows our numerical band dispersion with $k(a+b) \in (0,\pi)$, where $k \in (0, \frac{\pi}{a+b})$ defines the first Brillouin zone.[6] One sees that E_k for all the even bands, i. e., bands 2, 4 and 6, decrease with k, and all the odd bands increase. This is because $F(\varepsilon)$ is oscillatory with ε. If we want to express the band dispersion in the extended Brillouin zone, we need to shift the crystal momentum k as follows. We first note that, when we search roots of $F(\varepsilon)$ from -1 to 1, $k(a+b)$ only changes from π to 0 and does not exceed π. But, physically, $k(a+b)$ should start from π and end at 2π for the second band. The remaining even bands must have a similar change in their $k(a+b)$. We can correct this ordering of $k(a+b)$ by first changing $k(a+b)$ to $-k(a+b)$ and adding 2π for band 2, 4π for band 4, and 6π for band 6. For all the odd bands, we must shift by 2π for band 3 and 4π for band 5. The results are shown in Fig. 8.8(c). One sees that a gap is open at the Brillouin zone boundaries, with $k(a+b)$ being 1π, 2π, 3π, 4π, and so on. The KP model finally allows us to connect with the free-electron model and the infinite quantum-well model.

Exercise 8.5.6.

14. Explain how to shift bands in order to have a band structure like Fig. 8.8.

5 Our code `KP.f` uses the middle point method to find accurate values for ε.

6 Because k enters the equation through $\cos[k(a+b)]$, which is a periodic function of $k(a+b)$, shifting 1π produces the negative of $\cos[k(a+b)]$, and shifting 2π produces the same solution mathematically. But, physically, k represents a different wavevector. This contradiction must be corrected. We must manually extend our Brillouin zone, called the extended Brillouin zone in Chapter 9.

15. Find the roots of $F(\varepsilon)$, where $\mathcal{A} = 10$ and $r = 0.2$.
16. Choose $\mathcal{A} = 5$ and $r = 0.2$. (a) Use the code KP.f to compute and plot the band structure as a function of $k(a + b)$. (b) Investigate how the band gap changes with $\mathcal{A}$.

8.6 Problems

1. Ethylene has the chemical formula C_2H_4. The π molecular orbital can be represented by two p_z AOs at two carbon atoms that stick out from the main chain. The overlap integral S between two p_z AOs is $S_{12} = a\cos(\alpha)$, where α is the angle between two AOs and a is a constant. Assume $S_{11} = S_{22} = 1$. The Hamiltonian $H_{11} = H_{22} = \epsilon$ and $H_{12} = H_{21} = -t$. (a) Find the eigenvalues and eigenvectors. (b) Based on (a), describe what happens when we change α from 0 to π.
2. Butadiene (C_4H_6) has four carbon atoms forming a chain, whose Hamiltonian matrix is

$$\begin{pmatrix} E_0 & -t & 0 & 0 \\ -t & E_0 & -t & 0 \\ 0 & -t & E_0 & -t \\ 0 & 0 & -t & E_0 \end{pmatrix}.$$

   (a) Compute the eigenvalues. (b) Find the eigenvectors. (c) Assume that we have four π electrons occupying the lowest two MOs. Which MO is HOMO? Which one is LUMO? (d) Following (c), find the number of electrons at each C_i. (e) What happens to (d) if we only have three electrons?
3. Cyclobutadiene (C_4H_4) has four carbon atoms forming a ring, whose Hamiltonian matrix is

$$\begin{pmatrix} E_0 & -t & 0 & -t \\ -t & E_0 & -t & 0 \\ 0 & -t & E_0 & -t \\ -t & 0 & -t & E_0 \end{pmatrix}.$$

   (a) Compute the eigenvalues. (b) Find the eigenvectors. (c) Assume that we have four π electrons occupying the lowest two MOs. Which MO is HOMO? Which one is LUMO? (d) Following (c), find the number of electrons at each C_i. (e) What happens to (d), if we only have three electrons?
4. Benzene has six π electrons. Its Hamiltonian is $\hat{H} = \sum_{i=1}^{6} E_0 |i\rangle\langle i| - t\sum_{i=1}^{6} |i\rangle\langle i+1|$, where $i+1$ in the second summation is changed to 1 if $i = 6$. (a) Diagonalize this matrix to find the eigenvalues and eigenvectors. (b) find the number of electrons at each C_i.
5. Polyacetylene normally has dimerization due to the electron–lattice interaction, where one bond is longer with the hopping integral $-t_1$ and the other is shorter

with hopping integral $-t_2$. The on-site energy is E_0. (a) Write down the Hamiltonian. (b) Suppose the chain length is $L = N(a_1 + a_2)$, where a_1 and a_2 are the bond lengths, find the energy dispersion E_β.

6. Starting from our Kronig–Penney periodic potential, choose two different sets of $\mathcal{A}$, r to demonstrate that one can reproduce the results in the free electron model and the infinite quantum-well model.
7. In the Kronig–Penney potential, one can compute the linear momentum between two eigenstates at k_1 and k_2, $\langle k_1|\hat{p}_x|k_2\rangle$. (a) Show that this matrix is nonzero only if $(k_1 - k_2)a = 2n\pi$, where n is an integer. (b) Find $\langle k|\hat{p}_x|k\rangle$. Hint: Check Ref. [56].

9 Electrons in crystalline solids

Solids come in different sizes and shapes: Some are amorphous and some are crystallines. As briefly discussed in the previous chapter, larger systems are still off limits for QM. This chapter focuses on crystalline solids, where we are going to introduce some of the basic concepts in solid-state physics.

This chapter is grouped into five units:

- Unit 1, consisting of Sections 9.1 and 9.2, introduces the crystal structure, translational symmetry, the Bloch theorem, crystal momentum, and Brillouin zones.
- Unit 2, including Sections 9.3 and 9.4, presents the nearly free-electron model and the tight-binding model.
- Unit 3 includes Section 9.5 and introduces three major topics in solid state physics: Fermi surface, Fermi energy, and chemical potential.
- Unit 4 consists of Sections 9.6 and 9.7 covering quantum transports through conduction and transmission via the Landauer's formalism, and, finally, with application to scanning tunneling microscopy.
- Unit 5 (Section 9.8) introduces nuclear vibrations, the normal mode, and phonon concepts.

9.1 Crystal structure

Crystalline solids are a special group of solids, in which atoms form a regular pattern repeating itself infinitely in space. To understand the structures and properties of solids, it often suffices to look at these repeating patterns.

9.1.1 Unit cell and primitive cell

Figure 9.1(a) shows an example in a two-dimensional space, where two different types of atoms A (small circles) and B (big circles) are arranged on a chessboard. If we group two adjacent A and B atoms into a cell, we can completely cover the entire *solely by translating the cell.* This cell is called the primitive unit cell or the minimum cell. We can also choose ABAB as our cell. If we translate over a longer distance, we can again cover the entire crystal completely solely by translating the cell, so ABAB is also a unit cell, but contains two primitive unit cells and gains a different name, a supercell. One can have many different supercells, but only one primitive cell. So the ability to *completely cover the entire crystal solely by translating the cell* is a key characteristics of a unit cell. Cells such as ABA or BBA are not unit cells. All other symmetry operations, such as rotation or inversion, are not permitted when we define a unit cell.

https://doi.org/10.1515/9783110672152-009

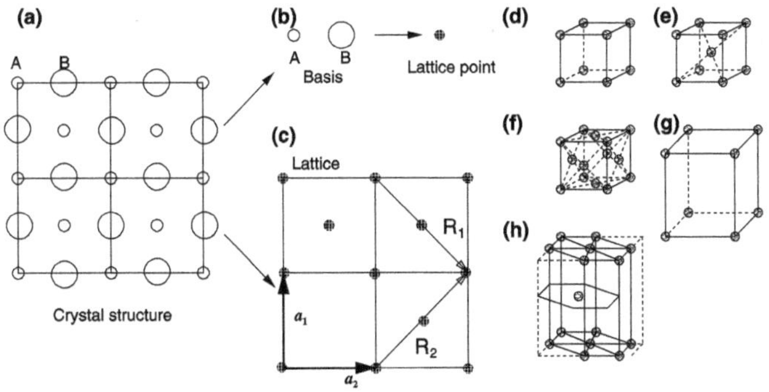

Figure 9.1: (a) A two-dimensional square lattice with two types of atoms: A (small circles) and B (big circles). (b) A basis consists of two different types of atoms, and is conceptualized to a single point, called the lattice point. (c) A crystal structure consists of basis and lattice. The two shortest vectors $\mathbf{a}_1$ and $\mathbf{a}_2$, primitive translational vectors, defines the lattice. (d) Simple cube. (e) Body-centered cube. (f) Face-centered cube. (g) Tetragonal lattice. (h) Hexagonal lattice.

Exercise 9.1.1.

1. Show that neither ABB nor ABA is not a unit cell, but AABB and BBAA are.

9.1.2 Basis and lattice points

A basis is a compartment that hosts a group of atoms. Figure 9.1(b) shows a basis that houses a primitive unit cell that has two atoms. There are many different ways of choosing a primitive cell. Mathematically, one conceptualizes the basis as a single point, the lattice point, by replacing every unit cell in a real crystal by a single lattice point (the grey circle). These lattice points form a scaffold, called the lattice, as shown in Fig. 9.1(c). The basis and the lattice point are the same thing, except that the former emphasizes the arrangement of the atoms inside each basis while the latter comprises rather abstract points. For this reason, one needs both the basis and the lattice to describe a crystal. The lattice provides lattice points to which a basis can be attached, while the basis specifies what atoms are there and how they are arranged: crystal structure = lattice + basis [22].

Exercise 9.1.2.

2. What is the main difference between the basis and lattice points?
3. To describe a crystal structure, what are the two key elements?

9.1.3 Primitive lattice vectors and lattice vectors

The concept of the lattice is formalized by the lattice vectors $\mathbf{R}$, pointing from one lattice point to another. The minimum lattice vectors are the primitive lattice vectors. Figure 9.1(c) shows two primitive vectors $\mathbf{a}_1$ and $\mathbf{a}_2$, so $\mathbf{R}$ can be written as $\mathbf{R} = l\mathbf{a}_1 + m\mathbf{a}_2$ for two-dimensional systems, or in three dimensions, $\mathbf{R} = l\mathbf{a}_1 + m\mathbf{a}_2 + n\mathbf{a}_3$, where l, m, and n are integers. The repeating pattern is mathematically realized by $\mathbf{R}$.

Exercise 9.1.3.

4. Figure 9.1(c) shows two vectors $\mathbf{R}_1$ and $\mathbf{R}_2$. Express them in terms of two primitive vectors $\mathbf{a}_1$ and $\mathbf{a}_2$.

9.1.4 Bravais lattice and lattice constant

Although there are many different crystal structures, there are only 14 unique lattice structures. These unique structures have a special name, Bravais lattices. Figures 9.1(d)–(h) show some of the most common structures, simple cubic (sc), body-centered cubic (bcc), face-centered cubic (fcc), tetragonal and hexagonal. All other lattice structures can be mapped onto 14 Bravais lattices. When we discuss lattice structures, we must use the Bravais lattices that enable us to define the lattice constant. The side lengths of Bravais lattices, *not of primitive cells*, are defined as the lattice constant. An exception is that sc, bcc, and fcc use the simple cube to define the lattice constant, even though their primitive cells are different. But a Bravais lattice, by definition, has no information about the basis. We often want to know both the Bravais lattice and the basis. For instance, NaCl and diamond both have a fcc Bravais lattice, but their bases are different. In diamond, the basis consists of two types of carbon atoms: one is $(0,0,0)$ and the other is at $(\frac{a}{4},\frac{a}{4},\frac{a}{4})$, but in NaCl, Na is at $(0,0,0)$ and Cl is at $(\frac{a}{2},\frac{a}{2},\frac{a}{2})$. Here, a is the lattice constant. This basis difference manifests itself when we carry out the symmetry operation such as rotation, inversion, and mirroring of the basis. These symmetry operations are not allowed when we define a lattice, but they are allowed when we define a basis. Fourteen Bravais lattices, together with many more types of basis, introduce 230 space groups [58]. This is the mathematical realization of our conclusion that crystal structure = lattice + basis.

9.2 Translational symmetry and Bloch theorem

9.2.1 Bloch theorem, Born–von Karman boundary condition and Bloch waves

A system property is determined by the Hamiltonian. Since the kinetic energy operator $-\frac{\hbar^2\nabla^2}{2m_e}$ depends only on the derivative of the coordinate, a rigid shift in position has no effect on it. So the translational symmetry is reflected in the potential energy. If the potential remains the same after shifting the potential by a lattice vector **R**,

$$V(\mathbf{r}+\mathbf{R}) = V(\mathbf{r}), \tag{9.2.1}$$

then the Hamiltonian is said to have the translational symmetry, $\hat{H}(\mathbf{r}+\mathbf{R}) = \hat{H}(\mathbf{r})$.

We introduce the translation operator $\hat{T}_{-\mathbf{R}}$ that shifts any operator $\hat{O}(\mathbf{r})$ or function $f(\mathbf{r})$ to $\hat{O}(\mathbf{r}+\mathbf{R})$ or $f(\mathbf{r}+\mathbf{R})$, i. e., $\hat{T}_{-\mathbf{R}}\hat{O}(\mathbf{r}) = \hat{O}(\mathbf{r}+\mathbf{R})$, $\hat{T}_{-\mathbf{R}}f(\mathbf{r}) = f(\mathbf{r}+\mathbf{R})$. Note that the result with $\hat{T}_{\mathbf{R}}$ is equivalent because it shifts **r** to **r** – **R** (recall Section 4.5.3). We apply $\hat{T}_{-\mathbf{R}}$ to $\hat{H}\psi(\mathbf{r})$, where ψ is an arbitrary wavefunction, to obtain

$$\hat{T}_{-\mathbf{R}}\hat{H}\psi(\mathbf{r}) = \hat{H}(\mathbf{r}+\mathbf{R})\psi(\mathbf{r}+\mathbf{R}) = \hat{H}(\mathbf{r}+\mathbf{R})\hat{T}_{-\mathbf{R}}\psi(\mathbf{r}) = \hat{H}(\mathbf{r})\hat{T}_{-\mathbf{R}}\psi(\mathbf{r}). \tag{9.2.2}$$

Since ψ is arbitrary, comparing the first and last terms of eq. (9.2.2) shows that $\hat{T}_{-\mathbf{R}}$ and $\hat{H}$ commute, $[\hat{T}_{-\mathbf{R}}, \hat{H}] = 0$, so they share common eigenfunctions. By examining the eigenfunctions of $\hat{T}_{-\mathbf{R}}$, we can gain insights into the crystal eigenequation,

$$\hat{H}(\mathbf{r})\psi(\mathbf{r}) = E\psi(\mathbf{r}), \tag{9.2.3}$$

where $\psi(\mathbf{r})$ is an eigenstate, not an arbitrary function. We apply $\hat{T}_{-\mathbf{R}}$ to eq. (9.2.3) and then use $\hat{H}(\mathbf{r}+\mathbf{R}) = \hat{H}(\mathbf{r})$ to find

$$\hat{H}(\mathbf{r}+\mathbf{R})\psi(\mathbf{r}+\mathbf{R}) = E\psi(\mathbf{r}+\mathbf{R}) \rightarrow \hat{H}(\mathbf{r})\psi(\mathbf{r}+\mathbf{R}) = E\psi(\mathbf{r}+\mathbf{R}),$$

which proves that $\psi(\mathbf{r}+\mathbf{R})$ is also an eigenstate with the same eigenenergy E. One can see $\psi(\mathbf{r}+\mathbf{R})$ must be related to $\psi(\mathbf{r})$.

Because $\hat{T}_{-\mathbf{R}}$ and $\hat{H}$ share common eigenfunctions $\psi(\mathbf{r})$, we must have

$$\hat{T}_{-\mathbf{R}}\psi(\mathbf{r}) = C(\mathbf{R})\psi(\mathbf{r}), \tag{9.2.4}$$

where $C(\mathbf{R})$ is the eigenvalue of $\hat{T}_{-\mathbf{R}}$. On the other hand, by virtue of the definition of the translation operator $\hat{T}_{-\mathbf{R}}$, we must have

$$\hat{T}_{-\mathbf{R}}\psi(\mathbf{r}) = \psi(\mathbf{r}+\mathbf{R}). \tag{9.2.5}$$

Comparing these two equations yields

$$\psi(\mathbf{r}+\mathbf{R}) = C(\mathbf{R})\psi(\mathbf{r}).$$

To find $C(\mathbf{R})$, we apply $T_{-\mathbf{R}_1}$ and $T_{-\mathbf{R}_2}$ consecutively to $\psi(\mathbf{r})$ to get

$$\hat{T}_{-\mathbf{R}_1}\hat{T}_{-\mathbf{R}_2}\psi(\mathbf{r}) = \hat{T}_{-\mathbf{R}_1}C(\mathbf{R}_2)\psi(\mathbf{r}) = C(\mathbf{R}_1)C(\mathbf{R}_2)\psi(\mathbf{r}),$$

which must be equal to a single translation $\hat{T}_{-\mathbf{R}_1-\mathbf{R}_2}\psi(\mathbf{r}) = C(\mathbf{R}_1+\mathbf{R}_2)\psi(\mathbf{r})$ (eq. (9.2.5)). Comparing both sides, we obtain

$$C(\mathbf{R}_1 + \mathbf{R}_2) = C(\mathbf{R}_1)C(\mathbf{R}_2). \tag{9.2.6}$$

If we let $\mathbf{R}_1 = 0$, then $C(\mathbf{R}_2) = C(0)C(\mathbf{R}_2)$ and $C(0) = 1$. If $\mathbf{R}_1 = \mathbf{R}_2 = \mathbf{R}$, $(C(\mathbf{R}))^2 = C(2\mathbf{R})$. If $\mathbf{R}_1 = -\mathbf{R}_2$, $C(\mathbf{R}_1) = (C(-\mathbf{R}_1))^{-1}$. This shows $C(\mathbf{R})$ has some special properties.

Next, following [20], we start from eq. (9.2.6) to derive an expression for $C(\mathbf{R})$. We first take the logarithm of both sides,

$$\ln C(\mathbf{R}_1 + \mathbf{R}_2) = \ln C(\mathbf{R}_1) + \ln C(\mathbf{R}_2). \tag{9.2.7}$$

We denote $\ln C(\mathbf{R})$ by $g(\mathbf{R})$, $g(\mathbf{R}_1) + g(\mathbf{R}_2) = g(\mathbf{R}_1 + \mathbf{R}_2)$. Then we take the derivative with respect to $\mathbf{R}_1$ to find $g'(\mathbf{R}_1) = g'(\mathbf{R}_1 + \mathbf{R}_2)$ and take another derivative with respect to $\mathbf{R}_2$ to find $g''(\mathbf{R}_1 + \mathbf{R}_2) = 0$. This proves that g must be a linear function of $\mathbf{R}$, $g(\mathbf{R}) = \mathbf{B}\cdot\mathbf{R} + D$. Because $C(0) = 1$ and $g(0) = 0$, $D = 0$. Then, $g(\mathbf{R}) = \mathbf{B}\cdot\mathbf{R} = \ln C(\mathbf{R})$, or $C(\mathbf{R}) = e^{\mathbf{B}\cdot\mathbf{R}}$. Hence, the eigenvalue of $\hat{T}_{-\mathbf{R}}$ is $e^{\mathbf{B}\cdot\mathbf{R}}$ and the eigenfunction is $\psi(\mathbf{r})$, which satisfies eq. (9.2.4),

$$\psi(\mathbf{r} + \mathbf{R}) = e^{\mathbf{B}\cdot\mathbf{R}}\psi(\mathbf{r}). \tag{9.2.8}$$

This shows that there is no need to assume the form of $C(\mathbf{R})$, and $\psi(\mathbf{r})$ must be identified with a free parameter $\mathbf{B}$. Now, we introduce an extra constraint to determine $\mathbf{B}$.

9.2.1.1 Born–von Karman boundary condition

Consider a crystal consisting of N_1, N_2, and N_3 cells along its three primitive basis vectors, $\mathbf{a}_1$, $\mathbf{a}_2$, and $\mathbf{a}_3$. We assume that the wavefunction remains the same after a rigid shift of the entire crystal by $N_j\mathbf{a}_j$, the Born–von Karman periodic boundary condition,

$$\psi(\mathbf{r} + N_j\mathbf{a}_j) = \psi(\mathbf{r}). \tag{9.2.9}$$

This periodic boundary condition has nothing to do with the translational symmetry. It is an extra constraint for us to fix $\mathbf{B}$. We specialize $\hat{T}_{-\mathbf{R}}$ to $\hat{T}_{-N_j\mathbf{a}_j}$, with translation along the $\mathbf{a}_j$ direction by N_j times, i. e., $\mathbf{R} = N_j\mathbf{a}_j$. Equation (9.2.8) becomes

$$\psi(\mathbf{r} + N_j\mathbf{a}_j) = e^{N_j\mathbf{B}_j\cdot\mathbf{a}_j}\psi(\mathbf{r}), \tag{9.2.10}$$

where we have added a subscript j to $\mathbf{B}$ since our translation is specialized along $\mathbf{a}_j$. Comparing eqs. (9.2.9) and (9.2.10), we find $e^{N_j\mathbf{B}_j\cdot\mathbf{a}_j} = 1$. If we choose $\mathbf{B}_j$ to be real, $\mathbf{B}_j$ must be zero since N_j and $\mathbf{a}_j$ cannot be zero. This is trivial. If $\mathbf{B}_j$ is imaginary, denoted as $i\mathbf{k}_j$, then $e^{iN_j\mathbf{k}_j\cdot\mathbf{a}_j} = 1$. Because $iN_j\mathbf{k}_j\cdot\mathbf{a}_j$ must be dimensionless, $\mathbf{k}$ has the dimension of the inverse of length, so it is a wavevector, and $\hbar\mathbf{k}$ is called the crystal momentum. To identify ψ with $\mathbf{k}$, we write it as $\psi_{\mathbf{k}}$, where $\mathbf{k}$ plays the same role as a quantum number. It allows us to solve $N_1N_2N_3$ Schrödinger equations separately, one $\mathbf{k}$ at a time, instead of a coupled giant equation, with the dimension at least of $N_1N_2N_3 \times N_1N_2N_3$ in terms of a matrix. It is also similar to a Fourier transform, but the transform is only limited to the Bravais lattice $\mathbf{R}$, but not $\mathbf{r}$, so the wavefunction is in a mixed representation $(\mathbf{k}, r)$.

9.2.1.2 Bloch theorem

Now, we can replace **B** by $i\mathbf{k}$ in eq. (9.2.8) to have

$$\hat{T}_{-\mathbf{R}}\psi_{\mathbf{k}}(\mathbf{r}) = \psi_{\mathbf{k}}(\mathbf{r}+\mathbf{R}) = e^{i\mathbf{k}\cdot\mathbf{R}}\psi_{\mathbf{k}}(\mathbf{r}), \tag{9.2.11}$$

which is the Bloch theorem. $\psi_{\mathbf{k}}$ is called the Bloch wavefunction. A different $\mathbf{k}$ produces a different phase factor. Because $\hat{T}$ and $\hat{H}$ share the common eigenstate, $\psi_{\mathbf{k}}(\mathbf{r})$ is an eigenstate of $\hat{H}$. Equation (9.2.11) connects the wavefunction at lattice point $\mathbf{R}$ with that at $\mathbf{R} = 0$. We note that although the system's crystal structure is periodic at every $\mathbf{R}$, its wavefunction is not; instead, every shift $\mathbf{R}$ produces a phase factor $e^{i\mathbf{k}\cdot\mathbf{R}}$ for the wavefunction. To isolate this phase factor, we introduce

$$\psi_{\mathbf{k}}(\mathbf{r}) = e^{i\mathbf{k}\cdot\mathbf{r}}u_{\mathbf{k}}(\mathbf{r}), \tag{9.2.12}$$

where $u_{\mathbf{k}}(\mathbf{r})$ is now periodic in $\mathbf{R}$, $u_{\mathbf{k}}(\mathbf{r}+\mathbf{R}) = u_{\mathbf{k}}(\mathbf{r})$, and $\psi_{\mathbf{k}}(\mathbf{r})$ obeys the Bloch theorem (see the exercise). We used $u(\mathbf{r})e^{i\mathbf{k}\cdot\mathbf{r}}$ in Chapter 1 to demonstrate the wave–particle duality. Finally, we should emphasize that not all Bloch wavefunctions can be written as eq. (9.2.12) (see eq. (9.4.13), but all Bloch wavefunctions can be written as $\psi_{\mathbf{k}}(\mathbf{r}+\mathbf{R}) = e^{i\mathbf{k}\cdot\mathbf{R}}\psi_{\mathbf{k}}(\mathbf{r})$. This is a consequence of the translational symmetry. A complication arises if we have a $\mathbf{k}'$ that differs from $\mathbf{k}$ by $\mathbf{G}$, i. e., $\mathbf{k}' = \mathbf{k}+\mathbf{G}$, and $\mathbf{G}\cdot\mathbf{R} = 2m\pi$, with m being an integer. Then our phase factor is the same $e^{i\mathbf{k}\cdot\mathbf{R}} = e^{i\mathbf{k}'\cdot\mathbf{R}}$, so is $\psi_{\mathbf{k}} = \psi_{\mathbf{k}+\mathbf{G}}$. This $\mathbf{G}$ is called the reciprocal lattice vector, which will be discussed in the next section.

Exercise 9.2.1.

5. A translational operator in space, $\hat{T}_\epsilon$, translates a one-dimensional function $\psi(x)$ to $\psi(x-\epsilon)$, $\hat{T}_\epsilon\psi(x) = \psi(x-\varepsilon)$. Show that, if ε is an infinitesimal number, $\hat{T}$'s corresponding operator is $\hat{p}_x$.
6. (a) Prove $\psi_{\mathbf{k}}(\mathbf{r}) = e^{i\mathbf{k}\cdot\mathbf{r}}u_{\mathbf{k}}(\mathbf{r})$ obeys the Bloch theorem. (b) Prove $u_{\mathbf{k}+\mathbf{G}}(\mathbf{r}) = e^{-i\mathbf{G}\cdot\mathbf{r}}u_{\mathbf{k}}(\mathbf{r})$.
7. (a) From eq. (9.2.11), prove $|\psi_{\mathbf{k}}(\mathbf{r}+\mathbf{R})|^2 = |\psi_{\mathbf{k}}(\mathbf{r})|^2$. (b) Explain this result.

9.2.2 Reciprocal lattice vectors and Brillouin zones

The previous subsection shows the translation property of the Bloch wavefunction. Here, we take a one-dimensional chain as an example to reveal further insights into the reciprocal lattice and, more importantly, introduce the Brillouin zone. Here $\mathbf{k}$ is just k for one dimension.

We consider a chain of length $L = Na$, which has N lattice sites and lattice constant a. As required by the Born–von Karman periodic boundary condition, $\psi_k(x+L) = \psi_k(x)$. On the other hand, the Bloch theorem requires $\psi_k(x+L) = e^{ikL}\psi_k(x)$. Comparing these two equations leads to $e^{ikL} = 1$, so $k = \frac{2\pi n}{L} = \frac{2\pi n}{Na}$, where n is an integer. If N is even, n starts from $-\frac{N}{2}+1$ and ends at $\frac{N}{2}$, so there are N different ks. If N is odd, $-\frac{N-1}{2} \le n < \frac{N-1}{2}$. The number of k points is the same as the number of lattice points. A positive (negative) k refers to the wave propagating along the $+x$ $(-x)$ direction.

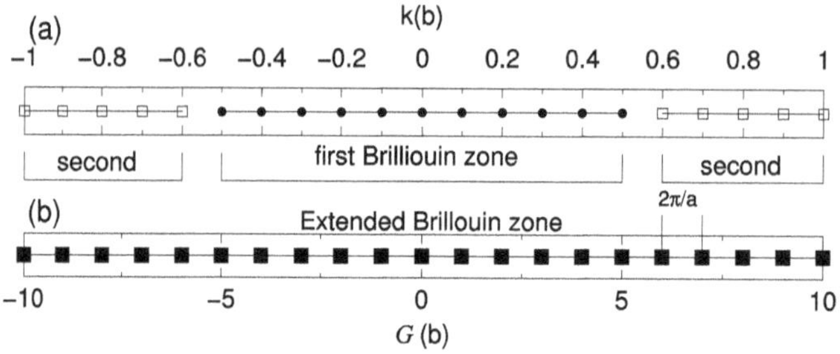

Figure 9.2: (a) The first (filled circles) and second (empty boxes) Brillouin zone with $N = 10$. $b = 2\pi/a$. (b) The extended Brillouin zone is identified by G. It is customary that, if the wavevector is smaller than b, we use k; but if larger than b, we use G. Once we combine k with G, $k + G$ can cover any wavevectors.

Among all possible ks, the smallest $k_{\min}$ is $\pm\frac{2\pi}{Na}$. Because the wavevector and the wavelength are reciprocal to each other, $\lambda = \frac{2\pi}{k}$, the longest wavelength is $\lambda = Na = L$, which is the system length. The largest $k_{\max} = \pm\frac{2\pi}{a}$ is defined as the primitive reciprocal lattice vector $b = 2\pi/a$, $ab = 2\pi$. We can use b as units of k, and then $k_{\max} = \pm b$. If k falls within $[-b/2, b/2]$, this region is called the first Brillouin zone (BZ) (see the filled circles in Fig. 9.2(a)). Here we set $N = 10$. The second BZ has two segments, $[-b, -b/2]$ and $[b/2, b]$ (see the empty boxes), but the length of the second BZ is the same as the first BZ, always b.

So far, we have only discussed the influence of the lattice structure on k. Now, we consider what happens if we introduce electrons into the system. An electron may have a higher energy and take a wavevector larger than k which is capped at b by the Born–von Karman condition. To go beyond b, we introduce a non-primitive reciprocal lattice vector G, where $G = mb$ and m is an integer, so we extend the reciprocal zone from $|k| \leq b$ to $k + G$, called extended zones. Figure 9.2(b) shows those extended zones. To see how this changes the wavefunction, consider a free electron, with its wavefunction (unnormalized) e^{ikx}. Then, in solids, we write it as $e^{i(k+G)x}$, where $|k| \leq b$ and $G = mb$, and m can be an integer.

In summary, in solid-state physics, k runs from $-b/2$ to $b/2$ in units of $\frac{b}{N}$, and G runs from b to mb in units of b. In three dimensions, a regular reciprocal-lattice vector $\mathbf{G} = l\mathbf{b}_1 + m\mathbf{b}_2 + n\mathbf{b}_3$ plays the same role as $\mathbf{R} = l\mathbf{a}_1 + m\mathbf{a}_2 + n\mathbf{a}_3$ in real space.

9.3 Nearly free-electron model

The previous section presents the framework for a solid. Now, we are ready to explore the nature of electronic states. The simplest model is the one-dimensional nearly free-electron model, where an electron experiences a periodic potential (see Fig. 9.3) that is contributed to by all the atoms. This potential energy is assumed weaker than the kinetic energy, so the electron still retains its free-electron character, which is the reason why we call it a nearly free-electron model. To simplify our problem further, we assume that

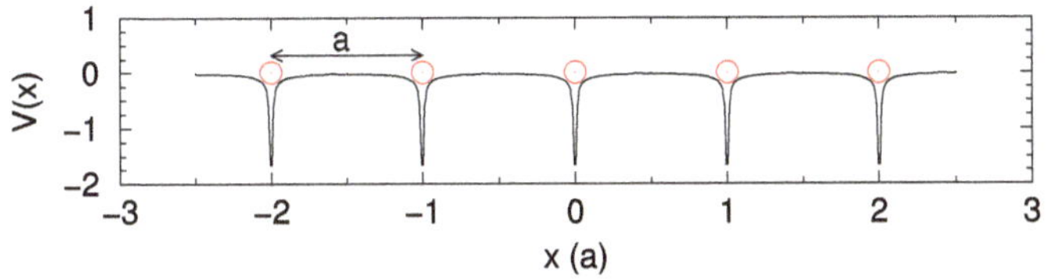

Figure 9.3: Periodic potential $V(x)$ in a one-dimensional nearly free-electron model of length $L = Na$. The lattice constant is a. We assume that a primitive cell contains only one atom.

the primitive cell contains only one atom, and the system has a length $L = Na$. N is the total number of lattice sites, and a is the lattice constant. The crystal Hamiltonian can be written as

$$\hat{H} = -\frac{\hbar^2}{2m_e}\frac{d^2}{dx^2} + V(x), \tag{9.3.1}$$

where we have changed the partial derivative with respect to x to the full derivative because both the kinetic energy and potential energy $V(x)$ are a function of x.

9.3.1 Nondegenerate perturbation treatment

The crystal potential (Fig. 9.3) differs from the atomic one in two aspects. First, the potential at every atom is contributed by potentials from all the atoms. Second, the potential does not decay to zero and instead repeats itself, $V(x+na) = V(x)$, at every lattice point. For these reasons, both bound and unbound states are present.

The Bloch wavefunction ψ_k is characterized by k, meaning that the solutions at different k are decoupled, in contrast to Section 8.2. Since $V(x)$ in eq. (9.3.1) is not expressed in the k space, we need to transform it to the k-space. Since the period of $V(x)$ is a, its Fourier transform is defined only at discrete lattice points as

$$V(x) = \sum_n V_n e^{inbx} = V_0 + \sum_{n\neq 0} V_n e^{inbx}, \tag{9.3.2}$$

where b is the reciprocal lattice vector ($b = 2\pi/a$), V_0 is the average potential, and $V_n = \frac{1}{Na}\int_0^{Na} e^{-inbx}V(x)dx$. We separate the Hamiltonian into the unperturbed $\hat{H}_0$ and perturbed $\hat{H}_I$ as

$$\hat{H} = \hat{H}_0 + \hat{H}_I, \tag{9.3.3}$$

where $\hat{H}_0 = -\frac{\hbar^2\partial^2}{m_e\partial x^2} + V_0$ and $\hat{H}_I = \sum_{n\neq 0} V_n e^{inbx}$. V_n is assumed to be so small that the problem can be treated by perturbation theory, but how small it is will become clear.

9.3.1.1 Zeroth-order perturbation

Perturbation theory only targets a state k in a system of length $L = Na$. At the zeroth order, we have only $\hat{H}_0$, and our Schrödinger equation is

$$-\frac{\hbar^2}{2m_e}\frac{d^2\phi_k}{dx^2} + V_0\phi_k = E_k^{(0)}\phi_k,$$

which has a known solution,

$$E_k^{(0)} = \frac{\hbar^2 k^2}{2m_e} + V_0, \quad \phi_k = \frac{1}{\sqrt{L}}e^{ikx}.$$

Here, the electron wavefunction is just like a regular plane wave of a free electron, except that its energy is shifted by V_0. So, the zeroth-order wavefunction is $\psi_k^{(0)} = \phi_k$.

9.3.1.2 First-order perturbation

The first-order energy correction is

$$E_k^{(1)} = \langle\phi_k|\hat{H}_I|\phi_k\rangle = \sum_{n\neq 0} V_n\langle\phi_k|e^{inbx}|\phi_k\rangle = \sum_{n\neq 0} V_n \int_0^{Na} \frac{1}{\sqrt{L}}e^{-ikx}e^{inbx}\frac{1}{\sqrt{L}}e^{ikx}dx,$$

whose integral is

$$\int_0^{Na} e^{inbx}dx = \frac{1}{inb}e^{inbx}\Big|_{x=0}^{x=Na} = \frac{e^{in\frac{2\pi}{a}Na} - 1}{inb} = 0.$$

So, $E_k^{(1)} = 0$, i. e., no correction at the first order.

The first-order wavefunction correction is

$$\psi_k^{(1)} = \sum_{q\neq k} \frac{\langle\phi_q|\hat{H}_I|\phi_k\rangle}{E_k^{(0)} - E_q^{(0)}}|\phi_q\rangle, \tag{9.3.4}$$

where $E_q^{(0)} = \frac{\hbar^2 q^2}{2m_e} + V_0$. There are two ways to compute $\langle\phi_q|\hat{H}_I|\phi_k\rangle$. One is to directly integrate it (see the homework exercise). The other is to break the integral over x in

$$\langle\phi_q|\hat{H}_I|\phi_k\rangle = \sum_{n\neq 0} V_n\frac{1}{L}\int_0^{Na} e^{-iqx}e^{inbx}e^{ikx}dx \tag{9.3.5}$$

into N segments of equal distance of a, i. e., N unit cells,

$$\int_0^{Na} e^{-iqx}e^{inbx}e^{ikx}dx = \int_0^{a} + \int_a^{2a} + \cdots + \int_{(N-1)a}^{Na}. \tag{9.3.6}$$

As an example, we choose a generic one

$$\int_{ma}^{(m+1)a} e^{(-iq+ik+inb)x}dx \tag{9.3.7}$$

and shift x, including its limits, by ma to get

$$\int_{0}^{a} e^{(-iq+inb+ik)(ma+x)}dx = e^{i(-q+nb+k)ma}\int_{0}^{a} e^{i(-q+nb+k)x}dx,$$

which reveals that each term in eq. (9.3.6) has a common integral multiplied by a term depending on m. Thus we can rewrite eq. (9.3.6) as the summation over m from 0 to $N-1$:

$$\sum_{m=0}^{N-1} e^{im(-q+nb+k)a}\int_{0}^{a} e^{i(k-q+nb)x}dx, \tag{9.3.8}$$

where the sum over all the lattice points m is called the Bloch sum. If $(k-q+nb)=0$, the sum is N and the integral is a, so we have $Na=L$ for eq. (9.3.6). If $(k-q+nb)\neq 0$, the sum is zero (see the following exercise). So, we can write eq. (9.3.8) as $\delta_{k-q+nb,0}$.

Inserting $\delta_{k-q+nb,0}$ into eq. (9.3.5), we get

$$\langle\phi_q|\hat{H}_I|\phi_k\rangle = \sum_{n\neq 0} V_n\frac{1}{L}Na\delta_{k-q+nb,0} = \sum_{n\neq 0} V_n\frac{1}{L}L\delta_{k-q+nb,0} = V_n\delta_{k-q+nb,0},$$

where n, q, k are linked by $n=\frac{q-k}{b}$ or $q=nb+k$, so the summation over q can be converted to the summation over n in eq. (9.3.4), or vice versa. So, our first-order wavefunction correction (eq. (9.3.4)) is

$$\psi_k^{(1)} = \sum_{q\neq k}\frac{\langle\phi_q|\hat{H}_I|\phi_k\rangle}{E_k^{(0)}-E_q^{(0)}}|\phi_q\rangle = \sum_{q\neq k}\frac{V_n}{E_k^{(0)}-E_q^{(0)}}\frac{e^{i(k+nb)x}}{\sqrt{L}}, \tag{9.3.9}$$

where we have converted e^{iqx} to $e^{i(k+nb)x}$, and the summation over q in the second equation must be understood as being over n as well.

The unnormalized ψ_k, up to first order, is

$$\psi_k = \frac{1}{\sqrt{L}}\left(1+\sum_{q\neq k}\frac{V_n}{E_k^{(0)}-E_q^{(0)}}e^{inbx}\right)e^{ikx} \equiv u_k(x)e^{ikx}, \tag{9.3.10}$$

where the summation over q is also over n because of $n=\frac{q-k}{b}$, and the periodic function $u_k(x)$ in eq. (9.2.12) is defined by those terms before e^{ikx} in eq. (9.3.10).

To reveal some physical insights concerning this new wavefunction, we choose a single term from the summation in eq. (9.3.10) (unnormalized),

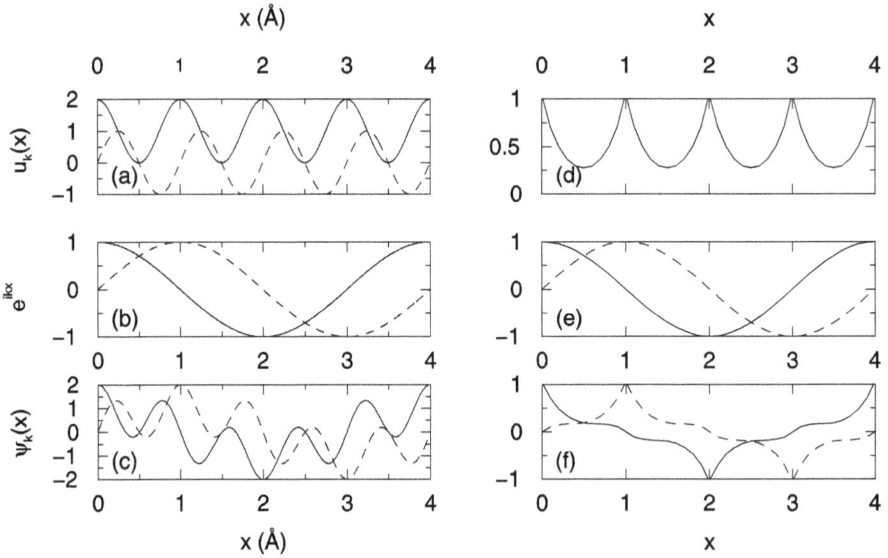

Figure 9.4: (a)–(c) are for the nearly free electron model. (d)–(f) are for the tight-binding model, with the horizonal axis in units of a. (a) and (d) The periodic function $u_k(x)$. In (d), it is the s orbital. The solid line denotes the real part and the dashed line is the imaginary part. (b) and (e) The Bloch envelope. (c) and (f) The Bloch wavefunction.

$$\psi_k = \frac{e^{ikx}}{\sqrt{L}}\left(1 + \frac{V_n}{E_k^{(0)} - E_q^{(0)}} e^{inbx}\right).$$

Figure 9.4(a) shows $u_k(x)$ is indeed periodic at every lattice site, where $a = 1\,\text{Å}$, $n = 1$, $b = \frac{2\pi}{a}$, $N = 4$, $L = Na$, and $k = \frac{2\pi}{Na}$. The Bloch envelope function e^{ikx} uses the smallest k (Fig. 9.4(b)). The final Bloch wavefunction is a product of the above two functions. Figure 9.4(c) reveals how complicated ψ_k is even for a nearly-free electron. This demonstrates the necessity of the Bloch wave. After normalization, we can compute the expectation value of the momentum $\hat{p}_x$ as

$$\langle \psi_k | \hat{p}_x | \psi_k \rangle = \hbar k + \frac{n\hbar b |Q_n|^2}{1 + |Q_n|^2},$$

where $Q_n = \frac{V_n}{E_k^{(0)} - E_q^{(0)}}$. One sees that the electron momentum in a solid consists of two parts [58]. One is the crystal momentum $\hbar k$, and the other is from the potential V_n. Only in the limit $V_n = 0$ are they the same.

The second-order energy correction is

$$E_k^{(2)} = \sum_{q \neq k} \frac{|\langle \phi_q | \hat{H}_I | \phi_k \rangle|^2}{E_k^{(0)} - E_q^{(0)}} = \sum_{n \neq 0} \frac{|V_n|^2}{\frac{\hbar^2 k^2}{2m_e} - \frac{\hbar^2}{2m_e}(k + nb)^2}.$$

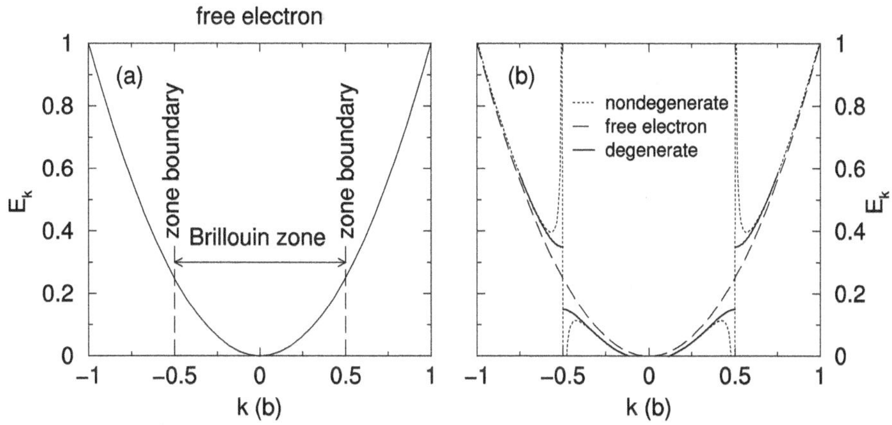

Figure 9.5: (a) Band dispersion of the free electron as a function of k. (b) Nearly-free electron band dispersion. The dotted line is from nondegenerate perturbation theory; the solid line is from degenerate perturbation theory; and the long-dashed line reproduces the free electron band. Energy gaps are open at $\pm b/2$.

So, the energy up to the second order is

$$E_k = E_k^{(0)} + E_k^{(2)} = \frac{\hbar^2 k^2}{2m_e} + \sum_{n\neq 0} \frac{|V_n|^2}{\frac{\hbar^2 k^2}{2m_e} - \frac{\hbar^2}{2m_e}(k+nb)^2}. \tag{9.3.11}$$

This shows that the crystal field adds an extra term to the otherwise free-electron energy. Figure 9.5(a) plots the energy dispersion for the free electron as a function of k. The dotted line in Fig. 9.5(b) reveals that if the denominator in eq. (9.3.11) is large, this extra term is small, so we have a nearly-free electron which follows the free-electron energy (long-dashed line). But as k approaches $\pm b/2$, there is a huge spike because, if $k^2 = (k + nb)^2$, or $k = -\frac{nb}{2}$, the energy computed from eq. (9.3.11) diverges. Figure 9.5(b) shows that the divergence occurs at $k = \pm\frac{b}{2}$, which is exactly at the Brillouin zone boundaries. This unphysical feature must be removed.

Exercise 9.3.1.

8. Prove the Bloch sum $\sum_{m=0}^{N-1} e^{im(-q+nb+k)a}$ is N if $(-q + nb + k) = 0$ and is zero if $(-q + nb + k) \neq 0$, where q and k are crystal momenta, b is the primitive reciprocal lattice vector, $b = 2\pi/a$, and N is the number of lattice sites.
9. Show u_k in eq. (9.3.10) is indeed periodic, meaning $u_k(x + na) = u_k$.
10. Starting from eq. (9.3.2), show $V(x)$ indeed satisfies $V(x) = V(x + na)$.
11. $V(x)$ is a periodic function, $V(x+a) = V(x)$. Find its Fourier transform in the reciprocal k space and compare your result with eq. (9.3.2).

9.3.2 Degenerate perturbation and energy band gap

The nondegenerate perturbation theory allowed us to address the band dispersion, but left an unresolved divergence. Equation (9.3.11) diverges if $k^2 = (k+nb)^2$, i. e., $k = -\frac{nb}{2}$. Since $n = \frac{q-k}{b}$, we find $q = +\frac{nb}{2}$, so $E_k^{(0)} = E_q^{(0)}$, i. e., q and k have degenerate energies. This divergence occurs at every $nb/2$. This is the reason why our Brillouin zones are defined at $\pm b/2$ and its multiples in Fig. 9.2. Having understood the origin of divergence, the solution is straightforward. We adopt the degenerate perturbation theory.

In the following, we choose two points, $k = \frac{nb}{2}(\Delta - 1)$ and $q = \frac{nb}{2}(\Delta + 1)$, that satisfy $k - q = nb$, as required by the Bloch sum. Here, Δ is a small dimensionless number and allows us to investigate the k-dependence of both the wavefunction and the eigenenergy around the BZ boundaries $\frac{nb}{2}$. Our Hamiltonian is the same as before, $\hat{H} = \hat{H}_0 + \hat{H}_I = -\frac{\hbar^2}{2m_e}\frac{d^2}{dx^2} + V_0 + \sum_{n\neq 0} V_n e^{inbx}$. The degenerate perturbation theory takes the unperturbed eigenstates $|\phi_k\rangle$ and $|\phi_q\rangle$ as the basis functions for $\hat{H}$.

The Hamiltonian matrix elements are $\langle\phi_k|\hat{H}|\phi_k\rangle = \frac{\hbar^2k^2}{2m_e} + V_0$, $\langle\phi_q|\hat{H}|\phi_q\rangle = \frac{\hbar^2q^2}{2m_e} + V_0$, $\langle\phi_k|\hat{H}|\phi_q\rangle = V_n$, and $\langle\phi_q|\hat{H}|\phi_k\rangle = V_n^*$. So our eigenequation is

$$\begin{pmatrix} \frac{\hbar^2k^2}{2m_e} + V_0 & V_n \\ V_n^* & \frac{\hbar^2q^2}{2m_e} + V_0 \end{pmatrix} \begin{pmatrix} C_1 \\ C_2 \end{pmatrix} = E \begin{pmatrix} C_1 \\ C_2 \end{pmatrix}, \tag{9.3.12}$$

which has nontrivial solutions only if its determinant is zero, i. e.,

$$E^2 - \left(\frac{\hbar^2k^2}{2m_e} + \frac{\hbar^2q^2}{2m_e} + 2V_0\right)E + \left(\frac{\hbar^2k^2}{2m_e} + V_0\right)\left(\frac{\hbar^2q^2}{2m_e} + V_0\right) - |V_n|^2 = 0.$$

Solving it produces two roots:

$$E_\pm = \frac{1}{2}\left(\frac{\hbar^2k^2}{2m_e} + \frac{\hbar^2q^2}{2m_e}\right) + V_0 \pm \sqrt{\left(\frac{\hbar^2k^2}{2m_e} - \frac{\hbar^2q^2}{2m_e}\right)^2/4 + |V_n|^2}. \tag{9.3.13}$$

9.3.2.1 $\Delta = 0$

If $\Delta = 0$, $k = -q = -\frac{nb}{2}$ and $E_\pm = \frac{\hbar^2k^2}{2m_e} + V_0 \pm |V_n|$. We see that $V(x)$ splits the initial degenerate band into two bands and opens a gap $\Delta E = E_+ - E_- = 2|V_n|$ at every $nb/2$. To see why this happens, we compute the wavefunction.

We choose a lower energy state, $E_- = \frac{\hbar^2k^2}{2m_e} + V_0 - |V_n|$, and substitute it into eq. (9.3.12) to find $C_2 = -\frac{|V_n|}{V_n}C_1$. Although one may wonder whether V_n is positive or negative [58], in the case of nearly free electrons, V_n is always negative because of the ion-core attraction, leading to $C_1 = C_2 = \frac{1}{\sqrt{2}}$. The wavefunction is $\psi_- = \frac{1}{\sqrt{2L}}(e^{ikx} + e^{-ikx}) = \sqrt{\frac{2}{L}}\cos(\frac{nbx}{2})$, a standing wave with a wavelength $\frac{2\pi}{nb/2} = \frac{2a}{n}$. The wavefunction has the maximum value and a zero slope at $x = ja$, characteristic of a bonding state. It has zero momentum, $\langle\psi_-|\hat{p}_x|\psi_-\rangle = 0$. In the presence of $V(x)$, the incident wave e^{ikx} is completely reflected from the lattice site once k matches $nb/2$, so the reflected and incident waves are counter-propagating to form a standing wave.

Similarly, the other eigenstate, $\psi_+ = \sqrt{\frac{2}{L}} i \sin kx = i\sqrt{\frac{2}{L}} \sin \frac{nbx}{2}$, is also a standing wave, an antibonding state, with zero momentum. Between E_+ and E_-, no state exists, so an energy gap is open at the BZ boundaries. This is the same quantization of energy as PIB, but here it is due to $V(x)$ and occurs once k matches $nb/2$. Away from the Brillouin zone boundaries, energy is continuous (as will be seen).

9.3.2.2 $\Delta \neq 0$

We substitute $k = \frac{-nb}{2}(1-\Delta)$ and $q = \frac{nb}{2}(1+\Delta)$ into eq. (9.3.13) to obtain

$$E_\pm = V_0 + T_n\left(\Delta^2 + 1 \pm \sqrt{4\Delta^2 + \frac{|V_n|^2}{T_n^2}}\right), \tag{9.3.14}$$

where $T_n = \frac{\hbar^2}{2m_e}(\frac{nb}{2})^2$. If we take the derivative of $E\pm$ with respect to Δ, $\frac{\partial E\pm}{\partial \Delta}$ is zero as $\Delta \to 0$ for both E_- and E_+. This means that $E_\pm$ is always perpendicular to the Brillouin zone boundary. Figure 9.6(a) shows that $E\pm$ both approach BZ at the right angle (solid lines), where E_+ is concave up. But, how E_- behaves depends on $|V_n|$ and T_n. We take the second derivative $\frac{\partial^2 E_\pm}{\partial \Delta^2}$ at $\Delta = 0$, with the details relegated to the following exercise:

$$\left.\frac{\partial^2 E_\pm}{\partial \Delta^2}\right|_{\Delta=0} = T_n\left(2 \pm \frac{4T_n}{|V_n|}\right).$$

So, if $|V_n| < 2T_n$, E_- is concave down. For a practical material, this condition can easily be met. This constitutes the condition for the nearly-free electron model: The electron's kinetic energy is larger than half of the potential energy $T_n > |V_n|/2$. Because E_+ curves up, while E_- curves down, they open a gap at the Brillouin zone boundary (Fig. 9.6(a)). Our treatment avoids the complication of the expansion of the square root in E in eq. (9.3.14) because it is difficult to justify whether $4\Delta^2$ or $\frac{|V_n|^2}{T_n^2}$ is a small quality, without stipulating the values of Δ, $|V_n|$ and T_n.

9.3.2.3 Momentum

The group momentum is defined as $\frac{m_e}{\hbar}\frac{\partial E_k}{\partial k}$. Since E_k is even with k, $\frac{\partial}{\partial k}$ is odd, and $p_x(-k) = -p_x(k)$ is an odd function of k. If we divide $p_x(k)$ by m_e, we get velocity $v_x(-k) = -v_x(k)$. Figure 9.6(b) shows that, away from the $nb/2$, the velocity is similar to that of the free electron (long-dashed line), but close to $nb/2$, it drops to zero. This means that the electron stops at the BZ boundary.

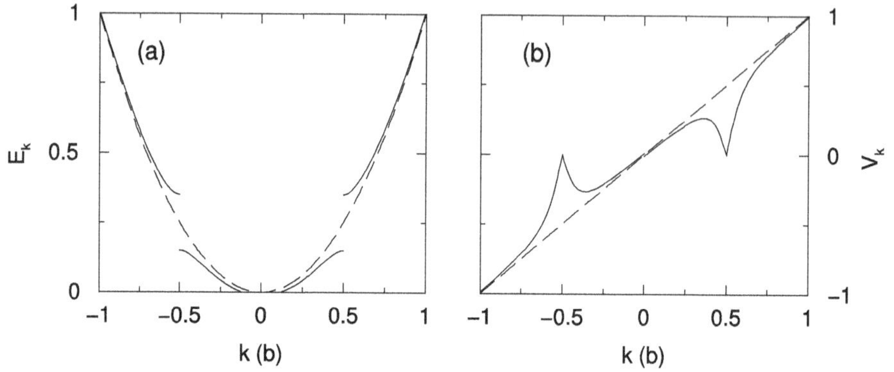

Figure 9.6: (a) Band structure of the nearly-free electron model (solid line). The band splits into two at $nb/2$. The long-dashed line is for the free electron. (b) Velocity dispersion. The velocity is in units of $\frac{n\hbar b}{2m_e}$.

We can also compute the expectation value of momentum $\hat{p}_x$ as well as $\langle\psi_+|\hat{p}_x|\psi_-\rangle$ and $\langle\psi_-|\hat{p}_x|\psi_+\rangle$ (see the following exercise), from which we can obtain the susceptibility

$$\chi_{xx}^{(1)}(\omega) = \frac{\hbar e^2}{m_e^2\epsilon_0\Omega}\sum_k \frac{\langle\psi_+|p_x|\psi_-\rangle\langle\psi_-|p_x|\psi_+\rangle}{\hbar\omega - (E_+ - E_-) + i\Gamma}\frac{f_+ - f_-}{E_+ - E_-}, \tag{9.3.15}$$

where the summation is over all the k points, k index on E and ψ is omitted, Ω is the length of the system (same as L, but used to be consistent with the three-dimensional formula), Γ is a small damping, and f is the occupation. We note that the conductivity and susceptibility are linked through, $\sigma = -i\omega\epsilon_0\chi^{(1)}$ in SI units.

9.3.2.4 Effective mass

The inverse of the effective mass $1/m^*$ is defined as $\frac{1}{\hbar}\frac{\partial v_k}{\partial k}$

$$\frac{1}{m^*} = \frac{1}{\hbar}\frac{\partial v_k}{\partial k} = \frac{1}{\hbar m_e}\frac{\partial p_x}{\partial k} = \frac{1}{\hbar m_e}\frac{\partial}{\partial k}\left(\frac{m_e}{\hbar}\frac{\partial E_k}{\partial k}\right) = \frac{1}{\hbar^2}\frac{\partial^2 E_k}{\partial k^2},$$

which is generic. For instance, for a free-electron $E_k = \frac{\hbar^2k^2}{2m_e}$, $\frac{1}{m^*} = \frac{1}{m_e}$. For our nearly free-electron model,

$$\frac{1}{m^*} = \frac{1}{m_e}\left(1 \pm \frac{1}{2}\frac{|V_n|^2/T_n^2}{\sqrt{[\Delta^2 + |V_n|^2/T_n^2]^3}}\right), \quad \text{valid for } \Delta \to 0.$$

The periodic potential V_n changes the effective mass as the electron moves through the crystal. At $\Delta = 0$, its mass is renormalized to $m^* = \frac{|V_n|}{|V_n| \pm \frac{T_n}{2}}m_e$. For the upper band, m^* is reduced; for the lower band, m^* is negative because $|V_n| < \frac{T_n}{2}$. This means that the electron has a negative acceleration even if a positive force applies on it. This is the consequence of the lattice potential.

Exercise 9.3.2.

12. Find the Hamiltonian matrix elements in eq. (9.3.12).
13. Show that, as far as $k - q = nb$, $\langle \phi_k | \hat{H} | \phi_q \rangle = V_n$ and $\langle \phi_q | \hat{H} | \phi_k \rangle = V_n^*$, are independent of Δ.
14. Starting from eq. (9.3.14): (a) prove $\frac{\partial E_\pm}{\partial \Delta}$ is zero for $\Delta \to 0$ for both E_- and E_+; and (b) prove $\left.\frac{\partial^2 E_\pm}{\partial \Delta^2}\right|_{\Delta=0} = T_n(2 \pm \frac{4T_n}{|V|})$.
15. At $\Delta \neq 0$, compute the momentum matrix elements $\langle \psi_+ | p_x | \psi_- \rangle$ between two eigenstates ψ_- and ψ_+.
16. Show at $\Delta \to 0$ that $m^* = \frac{|V_n|}{|V_n| \pm \frac{T_n}{2}} m_e$.

9.4 Tight-binding model

The nearly-free electron model reveals the rich physics of a crystal: band structure, Brillouin zone, and energy gap. But the model itself is more suitable for mobile electrons in metals that move around different atoms. The tight-binding model (TB) does the opposite, where the electron is assumed to be tightly bound to its parent atom and only hops between neighboring atoms, simulating semiconductors and insulators. We encountered the tight-binding model in Section 8.3. In chemistry, TB is also called the Hückel model. The words "tight" and "binding" refer to the fact that electron wavefunctions are tightly-bound to their parent atoms, with only a small overlap with neighboring atoms. The TB model is often used jointly with the linear combination of atomic orbitals (LCAO) (see Section 8.1.3). Here, we extend it to many atoms, so we can describe solids. Because of its flexibility, the TB model can simulate millions of atoms.

9.4.1 Three-dimensional systems

We consider a simple cubic structure (Fig. 9.7(a)), where the primitive cell has one atom. We place a single atom l at a lattice point $\mathbf{R}_l$, where $\mathbf{R}_l$ is the lattice vector, $\mathbf{R}_l = \ell_1 a\hat{x} + \ell_2 a\hat{y} + \ell_3 a\hat{z}$, a is the lattice constant, ℓ_1, ℓ_2 and ℓ_3 are integers, and $\hat{x}, \hat{y}$, and $\hat{z}$ are the unit vectors along the x-, y-, and z-axes, respectively. We further assume that only one atomic orbital, ϕ_s^{at}, an s orbital, participates in bonding. If we have more than one orbital, we can change the subscript s to a more general index.

The tight-binding approximation starts from the isolated atomic Schrödinger equation for atom l, which is

$$\hat{H}_l^{\mathrm{at}} \phi_s^{\mathrm{at}}(\mathbf{R}_l) = E_s^{\mathrm{at}} \phi_s^{\mathrm{at}}(\mathbf{R}_l), \tag{9.4.1}$$

or more specifically,

$$\hat{H}_l^{\mathrm{at}} \phi_s^{\mathrm{at}}(\mathbf{R}_l) = \left[-\frac{\hbar^2 \nabla_l^2}{2m_e} + V^{\mathrm{at}}(\mathbf{r} - \mathbf{R}_l) \right] \phi_s^{\mathrm{at}}(\mathbf{r} - \mathbf{R}_l) = E_s^{\mathrm{at}} \phi_s^{\mathrm{at}}(\mathbf{r} - \mathbf{R}_l), \tag{9.4.2}$$

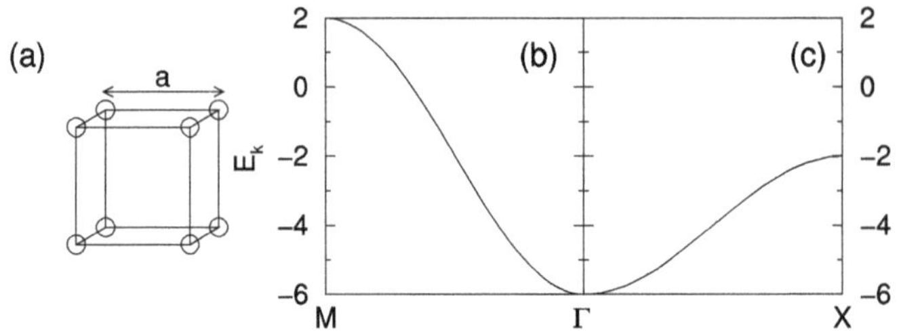

Figure 9.7: (a) Simple cubic structure, where the primitive cell contains one atom. a is the lattice constant. (b) Band dispersion along the $\Gamma - M$ direction. Here, we set $E^{at} - t_0$ to zero, and the energy unit is in t_1. (c) Band dispersion along the $\Gamma - X$ direction.

where E_s^{at} is the atomic orbital energy of eigenstate ϕ_s^{at} and $\hat{H}_l^{at}$ is the atomic Hamiltonian. For atoms $l_1, l_2, \ldots, l_N$, we have a similar equation, with $\mathbf{R}_l$ in the equation changed to $\mathbf{R}_{l_1}, \mathbf{R}_{l_2}, \ldots, \mathbf{R}_{l_N}$.

When N atoms form a solid, we have a new Hamiltonian at the lattice point $\mathbf{R}_l$,

$$\hat{H}_l = -\frac{\hbar^2 \nabla_l^2}{2m_e} + U(\mathbf{r} - \mathbf{R}_l) = \hat{H}_l^{at} + U(\mathbf{r} - \mathbf{R}_l) - V^{at}(\mathbf{r} - \mathbf{R}_l), \tag{9.4.3}$$

where $\hat{H}_l^{at}$ in the last equation is recovered by adding and then subtracting $V^{at}(\mathbf{r} - \mathbf{R}_l)$. To avoid confusion with $V^{at}(\mathbf{r} - \mathbf{R}_l)$, we denote the crystal potential by $U(\mathbf{r} - \mathbf{R}_l)$. $U(\mathbf{r} - \mathbf{R}_l)$ contains both the atomic potential $V^{at}(\mathbf{r} - \mathbf{R}_l)$ and the interatomic interactions. If we apply $\hat{H}_l$ to $\phi_l^{at}(\mathbf{r} - \mathbf{R}_l)$, we obtain

$$\hat{H}_l \phi_s^{at}(\mathbf{r} - \mathbf{R}_l) = E_s^{at} \phi_s^{at}(\mathbf{r} - \mathbf{R}_l) + \left[U(\mathbf{r} - \mathbf{R}_l) - V^{at}(\mathbf{r} - \mathbf{R}_l)\right] \phi_s^{at}(\mathbf{r} - \mathbf{R}_l), \tag{9.4.4}$$

where we have used eq. (9.4.1). It is clear that $\phi_s^{at}(\mathbf{r} - \mathbf{R}_l)$ is not an eigenstate of $\hat{H}_l$. In fact, other atoms have a similar equation, with the same E_s^{at} but its wavefunction is centered at different lattice sites. $U(\mathbf{r} - \mathbf{R}_l) - V^{at}(\mathbf{r} - \mathbf{R}_l)$ in eq. (9.4.3) constitutes the small perturbation $\hat{H}_I$ to $\hat{H}_l^{at}$. Because every $\{\phi_s^{at}\}$ corresponds to the same E_s^{at}, we have a degenerate problem and can solve the problem using the degenerate perturbation theory, where the eigenstate $\psi_{\mathbf{k}}(\mathbf{r})$ of $\hat{H}_l$, up to the first-order correction, is a linear combination of atomic orbitals $\{\phi_s^{at}\}$,

$$\psi_{\mathbf{k}}(\mathbf{r}) = \sum_l C_l^{\mathbf{k}} \phi_s^{at}(\mathbf{r} - \mathbf{R}_l), \tag{9.4.5}$$

where $C_l^{\mathbf{k}}$ is to be found.

9.4.1.1 Eigenstates

Because $\psi_{\mathbf{k}}(\mathbf{r})$ is a Bloch wavefunction, it must satisfy the Bloch theorem, $\psi_{\mathbf{k}}(\mathbf{r}+\mathbf{R}_l) = e^{i\mathbf{k}\cdot\mathbf{R}_l}\psi_{\mathbf{k}}(\mathbf{r})$, where $\mathbf{k}$ is the crystal momentum. This theorem itself demands that $C_l^{\mathbf{k}} = e^{i\mathbf{k}\cdot\mathbf{R}_l}C_0$ (see the next exercise), where C_0 is the normalization constant. This shows that we can find $C_l^{\mathbf{k}}$ without solving the crystal Schrödinger equation.

Now we can find the eigenenergy E from the crystal Schrödinger equation

$$\left[-\frac{\hbar^2}{2m_e}\nabla^2 + U(\mathbf{r})\right]\psi_{\mathbf{k}}(\mathbf{r}) = E\psi_{\mathbf{k}}(\mathbf{r}). \tag{9.4.6}$$

We substitute $\psi_{\mathbf{k}}(\mathbf{r})$ of eq. (9.4.5) into the left side of eq. (9.4.6) and use eq. (9.4.4) to get

$$E_s^{\mathrm{at}}\sum_l C_l^{\mathbf{k}}\phi_s^{\mathrm{at}}(\mathbf{r}-\mathbf{R}_l) + \sum_l C_l^{\mathbf{k}}[U(\mathbf{r}) - V^{\mathrm{at}}(\mathbf{r}-\mathbf{R}_l)]\phi_s^{\mathrm{at}}(\mathbf{r}-\mathbf{R}_l),$$

which equals the right side $E\sum_l C_l^{\mathbf{k}}\phi^{\mathrm{at}}(\mathbf{r}-\mathbf{R}_l)$. After some arrangement, we obtain

$$(E_s^{\mathrm{at}} - E)\sum_l C_l^{\mathbf{k}}\phi_s^{\mathrm{at}}(\mathbf{r}-\mathbf{R}_l) + \sum_l C_l^{\mathbf{k}}[U(\mathbf{r}) - V^{\mathrm{at}}(\mathbf{r}-\mathbf{R}_l)]\phi_s^{\mathrm{at}}(\mathbf{r}-\mathbf{R}_l) = 0. \tag{9.4.7}$$

Next, we multiply $\phi_s^{\mathrm{at}*}(\mathbf{r}-\mathbf{R}_m)$ by both sides and integrate over $\mathbf{r}$, so we have two integrals. The first is called the overlap integral,

$$\int \phi_s^{\mathrm{at}*}(\mathbf{r}-\mathbf{R}_m)\phi_s^{\mathrm{at}}(\mathbf{r}-\mathbf{R}_l)d\mathbf{r} \approx \delta_{lm}, \tag{9.4.8}$$

where we have made the tight-binding approximation by setting the overlap integral to zero, except for $l = m$ on the same lattice site. This simplifies the integral of the first $\sum_l$ in eq. (9.4.7) to $(E_s^{\mathrm{at}} - E)C_m^{\mathbf{k}}$.

The second is called the hopping integral. We do it differently[1] by keeping the hopping integral up to the nearest neighbors of $\mathbf{R}_m$ in the summation over l.

$$\int \phi_s^{\mathrm{at}*}(\mathbf{r}-\mathbf{R}_m)[U(\mathbf{r}) - V^{\mathrm{at}}(\mathbf{r}-\mathbf{R}_l)]\phi_s^{\mathrm{at}}(\mathbf{r}-\mathbf{R}_l)d\mathbf{r}. \tag{9.4.9}$$

Because $U(\mathbf{r}) = U(\mathbf{r}-\mathbf{R}_l)$, we can rewrite it as $U(\mathbf{r}-\mathbf{R}_l)$ and then shift all $\mathbf{r}$ by $\mathbf{R}_l$, so the integral only depends on $\mathbf{R}_m - \mathbf{R}_l$,

$$\int \phi_s^{\mathrm{at}*}(\mathbf{r}-(\mathbf{R}_m-\mathbf{R}_l))[U(\mathbf{r}) - V^{\mathrm{at}}(\mathbf{r})]\phi_s^{\mathrm{at}}(\mathbf{r})d\mathbf{r} \equiv -t(\mathbf{R}_m-\mathbf{R}_l) = -t_{m,l}, \tag{9.4.10}$$

where $t_{m,l}$ is called the hopping integral. Because $U(\mathbf{r}) - V^{\mathrm{at}}(\mathbf{r})$ is negative, $t_{m,l}$ is positive and $t_{m,l} = t_{l,m}$. If $m = l$, $t_{m,l} \equiv t_0$ is the atomic energy shift due to the difference between the crystal and atomic potentials $U(\mathbf{r}) - V^{\mathrm{at}}(\mathbf{r})$.

---

1 We cannot limit it to $l = m$ because only doing so returns to a group of isolated atoms, not a solid.

Equation (9.4.7) becomes

$$(E_s^{at} - E)C_m^{\mathbf{k}} - \sum_l t(\mathbf{R}_m - \mathbf{R}_l)C_l^{\mathbf{k}} = 0, \tag{9.4.11}$$

which is a secular equation if we write the Hamiltonian matrix out completely,

$$\begin{pmatrix} E_s^{at} - t_0 - E & -t_{1,2} & 0 & \cdots & 0 & -t_{1,N} \\ -t_{2,1} & E_s^{at} - t_0 - E & -t_{2,3} & 0 & \cdots & 0 \\ \cdots & \cdots & \cdots & \cdots & \cdots & 0 \\ \cdots & \cdots & \cdots & \cdots & \cdots & -t_{N-1,N} \\ -t_{N,1} & 0 & \cdots & 0 & -t_{N,N-1} & E_s^{at} - t_0 - E \end{pmatrix} \begin{pmatrix} C_1^{\mathbf{k}} \\ C_2^{\mathbf{k}} \\ \cdots \\ C_{N-1}^{\mathbf{k}} \\ C_N^{\mathbf{k}} \end{pmatrix} = 0. \tag{9.4.12}$$

Its eigenvector is just $\{C_l^{\mathbf{k}}\}$ as previously given, $C_l^{\mathbf{k}} = C_0 e^{i\mathbf{k}\cdot\mathbf{R}_l}$.

To determine the normalization constant C_0, we start from our wavefunction $\psi_{\mathbf{k}}(\mathbf{r}) = C_0 \sum_l e^{i\mathbf{k}\cdot\mathbf{R}_l}\phi_s^{at}(\mathbf{r} - \mathbf{R}_l)$. We invoke the normalization $\langle\psi_{\mathbf{k}}(\mathbf{r})|\psi_{\mathbf{k}}(\mathbf{r})\rangle = 1$,

$$\langle\psi_{\mathbf{k}}(\mathbf{r})|\psi_{\mathbf{k}}(\mathbf{r})\rangle = C_0^2 \sum_{l_1 l_2} e^{i\mathbf{k}\cdot(\mathbf{R}_{l_2} - \mathbf{R}_{l_1})}\langle\phi_s^{at}(\mathbf{r} - \mathbf{R}_{l_1})|\phi_s^{at}(\mathbf{r} - \mathbf{R}_{l_2})\rangle = C_0^2 N_1 N_2 N_3 = 1,$$

so $C_0 = 1/\sqrt{N_1 N_2 N_3}$, where N_i is the number of cells along three directions. Our final normalized wavefunction is

$$\psi_{\mathbf{k}}(\mathbf{r}) = \frac{1}{\sqrt{N_1 N_2 N_3}} \sum_l e^{i\mathbf{k}\cdot\mathbf{R}_l}\phi_s^{at}(\mathbf{r} - \mathbf{R}_l). \tag{9.4.13}$$

This shows that, at every lattice site, the wavefunction is modulated by the phase factor. Our periodic function $u(\mathbf{r})$ in the standard Bloch wavefunction (eq. (9.2.12)) is just the atomic function $\phi_s^{at}(\mathbf{r} - \mathbf{R}_l)$. $\psi_{\mathbf{k}}(\mathbf{r})$ does not have $e^{i\mathbf{k}\cdot\mathbf{r}}$, in contrast to the nearly free-electron model (eq. (9.3.10)).

9.4.1.2 Band energy

Once we have the wavefunction, we can find the band energy. We insert $C_l^{\mathbf{k}}$ into eq. (9.4.11) to get

$$E_{\mathbf{k}} = E_s^{at} - \sum_l t(\mathbf{R}_m - \mathbf{R}_l)e^{i\mathbf{k}\cdot(\mathbf{R}_l - \mathbf{R}_m)},$$

where we have added a subscript $\mathbf{k}$ to E to remind us that the energy disperses with $\mathbf{k}$. Since the summation depends on $\mathbf{R}_m - \mathbf{R}_l$, we can set $\mathbf{R}_m$ to zero,

$$E_{\mathbf{k}} = E_s^{at} - \sum_l t(\mathbf{R}_l)e^{i\mathbf{k}\cdot\mathbf{R}_l}. \tag{9.4.14}$$

If we only include terms up to the nearest neighbors and denote $t(\mathbf{R}_l = 0) = t_0$ and $t(\mathbf{R}_l) = t_1$ (for the nearest neighbor), we have

$$E_{\mathbf{k}} = E_s^{\text{at}} - t_0 - t_1 \sum_{l(\text{nearest neighbor})} e^{i\mathbf{k}\cdot\mathbf{R}_l}, \tag{9.4.15}$$

where t_0 is the onsite energy shift with respect to E_s^{at}.

Example 1 (Band structure in a simple cubic structure). Use eq. (9.4.15) to calculate the band dispersion in a simple cubic structure up to the first neighbors along two directions, one from Γ to M and the other from Γ to X. Here, the Γ point is at coordinate $(0,0,0)\frac{2\pi}{a}$, M at $(\frac{1}{2},\frac{1}{2},0)\frac{2\pi}{a}$, and X at $(\frac{1}{2},0,0)\frac{2\pi}{a}$.

We directly use eq. (9.4.15). We choose the origin at $(0,0,0)$, and then six neighbors are located at $(\pm a,0,0)$, $(0,\pm a,0)$ and $(0,0,\pm a)$, respectively. The summation over $(\pm a,0,0)$ is $-t_1 e^{ik_x(+a)} - t_1 e^{ik_x(-a)} = -2t_1\cos k_x a$. The other two can be found similarly, $-2t_1\cos k_y a$ and $-2t_1\cos k_z a$. Then, the energy dispersion is $E_{\mathbf{k}} = E^{\text{at}} - t_0 - 2t_1(\cos k_x a + \cos k_y a + \cos k_z a)$.

In solid state physics, high symmetry **k** points are denoted by special letters. Different crystal structures use different labels, except the Γ point whose coordinate is always $(0,0,0)\frac{2\pi}{a}$. For a simple cubic structure, X has the coordinate $(\frac{1}{2},0,0)$ in units of $\frac{2\pi}{a}$, with its six equivalent points $(\pm\frac{1}{2},0,0)$, $(0,\pm\frac{1}{2},0)$, and $(0,0,\pm\frac{1}{2})$ in units of $\frac{2\pi}{a}$. We have $E(\Gamma) = E^{\text{at}} - t_0 - 6t_1$, $E(X) = E^{\text{at}} - t_0 - 2t_1$. The band dispersion is $E_{\mathbf{k}} = E^{\text{at}} - t_0 - 2t_1[2 + \cos(k_x a)]$ for the Γ-X line, where k_x runs from 0 to $b/2$. Since M is at $(\frac{1}{2},\frac{1}{2},0)\frac{2\pi}{a}$, $E_k = E^{\text{at}} - t_0 - 2t_1[1 + 2\cos(ka)]$ for $\Gamma - M$, where k runs from 0 to $b/2$.

Figure 9.7(b) shows the band dispersion along the Γ-M direction, where we have set $E^{\text{at}} - t_0$ to 0. It starts from −6 (in units of t_1) to 2. However, along the Γ-X direction, Fig. 9.7(c) shows the band disperses from −6 to −2, weaker than that of the Γ-M direction, because the M point is at $(\frac{1}{2},\frac{1}{2},0)\frac{2\pi}{a}$.

Equation (9.4.15) is our key result. It shows that, if we had no neighbor, then $E_{\mathbf{k}}$ is just the atomic energy plus an average potential $-t_0$, just a single energy level. With the neighboring atoms, a single energy level broadens into a band, a continuous energy regime, called the energy band. For different **k**, $E_{\mathbf{k}}$ disperses differently.

For other structures, we can carry out a similar calculation. The only difference is that the number of neighbors and their coordinates can be different. In the exercise, one uses a body-centered cubic structure and the other uses a face-centered cubic structure. The validity of tight-binding approximation depends on the overlap of the atomic orbital with neighboring atoms. If the overlap is small, this approximation is very good.

Exercise 9.4.1.

17. Starting from $\psi_{\mathbf{k}}(\mathbf{r}) = \sum_l C_l^{\mathbf{k}} \phi_s^{\text{at}}(\mathbf{r} - \mathbf{R}_l)$, show that if $\psi_{\mathbf{k}}(\mathbf{r})$ is a Bloch wave, then $C_l^{\mathbf{k}}$ must take the form of $e^{i\mathbf{k}\cdot\mathbf{R}_l}$.

18. The body-centered cubic lattice has eight nearest neighbors. (a) Starting from eq. (9.4.14), show its band dispersion $E_{\mathbf{k}} = E_s^{\text{at}} - t_0 - t_1\cos(k_x a/2)\cos(k_y a/2)\cos(k_z a/2)$. (b) Choose the Γ-X line to plot the band.

9.4.2 One-dimensional ring

In polyacetylene, we employed a simple model to describe its band structure. Here, we adopt a similar model by considering a one-dimensional ring, with a periodic boundary condition. Our Hamiltonian is

$$\hat{H} = \sum_s t_0|s\rangle\langle s| - t\sum_s(|s+1\rangle\langle s| + |s\rangle\langle s+1|), \tag{9.4.16}$$

where s is the site index and runs from 1 to N and $|s\rangle$ is the orthonormalized atomic orbital, $\langle s_1|s_2\rangle = \delta_{s_1,s_2}$. The periodic boundary condition (PBC) means that, if $s = N+1$, this site is interpreted as $s = 1$; if $s = 0$, this is $s = N$. Similar to eq. (9.4.5), the translational symmetry ensures that the eigenstate must be a Bloch wave (unnormalized), $|\psi_k\rangle = \sum_s e^{iska}|s\rangle$, where k replaces β in Section 8.4.

Next, we apply $\hat{H}$ to ψ_k to test whether ψ_k is an eigenstate of $\hat{H}$. First, we change s to s_1 in $|\psi_k\rangle = \sum_{s_1} e^{is_1ka}|s_1\rangle$ and s in $\hat{H}$ in eq. (9.4.16) to s_2.

$$\hat{H}|\psi_k\rangle = \sum_{s_2,s_1} t_0|s_2\rangle\langle s_2|e^{is_1ka}|s_1\rangle - t\sum_{s_2,s_1}(|s_2+1\rangle\langle s_2|s_1\rangle e^{is_1ka} + |s_2\rangle\langle s_2+1|s_1\rangle e^{is_1ka}).$$

Using $\langle s_2|s_1\rangle = \delta_{s_1s_2}$ and changing s_1 and s_2 to s simplify the equation to

$$\hat{H}\psi_k = \sum_s t_0e^{iska}|s\rangle - t\sum_s(e^{iska}|s+1\rangle + e^{i(s+1)ka}|s\rangle),$$

where the second term can be written as $\sum_s |s\rangle e^{i(s-1)ka}$. Then, we combine the last two terms to get

$$-t\sum_s(e^{i(s-1)ka} + e^{i(s+1)ka})|s\rangle = -2t\cos ka\sum_s e^{iska}|s\rangle = -2t\cos ka|\psi_k\rangle.$$

Putting all the terms together yields $\hat{H}\psi_k = (t_0 - 2t\cos ka)\psi_k \equiv E_k\psi_k$, which proves that ψ_k is indeed an eigenstate of $\hat{H}$ and E_k is the eigenenergy.

9.4.2.1 Group velocity

We can compute the group velocity of the band,

$$v_k = \frac{1}{\hbar}\frac{\partial E_k}{\partial k} = \frac{2ta\sin ka}{\hbar}.$$

If $k = \frac{nb}{2}$, $ka = \frac{2n\pi}{2a}a = n\pi$, and v_k is zero, where the band is perpendicular to the Brillouin zone edge, same as the nearly free-electron model.

9.4.2.2 Wavefunction, position matrix elements and Berry connections

To normalize ψ_k, we compute $\langle\psi_k|\psi_k\rangle = \sum_{s_1 s_2} e^{-is_1 ka} e^{is_2 ka}\langle s_1|s_2\rangle = \sum_{s_1=1}^{N} 1 = N$, so ψ_k must be divided by $\sqrt{N}$,

$$\psi_k = \frac{1}{\sqrt{N}}\sum_s e^{iska}|s\rangle. \tag{9.4.17}$$

We notice that the wavefunction of the ring is modulated through the crystal momentum k. k is different from β (Section 8.4): k is $\frac{lb}{N} = \frac{2\pi l}{Na} = \frac{2\pi l}{L}$, but $\beta_l = \frac{\pi l}{L}$. This is because in a chain with an open boundary condition, the waves are standing waves, but, in a chain with a periodic boundary, the waves are traveling waves, and at least one cycle of the wave must fit inside the system. Figures 9.4(d)-(f) show the atomic periodic wavefunction, the Bloch enevelope and Bloch wavefunction in a system with four lattice sites and with the smallest k_{min}.

With the wavefunction, we can compute various properties. We take the position operator as an example, $\hat{x} = \sum_s sa|s\rangle\langle s|$, whose expectation value in $|\psi_k\rangle$ is

$$\langle\psi_k|\hat{x}|\psi_k\rangle = \frac{1}{N}\sum_{s,s_1,s_2} e^{-is_1 ka} e^{is_2 ka} sa\langle s_1|s\rangle\langle s|s_2\rangle. \tag{9.4.18}$$

To compute this, we take the derivative of ψ_k (eq. (9.4.17)) with respect to k,

$$\frac{\partial|\psi_k\rangle}{\partial k} = \frac{1}{\sqrt{N}}\sum_s (isa)e^{iska}|s\rangle.$$

Then, we multiply $-i\langle\psi_k|$ from the left and obtain

$$-i\langle\psi_k|\frac{\partial\psi_k}{\partial k}\rangle = \frac{1}{\sqrt{N}}\frac{1}{\sqrt{N}}\sum_{s_1 s_2} e^{-is_1 ka}(s_2 a)e^{is_2 ka}\langle s_1|s_2\rangle = \frac{a}{N}\sum_s s = \frac{(N+1)a}{2}, \tag{9.4.19}$$

which is called the Berry connection. Equation (9.4.19) is exactly the same as eq. (9.4.18), the expectation value of the position operator. Since our chain starts at a and ends at Na, one sees that the expectation value is just at the middle of the system. This is well defined, in contrast to the common belief in the literature [59]. If $N \to \infty$, $\langle\psi_k|\hat{x}|\psi_k\rangle$ goes to infinity, but this never happens because a solid is always finite.

We can also compute the transition matrix element between ψ_{k_1} and ψ_{k_2},

$$\langle\psi_{k_1}|\hat{x}|\psi_{k_2}\rangle = \frac{a}{1-e^{-i\Delta ka}} = \frac{a}{2} - \frac{ai}{2}\cot\frac{\Delta ka}{2}, \tag{9.4.20}$$

where $\Delta k = k_1 - k_2$. Equation (9.4.20) never approaches eq. (9.4.19) at $\Delta k \to 0$. If $\Delta k = 0$, we use eq. (9.4.19). In [60, 61], the continuous crystal momentum is used, which leads to the δ function and thus a highly singular expression $\partial\delta(\mathbf{k}-\mathbf{q})/\partial k_x$, where $\mathbf{k}$ and $\mathbf{q}$ are two different wavevectors. This hides the beauty of the orthogonalization between two wavefunctions. Here, in the discrete crystal momentum, which represents an actual crystal of finite size, such a singular term does not occur. Instead, we obtain a well-defined matrix element. This demonstrates the position operator in crystals is a valid operator, in contrast to the literature [59, 62].

Exercise 9.4.2.
19. Three atoms form a ring. It is known that the site energy is E_0, and the nearest-neighbor integral is $-t$, where t is positive. Find its eigenstates and eigenenergies.
20. Suppose $\psi_k(s) = e^{iska}|s\rangle$. Show $\psi_k(s+1) = e^{ika}\psi_k(s)$. Hint: Use $|s\rangle = |s+1\rangle$.
21. Without using the Berry connection, use the normalized wavefunction (eq. (9.4.17)) to directly compute the expectation value of the position operator.

9.5 Occupying the energy band: Fermi surface and Fermi velocity

The previous section explained how the band is formed. This section explains how the electrons fill up energy bands in the crystal momentum space. Therefore, we must first find out the number of **k** points, N_k, and then the number of bands per **k**. For a three-dimensional system, $N_1\mathbf{a}_1 \times N_2\mathbf{a}_2 \times N_3\mathbf{a}_3$, $N_k = N_1N_2N_3$, where $\mathbf{a}_i$ is the primitive lattice vector.

9.5.1 Fermi energy in metals, semiconductors and insulators

According to the Pauli exclusion principle, each state takes only two electrons at maximum, one for a spin-up and the other for a spin-down channel. Assuming that each **k** point has a single band such as s-orbitals,[2] we have $2N_k$ states available.

To be definitive, we consider only a case at $T = 0$ K, with N_e electrons filling these band states from a lower to a higher energy state. The highest occupied band state has a special significance because electrons in this state are most energetic and are most likely to interact with external fields and electrons, responsible for an array of chemical and physical properties.[3] The energy of this state is defined as the Fermi energy E_F, which separates the occupied from the unoccupied states. Figure 9.8 shows three examples. Figure 9.8(a) shows a case for metals, where the number of electrons N_e is fewer than the number of available states $2N_k$. Electrons only partially occupy a band (filled circles), leaving some part of the band unoccupied (empty circles). E_F cuts through the band. If we apply an electric field on the system, the electron can easily move to the next k and occupy a state that is slightly above the Fermi level, leaving a state unoccupied, i. e., a hole, behind. This explains why metals can conduct electricity.

2 If we have a p- or d-orbital, each **k** point has 3 and 5 bands, respectively.

3 In chemistry, in molecules these orbitals are called frontier orbitals because those orbitals are at the front to react.

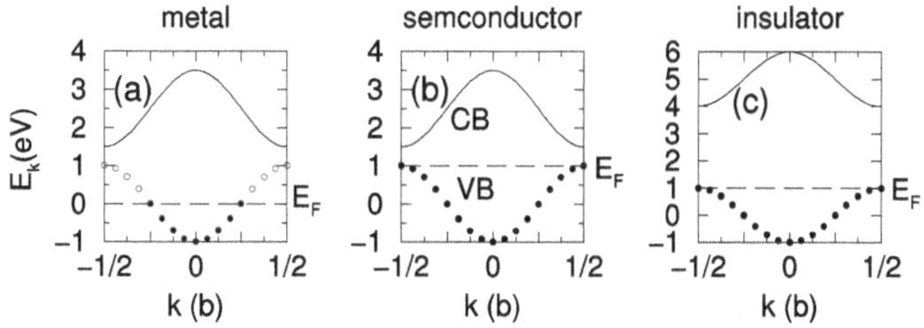

Figure 9.8: (a) Band dispersion in metals, where the number of electrons is less than the available states provided by the lattice. The band is only partially filled (filled circles). The dashed line denotes E_F. Electrons are ready to move into an unoccupied state (unfilled circles). (b) In semiconductors, the number of electrons is the same as the number of available states. The valence band is filled completely, but the unfilled conduction band is only 1–2 eV above the valence band, i. e., a small band gap, E_g. (c) Insulators are similar to semiconductors, but the band gap E_g is much larger, typically above 5 eV.

If we have more electrons, E_F increases. If the number of electrons N_e is equal to the number of available states $2N_k$ for a low-lying band, the band is completely occupied. These occupied bands are called valence bands (VBs in Fig. 9.8(b)). Figure 9.8(b) shows that E_F is now at the top of the highest VB. In the meantime, the unoccupied band is 1–2 eV above the highest VB. This is a semiconductor. All unoccupied bands are called conduction bands (CBs). The energy gap between the highest VB and the lowest CB is defined as the band gap E_g. This can be compared with HOMO and LUMO in Section 8.3.1. Semiconductors conduct electricity not as well as a metal, but one often uses thermal, electric, and optical fields to excite electrons from VB to CB and conduct electricity. This property underlines the semiconductor industry. Figure 9.8(c) shows the band structure for insulators. They are similar to semiconductors, but with a larger E_g, above 5 eV. Both E_g and E_F are determined by band structures of a material. In the following, we show one example of E_F.

Example 2 (Fermi energy and Fermi wavevector). N_e free electrons are sealed inside a chain of length $L = Na$, where N_e is even, a is the lattice constant, and N is the number of lattice sites. The electron has energy dispersion $E_k = \frac{\hbar^2 k^2}{2m_e}$, where k is the crystal momentum. (a) Find the Fermi wavevector k_F. (b) Find the Fermi energy as a function of electron line density $n_e = N_e/L$.

(a) We start with the energy dispersion $E_k = \frac{\hbar^2 k^2}{2m_e}$. At E_F, it is $E_F = \frac{\hbar^2 k_F^2}{2m_e}$, where k_F is the Fermi wavevector. First, we need to see what k_F can take. According to the Born–von Karman boundary condition (Section 9.2.1), $e^{ikL} = 1$, so $k = 2\pi n/L = \frac{2\pi}{Na}n$, where n is an integer, $1, \dots, N$. So, the first k_1 is $\frac{2\pi}{Na}$. If we had one electron, then k_F would be k_1; two electrons, k_F is still k_1 since every k takes two electrons. So, all we need to do is to find how many states are available between the Fermi wavevectors $-k_F$ and k_F, and what the minimum gap Δk is between two adjacent k points. Δk is just $\frac{2\pi}{Na}$. So, now the number of available states is $\frac{2k_F}{\Delta k}$. Because each state can accommodate two electrons, spin-up and spin-down, the total number of electrons N_e must be equal to [50], $N_e = 2 \times \frac{2k_F}{\Delta k} \to k_F = \frac{N_e \pi}{2L} = \frac{n_e \pi}{2}$, showing that the Fermi wavevector k_F is directly related to the electron line density.

(b) Using the energy dispersion, we can find $E_F = \frac{\hbar^2 k_F^2}{2m_e} = \frac{\hbar^2 \pi^2 n_e^2}{8m_e}$.

Exercise 9.5.1.
22. What are the band structure differences among metals, semiconductors, and insulators?
23. Which metal, sodium or potassium, has a higher Fermi energy?
24. A one-dimensional system has the band dispersion $E_k = E_0 \cos(ka)$, where E_0 is an energy constant, k is the wavevector, and a is the lattice constant. Suppose we have N_e electrons. (a) Find the Fermi wavevector. (b) Find the Fermi energy.

9.5.2 Fermi surface

The Fermi level separates the occupied from unoccupied bands in the band structure along a single-crystal momentum direction. A surface that separates the occupied from the unoccupied states in the **k** space is called the Fermi surface, where every point has E_F. In one dimension, the Fermi surface is just two dots, defined by $\pm k_F$, the Fermi wavevector. In two dimension, the Fermi surface is a two-dimensional curve, and, in three dimensions, we have a surface. Within the surface, all the states are occupied. The Fermi surface provides the band structure information in the reciprocal space. We have only the coordinates of Fermi vector $\mathbf{k}_F$, and the Fermi energy and the number of electrons become our input parameters to construct such a surface.

We take the free-electron gas an example. A free electron has energy $E_\mathbf{k} = \frac{\hbar^2(k_x^2+k_y^2+k_z^2)}{2m_e}$. If we fix $E_\mathbf{k}$ at E_F, then k_x, k_y and k_z form a sphere of equal energy centered at $(0,0,0)$. The surface of this sphere is the Fermi surface. Only the free electron has such a sphere. The rest have some distortions.

We take the band structure of the tight-binding model in the previous section (eq. (9.4.15)) as our second example. We slightly modify the band dispersion to

$$(E_\mathbf{k} - E^{\mathrm{at}} + t_0)/(2t_1) = \cos k_x a + \cos k_y a + \cos k_z a.$$

For an easy illustration in the two dimension, we let $\cos k_z a = 0$ or $k_z a = \frac{\pi}{2}$, and denote the term on the left side as a dimensionless $-\epsilon_\mathbf{k}$, so we have

$$-\varepsilon_\mathbf{k} = \cos k_x a + \cos k_y a, \tag{9.5.1}$$

which lies between −2 and +2. E_F is also dimensionless. In principle, we should choose N_e and then find E_F, but this is tedious. Instead, we choose a E_F, which is equivalent to choosing a N_e, and then set $-\varepsilon_\mathbf{k} = E_F$.

To appreciate how the Fermi surface changes, we start with a small E_F and $\mathbf{k}_F$, i. e., very few electrons. To be more precise, we assume $E_F = -1.95$. In this case, our k is small, so we can expand $\cos k_x a$ and $\cos k_y a$ around $\mathbf{k} = 0$ as

$$-\varepsilon_\mathbf{k} = 1 - \left(\frac{k_x a}{2}\right)^2 + 1 - \left(\frac{k_y a}{2}\right)^2 + \cdots.$$

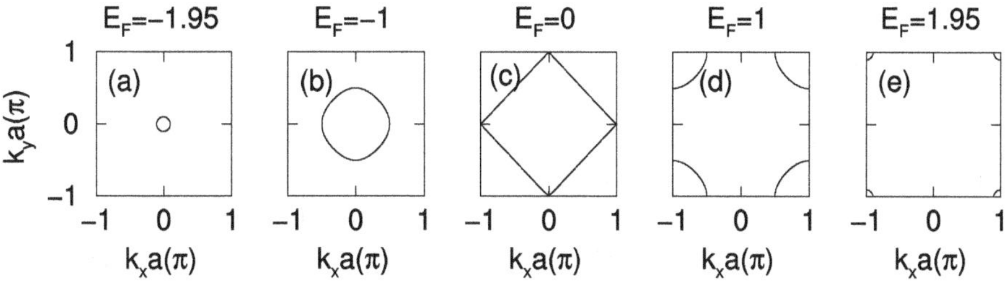

Figure 9.9: Fermi surfaces at five different Fermi energies (thin lines): (a) $E_F = -1.95$, (b) $E_F = -1$, (c) $E_F = 0$, (d) $E_F = 1$, and (e) $E_F = 1.95$. E_F is dimensionless due to eq. (9.5.1). All the states are filled from the Γ point ($k_x = 0, k_y = 0$) to the Fermi surface. These surfaces are created using the `fermisurface.f`.

Ignoring higher-order terms, we find

$$\left(\frac{k_x a}{2}\right)^2 + \left(\frac{k_y a}{2}\right)^2 = 2 + \varepsilon_{\mathbf{k}}.$$

On the Fermi surface, $\varepsilon_{\mathbf{k}}$ is fixed at E_F, so k_x and k_y form a circle with radius $\frac{2\sqrt{2+\varepsilon_{\mathbf{k}}}}{a}$. The pair of (k_x, k_y) represents a Fermi wavevector. Figure 9.9(a) shows all the states inside the circle are filled with electrons, and outside they are unfilled. This means that, if E_F is small, the highest occupied $\varepsilon_{\mathbf{k}}$ is small as well.

Our actual Fermi surface is computed numerically from eq. (9.5.1) where we set E in the code `fermisurface.f` to −1.95. Then, we scan k_x from $-\pi/a$ to π/a, i. e., from $-b/2$ to $b/2$. For every k_x, we solve for k_y from eq. (9.5.1).[4]

If we have more electrons, E_F becomes larger, and so is ε_k. At $E_F = -1$, our Fermi surface is distorted. Figure 9.9(b) shows the distortion is stronger as k_x and k_y approach the BZ boundaries. At $\varepsilon_{\mathbf{k}} = 0$, the distortion is so strong that we cannot use the same expansion anymore. Instead, we can directly solve $\cos k_x a + \cos k_y a = 0$ to get $k_y a = \pm k_x a \pm \pi$, which produces four line equations. It is extraordinary that the Fermi surface collapses into four lines (see Fig. 9.9(c)).

For a larger ka close to π, we need to expand $\cos ka$ around π,

$$\begin{aligned}-\varepsilon_{\mathbf{k}} = \cos(\pi) &+ \left.\frac{d\cos(k_x a)}{dk_x a}\right|_{k_x a=\pi}(k_x a - \pi) + \frac{1}{2!}\left.\frac{d^2\cos(k_x a)}{d(k_x a)^2}\right|_{k_x a=\pi}(k_x a - \pi)^2 + \cdots \\ + \cos(\pi) &+ \left.\frac{d\cos(k_y a)}{dk_y a}\right|_{k_y a=\pi}(k_y a - \pi) + \frac{1}{2!}\left.\frac{d^2\cos(k_y a)}{d(k_y a)^2}\right|_{k_y a=\pi}(k_y a - \pi)^2 + \cdots,\end{aligned}$$

which can be reorganized into a simple form by dropping higher-order terms,

4 In the code, we denote k_x by x and denote k_y by y.

$$(k_x a - \pi)^2 + (k_y a - \pi)^2 = 2(2 - \varepsilon_{\mathbf{k}}) \rightarrow \left(k_x - \frac{\pi}{a}\right)^2 + \left(k_y - \frac{\pi}{a}\right)^2 = \frac{2(2 - \varepsilon_{\mathbf{k}})}{a^2}.$$

So, this is a circle centered around $\frac{\pi}{a}$ with radius of $\frac{\sqrt{2(2-\varepsilon_{\mathbf{k}})}}{a}$, which is plotted in Fig. 9.9(d). Caution must be taken that those four corner quarter circles are unoccupied regions. As E_F is increased to 1.95, almost the entire zone is filled, with four tiny unfilled quarter-circles at four corners (Fig. 9.9(e)). Finally, we should note again that, although we increase E_F here, what is really increased is the number of electrons.

Exercise 9.5.2.

25. Use the code `fermisurface.f` to produce the entire set of Fermi surfaces as shown in Fig. 9.9(e).
26. Emulating the previous case, find the coordinates of (k_x, k_y).

9.5.3 Chemical potential

To this end, we limit ourselves to $T = 0$ K, where we can define the Fermi energy E_F, the Fermi wavevector $\mathbf{k}_F$ and the velocity is called the Fermi velocity $v_F = \frac{1}{\hbar}\frac{\partial E_k}{\partial k}|_{k=k_F}$. Figure 9.10(a) reviews a typical band structure for a metal, where the Fermi energy E_F is denoted by a horizontal line.

When $T \neq 0$ K, the Fermi energy must be replaced by the chemical potential because electron occupation follows the Fermi–Dirac distribution f_{FD}. For the single-spin channel, the distribution is given by

$$f_{\text{FD}}(E_{\mathbf{k}}) = \frac{1}{e^{\beta(E_{\mathbf{k}}-\mu)} + 1}, \tag{9.5.2}$$

where $E_{\mathbf{k}}$ is the band energy, $\beta = 1/k_B T$, k_B is the Boltzmann constant, and μ is the chemical potential. Figure 9.10(b) is the Fermi distribution (eq. (9.5.2)) for $k_B T = 0.05$ eV. For a single energy state such as Fig. 9.10(b), μ is equal to the energy with $f_{\text{FD}} = 1/2$.

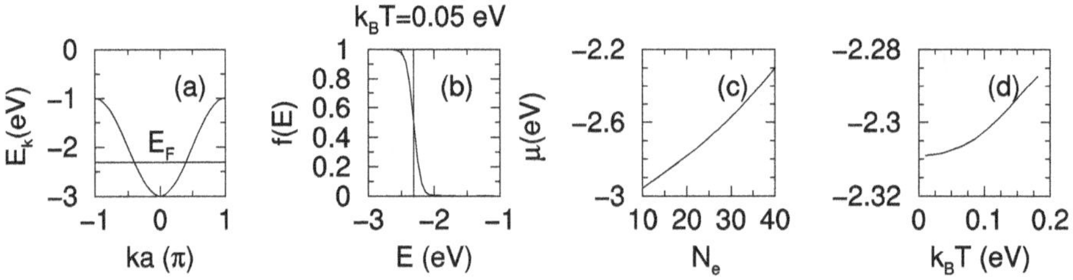

Figure 9.10: (a) Band dispersion with the Fermi energy E_F highlighted. (b) Fermi–Dirac distribution $f(E)$. (c) and (d) Chemical potential μ as a function of the number of electrons N_e and $k_B T$, respectively.

To see why we call μ a chemical potential, consider a specific system with N_e electrons. Then, the sum of f_{FD} over $E_{\mathbf{k}}$ must equal N_e,

$$\sum_k f_{\mathrm{FD}}(E_{\mathbf{k}}) = \sum_{\mathbf{k}} \frac{1}{e^{\beta(E_{\mathbf{k}}-\mu)}+1} = N_e .$$

For a given T and N_e, μ is the only variable to be adjusted until the sum is N_e. Different N_e produces a different μ, so we call it the chemical potential.

In the following, we take a simple cubic structure as an example. Its band dispersion is $E_{\mathbf{k}} = E^{\mathrm{at}} - t_0 - 2t_1 \cos k_x a$, where we drop the terms $\cos k_y a$ and $\cos k_z a$ for simplicity. We set $E^{\mathrm{at}} - t_0$ to -3 eV and t_1 to 1 eV. $k_x a$ runs from $-\pi$ to π in steps of $\pi/100$, so there are 100 k points. First, we fix $k_B T = 0.05$ eV and change N_e from 10 to 40.[5] The result is computed using code `fermi200.f` in the Appendix 11.16. We use the middle point method to find μ. The chemical potential sensitively depends on N_e (see Fig. 9.10(c)). A larger N_e pushes μ upward. The dependence of μ on $k_B T$ is shown in Fig. 9.10(d), and, if we extrapolate μ to $k_B T = 0$, we get E_F.

Exercise 9.5.3.

27. Use the code `fermi200.f` to reproduce data shown in Figs. 9.10(c) and (d).

28. Use the same code `fermi200.f`, but change the band dispersion from $\cos k_x a$ to a free-electron one $\hbar^2 k^2/(2m_e)$. (a) Plot the chemical potential as a function of N_e. (b) Extrapolate the chemical potential to $T = 0$ just as Figs. 9.10(c) and (d).

9.6 Transport on a nanometer scale

Chemical potential introduced in the previous section is important for transport on a nanometer scale. Since the 1980s, the electronic device size shrunk to a few hundred nanometers, where the quantum effects of electrons become very important. We enter the quantum transport era, which is fueled by the scanning tunneling microscope, the discovery of C_{60}, the quantum Hall effect, giant magnetoresistance, and tunneling magnetoresistance, each awarded a Nobel prize. To our knowledge, very few QM textbook has introduced quantum transport, partly because an open-system is too difficult to describe with the Schrödinger equation. A detailed treatment of quantum transport involves several advanced topics [63, 64]. We find a simple way to introduce this exciting topic to the reader through the transmission coefficients and Landauer's formalism.

---

5 It is important to note that we cannot arbitrarily choose N_e for two reasons. First, to accommodate N_e electrons, the number of k points must be larger than $N_e/2$ for a single band. Second, our band dispersion E_k has a limited energy range. If we choose N_e too large, the top occupied level may be close to the maximum band energy, so f_{FD} abruptly stops at the maximum energy. This limitation is due to our band structure in the tight-binding model. In reality, this should not happen as there are always unoccupied bands available.

9.6.1 Minimum conductance and von Klitzing resistance constant

We consider a special case of direct current through our device. Figure 9.11(a) shows our device (the central part) connected with two leads (contacts), source and drain. In general, the device should also be connected with a gate, which is ignored here. We start with our current continuity equation,

$$\frac{\partial \rho}{\partial t} = -\nabla \cdot \mathbf{J} + \frac{U - U^*}{i\hbar}\rho, \tag{9.6.1}$$

where ρ is the density pertaining to our device or the channel[6] in the device. The density in the source and drain is not considered, different from our prior presentations. We purposely use U to represent the potential energy to avoid confusion with voltage V. The effect of the source and drain enters through U.[7]

Different from the closed system (see eq. (2.1.13)), the open system must keep the second term in eq. (9.6.1). In particular, for DC, where $\nabla \cdot \mathbf{J} = 0$,[8] the second term is the only source for the density change. We rewrite U as $U = e(V_r + iV_D/2)$, where V_r is the real part of the potential and $V_D/2$ is the imaginary part, representing the voltage at one lead. We use $\frac{1}{2}$, so $U - U^*$ depends only on the potential difference between two leads, $U - U^* = ieV_D$. To find the electric current I, we multiply the equation by $-e$ and integrate over space to get

$$I = \left| -e \int \frac{\partial \rho}{\partial t} d\mathbf{r} \right| = \left| \frac{e^2 V_D}{\hbar} \int \rho d\mathbf{r} \right| = \frac{e^2}{\hbar} M V_D. \tag{9.6.2}$$

The integration over space $\int \rho d\mathbf{r}$ is a charge-carrier number M. Different from the normalization condition in a closed system in eq. (1.4.2), this is not the total number of charge carriers in the device; instead, it represents the number of charge carriers that can physically move through the device from the source to the drain (Fig. 9.11(a)) to contribute to the current. To extract the crucial physics, we replace $\int \rho d\mathbf{r}$ in eq. (9.6.2) by the number of channels M in the device.[9] The carts in Fig. 9.11(b) represent the charge carriers. Inside the device, we have five carts, but only one in channel B can move through. The bottom four carts in channel C cannot move easily because their energy level is lower than the energy level in the drain. Channel A has no contribution since there is

6 It is common that one uses “channel” instead of “energy level” in transport, although they are exactly the same thing. Using channels highlights the transport nature.

7 In the advanced transport theory, only small portions of the source and drain are included to improve the wavefunction accuracy on the interfaces between the device and the source/drain [63].

8 Even if we do not make this assumption, we can drop off the first term because, experimentally, without bias the current must be zero. This means only the second term remains in the equation.

9 This avoids the problem that we have to specify how many electrons are in the channel.

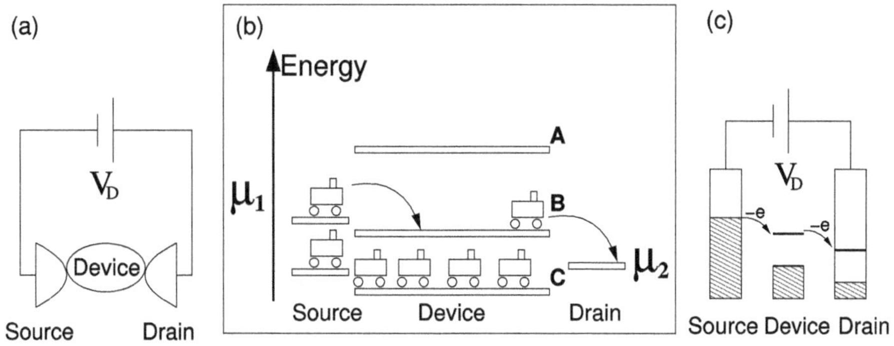

Figure 9.11: (a) A typical quantum transport device is connected to the power supply through the source on the left and drain on the right. (b) The current flow depends on $\mu_1 - \mu_2$ and the number of available channels in the device. The small carts in the figure represent the charge carriers such as electrons. The quantization of energy level in the nanodevice represents a resistance to carriers at contacts between the device and source/drain, while in the channel there is no resistance, just like bullets, ballistic transport. (c) Energy diagram across the source, device, and drain, where the shaded regions are filled states according to the Fermi distribution.

no charge carrier. Figure 9.11(c) highlights three conditions for channel B to be operative. First, the chemical potential in the source is higher than that in channel B of the device, while the drain chemical potential is lower than channel B. Finally, the device still has enough states to be occupied [63]. These three conditions allow the device to discharge the old charge carrier to the drain and ready to receive new charge carries from the source. If for some reason one of three conditions is not fulfilled, the current flow stops, leading to spatial charge accumulation.

Figure 9.11(c) has one channel, $M = 1$,

$$I = \frac{e^2}{\hbar} V_D \to G_0 = \frac{I}{V_D} = \frac{e^2}{h} = 38.7\,\mu\text{S}, \tag{9.6.3}$$

which is the maximum conductance (h without 2π). One can see that the conductance is not infinite, in contrast to the classical physics prediction. Its reciprocal is the minimum resistance, the von Klitzing constant, $R_K = \frac{1}{G_0} = 25.8\,\text{K}\Omega$.[10] This resistance cannot be described by Ohm's law. It is purely due to the quantum nature of the device at a small scale. Because the channel length is so short, charge carriers inside the channel of the device travel like a bullet, i. e., ballistic transport, without resistance, but when they enter the device from the source and when they leave the channel, they encounter resistance because the energy level quantization restricts them to a few channels. This is similar to the situation that when 100 students enter the classroom through one door, and every-

10 In 1985, Klaus von Klitzing received the Nobel prize for the discovery of the quantum Hall effect.

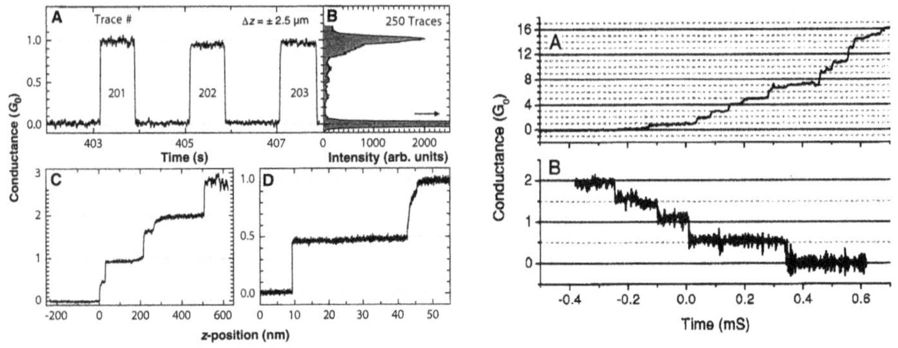

Figure 9.12: (Left) Conductance in carbon nanotube [65]. (A) Conductance change as a tube moves in and out of mercury. (B) Statistics. (C) Conductance plateaus as a function of position z. (D) Tip effects. Note their G_0 is $2e^2/h$. Used with permission from AAAS. (Right) (A) Conductance-time curves in Cu nanowires taken at room temperature [66]. (B) The half-step size of G_0 shows the broken spin degeneracy. Used with permission from AIP.

one experiences resistance. In nanodevices, such as our cell phones, the heating occurs at contacts between the device and source/drain.

In nanowires, one can directly measure G_0. Franck et al. [65] moved their nanotube in and out of mercury and found that the conductance shows a short jump (left figure in Fig. 9.12). Gillingham et al. [66] used a much simpler method. They employed two copper nanowires and measured the conductance by a Tektronix TDS430A digital oscilloscope at room temperature. They detected jumps in conductance as they vibrated the wires at 5–10 Hz. The right figure in Fig. 9.12 (Fig. A) is their experimental results, and Fig. 9.12 (Fig. B) shows that the broken spin degeneracy (see Chapter 6) leads to the half steps. If we have M channels, we have M times more current. This is the reason why we use M instead. Ohm's law can be recovered if the device is much longer and the diffusive resistance in the channel dominates over the resistance at contacts.

Exercise 9.6.1.

29. Another way to derive G_0 is to use the group velocity of the charge carrier, υ_x. This, however, slightly mixes the quantum and classical treatments [63]. Suppose the charges move along the $+x$-direction. The current is $I = -\frac{e}{L}\sum_{\upsilon_x} \upsilon_x$, where the summation is over all the contributing charges, L is the length of the device, and υ_x is their velocity. Using the relationship between υ_x and energy E, $\upsilon_x = -\frac{1}{\hbar}\frac{\partial E}{\partial k_x}$, where k_x is the wavevector, and converting $\sum_{\upsilon_x} 1/L$ to $\int \frac{dk}{2\pi}$, show $I = \frac{e^2}{h}(V_1 - V_2)$. Here V_1 and V_2 are voltages at two contacts.

9.6.2 Landauer theory: conductance is transmission

To understand quantum transport brings us back to our source and drain introduced in the previous subsection. We recall that chemical potential μ changes with the number of electrons N. The source, which is connected to the negative and electron rich electrode,

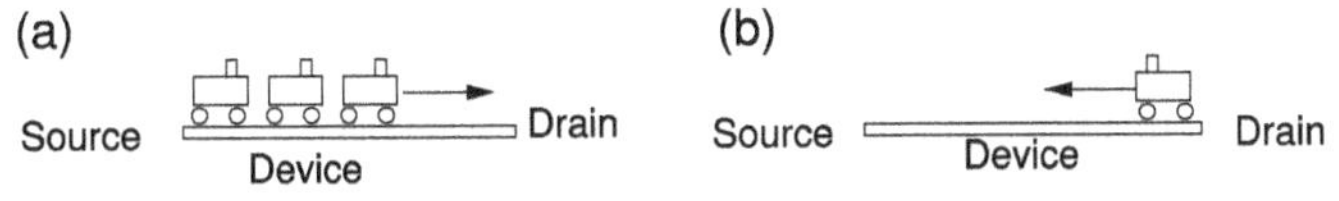

Figure 9.13: (a) The source tries to move electrons from left to right. (b) The drain tries to move electrons from right to left. The competition between them gives the net current. This is the origin of the subtraction between two Fermi functions.

has a higher chemical potential μ_1 than μ_2 of the drain, i. e., $\mu_1 > \mu_2$.[11] The difference is $\mu_1 - \mu_2 = e(V_1 - V_2) = eV_D$, where V_D is the power supply's voltage. In Fig. 9.13, we schematically illustrate this connection. Figure 9.13(a) illustrates that the source pushes the electrons (carts) to the right, while Fig. 9.13(b) shows that the drain pushes the electron to the left. A competition between these two gives the current. We should emphasize several theoretical difficulties here. A single chemical potential is well defined for a single material, but, when our system has three parts, the device, source, and drain, they cannot have the same chemical potential.[12] Second, the entire system cannot be treated in the crystal momentum space, as we did previously using Bloch wavefunctions, because there is no translational symmetry among the device, source, and drain. Thus, we are forced to work in real space, not reciprocal space, where Schrödinger equation even for a simplest case is unsolvable. Third, this is a truly nonequilibrium process.

In 1957, R. Landauer [67] pioneered the idea that conductance is transmission. It allows us to connect the current with transmission through [63],

$$I = \frac{e}{h}\int_{-\infty}^{\infty} dET(E)(f_1(E) - f_2(E)), \tag{9.6.4}$$

where $T(E)$ is the transmission of the device and is dimensionless. The integration is over energy states of the device. The source and drain enter the equation through their respective Fermi functions, f_1 and f_2. The competition between the source and drain as seen in Fig. 9.13 is now realized mathematically by the Fermi function difference $(f_1(E) - f_2(E))$, where $f_1 = \frac{1}{e^{(E-\mu_1)/k_B T}+1}$ and $f_2 = \frac{1}{e^{(E-\mu_2)/k_B T}+1}$. One can see, if $\mu_1 = \mu_2, I = 0$. If $f_1 > f_2$, $I > 0$; otherwise, $I < 0$. This shows that the current does change sign if we switch the polarity of the power supply. If μ_1 is very close to $\mu_2, f_1 - f_2$ is

$$f_1 - f_2 \approx \frac{\partial f}{\partial \mu}(\mu_1 - \mu_2) = -\frac{\partial f}{\partial \mu}e(V_1 - V_2), \tag{9.6.5}$$

11 Rigorously speaking, the chemical potential energy becomes more negative once we have more electrons. Here, in order to simplify our explanation, we make a compromise: A higher chemical potential leads to a higher chemical potential energy by assuming the electron carries a *positive* charge. It is in this spirit that Fig. 9.11(b) has a higher μ_1 on the source side and lower μ_2 on the drain side.

12 If they do, there is no current.

and then the current is

$$I = \frac{-e^2}{h} \int_{-\infty}^{\infty} dE \frac{\partial f}{\partial \mu} (V_1 - V_2) T(E). \tag{9.6.6}$$

Since $\frac{\partial f}{\partial \mu} = -\frac{\partial f}{\partial E}$, whose proof is left as a homework assignment, we have

$$I = +\frac{e^2}{h} \int_{-\infty}^{\infty} dE \frac{\partial f}{\partial E} T(E)(V_1 - V_2). \tag{9.6.7}$$

If the transmission $T(E) = 1$, which is called the ballistic transport since any electrons move through without reflection, just like a bullet, then the integral of $\frac{\partial f}{\partial E}$ over E is just $f(+\infty) - f(-\infty)$, which is -1. We take the absolute value and have $I = \frac{e^2}{h}(V_1 - V_2)$, which gives conductance $G_0 = \frac{e^2}{h}$. This proof is more rigorous than we just did.

Here, the Fermi function f is just for one channel, i. e., one state. If we have M degenerate states, i. e., M channels, we have to sum over all the these channels, but, since each channel contributes the same amount of current, our current will be M times larger,

$$I = M \frac{e^2}{h} (V_1 - V_2). \tag{9.6.8}$$

If $T(E) < 1$, the conductance G is smaller than G_0. This is called the scattered transport, where not all the electrons can go through the barrier. This connects with the quantum tunneling of Chapter 2. The actual dependence of I on V depends on the transmission $T(E)$.

In the following, we present a numerical example to reveal all the details in eq. (9.6.4), without making a low-bias approximation. At zero bias, we set the chemical potential for two contacts, source, and drain, to $\mu_1 = \mu_2 = -1$ eV. Figure 9.14(a) shows the Fermi function in the source or the drain, with $\beta = \frac{1}{0.025\ \text{eV}}$.

If the bias voltage V is applied, μ_1 is changed to $\mu_1 + eV/2$ for the source, and μ_1 is changed to $\mu_1 - eV/2$ for the drain, so the Fermi functions are

$$f_1 = \frac{1}{e^{\beta(E+\mu_1+eV/2)} + 1}, \quad f_2 = \frac{1}{e^{\beta(E+\mu_2-eV/2)} + 1}.$$

Figure 9.14(b) shows the difference $f_1 - f_2$ as a function of E for two biases. As we increase V, the nonzero window widens. We again take the ballistic limit, with $T(E) = 1$, and numerically integrate eq. (9.6.4) over E to find the current I. The results are shown in Fig. 9.14(c). The current strictly linearly increases with V. The slope is our conductance $G_0 = 38.7\ \mu$S. In practice, one often computes the conductance by dI/dV, which is shown in Fig. 9.14(d). This conductance is independent of V. This means our current $I = \frac{e^2}{h}(V_1 - V_2)$, even without the low-bias approximation and even at a finite temperature. This explains why experimentally Fig. 9.12 shows steps at G_0. However, mathematically it is

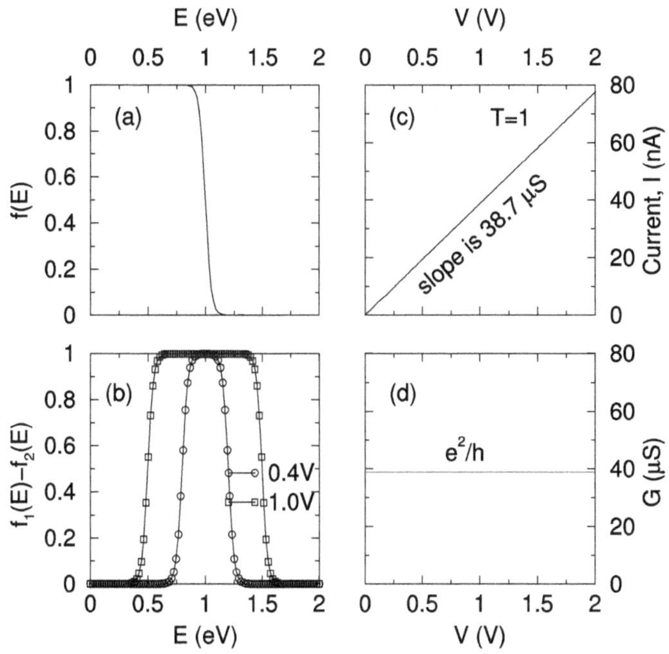

Figure 9.14: (a) Fermi distribution in the device alone. (b) Difference between two Fermi functions at two biases of 0.4 (circles) and 1.0 V (squares). The temperature is fixed at $\beta = 1/(0.025\text{ eV})$. (c) I-V curve with transmission $T = 1$ (ballistic limit) (d) Conductance does not change with V for the ballistic transport.

not obvious why, $I = \frac{e}{h}\int_{-\infty}^{\infty} dE(T(E) = 1)(f_1(E) - f_2(E)) = \frac{e^2}{h}(V_1 - V_2)$, whose proof is rendered as a homework exercise.

Exercise 9.6.2.

30. Show (a) for the Fermi function f, $\frac{\partial f}{\partial \mu} = -\frac{\partial f}{\partial E}$; (b) if $T(E) = 1$, use this relationship to prove $I = \frac{e^2}{h}V$.

31. Consider a ballistic transport with the transmission $T = 1$. Prove the current is independent of temperature and can be written as

$$I = \frac{e}{h}\int_{-\infty}^{\infty}\big(f_1(E + eV_1) - f_2(E + eV_2)\big)dE = \frac{e^2}{h}(V_1 - V_2),$$

where f_1 and f_2 are two Fermi functions.

9.6.3 Device's density of states (DOS)

So far, we have investigated only how the voltage bias affects the I-V curve through the source (f_1) and drain (f_2), but the details of the device information are absent. In particular, there is no information on the eigenstates, i. e., channels, of the device. In fact, Landauer's formula (eq. (9.6.4)) in its present form does not allow to include eigenstates, such as those illustrated in Fig. 9.11(c). If the device has 1 or 10 eigenstates per unit energy, this integration stays the same. So, we need a quantity to tell the difference between 1 or 10 eigenstates per unit energy.[13] This quantity is called the density of states (DOS), the number of states at energy E per unit energy, $\rho(E)$,[14]

$$\rho(E) = A \sum_n \frac{1}{\sqrt{(E - E_n)^2 + \delta^2}} \quad \text{(states/energy)}, \tag{9.6.9}$$

where the summation is over the eigenenergy E_n of the channel device and δ is a broadening to avoid divergence. A is a constant to ensure that the normalization of $\rho(E)$ is equal to the number of electrons in the device, N,

$$\int_{-\infty}^{\infty} \rho(E)dE = N.$$

With $\rho(E)$, we now rewrite eq. (9.6.4) as

$$I = \frac{e}{h} \int_{-\infty}^{\infty} dET(E)\rho(E)(f_1(E) - f_2(E)). \tag{9.6.10}$$

In the following, as an example we suppose that the device only has one electron and one eigenenergy $E_n = 2\,\text{eV}$. Our next example uses $E_n = 1\,\text{eV}$. Figure 9.15(a) is our normalized DOS, which peaks at E_n. The normalization is done within $[-20\,\text{eV}, 20\,\text{eV}]$. The channel restriction with a single state reduces the current significantly. Figure 9.15(b) shows that, at $V = 1\,\text{V}$, $I = 0.071\,\text{nA}$, but in our ballistic case (Fig. 9.14(b)), $I = 3.87\,\text{nA}$. It saturates with V (see the circles in Fig. 9.15(b)). What is more interesting is how G changes. It peaks at 2 V, which exactly corresponds to $E_n = 2\,\text{eV}$ in our DOS. This explains why the current flows only if there is a channel between two chemical potentials of the drain and source. G peaks whenever there is such a channel.

To this end, our calculation is limited to the ballistic transport $T(E) = 1$. Next, we consider a case with $E_n = 1\,\text{eV}$. Figure 9.15(c) shows our DOS, which is normalized between 0 and 2 eV. We now see that it peaks at 1 eV, in contrast to Fig. 9.15(a). In order to see how transmission $T(E)$ affects transport, we use the transmission from Chapter 2, where the finite barrier height V_0 is 2 eV and the barrier width is 3 Å. The current and conductance are given in Fig. 9.15(d). One sees that the current is further reduced since $T(E)$ becomes smaller. The current has a plateau after 2 eV because there are no more

13 M in eq. (9.6.2) gives us good guidance.

14 Unfortunately, ρ has been overly used, but it is mostly related to density.

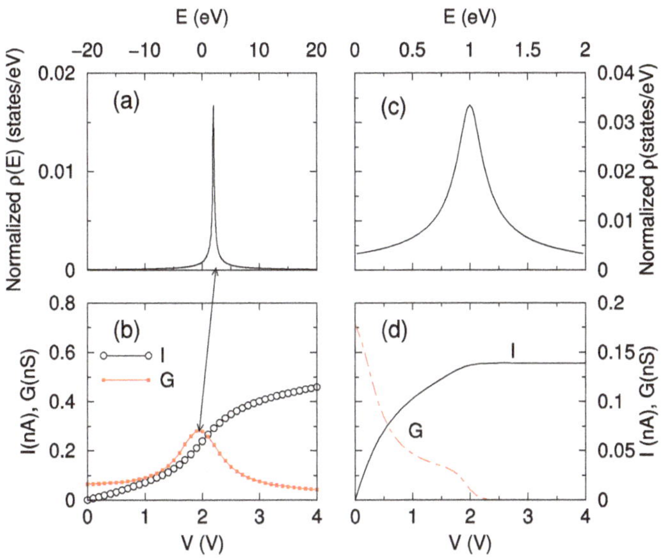

Figure 9.15: (a) The normalized density of states $\rho(E)$ as a function of E within [−20 eV, 20 eV]. Only a single eigenstate at 2 eV is included. The broadening δ is 0.1 eV. (b) Current-voltage curve (empty circles) and conductance–voltage curve (filled circles) under a single eigenvalue limit and ballistic limit. (c) The normalized density of states with $E_n = 1$ eV. (d) I-V (solid) and G-V (dashed) curves with a realistic $T(E)$ from quantum tunneling.

states larger than 2 eV that we set. The conductance no longer peaks at E_n due to the influence of $T(E)$. This shows that in a practical device, I-V and G-V curves sensitively depend on both the transmission and DOS of the device.

What we have presented so far is an introduction to quantum transport. One might wonder why we take the Landauer approach, instead of the time-dependent Schrödinger equation. The answer is that we cannot solve the Schrödinger equation, even numerically, because we are dealing with three subsystems, two semi-infinite subsystems, source and drain, and one finite device. These semi-infinite systems cannot be treated in the momentum space either, as solid-state physics would. Essentially, we are forced to compute everything in real space. Landauer's idea is to use a transmission coefficient as a way to investigate transport properties. The advanced reader may consult several excellent books on this topic [63].

Exercise 9.6.3.

32. Suppose that $E_n = 0.5$ eV and $\delta = 0.01$ eV. Compute the I-V curve as a function of voltage for the transmission (a) $T(E) = 1$ and (b) $T(E) = 0.5$. (c) Compare the results from (a) and (b) and explain the difference.

9.7 Scanning tunneling microscope

In 1981, Gerd Binnig and Heinrich Rohrer [68] invented the scanning tunneling microscope (STM), for which they were awarded a Nobel Prize in Physics in 1986. STM for the first time enables one to "visualize" atoms on atomic scales. It represents a triumph of quantum tunneling in real applications from surface structure characterization to surface reconstructions. A subsequent discovery of the atomic force microscope (AFM) enables one to investigate insulating samples, such as DNA and polymers. This section is devoted to some of its basic principles and operations.

9.7.1 Principle of the STM

A typical STM consists of a metallic tip, which scans across the surface of a sample, a power supply, and a current detection system, which may include a feedback loop. Figure 9.16(a) shows a typical scheme. Three piezodrives, made of a piezomaterial, move the tip along the x-, y-, and z-axes. The piezodrives are calibrated by capacitor plates. A typical sensitivity is $2.0 \pm 0.2\,\text{Å/V}$. This means, for every volt, the tip moves $2\,\text{Å}$. One brings the tip close to the sample's surface, without touching the surface. Once a bias is applied between the tip and sample, electrons can tunnel through a vacuum barrier and generate a current of nA or pA. Experimentally, one measures the current as a function of position x, y and z. The current change encodes the electron density and its spatial profile that reflect the underlying atomic arrangement. The tip is usually very sharp. One can use a pair of scissors to cut a wire at 90°, post-processed by chemical etching.

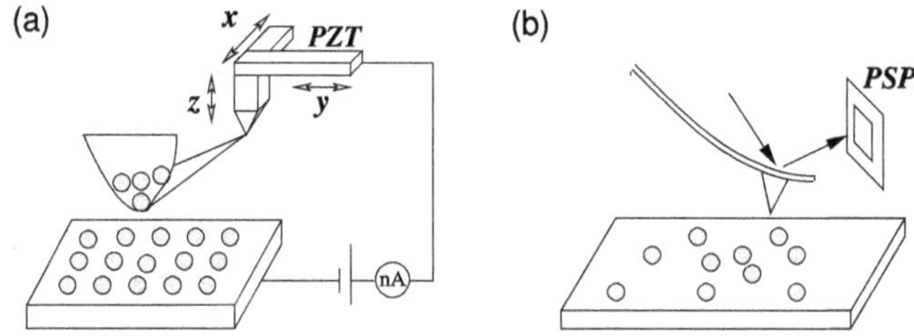

Figure 9.16: (a)Typical scanning tunneling microscope (STM) setup. A sharp tip, usually mounted on a piezomaterial, such as Pb(Ti, Zr)O_2 (PZT), is brought close to a conductor surface. Electrons tunnel through the gap between the tip and the surface, and under an external voltage bias, they contribute a current. The current sensitively depends on the distance between the tip and surface. The image in Fig. 1.2 is produced by STM. (b) For insulating surfaces such as DNA, one detects the attractive force between the tip and surface by measuring the reflected light from the back of the tip. The reflected signal is recorded by a position-sensitive photodetector.

9.7.2 How STM gains atomic resolution and the Binnig–Rohrer factor

It is the quantum tunneling that enables STM to obtain atomic resolution. In Section 2.7.2, we showed that transmission exponentially depends on the barrier width. From Section 9.6, we see that the current is directly proportional to the transmission.

Suppose the distance between the atoms on a sample surface and those on the tip is z, which is the same as a in Section 2.7.2, and the transmission is

$$T(E) = g(E)e^{-\frac{2z}{\hbar}\sqrt{2m_e(V_0-E)}}, \tag{9.7.1}$$

where $g(E)$ is a function of E and V_0 and is less than 1. To simplify our notation, we define $\frac{2}{\hbar}\sqrt{2m_e}$ as the Binnig–Rohrer factor,

$$S_{\mathrm{BR}} = \frac{2}{\hbar}\sqrt{2m_e} = 1.025\,\text{Å}^{-1}\text{eV}^{-1/2}. \tag{9.7.2}$$

Assume that we only apply a low bias, so E can be approximated by the Fermi energy E_f. V_0 is the barrier potential due to the surface confinement, where electrons cannot escape from the surface, and $V_0 - E_f$ is defined as the work function, ϕ. We caution again that V_0 is a potential energy. Under the low-bias limit (see Section 9.6), the current is proportional to $T(E_f)$:

$$I(z) = \frac{e^2}{h}g(E_f)e^{-S_{\mathrm{BR}}z\sqrt{\phi}}\Delta V, \tag{9.7.3}$$

which is found from eq. (9.6.7). If we have two barrier widths, z_1 and z_2, and if we further assume the charge density is the same, then the ratio of two currents I_1 and I_2 is

$$\frac{I(z_1)}{I(z_2)} = e^{-S_{\mathrm{BR}}(z_1-z_2)\sqrt{\phi}}. \tag{9.7.4}$$

Its logarithm is

$$\ln\frac{I(z_1)}{I(z_2)} = -S_{\mathrm{BR}}\sqrt{\phi}(z_1 - z_2). \tag{9.7.5}$$

So, the slope of $\ln I$ is a direct measurement of $\sqrt{\phi}$. If z is in Å, then ϕ is in eV. The reduction of current is exponentially dependent of z. So any atoms, slightly away from the tip, have a negligible contribution to the total current. This provides the atomic sensitivity.

In experiments, one can fix the current and change z, while scanning along the xy-plane. This is called the constant current mode, by adjusting z. This can be done by changing the voltage to the piezodrive along the z-axis. If the surface is electronically homogeneous, the constant current means the same height z everywhere. Then, the current is a function of x, y, z, or $I(x, y, z)$, which encodes the surface information of a sample.

STM is restricted to conducting surfaces and does not work for insulating materials. The atomic force microscope (AFM) is used instead. It uses the force between the tip and sample. The cantilever deflects up or down, depending on the force. Figure 9.16(b) shows the setup. Modern STM and AFM are much more sophisticated and widely used in research and industry.

Exercise 9.7.2.

33. Show the Binnig–Rohrer factor $S_{BR} = \frac{2}{\hbar}\sqrt{2m_e} = 1.025\,\text{Å}^{-1}\text{eV}^{-\frac{1}{2}}$.

34. The tunnel current is roughly of the form [68] $I = f(V)\exp(-\sqrt{\bar{\phi}}\cdot z)$, where $f(V)$ is a weighted joint local density of states of tip and object, $\bar{\phi}$ is the average tunnel barrier height in eV, and z is the separation of two metals in Å. (a) It is clear that the unit of the exponential function is not correct. What are missing? Hints: Check eq. (2.7.2). (b) Given $\bar{\phi} = 5$ eV, after correction of the units, compute the current ratio at $z_1 = 1$ Å and $z_2 = 1$ Å.

35. Tungsten tips are used for STM. The work function of tungsten is 4.5 eV. When the barrier width z is 1 Å between the tungsten tip and a sample, the current is 2 nA. Find the currents at $z = 2$ Å and 4 Å.

9.8 Lattice vibrations

The harmonic oscillator is a model for a single atom. In real materials, we have many atoms, often coupled to each other. For instance, in CO_2 the carbon is bonded to two oxygen atoms. In C_{60}, each carbon has three neighboring carbon atoms. We have 174 vibrational modes. In fact, for a system with N atoms, the total number of the degrees of freedom is $3N$. We subtract 3 translational modes and 3 rotational modes, and we have $3N-6$ normal modes of nonzero frequency and $3N-6$ normal modes of molecular vibrations. This section starts with classical physics and then introduces the quantum concept: phonon.

9.8.1 Normal modes

We consider a chain of N identical atoms along the x-axis, situated at $x_1, x_2, \ldots x_N$ (see Fig. 9.17(a)). The spatial separation between two neighboring atoms is a, so $x_l = la$, where a is the lattice constant. Every atom has two neighbors. Consider atom l, with position x_l, which has two neighbors: atom $l+1$ and atom $l-1$. The interaction between atoms is modeled by an ideal harmonic oscillator, with spring constant C.[15] The force from atom $l+1$ on atom l is $C[(x_{l+1}-x_l)-(x^0_{l+1}-x^0_l)]$ and that from $l-1$ is $C[(x_{l-1}-x_l)-(x^0_{l-1}-x^0_l)]$, where x^0_l is the equilibrium position of atom l, so the total force on l is

$$\begin{aligned} F_l &= C[(x_{l+1}-x^0_{l+1})-(x_l-x^0_l)] + C[(x_{l-1}-x^0_{l-1})-(x_l-x^0_l)] \\ &\equiv C(u_{l+1}-u_l) + C(u_{l-1}-u_l) = C(u_{l+1}+u_{l-1}-2u_l), \end{aligned} \tag{9.8.1}$$

15 We do not use k here since k is reserved for a wavevector.

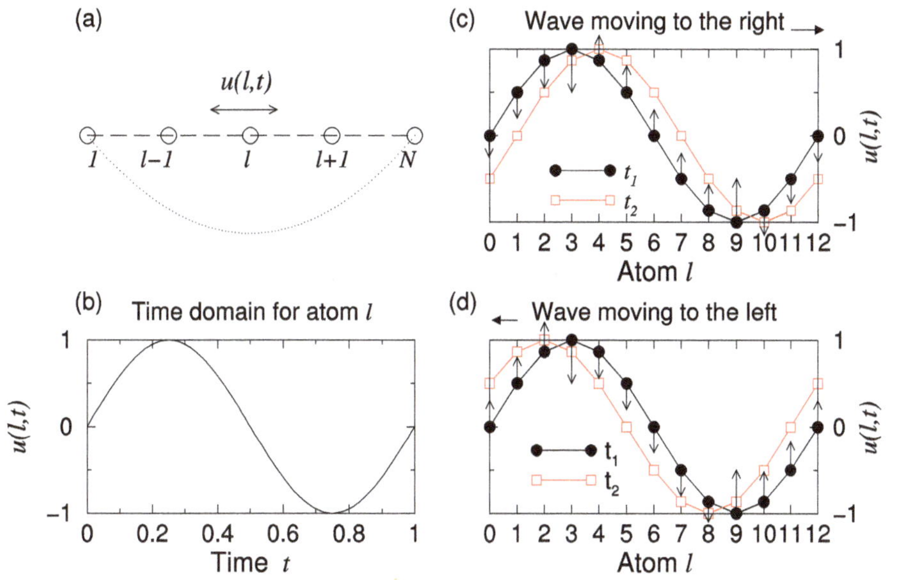

Figure 9.17: (a) A chain of atoms are arranged along the x-axis. Each atom l has two neighbors, $l-1$ and $l+1$. At the end of the chain, we connect atom 1 and atom N. (b) The displacement $u(l,t)$ as a function of time t for atom l. (c) Collective vibration in a chain of 12 atoms at time t_1 (filled circles) and t_2 (empty boxes), where $t_1 < t_2$. The displacement $u(l,t)$ is plotted as a function of atom index l. The wave moves to the right. (d) The same as (c), but the wave moves to the left.

where $u_l = x_l - x_l^0$ is the displacement of atom l. Equation (9.8.1) works for any atom away from two ends of the chain, but, for $l = 1$ and $l = N$, the atom only has one neighbor, so the force cannot be balanced, which is unphysical. To resolve this issue, one either manually fixes these two end atoms or one links the $l = 1$ atom to the $l = N$ atom. This second method is called the periodic boundary condition, which is used below. The dashed line in Fig. 9.17(a) shows this periodic condition.

Suppose that all the atoms have the same mass m. Equation (9.8.1) provides an acceleration for atoms, and the Newtonian equation of motion is

$$m\frac{\partial^2 u_l}{\partial t^2} = C(u_{l+1} + u_{l-1} - 2u_l). \tag{9.8.2}$$

Given an initial condition, this equation describes various dynamics, where u depends on both time and space. To simplify our calculation, we examine only a normal-mode solution, where $u_l(t=0)$ contains a single frequency, ω, and $u(l,t) = u_l(t=0)e^{i\omega t}$. Figure 9.17(b) shows an example of u_l for one period of time for atom l. Figure 9.17(c) shows $u(l,t)$ at a fixed time t_1 (filled dots) as a function of atom l for a chain of 12 atoms for the wave traveling to the $+x$-axis, or the wavevector $k > 0$. This is a snapshot of all the atoms in space, so one sees that a wave across all the atoms appears. The vertical arrows in Figs. 9.17(c) and 9.17(d) denote moving directions of the atoms at the next time

t_2 (empty boxes). A simple way to find those directions is: (i) Locate the atoms at their maximum positive and negative displacements, i. e., crests or troughs, here, $l = 3$ and $l = 9$; (ii) draw two arrows toward the $u = 0$ line; and (iii) check which direction the collective wave propagates. In our case, it is to the right. All the atoms behind these extreme atoms, i. e., $l = 3$ and $l = 9$, have the same moving direction until one hits on the next extreme atom. Figure 9.17(d) shows a wave propagating to the left, or $k < 0$.

The normal-mode solution decouples the spatial dependence of u_l from its time dependence, i. e., $u_l(t) = u_l(t = 0)e^{i\omega t}$. Substituting $u_l(t)$ into eq. (9.8.2) yields

$$-m\omega^2 u_l(t = 0) = C(u_{l+1}(t = 0) + u_{l-1}(t = 0) - 2u_l(t = 0)), \tag{9.8.3}$$

where the time dependence of $u_l(t)$ is eliminated. Because the expression $u_{l+1}(t = 0) + u_{l-1}(t = 0) - 2u_l(t = 0)$ is a finite difference form of $\frac{d^2u}{dx^2}$ if u were continuous, eq. (9.8.3) is similar to the differential equation

$$C\frac{d^2u_l}{dx^2}(\Delta x)^2 + m\omega^2 u_l = 0,$$

which suggests a solution of $u_l \sim e^{ikx}$. In our case, $u_l(t = 0)$ must take a traveling wave solution, $u_l(t = 0) = u_0e^{ikla}$, since we are interested in wave propagation. Here, u_0 is a constant and determined by the initial boundary condition. In Fig. 9.17, u is written as $u(l, t)$ to highlight the fact that u depends on the atom position l and time t.

To see whether $u_l = u_0e^{ikla}$ is indeed a solution, we substitute it into eq. (9.8.3),

$$-m\omega^2 u_0e^{ilak} = C(u_0e^{i(l+1)ak} + u_0e^{i(l-1)ak} - 2u_0e^{ilak}). \tag{9.8.4}$$

Canceling the factor u_0e^{ilak} on both sides yields

$$-m\omega^2 = C(2\cos(ak) - 2) = 2C(\cos(ak) - 1) = 2C(-2\sin^2(ka/2)),$$

$$\omega^2 = \frac{4C}{m}\sin^2(ka/2) \to \omega_k = 2\sqrt{\frac{C}{m}}|\sin(ka/2)|.$$

This demonstrates that, if we choose a proper ω, both sides have no dependence on l anymore, so that our trial solution is a correct solution. We purposely add a subscript k to ω to denote that ω is a function of k. For this reason, we call ω_k a dispersion relationship of lattice vibrations. Figure 9.18 shows this dispersion. Between $-\pi/a$ and π/a is called the first Brillouin zone, as known from the prior sections. Although the dispersion repeats itself outside this region, they are unphysical. This is because, physically, the longest wavelength of $\lambda = Na$ covers all the atoms, and the shortest wavelength of $\lambda = 2a$ only spans two atoms. Shorter than $2a$ is unphysical. This implies that the largest $|k|$ is π/a, and the smallest k is $2\pi/Na$, which is also the step size, Δk. To match the periodic boundary condition, k must take the integer multiple, $n\Delta k = n(2\pi)/(Na)$, where n runs from $-N/2$ to $N/2$, excluding $n = 0$. The solid line in Fig. 9.18 is an ideal case with

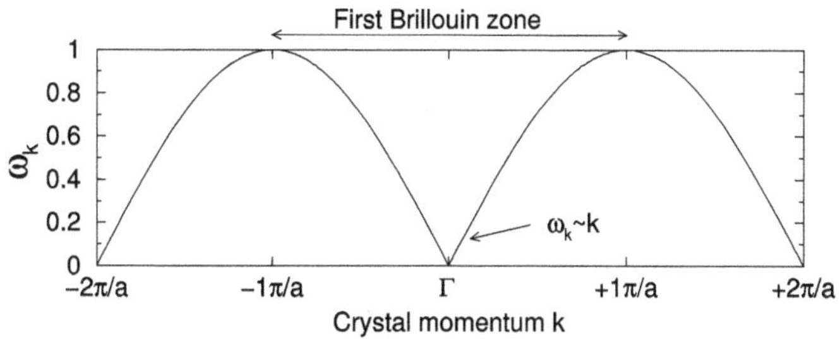

Figure 9.18: Vibrational frequency dispersion ω_k, i. e., phonon dispersion as a function of crystal momentum k. The crystal momentum range within $[-\pi/a : \pi/a]$ defines the first Brillouin zone. Although ω_k is periodic in k, it does not mean that k can take values beyond the Brillouin zone, which is different from the electron band structure. The unit of ω_k is $2\sqrt{C/m}$, where C is the spring constant and m is the mass of the particle.

$N \to \infty$, which never exists in a real sample, so the $k = 0$ point must be excluded. For 12 atoms there are only 12 k's. If $k < 0$, this means that our wave moves along the $-x$-axis, which is illustrated in Fig. 9.17(d). The wave nature of the collective motion involving all the atoms is better reflected with a longer λ or smaller k. This corresponds to the linear portion of the dispersion in Fig. 9.18, where for a small k, $\omega_k \approx \sqrt{\frac{C}{m}}ka$, so ω_k is linearly proportional to k.

For the same reason, we write u_l as $u_l^k = u_0 e^{ikla}$. The displacement of neighboring atoms forms a continuous wave, with a fixed phase difference,

$$\frac{u_{l+1}^k}{u_l^k} = \frac{u_0 e^{ia(l+1)k}}{u_0 e^{ia(l)k}} = e^{iak},$$

which is independent of l. When the next moment comes, the wave simply shifts toward the propagation direction. This is the collective motion of all the atoms in the chain.

In molecules and three-dimensional solids, one normally computes the normal mode frequency by diagonalizing dynamic matrices. C_{60} is one example. Sixty carbon atoms form a Buckminister fullerene cage structure with the highest point group symmetry I_h. Table 9.1 shows the frequencies of vibrational modes in C_{60} for both the Raman and infrared active modes.

Exercise 9.8.1.

36. A chain of 60 carbon atoms has a phonon frequency of 5 THz. Compute the spring constant C in the units of eV/Å.

37. C_{60} has a mode of frequency 481 cm^{-1}. The units are in wavenumbers, i. e., the number of waves per cm. Use $\nu = \frac{c}{\lambda}$ to convert it to the units of THz.

38. For a small k, show $\omega_k \approx \sqrt{\frac{C}{m}}ka$, where ω_k is linearly proportional to k.

Table 9.1: Main normal mode frequencies in C_{60} computed with our parameters and comparison with the experiments [69, 70]. The agreement with theory [71] is excellent, where the difference is less than 1%, so their numerical results are not listed. The frequencies are in the unit of cm^{-1}.

Raman modes	$A_g(1)$	$A_g(2)$	H_g	H_g	H_g	H_g	H_g	H_g	H_g	H_g
ν (theory)	481	1476	245	374	479	772	1184	1410	1545	1817
ν (experiment)	496	1470	273	437	710	774	1099	1250	1428	1575

IR modes	T_{1u}	T_{1u}	T_{1u}	T_{1u}
ν (theory)	427	583	1326	1498
ν (experiment)	528	577	1183	1429

9.8.2 Phonons

If we quantize collective vibrational excitations in a molecule or solid, we have phonons. Then, the dispersion shown in Fig. 9.18 is called the phonon dispersion. The phonon frequency is typically on the order of THz (= 10^{12} Hz). The phonon's Hamiltonian is

$$\hat{H} = \sum_k \left(\hat{a}_k^\dagger \hat{a}_k + \frac{1}{2} \right) \hbar\omega_k,$$

where the ladder operators $\hat{a}_k^\dagger$ and $\hat{a}_k$ are now called creation and annihilation operators, respectively, with a subscript, k. The number operator is $\hat{n}_k = a_k^\dagger a_k$. The eigenenergy is $E_k = (n_k + \frac{1}{2})\hbar\omega_k$. The connection between the number of phonons and the displacement of the nuclear vibration is also the same as the harmonic oscillator. A larger displacement corresponds to many excited phonons.

Phonons and photons are bosons. A single energy level can have many phonons. They follow the Bose–Einstein distribution,

$$\langle \hat{n}_k \rangle = f_{\mathrm{BE}} = \frac{1}{e^{\beta E_k} - 1}, \tag{9.8.5}$$

where $\beta = 1/k_B T$. Because f_{BE} must be positive, E_k for bosons must be positive. If E_k is small, $f_{\mathrm{BE}} \propto 1/(\beta E_k)$ can be a huge number. If lots of phonons occupy the same level, we have Bose–Einstein condensation. This happens in cold atoms and superconductors. Figure 9.19 compares three distributions: Maxwell–Boltzmann, Fermi–Dirac, and Bose–Einstein distributions.

Quantization of the phonon energy does not only explain the blackbody radiation as seen in Chapter 3 but also, more importantly, explains the specific heat change with temperature. The specific heat c is defined as the amount of heat energy needed to raise the temperature of a matter per unit of mass,

$$c = \frac{dE}{dT}, \tag{9.8.6}$$

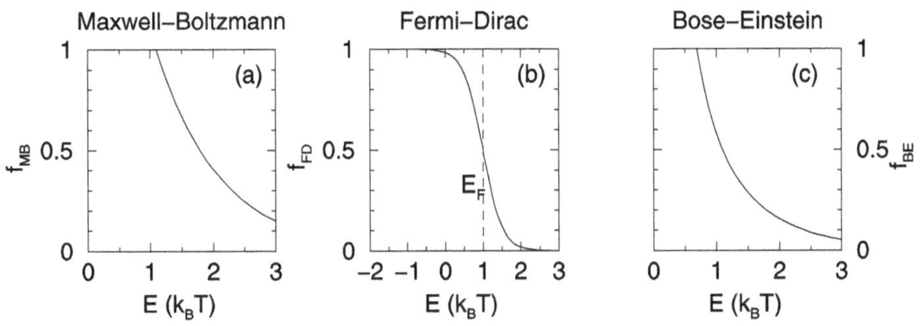

Figure 9.19: (a) Maxwell-Boltzmann, (b) Fermi–Dirac and (c) Bose–Einstein distributions.

where E is the heat energy and T is the temperature. In SI units, c is in J/K. Classical Maxwell–Boltzmann distribution predicts the thermal energy for a three-dimensional oscillator is $3k_BT$. Because one over the unit of mass (in grams) is the Avogadro number N_A, the total heat energy is $E = 3N_Ak_BT$, which leads to $c = 3N_Ak_B$. This means that c is independent of temperature, which contradicts the experimental observation. Experimentally, it is observed that c goes to zero as $T \to 0$ and approaches a constant as T becomes very high. In 1907, Einstein adopted the quantum concept and finally resolved this issue.

Exercise 9.8.2.

39. C_{60} has a mode of 1817 cm^{-1}. Find its corresponding energy E in eV. (Answer: 0.225 eV).
40. According to eq. (9.8.5), find E/k_BT if $f_{BE} = 1000$.
41. Einstein used the thermal energy per mole $E = 3h\nu N_A/(e^{h\nu/k_BT} - 1)$. (a) Compute c. (b) Plot c as a function of T for $\nu = 0.1$ eV and $\nu = 1$ eV. What do you find?

9.9 Problems

1. Consider an operator, $\hat{T}$, which transforms a wavefunction $\psi(t)$ of a system from time t to $t + \Delta t$, where Δt is infinitesimal. If $\psi(t + \Delta t) = e^{-i\omega\Delta t}\psi(t)$, show the operator that can perform such change is a Hamiltonian operator.
2. (a) Directly integrate

$$\frac{1}{L}\int_0^{Na} e^{-iqx}e^{inbx}e^{ikx}dx,$$

where $L = na$, b is the primitive reciprocal lattice vector, $b = 2\pi/a$, q, and k is the crystal momentum. (b) Consider two cases: (i) $(-q + nb + k) = 0$ and (ii) $(-q + nb + k) \neq 0$.

3. In the nearly free-electron model, at $\Delta \neq 0$, (a) compute the expectation value of $\hat{p}_x$ from E_k using the Hellmann–Feynman theorem $\langle\psi_k|\hat{p}_x|\psi_k\rangle = \frac{m_e}{\hbar}\frac{\partial E_k}{\partial k}$. (b) Compute $\langle\psi_k|\hat{p}_x|\psi_k\rangle$ using the eigenfunction to prove that they are exactly the same. (c) Compute the velocity v_x. (d) Show the expectation value of $\langle\psi_-|\hat{p}_x|\psi_-\rangle = (\Delta - \frac{1}{2})(1 - \frac{1}{\sqrt{(\Delta-\frac{1}{2})^2+V_n^2}})$, which matches $m\frac{\partial E_-}{\hbar\partial k}$.
4. Starting from eq. (9.3.10), (a) normalize it; (b) compute the expectation of $\hat{p}_x$; (c) compute $\langle\psi_k|x|\psi_k\rangle$ (Answer: $\langle\psi_k|x|\psi_k\rangle = \frac{L}{2}$); (d) compute $\langle\psi_{k_1}|x|\psi_{k_2}\rangle$, where k_1 and k_2 are different.
5. According to Section 9.4.1.1, prove the Berry connection $-i\langle\psi_k|\frac{\partial\psi_k}{\partial k}\rangle$ is equal to the lattice site expectation value $\frac{1}{N}\sum_l R_l$.
6. (a) Starting from eq. (9.4.17), prove $\psi_{k+nb} = \psi_k$. (b) Directly starting from two eigenstates $|\psi_{k_1}\rangle$ and $|\psi_{k_2}\rangle$, show that the transition matrix elements $\langle\psi_{k_1}|\hat{x}|\psi_{k_2}\rangle$ are given by eq. (9.4.20).
7. (a) Starting from eq. (9.4.14), find the energy dispersion E_k of the face-centered cubic. (b) Choose the Γ-X line to plot the band.
8. The body-centered cubic has the energy dispersion $E_k = -t\cos(k_x a)\cos(k_y a) \times \cos(k_z a)$. Setting $k_z a = 0$, plot the Fermi surfaces at several different Fermi energies, $E_F = -0.9t, -0.5t, 0, 0.5t, 0.9t$.
9. (a) Show that a free-electron gas model in a two-dimensional square lattice has the Fermi wavevector of $k_F = \sqrt{2\pi n_e}$, where n_e is the electron surface density $n_e = N_e/(Na)^2$, N is the number of lattice sites, a is the lattice constant, and N_e is the number of electrons. (b) Show that in a free-electron gas model in a three-dimensional cubic lattice, the Fermi wavevector is $k_F = \sqrt[3]{3\pi^2 n_e}$, where $n_e = N_e/(Na)^3$.
10. A one-dimensional system has band dispersion $E_k = E_0\cos(ka)$, where E_0 is an energy constant, k is the wavevector, and a is the lattice constant. Suppose we have N_e electrons. Draw the Fermi surfaces by choosing several different N_e.
11. A one-dimensional system has the band structure $E_k = \frac{\hbar^2}{ma^2}(1-\cos(ka)+2\cos(2ka))$. (a) Find the energy bandwidth (defined as the difference between the band maximum and minimum). (b) Calculate the group velocity. (c) Compute the effective mass of the electron at Γ point.
12. A one-dimensional periodic potential is given as $V(x) = -V_0\sum_{n=1}^{N}\delta(x-na)$, where a is the lattice constant, δ is the Dirac delta function, and V_0 is a constant. Before forming a solid, atoms are in the atomic one-dimensional s orbital, $\phi^{\text{at}}(x-na) = \sqrt{c}e^{-c|x-na|}$. (a) Using the tight-binding approximation, compute the energy band E_k. Hint: Follow eqs. (9.4.9), (9.4.10), and (9.4.14). You can set the atomic energy to E_s^{at}. (Answer: $E_k = E_s^{\text{at}} - 2V_0 ce^{-ca}\cos(ka)$). (b) Using the variational principle to minimize E_k to find c.

10 Special topics: many-body systems, magnetism, and quantum information

This chapter introduces some advanced topics: electron correlation, magnetism, the Berry phase, and quantum information processing. Except for the Berry phase, the remaining sections heavily rely on many-body wavefunctions, so it is necessary to first cover Unit 1. The Berry phase is included since it finds many applications in condensed matter and provides some additional insights into QM.

This chapter consists of four units:

- Unit 1, containing Sections 10.1 and 10.2, introduces several approximations for many-body problems and then focuses on particle indistinguishability and many-body wavefunctions of bosons and fermions.
- Unit 2 includes Sections 10.3, 10.4, and 10.5. It introduces the spin singlet and spin triplet, the Coulomb and exchange correlation, and explains where magnetism comes from. The Heisenberg exchange spin model is also introduced.
- Unit 3 contains Section 10.6. It addresses quantum information, where quantum bits, gates and the hardware for quantum computers are introduced.
- Unit 4 contains Section 10.7. It starts with the adiabatic approximation, leading to the geometrical phase. Then it introduces the Berry connection and the Berry phase.

10.1 Many-body systems under various approximations

Molecules and solids contain lots of electrons and nuclei, given by the Avogadro number $N_A = 6.023 \times 10^{23}$ per mole. Note N_A is the inverse of the atomic mass unit m_u (in grams): $N_A \times m_u = 1$ (chemistry × physics = 1). In a system of N nuclei and M electrons, even if we treat them all classically, this is not an easy task. The normal-mode nuclear vibration and its quantum, the phonon, in molecules and solids discussed in Section 9.8 are a way to describe their collective interactions. If treated fully quantum mechanically, the total wavefunction is $\Psi(\mathbf{R}_1, \ldots, \mathbf{R}_N; \mathbf{r}_1, \ldots, \mathbf{r}_M)$, where $\mathbf{R}_i$ is the nuclear coordinates, and $\mathbf{r}_i$ is the electron's coordinates. A full quantum mechanical treatment is nearly impossible except for a few small systems. Various levels of theory are developed.

10.1.1 Born–Oppenheimer approximation

The first approximation is called the Born-Oppenheimer approximation. Because nuclei are at least four orders of magnitude heavier than the mass of electrons, one can approximate $\Psi(\mathbf{R}_1, \ldots, \mathbf{R}_N; \mathbf{r}_1, \ldots, \mathbf{r}_M)$ by a product of the nuclear $\Psi_{\mathrm{nu}}(\mathbf{R}_1, \ldots, \mathbf{R}_N)$ and electronic wavefunctions $\Psi_{\mathrm{el}}(\mathbf{r}_1, \ldots, \mathbf{r}_M)$, i. e.,

https://doi.org/10.1515/9783110672152-010

$$\Psi(\mathbf{R}_1,\ldots,\mathbf{R}_N;\mathbf{r}_1,\ldots,\mathbf{r}_M) \approx \Psi_{\rm nu}(\mathbf{R}_1,\ldots,\mathbf{R}_N)\Psi_{\rm el}(\mathbf{r}_1,\ldots,\mathbf{r}_M).$$

Here, Ψ represents a many-body (many nuclei or many electrons) wavefunction.

We denote the nuclear Hamiltonian as $\hat{H}_{\rm nu}$, which includes nucleus–nucleus interaction, nucleus–electron interaction as $\hat{H}_{\rm ne}$, and the electron Hamiltonian as $\hat{H}_{\rm el}$, which includes the electron–electron interaction. So, the total Hamiltonian is $\hat{H} = \hat{H}_{\rm nu} + \hat{H}_{\rm ne} + \hat{H}_{\rm el}$. Our approximate Schrödinger equation is

$$\hat{H}\Psi = E\Psi. \tag{10.1.1}$$

To find the effective Schrödinger equation for electrons, we left multiply the nuclear wavefunction $\langle\Psi_{\rm nu}|$ and integrate over the nuclear coordinates to get

$$\left[\langle\Psi_{\rm nu}|\hat{H}_{\rm nu}|\Psi_{\rm nu}\rangle + \langle\Psi_{\rm nu}|\hat{H}_{\rm ne}|\Psi_{\rm nu}\rangle + \hat{H}_{\rm el}\right]|\Psi_{\rm el}\rangle = E|\Psi_{\rm el}\rangle. \tag{10.1.2}$$

One sees that the electrons move in the effective potential created by the nuclei. The first term is a constant and represents the nuclear energy. The second term is still an operator because the electronic part is not integrated over. The last term is the pure electron Hamiltonian. Similarly, we can find the effective Schrödinger equation for nuclei by left multiplying $\langle\Psi_{\rm el}|$ and integrating over the electronic coordinates,

$$\left[\hat{H}_{\rm nu} + \langle\Psi_{\rm el}|\hat{H}_{\rm ne}|\Psi_{\rm el}\rangle + \langle\Psi_{\rm el}|\hat{H}_{\rm el}|\Psi_{\rm el}\rangle\right]|\Psi_{\rm nu}\rangle = E|\Psi_{\rm nu}\rangle. \tag{10.1.3}$$

The time-dependent Schrödinger equations are similar. Although we separate the nuclear and electronic wavefunctions, this does not mean that nuclei and electrons are independent of each other. In fact, the solution in eq. (10.1.2) affects the solution in eq. (10.1.3), and vice versa. Therefore, one has to solve them self-consistently. Essentially, when we solve electronic wavefunctions, we assume that the nuclei coordinates do not change. For this reason, this approximation is also called the adiabatic approximation.

10.1.2 Independent electron approximation

To further simplify our equation, we take another approximation: independent particle approximation, where particles do not interact each other. We use M electrons as an example. The total Hamiltonian is

$$\hat{H}_{\rm el} = \hat{H}_1 + \hat{H}_2 + \cdots + \hat{H}_M, \tag{10.1.4}$$

where $\hat{H}_1$ only acts upon electron 1, $\hat{H}_2$ on 2, and so on. Assuming their eigenfunction is $|\phi_i\rangle$ for electron i, $\hat{H}_i|\phi_i\rangle = E_i|\phi_i\rangle$, the eigenfunction of $\hat{H}_{\rm el}$ must be

$$\Psi_{\rm el}(1,2,\ldots,M) = \phi_1\phi_2\cdots\phi_M = \prod_i^M \phi_i, \tag{10.1.5}$$

which can be checked by directly substituting $\Psi_{\mathrm{el}}(1,2,\ldots,M)$ into $\hat{H}_{\mathrm{el}}\Psi_{\mathrm{el}} = E\Psi_{\mathrm{el}}$. And the total eigenenergy E is

$$E = E_1 + E_2 + \cdots E_M = \sum_i^M E_i. \tag{10.1.6}$$

This shows two crucial features of the independent electron approximation: (1) The total energy is the sum over all the individual eigenenergies (eq. (10.1.6)), but (2) the total eigenfunction is the product of individual eigenfunctions (eq. (10.1.5)). A wavefunction of this type is called an uncorrelated (in correlated electron physics), or unentangled (in quantum information), wavefunction or state.

The adiabatic and independent-electron approximations have been very successful in describing various phenomena in chemistry and physics and explain, why in atoms, molecules, and solids we can add the energies of each electron to get an approximate total energy. This also includes the band structure in solids and molecular orbitals in molecules.

10.1.3 Hartree–Fock approximation and density functional theory

To this end, our QM treatment is limited to the single independent-particle level. Both the molecular orbital theory and the band-structure theory in prior chapters are based on this picture. Once we include the electron–electron interaction, our electronic Hamiltonian is $\hat{H}_e + \hat{H}_{\mathrm{ee}}$, where $\hat{H}_e$ is the single-electron Hamiltonian and $\hat{H}_{\mathrm{ee}}$ is the electron-electron interaction.

We consider two electrons, with $\hat{H}_{\mathrm{el}} = \hat{H}_{e1} + \hat{H}_{e2} + \hat{H}_{\mathrm{ee}},$ and our many-electron Schrödinger equation is

$$\hat{H}\Psi(1,2) = E\Psi(1,2) \rightarrow (\hat{H}_{e1} + \hat{H}_{e2} + \hat{H}_{\mathrm{ee}})\Psi(1,2) = E\Psi(1,2), \tag{10.1.7}$$

where we have removed the subscript "el" on Ψ. If we assume $\Psi(1,2)$ can be written as a product of two single-electron wavefunctions ϕ_1 and ϕ_2, $\Psi(1,2) = \phi_1\phi_2$, this is called the Hartree–Fock approximation.

However, the independent-particle theory cannot explain the Pauli exclusion principle, magnetism, superconductivity, and other properties. This is because, in general, the two-electron wavefunction Ψ cannot be written as a product of two single-electron wavefunctions. The wavefunction of this type is called the correlated wavefunction, or the entangled wavefunction. It is this wavefunction that introduces magnetism, strongly correlated systems, quantum computing, and quantum cryptography. Modern theories are mostly based on the density-functional theory [72] and the Kohn–Sham equation [73], where the electron–electron interactions, Coulomb interaction, and exchange correlation are treated by approximate density functionals.

Exercise 10.1.3.
1. Use the independent-particle method to write down the total energy and eigenfunction for benzene in terms of E_{C_i-2s} and E_{C_i-2p}, where C_i refers to six carbon atoms and 2*s* and 2*p* are the electrons. Ignore the 1*s*-electron.
2. Consider a carbon atom. (a) Show $\Psi = \phi_n\phi_{1s}\phi_{2s}\phi_{2p}$ is not an eigenfunction of $\hat{H} = \hat{H}_n + \hat{H}_{1s} + \hat{H}_{2s} + \hat{H}_{2p} + \hat{H}_{n-1s} + \hat{H}_{n-2s} + \hat{H}_{n-2p}$. (b) Show the sum, $E \neq E_n + E_{1s} + E_{2s} + E_{2p}$, is not the eigenenergy of $\hat{H}$.
3. Derive eqs. (10.1.2) and (10.1.3).
4. Suppose $\Psi(1,2) = \phi_1\phi_2$ for a system of two electrons. Derive the Hartree–Fock equations for electrons 1 and 2, in a similar manner as eqs. (10.1.2) and (10.1.3).

10.2 Particle indistinguishability in many-body systems

In CM, particles are distinguishable by their shapes, colors, locations, or other attributes. For instance, two identical cars can be distinguished by their locations. In QM, we can also distinguish them in the same way as CM. For instance, we can tell the electron from the proton in the H atom by mass and charge. However, this is no longer possible if we have two identical particles. Two quantum particles are defined as *identical* or *indistinguishable* (i) if they have the same physical properties, such as mass, charge, spin, polarization, and frequency, *and* (ii) if they appear in the same location and at the same time. Here, the same location means two spatial wavefunctions are overlapped in space. In short, overlap ≡ indistinguishability ≡ entanglement [10]. When any identical particles exchange, they are indistinguishable from the viewpoint of probability density.

But, the wavefunction is always *distinguishable*. Suppose a system consists of N identical particles. Its wavefunction is a many-body wavefunction, denoted as $\Psi(\mathbf{r}_1, \mathbf{r}_2, \ldots, \mathbf{r}_N; s_1, s_2, \ldots, s_N; t)$, where $\mathbf{r}_i$ and s_i are the spatial and spin coordinates and t is time. If we exchange particle i with j, we have a new wavefunction

$$\mathcal{P}_{ij}\Psi(\mathbf{r}_1, \mathbf{r}_2, \ldots, \mathbf{r}_i, \ldots, \mathbf{r}_j, \ldots, \mathbf{r}_N; s_1, s_2, \ldots, s_i, \ldots, s_j, \ldots, s_N; t) \tag{10.2.1}$$

$$= \Psi(\mathbf{r}_1, \mathbf{r}_2, \ldots, \mathbf{r}_j, \ldots, \mathbf{r}_i, \ldots, \mathbf{r}_N; s_1, s_2, \ldots, s_j, \ldots, s_i, \ldots, s_N; t), \tag{10.2.2}$$

where $\mathcal{P}_{ij}$ is defined as the exchange operator that exchanges i with j. How this new wavefunction is related to the old wavefunction is the topic of this section.

10.2.1 Identical particles and entanglement

We start with a system with two identical particles at the same time and the same place, such as two electrons in H_2, but not one electron and one proton in the H atom. If we have two photons, they must have the same frequency, polarization, and other physical properties, and arrive at the same time. The indistinguishability depends on space and time, which underlies linear optical quantum computing.

We use $\Psi(1,2;t)$ to denote the many-body wavefunction (here the two-body wavefunction), where 1 and 2 denote particles 1 and 2, respectively, and t is the time that will be omitted in the following. We want to make this more explicit. Suppose particle 1 is at spatial coordinates $\mathbf{r}_1$ with spin s_1, where $\mathbf{r}_1$ refers to (x_1, y_1, z_1) and s_1 refers to a particular spin state such as $|\uparrow\rangle$ or $|\downarrow\rangle$ or any mixture of spin-up and spin-down. So, $\Psi(1,2)$ is really $\Psi(\mathbf{r}_1, s_1; \mathbf{r}_2, s_2)$.

When we exchange two particles $1 \leftrightarrow 2$, we move particle 1 to $\mathbf{r}_2$ and assign spin s_2 to it, while particle 2 to $\mathbf{r}_1$ with spin s_1. We define the exchange operator $\mathcal{P}_{12}$ as $\mathcal{P}_{12}|\Psi(1,2)\rangle = |\Psi(2,1)\rangle$, where $\mathcal{P}_{12} = \mathcal{P}_{21}$ by definition. It is easy to show $\mathcal{P}_{12}^2 = 1$. As seen in Chapter 1, in QM only the probability density ρ has a physical meaning. For a single-body wavefunction, $\psi(1)$, $\rho(1) = |\psi(1)|^2$, where we have already integrated over the spin wavefunction χ. For a two-body state, we define the density as $\rho(1,2) = \Psi^*(1,2)\Psi(1,2) = |\Psi(1,2)|^2$. Now, we exchange $1 \leftrightarrow 2$, $\mathcal{P}_{12}\rho(1,2) = \rho(2,1)$. If $\rho(1,2) = \rho(2,1)$, then we have two identical particles; otherwise we don't, such as one electron and one proton. The meaning of identical particles is realized mathematically through the density, i. e.,

$$\rho(1,2) = \rho(2,1) \longrightarrow |\Psi(1,2)|^2 = |\Psi(2,1)|^2 \longrightarrow \Psi(1,2) = e^{i\theta}\Psi(2,1),$$

where θ is a real phase. This means that every exchange in the identical particle indices 1 with 2 in the wavefunction $\Psi(1,2)$ produces a phase factor of $e^{i\theta}$. Now, if we do it again, we should get another phase factor as

$$\Psi(1,2) = e^{i\theta}\Psi(2,1) = e^{i\theta}e^{i\theta}\Psi(1,2) = e^{2i\theta}\Psi(1,2) \to e^{2i\theta} = 1 \to \theta = 0 \text{ or } \pi.$$

Before we discuss the phase θ, we should emphasize the general property of $\Psi(1,2)$: Ψ in general cannot be written as a product of the spin and spatial wavefunctions and even the spatial wavefunction alone cannot be rewritten as a product of two spatial single-particle wavefunctions. Therefore, a many-body state is entangled. The entanglement can be in space or in spin or both. These general cases are rather complicated for the beginner, so our following presentation is an approximate version of Ψ, where Ψ is approximated as a product of ψ and χ, i. e., $\Psi \approx \psi \otimes \chi$, with $\psi(\mathbf{r}_1, \mathbf{r}_2, \ldots, \mathbf{r}_N)$ being the spatial part and $\chi(s_1, s_2, \ldots, s_N)$ the spin part. $\otimes$ represents a product. As before, $\phi(\mathbf{r})$ denotes a single-body spatial wavefunction. Here, time t is omitted.

Exercise 10.2.1.

5. Show $\mathcal{P}_{12}^2 = 1$.
6. Suppose a system has two particles. (a) Show, if particle 1 is a proton and particle 2 is an electron, the density $\rho(1,2) \neq \rho(2,1)$. (b) If $\rho(1,2) \neq \rho(2,1)$, prove $\Psi(1,2)$ cannot differ from $\Psi(2,1)$ only by a phase factor.

10.2.2 Bosons

If we choose $\theta = 0$, we have $\Psi(1,2) = \Psi(2,1)$, where the wavefunction does not change signs when we exchange two particles. The wavefunction of this type is called symmetric, and its corresponding particles are called bosons. A photon and phonon (quantum of vibration) are bosons. Bosons have integer spin angular-momentum quantum numbers $s = 0, 1, \ldots$. If we have three bosons, the wavefunction does not change sign if we perform an exchange of any pairs of particles, $\Psi(1,2,3) = \Psi(2,1,3) = \Psi(1,3,2), \ldots$. The many-body wavefunction introduces entanglement, in general, $\rho(1,2) \neq \rho(1)\rho(2)$.

Example 1 (Many-body wavefunction for bosons). (a) A system has two bosons in the same single-particle normalized state ψ_a. Write down the wavefunction and check whether $\rho(1,2) = \rho(1)\rho(2)$. (b) If two bosons occupy two different states ψ_a and ψ_b, write down all the possible wavefunctions and check whether $\rho(1,2) = \rho(1)\rho(2)$.

(a) Since we only have a single-particle state, there is only one possible configuration, $\Psi(1,2) = \psi_a(1)\psi_a(2)$, where indices 1 and 2 refer to particles 1 and 2, respectively. Exchanging 1 with 2 does not change Ψ, so we say the bosonic wavefunction is symmetric. According to the definition, $\rho(1,2) = \Psi(1,2)\Psi^*(1,2) = \psi_a(1)\psi_a(2)\psi_a^*(1)\psi_a^*(2) = \rho(1)\rho(2)$.

(b) In this case, we have many possible wavefunctions. We can put both bosons into the same wavefunction as (a), $\Psi(1,2) = \psi_a(1)\psi_a(2)$ to get $\rho(1,2) = \rho(1)\rho(2)$. Next, if we want to put them in different wavefunctions, caution must be taken to make the wavefunction symmetric. Suppose we have $\psi_a(1)\psi_b(2)$, but this configuration is not symmetric because, if we exchange $1 \leftrightarrow 2$, we get $\psi_a(2)\psi_b(1)$, which is not the same as $\psi_a(1)\psi_b(2)$. This means that we need to add these two configurations up, $\Psi(1,2) = \psi_a(1)\psi_b(2) + \psi_a(2)\psi_b(1)$. Now, if we exchange $1 \leftrightarrow 2$, $\Psi(1,2) = \Psi(2,1)$. The normalized wavefunction is $\Psi(1,2) = \frac{1}{\sqrt{2}}(\psi_a(1)\psi_b(2) + \psi_a(2)\psi_b(1))$. $\rho(1,2) = \frac{1}{2}(\rho_a(1)\rho_b(2) + \rho_a(2)\rho_b(1) + \psi_a(1)^*\psi_b(1)\psi_b^*(2)\psi_a(2) + \psi_a(2)^*\psi_b(2)\psi_b^*(1)\psi_a(1))$, which shows that $\rho(1,2) \neq \frac{1}{2}(\rho_a(1)\rho_b(2) + \rho_a(2)\rho_b(1))$.

Exercise 10.2.2.

7. If a system has three bosons occupying two normalized single-particle states ψ_a and ψ_b, write down all the possible wavefunctions.

8. If a system has two bosons occupying three normalized single-particle states ψ_a, ψ_b, and ψ_c, write down all the possible wavefunctions.

9. Suppose at $t = 0$ the wavefunction is symmetric, $\Psi(1,2,t=0) = \Psi(2,1,t=0)$. Prove that, at time t, it is also symmetric, $\Psi(1,2,t) = \Psi(2,1,t)$. Hints: use time-dependent Schrödinger equation in Appendix 11.2.

10. A system consists of two identical particles, whose wavefunction is $\Psi(1,2) = (\psi_a(1)\psi_b(2) + \psi_a(2)\psi_b(1))/\sqrt{2}$, where both ψ_a and ψ_b are normalized. (a) What is the probability to find particle 1? (b) What is the probability to find particle 2?

10.2.3 Fermions and the Pauli exclusion principle

If we choose $\theta = \pi$, we have fermions, $\Psi(1,2) = -\Psi(2,1)$, where the wavefunction changes sign for each exchange of particles and is said to be antisymmetric. These particles are called fermions. Electrons and protons are fermions, and a single fermion has only spin half-integers, such as $\frac{1}{2}, \frac{3}{2}, \ldots$. We must emphasize that, to tell whether a particle is a fermion or not, we must use its single particle's spin quantum number. This complication arises from the fact that two fermions can form a boson. For instance, the ground state of Cd (see Section 6.5.2) has zero spin. In superconductors, the Cooper pair which has two electrons behaves like a composite boson. Any odd number of fermions still behave as a fermion.

If we have N particles, Ψ must be written as $\Psi(\mathbf{r}_1 s_1, \mathbf{r}_2 s_2, \ldots, \mathbf{r}_N s_N)$, where $\mathbf{r}_i$ and s_i are the spatial and spin indices. Before we provide an example on a many-body state Ψ, we recall that for a single fermion, such as electrons, its single-particle state ψ is a product of the spatial (orbital) ϕ and spin parts (α, β) as $\psi = \phi\alpha$ or $\psi = \phi\beta$ (see eq. (6.1.14)). Similar to bosons, in general, $\rho(1,2) \neq \rho(1)\rho(2)$.

Example 2 (Two-body wavefunctions for electrons). If a system has two electrons in the same single-particle normalized spatial orbital ϕ, write down the possible wavefunction.

We first notice that the problem provides a spatial orbital ϕ without specifying the spin state. This allows us to form two possible ψ: $\psi_a = \phi\alpha$ and $\psi_b = \phi\beta$. If we place both electrons in the same ψ_a, we have $\Psi(1,2) = \psi_a(1)\psi_a(2)$. If we exchange $1 \leftrightarrow 2$, $\Psi(2,1) = \psi_a(2)\psi_a(1) = \Psi(1,2)$. This is a symmetric wavefunction, so it violates the antisymmetric requirement, and we cannot use this one. What if we place them in two different orbitals? Suppose we start with $\Psi(1,2) = \psi_a(1)\psi_b(2)$. But if we exchange $1 \leftrightarrow 2$, we get $\psi_a(2)\psi_b(1)$. This shows that this new combination must be appended to $\psi_a(1)\psi_b(2)$. However, if we directly add them up, then we end up with a symmetric wavefunction again. The only choice is to subtract them from each other to get $\Psi(1,2) = \frac{1}{\sqrt{2}}(\psi_a(1)\psi_b(2) - \psi_a(2)\psi_b(1))$, which is a Slater determinant,

$$\Psi(1,2) = \frac{1}{\sqrt{2}}\begin{vmatrix} \psi_a(1) & \psi_a(2) \\ \psi_b(1) & \psi_b(2) \end{vmatrix} = \frac{1}{\sqrt{2}}\big(\psi_a(1)\psi_b(2) - \psi_a(2)\psi_b(1)\big).$$

Exchanging $1 \leftrightarrow 2$ produces a sign change for $\Psi(1,2) = -\Psi(2,1)$, antisymmetric. The density is $\rho(1,2) = \frac{1}{2}(\rho_a(1)\rho_b(2) + \rho_b(1)\rho_a(2) - \rho_{ab}^{ex}(1)\rho_{ba}^{ex}(2) - \rho_{ab}^{ex}(2)\rho_{ba}^{ex}(1)) \neq \rho_a(1)\rho_b(2)$, where $\rho_{ab}^{ex} = \psi_a^*\psi_b$ is also called the exchange charge.

The previous example prepares us to explain the Pauli exclusion principle: Each orbital can take a maximum of two electrons. If two electrons have the same spin orientation, i. e., both up or both down, we have to put them into two different orbitals. Here, we can show the reason behind this. Suppose that we want to put two electrons into the same 1s-orbital with the same spin α. Then, the single electron wavefunction is $\psi_a = \phi_{1s}\alpha$. The two-electron wavefunction is $\Psi(1,2) = \frac{1}{\sqrt{2}}\begin{vmatrix} \psi_a(1) & \psi_a(2) \\ \psi_a(1) & \psi_a(2) \end{vmatrix} = 0$. This is how mathematically the many-body wavefunction forbids such a configuration.

On the other hand, if two electrons are in the same 1s-orbital, but in two different spin states α and β, then we have two distinctive single-particle states $\psi_a = \phi_{1s}\alpha$ and $\psi_b = \phi_{1s}\beta$. The many-body state $\Psi(1,2)$ is allowed since

$$\Psi(1,2) = \frac{1}{\sqrt{2}}\begin{vmatrix} \psi_a(1) & \psi_a(2) \\ \psi_b(1) & \psi_b(2) \end{vmatrix} = \frac{1}{\sqrt{2}}\begin{vmatrix} \phi_{1s}(1)\alpha(1) & \phi_{1s}(2)\alpha(2) \\ \phi_{1s}(1)\beta(1) & \phi_{1s}(2)\beta(2) \end{vmatrix} \neq 0. \tag{10.2.3}$$

Exercise 10.2.3.

11. If a system has two electrons occupying two different normalized single-particle orbitals ϕ_p and ϕ_q, where the spin states for each electron are not specified, (a) write down all the possible wavefunctions; (b) find $\rho(1,2)$; and (c) show $\rho(1,2) \neq \rho(1)\rho(2)$.
12. If a system has two electrons occupying three normalized single-particle orbitals ϕ_p, ϕ_q, and ϕ_s, write down all the possible wavefunction; (b) find $\rho(1,2)$; and (c) show $\rho(1,2) \neq \rho(1)\rho(2)$.
13. If $\Psi(1,2,t=0) = -\Psi(2,1,t=0)$, prove that, at time t, $\Psi(1,2,t) = -\Psi(2,1,t)$. Hint: Use eqs. (11.5) and (3.5.2) to rewrite $\Psi(1,2,t)$ in terms of $\Psi(1,2,0)$.

10.3 Spin singlet and triplet and spatial entanglement

According to the combinatorial rules of spins (Section 6.5.1), two spins s_1 and s_2 can produce the total spin $S = s_1 + s_2, s_1 + s_2 - 1, \ldots, |s_1 - s_2|$. We use χ_{S,M_S} to denote a many-spin state, where S and M_S are the total spin quantum number and the magnetic spin quantum number, respectively. For two electrons with spin 1/2, $S = 0$ and 1. For $S = 0$, $M_S = 0$, we only have one single spin state, called the spin singlet. We denote its wavefunction as $\chi_{0,0}$, where the first subscript refers to S and the second to M_S. For $S = 1$, $M_S = 0, \pm 1$, called the spin triplet, with three eigenfunctions, $\chi_{1,\pm 1}$ and $\chi_{1,0}$.

10.3.1 Singlet and triplet

We use the previous example to demonstrate this clearly by rewriting eq. (10.2.3) as,

$$\Psi(1,2) = \frac{1}{\sqrt{2}}\phi_{1s}(1)\phi_{1s}(2)\begin{vmatrix}\alpha(1) & \alpha(2)\\ \beta(1) & \beta(2)\end{vmatrix} \equiv \phi_{1s}(1)\phi_{1s}(2)\chi(1,2) \tag{10.3.1}$$

which reveals that the spatial part of the wavefunction is symmetric, i. e., $\phi_{1s}(1)\phi_{1s}(2) = \phi_{1s}(2)\phi_{1s}(1)$, but the spin part is antisymmetric, $\chi(1,2) = -\chi(2,1)$. The whole wavefunction is still antisymmetric, $\Psi(1,2) = -\Psi(2,1)$.

The spin wavefunction is called the spin singlet because we only have one single spin state for the same spatial wavefunction. The spin singlet can have another type of spatial wavefunction, if we choose two different spatial wavefunctions, say ϕ_{1s} and ϕ_{2p}. The spatial symmetric wavefunction is $\frac{1}{\sqrt{2}}(\phi_{1s}(1)\phi_{2p}(2) + \phi_{1s}(2)\phi_{2p}(1))$.

$$\Psi(1,2) = \frac{1}{2}(\phi_{1s}(1)\phi_{2p}(2) + \phi_{1s}(2)\phi_{2p}(1))\begin{vmatrix}\alpha(1) & \alpha(2)\\ \beta(1) & \beta(2)\end{vmatrix}. \tag{10.3.2}$$

The spin singlet wavefunction has zero spin, so we rewrite it as $\chi_{0,0}(1,2)$,

$$\chi_{0,0}(1,2) = \frac{1}{\sqrt{2}}(\alpha(1)\beta(2) - \alpha(2)\beta(1)). \tag{10.3.3}$$

Example 3 (Spin singlet is an eigenfunction of the total spin). Prove that $\chi_{0,0}$ is an eigenfunction of both $\hat{S}_z$ and $\hat{\mathbf{S}}^2$.

We apply $\hat{S}_z = \hat{S}_{1z} + \hat{S}_{2z}$ to the first term in $\chi_{0,0}$ (eq. (10.3.3)),

$$(\hat{S}_{1z} + \hat{S}_{2z})\alpha(1)\beta(2) = \hat{S}_{1z}\alpha(1)\beta(2) + \hat{S}_{2z}\alpha(1)\beta(2) = \frac{\hbar}{2}\alpha(1)\beta(2) + \alpha(1)\frac{-\hbar}{2}\beta(2),$$

where $\hat{S}_{1z}$ only acts upon $\alpha(1)$ since it belongs to electron 1 and has no effect on $\beta(2)$. $\hat{S}_z\alpha(2)\beta(1)$ in eq. (10.3.3) can be worked out similarly. Finally, we have

$$(\hat{S}_{1z} + \hat{S}_{2z})\chi_{0,0} = (\hat{S}_{1z} + \hat{S}_{2z})\frac{1}{\sqrt{2}}\left[\alpha(1)\beta(2) - \alpha(2)\beta(1)\right] = 0\hbar\chi_{0,0},$$

which shows that $\chi_{0,0}$ is an eigenfunction with the eigenvalue of $0\hbar$ as expected.

Similarly, we can apply the total spin operator $\hat{\mathbf{S}}^2$ to $\chi_{0,0}$ and write out $\hat{\mathbf{S}}^2$ as

$$\hat{\mathbf{S}}^2 = (\hat{\mathbf{S}}_1 + \hat{\mathbf{S}}_2)^2 = \hat{\mathbf{S}}_1^2 + \hat{\mathbf{S}}_2^2 + 2\hat{\mathbf{S}}_1 \cdot \hat{\mathbf{S}}_2 = \hat{\mathbf{S}}_1^2 + \hat{\mathbf{S}}_2^2 + 2(\hat{S}_{1x}\hat{S}_{2x} + \hat{S}_{1y}\hat{S}_{2y} + \hat{S}_{1z}\hat{S}_{2z}).$$

We use eqs. (6.1.13) and (6.1.11) to find

$$\hat{\mathbf{S}}^2\chi_{0,0}(1,2) = 0(0+1)\hbar^2\chi_{0,0}(1,2),$$

which proves that $\chi_{0,0}(1,2)$ is indeed an eigenfunction for the total spin.

Spin triplet wavefunctions are different from those in singlets. They are symmetric if we exchange two spins, $1 \leftrightarrow 2$. The spin triplet wavefunctions $\chi(1,2)$ are

$$\chi_{1,1} = \alpha(1)\alpha(2), \quad \chi_{1,0} = \frac{1}{\sqrt{2}}(\alpha(1)\beta(2) + \alpha(2)\beta(1)), \quad \chi_{1,-1} = \beta(1)\beta(2), \tag{10.3.4}$$

each of which is symmetric. $\chi_{0,0}$, $\chi_{1,0}$, and $\chi_{1,\pm1}$ are also eigenfunctions of $\hat{\mathbf{S}}^2$ and $\hat{S}_z$.

Exercise 10.3.1.

14. Show the three triplets in eq. (10.3.4) are the eigenfunctions of $\hat{\mathbf{S}}^2$ and $\hat{S}_z$.

10.3.2 Entanglement between spin and spatial spaces

Spin wavefunctions χ do not exist alone. It is incorrect to say that we can move two spins far apart because "far apart" refers to the spatial dimension. Neither can we say that we allow the spin going into a detector at a particular location because spin has no spatial dependence $\nabla\chi = 0$ (recall the Stern–Gerlach experiment). A correct way to understand spin wavefunctions is to include the spatial part of the wavefunction, true for single-electron and many-electron wavefunctions. Only the spatial part of the wavefunction can be identified with a particular location, namely, entanglement.

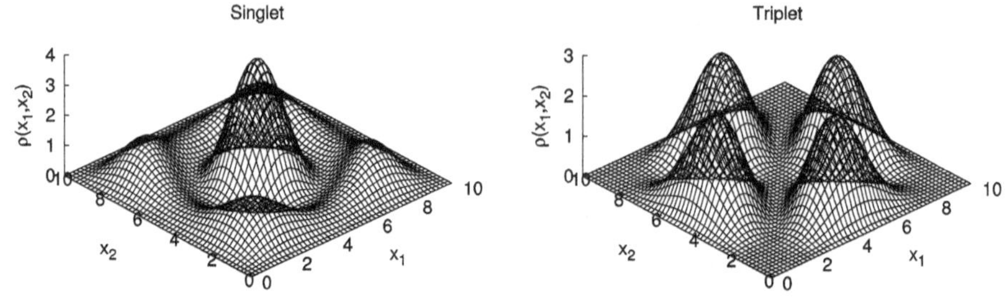

Figure 10.1: (Left) Two-body spatial density $\rho(x_1, x_2)$ for two electrons (x_1, x_2) in a spin singlet state in the infinite quantum well. The well starts at $x = 0$ and ends at $x = 10$. The center is at $x_1 = x_2 = 5$. (Right) Two-body density $\rho(x_1, x_2)$ of a spin triplet.

To be specific, we consider two electrons occupying two spatial eigenstates of the infinite quantum well, $\phi_a(x) = \sqrt{\frac{2}{a}}\sin(\frac{\pi x}{a})$ and $\phi_b(x) = \sqrt{\frac{2}{a}}\sin(\frac{3\pi x}{a})$.[1] First, we construct a spatial wavefunction for the singlet. Because $\chi_{0,0}$ is antisymmetric, the spatial part must be symmetric, so the only possible wavefunction is

$$\Psi(1,2) = \frac{1}{\sqrt{2}}(\phi_a(1)\phi_b(2) + \phi_a(2)\phi_b(1))\chi_{0,0} = \psi_{\text{singlet}}(1,2)\chi_{0,0},$$

whose spatial wavefunction $\psi_{\text{singlet}}(1,2)$ is

$$\psi_{\text{singlet}}(1,2) = \frac{\sqrt{2}}{a}\left[\sin\left(\frac{\pi x_1}{a}\right)\sin\left(\frac{3\pi x_2}{a}\right) + \sin\left(\frac{\pi x_2}{a}\right)\sin\left(\frac{3\pi x_1}{a}\right)\right].$$

Figure 10.1 (left) shows that its probability density $\rho(1,2) = |\psi_{\text{singlet}}(1,2)|^2$ has a maximum in the middle of the box. This reveals an important property: In spin singlets, two electrons spatially can be as close as possible, only subject to the Coulomb repulsion.

The spin triplets allow only an antisymmetric spatial wavefunction $\psi_{\text{triplet}}(1,2)$,

$$\begin{aligned}\psi_{\text{triplet}}(1,2) &= \frac{1}{\sqrt{2}}(\phi_a(1)\phi_b(2) - \phi_a(2)\phi_b(1)) \\ &= \frac{\sqrt{2}}{a}\left[\sin\left(\frac{\pi x_1}{a}\right)\sin\left(\frac{3\pi x_2}{a}\right) - \sin\left(\frac{\pi x_2}{a}\right)\sin\left(\frac{3\pi x_1}{a}\right)\right].\end{aligned}$$

Figure 10.1 (right) shows its probability density $\rho(1,2) = |\psi_{\text{triplet}}(1,2)|^2$. The density does not peak at the center; instead, it spreads around. Even before we consider the Coulomb repulsion, the electrons already repel each other due to the antisymmetry of the spatial

1 Choosing $\phi_a(x) = \sqrt{\frac{2}{a}}\sin(\frac{\pi x}{a})$ and $\phi_b(x) = \sqrt{\frac{2}{a}}\sin(\frac{2\pi x}{a})$ is not ideal because ϕ_a and ϕ_b do not share the same parity, and the difference between the antisymmetric and symmetric many-body wavefunctions is less obvious. Choosing $\phi_a = \phi_b$ is not an option either because it only provides a symmetric spatial wavefunction that cannot be used for a triplet.

wavefunction. It is qualitatively different from the spatial wavefunction for a spin singlet. This demonstrates the entanglement between the spin and spatial degrees of freedom. In the density functional theory, this antisymmetry creates a hole, i.e, an exchange hole, for other electrons if an electron is already present.

Exercise 10.3.2.

15. Why cannot we say a spin is located at a position in space?

16. Two electrons sealed inside a box with dimension $a \times a \times a$ are two eigenfunctions $|k_1\rangle = a^{-3/2}e^{i\mathbf{k}_1\cdot\mathbf{r}}$ and $|k_2\rangle = a^{-3/2}e^{i\mathbf{k}_2\cdot\mathbf{r}}$. (a) If both electrons have the same spin orientation, write down the many-body wavefunction, normalize it, and find $\rho(1,2)$. (b) If both electrons have opposite spin orientations, write down the many-body wavefunction and normalize it. (c) Find $\rho(1,2)$, and show that $\rho(1,2)$ cannot be written as a product of single densities $\rho(1)$ and $\rho(2)$.

10.4 Kinetic, Coulomb and exchange correlation energies

To appreciate new features of a many-body state, we consider single-body operators $\hat{O}(i)$ and two-body operators $\hat{O}(i,j)$.

10.4.1 Single-particle operators: kinetic energy

Since spin operators have just been examined, here we focus on the spatial part of many-electron states, $\psi(\mathbf{r}_1,\ldots,\mathbf{r}_N)$. We will omit the spin wavefunctions χ, but one must remember that the product of ψ and χ must be antisymmetric for electrons.

To be more definitive, we also consider a two-electron state $\psi(1,2)$ that consists of two different normalized single-particle states, ϕ_a and ϕ_b, $\langle\phi_i|\phi_i\rangle = 1$,

$$\psi(1,2) = \frac{1}{\sqrt{2}}(\phi_a(1)\phi_b(2) \pm \phi_b(1)\phi_a(2)) \equiv \frac{1}{\sqrt{2}}(A \pm B),$$

where $A = \phi_a(1)\phi_b(2)$, $B = \phi_a(2)\phi_b(1)$, and + and − denote the symmetric and antisymmetric wavefunctions, respectively.

The expectation value of the single-electron operator $\hat{O}(1)$ for electron 1 is,

$$\begin{aligned}\langle\psi|\hat{O}(1)|\psi\rangle &= \frac{1}{2}\int_1\int_2 (A^* \pm B^*)\hat{O}(1)(A \pm B)d\mathbf{r}_1 d\mathbf{r}_2\\ &= \frac{1}{2}(\langle A|\hat{O}(1)|A\rangle + \langle B|\hat{O}(1)|B\rangle \pm \langle A|\hat{O}(1)|B\rangle \pm \langle B|\hat{O}(1)|A\rangle),\end{aligned} \tag{10.4.1}$$

which contains four integrals over electron coordinates $\mathbf{r}_1$ and $\mathbf{r}_2$. We first compute the diagonal terms with AA^* and BB^*,

$$\langle A|\hat{O}(1)|A\rangle = \int_1\int_2 \phi_a^*(1)\phi_b^*(2)\hat{O}(1)\phi_a(1)\phi_b(2)d\mathbf{r}_1 d\mathbf{r}_2 = \int_1 \phi_a^*(1)\hat{O}(1)\phi_a(1)d\mathbf{r}_1,$$

and

$$\langle B|\hat{O}(1)|B\rangle = \int_1\int_2 \phi_b^*(1)\phi_a^*(2)\hat{O}(1)\phi_b(1)\phi_a(2)d\mathbf{r}_1 d\mathbf{r}_2 = \int_1 \phi_b^*(1)\hat{O}(1)\phi_b(1)d\mathbf{r}_1,$$

where we have used the normalization condition $\langle\phi_i|\phi_i\rangle = 1$. The crossing terms $\langle A|\hat{O}(1)|B\rangle$ and $\langle B|\hat{O}(1)|A\rangle$ are complicated since they depend on the overlap $\langle\phi_a|\phi_b\rangle$.

If $\langle\phi_a|\phi_b\rangle = 0$, then the crossing terms are zero. $\langle\psi|\hat{O}(1)|\psi\rangle$ is just the average value of two states, $\langle\psi|\hat{O}(1)|\psi\rangle = \frac{1}{2}(\langle A|\hat{O}(1)|A\rangle + \langle B|\hat{O}(1)|B\rangle)$. Take the position operator in PIB as an example. Since $\langle\phi_a(1)|x_1|\phi_a(1)\rangle = \frac{a}{2}$ and $\langle\phi_b(1)|x_1|\phi_b(1)\rangle = \frac{a}{2}$, the expectation value is $\langle\psi|x_1|\psi\rangle = \frac{a}{2}$. Whether $\psi(1,2)$ is symmetric or antisymmetric produces the same $\langle\psi|\hat{O}(1)|\psi\rangle$.

If $\langle\phi_a|\phi_b\rangle \neq 0$, then the crossing terms are nonzero, and whether $\psi(1,2)$ is symmetric or antisymmetric makes a big difference,

$$\begin{aligned}\langle\psi|\hat{O}(1)|\psi\rangle &= \frac{1}{2}(\langle\phi_a|\hat{O}(1)|\phi_a\rangle + \langle\phi_b|\hat{O}(1)|\phi_b\rangle)\\ &\pm \frac{1}{2}(\langle\phi_a|\hat{O}(1)|\phi_b\rangle\langle\phi_b|\phi_a\rangle \pm \langle\phi_b|\hat{O}(1)|\phi_a\rangle\langle\phi_a|\phi_b\rangle).\end{aligned} \tag{10.4.2}$$

For the single-electron operator $\hat{O}(2)$ for electron 2, we only need to change 1 to 2 in the eq. (10.4.2).

In summary, a key feature of the expectation values of single-body operators for the many-body wavefunctions is that, if the spatial wavefunction is a product of two single-particle wavefunctions, their expectation values are formed by their expectation values of the single-body wavefunction. In the following, we show two examples.

Example 4 (Single-particle probability in a many-body state). Two electrons are in two different single-particle normalized functions ϕ_a and ϕ_b. Their total wavefunction is $\psi(1,2) = \frac{1}{\sqrt{2}}(\phi_a(1)\phi_b(2) - \phi_a(2)\phi_b(1))$. (a) Find the probability density $\rho(1)$ to find electron 1. (b) Find the probability density $\rho(2)$ to find electron 2. (c) What can we conclude from (a) and (b)?

(a) There are two methods to find the single-particle probability from a many-particle state. One is to use the single-particle operator, and the other is to integrate over coordinates of the rest of particles. Here, we use the second method. $\rho(1) = \int_2 \psi^*(1,2)\psi^*(1,2)d\mathbf{r}_2 = \frac{1}{2}(|\phi_a(1)|^2 + |\phi_b(1)|^2 - \phi_a^*(1)\phi_b(1)\int_2 \phi_b^*(2)\phi_a(2)d\mathbf{r}_2 - \phi_b^*(1)\phi_a(1)\int_2 \phi_a^*(2)\phi_b(2)d\mathbf{r}_2)$.

(b) For electron 2, we only need to exchange 1 and 2, $\rho(2) = \int_1 \psi^*(1,2)\psi^*(1,2)d\mathbf{r}_1 = \frac{1}{2}(|\phi_a(2)|^2 + |\phi_b(2)|^2 - \phi_a^*(2)\phi_b(2)\int_1 \phi_b^*(1)\phi_a(1)d\mathbf{r}_1 - \phi_b^*(2)\phi_a(2)\int_1 \phi_a^*(1)\phi_b(1)d\mathbf{r}_1)$.

(c) If we compare $\rho(1)$ and $\rho(2)$, we see they are exactly the same, $\rho(1) = \rho(2)$. This means that they are identical particles.

Example 5 (Kinetic energy expectation in a many-body state). Suppose

$$\psi(1,2) = \frac{1}{\sqrt{2}}\big(\phi_a(1)\phi_b(2) \pm \phi_b(1)\phi_a(2)\big),$$

where ϕ_a and ϕ_b are normalized but not orthogonal. Find the total kinetic energy $\langle\psi(1,2)|(\hat{T}_1 + \hat{T}_2)|\psi(1,2)\rangle$, where $\hat{T}$ is the kinetic energy operator.

We can directly use eq. (10.4.2) to get $\langle\psi(1,2)|(\hat{T}_1 + \hat{T}_2)|\psi(1,2)\rangle$. Since the expression is too long, we denote $|\phi_a\rangle$ by $|a\rangle$ and $\langle\phi_a|\phi_b\rangle$ by $\langle a|b\rangle$. The rest is the same. $\langle\psi(1,2)|(\hat{T}_1 + \hat{T}_2)|\psi(1,2)\rangle$ is

$$\frac{1}{2}\sum_{i=1}^{2}\big(\langle a|\hat{T}_i|a\rangle + \langle b|\hat{T}_i|b\rangle \pm \langle a|\hat{T}_i|b\rangle\langle b|a\rangle \pm \langle b|\hat{T}_i|a\rangle\langle a|b\rangle\big),$$

which shows that the kinetic energy is also additive.

Exercise 10.4.1.

17. Two electrons are in two single-particle wavefunctions sealed inside a line segment between $x \in (-L/2, L/2)$, $\phi_a(x) = e^{ik_ax}/\sqrt{L}$ and $\phi_b(x) = e^{ik_bx}/\sqrt{L}$, where $k_a \neq k_b$. (a) Write down a symmetric two-body wavefunction and compute the expectation value of the position operator. (b) Write down an antisymmetric two-body wavefunction and compute the expectation value of the position operator.

18. Two electrons are confined in $x \in (-\frac{a}{2}, \frac{a}{2})$. Suppose they are in a wavefunction $\psi(1,2) = \frac{1}{\sqrt{2}}(\phi_a(1)\phi_b(2) \pm \phi_b(1)\phi_a(2))$, where $\phi_a(x) = \cos(\frac{\pi x}{a})$ and $\phi_b = e^{-\frac{|x|}{a}}$. (a) Normalize both ϕ_a and ϕ_b and find $\langle\phi_a|\phi_b\rangle$. (b) Compute the expectation value of the total kinetic energy. (c) Find the difference in kinetic energy between the symmetric and antisymmetric wavefunctions.

10.4.2 Coulomb and exchange correlation energies

A two-body operator $\hat{O}(1,2)$ has the hallmark of many-body states. The Coulomb repulsion between two electrons is such an operator, where in 3D $V(r_{12}) = \frac{e^2}{4\pi\epsilon_0 r_{12}}$. In CM, the Coulomb repulsion is the only force between two electrons. But in QM, the many-body state produces something extra.

Let's assume again the symmetric (+) and antisymmetric (−) wavefunctions $\psi(1,2) = \frac{1}{\sqrt{2}}(A \pm B)$, where $A = \phi_a(1)\phi_b(2)$ and $B = \phi_b(1)\phi_a(2)$. The expectation value of $V(r_{12})$ is

$$\langle\Psi|V|\Psi\rangle = \frac{1}{2}(\langle A|V|A\rangle + \langle B|V|B\rangle \pm \langle A|V|B\rangle \pm \langle B|V|A\rangle). \tag{10.4.3}$$

We examine the first two terms

$$\begin{aligned}\langle A|V|A\rangle = \langle B|V|B\rangle &= \frac{e^2}{4\pi\epsilon_0}\int_1\int_2 \phi_a^*(1)\phi_b^*(2)\phi_a(1)\phi_b(2)r_{12}^{-1}d\mathbf{r}_1 d\mathbf{r}_2 \\ &= \frac{e^2}{4\pi\epsilon_0}\int_1\int_2 \rho_a(1)\rho_b(2)r_{12}^{-1}d\mathbf{r}_1 d\mathbf{r}_2,\end{aligned} \tag{10.4.4}$$

which is the Coulomb repulsion between electrons, $\rho_a(1)$ and $\rho_b(2)$, called the Coulomb integral. $E_C = \frac{1}{2}(\langle A|V|A\rangle + \langle B|V|B\rangle) = \langle A|V|A\rangle)$. Here, we have introduced the charge density $\rho_a = \phi_a^*\phi_a$ and $\rho_b = \phi_b^*\phi_b$, Because $\rho > 0$, $E_C > 0$. Hund's second rule (Section 6.5.2.2) requires the sum of orbital angular-momentum quantum numbers to be the largest to reduce E_C. We provide one problem at the end to demonstrate this. But, the last two terms in eq. (10.4.3) have no classical analogue,

$$\begin{aligned} E_X = \langle A|V|B\rangle &= \frac{e^2}{4\pi\epsilon_0}\int_1\int_2 \phi_a^*(1)\phi_b^*(2)\phi_a(2)\phi_b(1)r_{12}^{-1}\mathbf{dr}_1\mathbf{dr}_2 \\ &= \frac{e^2}{4\pi\epsilon_0}\int_1\int_2 \rho_{\text{ex}}(1)\rho_{\text{ex}}(2)r_{12}^{-1}\mathbf{dr}_1\mathbf{dr}_2, \end{aligned} \tag{10.4.5}$$

where $\phi_a^*(1)\phi_b(1)$ is not a regular density, so we write it as $\rho_{\text{ex}} = \phi_a^*\phi_b$. We add a subscript "*ex*" to ρ to denote an exchange between ϕ_a and ϕ_b. $\langle A|V|B\rangle$ is called the exchange integral, E_X. It can be proven that E_X is also positive [74] (see the following exercise). Then, our total potential energy is

$$\langle\Psi|V|\Psi\rangle = E_C \pm E_X, \tag{10.4.6}$$

where $+(-)$ is due to the symmetric (antisymmetric) wavefunction. Because E_C and E_X are both positive, the symmetric spatial wavefunction (+) has a higher potential energy because the electrons in spin singlets tend to move closer to each other as seen in Fig. 10.1. This is the origin of Hund's first rule (Section 6.5.2.2), where we force the electrons into a spin symmetric state. If the potential energy were the only determining factor of a system, the system would always be in an antisymmetric state (−), where the spin wavefunction is symmetric with spins pointing in the same direction. In other words, we would never have the H_2 molecule.

The missing piece is the contribution of the kinetic energy. Following the second example in the last section, we have

$$\frac{1}{2}\sum_{i=1}^{2}(\langle a|\hat{T}_i|a\rangle + \langle b|\hat{T}_i|b\rangle \pm \langle a|\hat{T}_i|b\rangle\langle b|a\rangle \pm \langle b|\hat{T}_i|a\rangle\langle a|b\rangle).$$

If $\langle b|a\rangle$ and $\langle a|b\rangle$ are nonzero and $\langle a|\hat{T}_i|b\rangle$ and $\langle b|\hat{T}_i|a\rangle$ are negative, then the spatial symmetric wavefunction (+) has a lower energy. Because the total wavefunction must be antisymmetric, this means the spin wavefunction is antisymmetric in a singlet state. This is the case for the H_2 molecule, He, and other light elements in the first and second rows of the period table. Caution must be taken since actual situations are much more complicated, whenever electron correlation becomes strong [15].

Exercise 10.4.2.
19. Based on eq. (10.4.6), under what conditions can we ignore the difference between symmetric and antisymmetric wavefunctions?
20. $\psi(1,2) = \frac{1}{\sqrt{2}}(A \pm B)$, where $A = \phi_a(1)\phi_b(2)$ and $B = \phi_b(1)\phi_a(2)$. Show (a) $\langle A|V|A\rangle = \langle B|V|B\rangle$, and (b) $\langle A|V|B\rangle = \langle B|V|A\rangle$, where V is the potential.
21. The kinetic energy operator for a two-electron system is $-\frac{\hbar^2\nabla_1^2}{2m_e} - \frac{\hbar^2\nabla_2^2}{2m_e}$. Its wavefunction is $\psi(1,2) = \frac{1}{\sqrt{2}}[\phi_a(1)\phi_b(2) \pm \phi_a(2)\phi_b(1)]$. (a) Show, if $\langle\phi_a|\phi_b\rangle = 0$, the expectation value of the kinetic energy is the same for the symmetric and antisymmetric wavefunctions. (b) If $\langle\phi_a|\phi_b\rangle \neq 0$, show they are different.
22. Suppose $\psi(1,2) = \frac{1}{\sqrt{2}}(\phi_a(1)\phi_b(2) \pm \phi_b(1)\phi_a(2))$, where ϕ_a and ϕ_b are two eigenfunctions of PIB for $x \in (0,a)$, $\phi_a = \sqrt{\frac{2}{a}}\sin(\frac{\pi x}{a})$, $\phi_b = \sqrt{\frac{2}{a}}\sin(\frac{2\pi x}{a})$. Compute the expectation value of x_1x_2 for (a) the symmetric two-body state and (b) the antisymmetric two-body state.
23. Prove both the Coulomb and exchange integrals are positive. Hint for the exchange integral: Choose two arbitrary points A and B and then assume ρ_{ex} at A is negative and at B is positive (both positive or both negative always lead to a positive E_X). Then, compute $\rho_{ex}(1)\rho_{ex}(2)r_{12}^{-1}$ for four possible combinations for two electrons: both at A or B, and one at A and the other at B. Then, write down four expressions such as $\rho_{ex}(1A)\rho_{ex}(2B)r_{12}^{-1}$. Since r_{12}^{-1} is much larger when both electrons are at the same location than at two different locations, the positive term is much larger than the negative term.

10.5 Origin of magnetism and the Heisenberg model

Modern information technology depends on magnetism. Big data centers house enormous magnetic storage disks. Magnetism is quantum mechanical and is rooted in the electron-exchange interaction. The exchange interaction consists of two contributions: one is from the Coulomb interaction and the other is from the kinetic energy term. The exchange integral in eq. (10.4.4) represents part of the exchange interaction. Whether a system becomes magnetic depends on the competition between the kinetic energy and exchange interaction. If the exchange interaction wins, then we have a magnetic material. This stringent condition is difficult to meet in solids. We only have four element ferromagnets, Fe, Co, Ni, and Gd at room temperature, and Dy at 85 K, and one antiferromagnet Cr at 311 K. Most magnetic materials are alloys [75, 76]. More recently, nitrogen-vacancy (NV) in diamond has been found to be magnetic.

The simplest model to describe magnetism is the Heisenberg spin model. It ignores the kinetic energy term completely and retains the exchange term between two spins (see Fig. 10.2(a)):

$$\hat{H} = -\sum_{ij} J_{ij}\hat{\mathbf{S}}_i \cdot \hat{\mathbf{S}}_j, \tag{10.5.1}$$

where J_{ij} is called the Heisenberg exchange constant between atoms i and j, and $\hat{\mathbf{S}}_i$ has a site index i because J_{ij} depends on i. J_{ij} is related to E_X but contains other contributions,[2]

2 This is because E_X is always positive. To get a negative J_{ij}, one has to include other terms.

so it can be positive or negative. J_{ij} is typically on the order of meV. If J_{ij} is positive, we have a ferromagnet, where spins at each atomic site point in the same direction. If J_{ij} is negative, we have an antiferromagnetic coupling, where spins at two neighboring atoms point in two opposite directions. If J_{ij} is negative, and spins also point in two opposite directions, but the magnitudes of the spins at two neighboring atoms differ, then we have a ferrimagnet. There are more complicated spin configurations [76]. A direct calculation of J_{ij} is possible by using the Green function, or by the spin singlet and triplet energy difference.

In the following, we show an example of two spins located at two atomic sites. This identification of spin to a site is made possible because J_{ij} is spatially dependent. Our Hamiltonian is

$$\hat{H} = -J_{12}\hat{\mathbf{S}}_1 \cdot \hat{\mathbf{S}}_2, \tag{10.5.2}$$

where J_{12} is units of $J/\hbar^2$ since $\hat{\mathbf{S}}$ is units of $\hbar$. There are four possible spin configurations for two sites, denoted as $|a\rangle = \alpha(1)\alpha(2)$, $|b\rangle = \beta(1)\beta(2)$, $|c\rangle = \alpha(1)\beta(2)$, and $|d\rangle = \beta(1)\alpha(2)$. From eq. (10.3.4), we recognize that $|a\rangle$ is just $\chi_{1,1}$ and $|b\rangle$ is $\chi_{1,-1}$. Indeed, because of the decoupling between the spin and spatial dimensions, the knowledge from multiple spins can be directly used. This also applies to $|c\rangle$ and $|d\rangle$, but for the moment, they are not an eigenstate of $\hat{H}$. We can prove a linear combination of them can become an eigenstate. Without relying on eq. (10.3.4), first, we rewrite $\hat{H}$ as

$$\hat{H} = -J_{12}\left[\hat{S}_{1,z}\hat{S}_{2,z} + \frac{1}{2}(\hat{S}_{1,+}\hat{S}_{2,-} + \hat{S}_{1,-}\hat{S}_{2,+})\right]. \tag{10.5.3}$$

If we apply $\hat{H}$ to $|a\rangle$ and $|b\rangle$, respectively, we get

$$\hat{H}|a\rangle = -\frac{\hbar^2 J_{12}}{4}|a\rangle, \quad \hat{H}|b\rangle = -\frac{\hbar^2 J_{12}}{4}|b\rangle,$$

which shows that both $|a\rangle$ and $|b\rangle$ are the eigenstates of $\hat{H}$, with eigenvalue $-\frac{\hbar^2 J}{4}$. But, when we apply $\hat{H}$ to $|c\rangle$ and $|d\rangle$, we find

$$\hat{H}|c\rangle = \frac{J_{12}\hbar^2}{4}|c\rangle - \frac{J_{12}\hbar^2}{2}|d\rangle, \quad \hat{H}|d\rangle = \frac{J_{12}\hbar^2}{4}|d\rangle - \frac{J_{12}\hbar^2}{2}|c\rangle, \tag{10.5.4}$$

where neither $|c\rangle$ nor $|d\rangle$ is an eigenstate of $\hat{H}$. Instead, we obtain the matrix

$$\hat{H} = \begin{pmatrix} \frac{J_{12}\hbar^2}{4} & -\frac{J_{12}\hbar^2}{2} \\ -\frac{J_{12}\hbar^2}{2} & \frac{J_{12}\hbar^2}{4} \end{pmatrix}. \tag{10.5.5}$$

We can diagonalize it to find its eigenvalues and eigenvectors (see the following exercise). These two eigenvectors are exactly $\chi_{1,0}$ and $\chi_{0,0}$, whose respective eigenvalues are

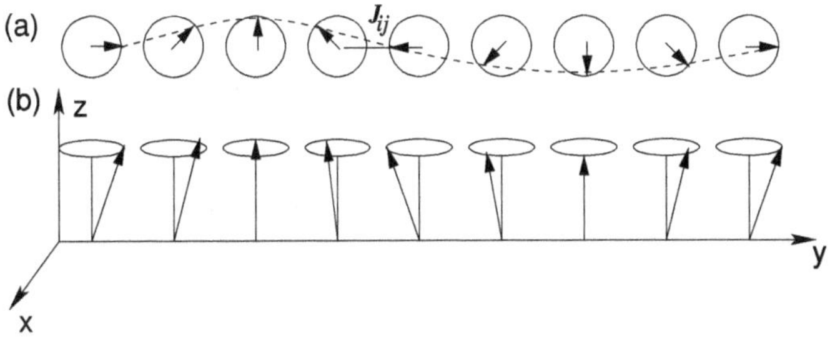

Figure 10.2: (a) Top view of spin precession. Spins at atomic sites are exchange coupled to each other and precess with time with a fixed phase difference between them. This forms a collective spin wave excitation, whose quantum is a magnon. (b) Three-dimensional view of spin waves.

$E_{1,0} = -\frac{\hbar^2 J_{12}}{4}$ and $E_{0,0} = \frac{3\hbar^2 J_{12}}{4}$. Now, the difference between the singlet and triplet is $\Delta E = E_{0,0} - E_{1,0} = \hbar^2 J_{12}$. This is how one can compute the exchange interaction J_{12}. If $J_{12} > 0$, three triplet states have a lower energy, so each site has spin pointing in the same direction, an ferromagnetic phase. On the other hand, if $J_{12} < 0$, the singlet has a lower energy, where the spins at different sites point in opposite directions. This is an antiferromagnetic phase.

Just as the nuclear vibrations show a collective behavior, spins at different lattice sites precess with a fixed phase with time. This is mediated by the exchange interaction. Figure 10.2(a) is a top view of spin waves. These collective spin excitations, once quantized, are called magnons. Magnons are bosons, like photons and phonons. Figure 10.2(b) shows the three-dimensional picture, where atoms are on the y-axis.

Exercise 10.5.0.

24. Show that $\hat{\mathbf{S}}_1 \cdot \hat{\mathbf{S}}_2$ can be written as $\hat{S}_{1,z}\hat{S}_{2,z} + \frac{1}{2}(\hat{S}_{1,+}\hat{S}_{2,-} + \hat{S}_{1,-}\hat{S}_{2,+})$.
25. Prove $\hat{H}|a\rangle = -\frac{\hbar^2 J_{12}}{4}|a\rangle$, $\hat{H}|b\rangle = -\frac{\hbar^2 J_{12}}{4}|b\rangle$.
26. Prove eqs. (10.5.4).
27. Find the eigenvalues and eigenvectors of eq. (10.5.5).

10.6 Quantum information technology

Quantum information technology has two major branches: quantum computing and quantum communication (quantum cryptography). Quantum computing employs the state superposition of single-body and many-body states. Quantum communication is based on the state entanglement in a correlated many-body state. For this reason, these two simple QM concepts are accessible to those who have little background in QM and are partly responsible for rapid dissemination of quantum technology into other fields. We will limit ourselves to quantum computing. Quantum computers have advantages over classical computers. For instance, the Shor factoring method is much faster, and the Grover search scheme provides a square-root speedup. But not just any quantum systems are suitable for a quantum computer. In 2000, DiVincenzo [77] proposed five criteria for quantum computation: (1) a scalable physical system with well characterized qubits; (2) the ability to initialize the state of the qubits to a simple fiducial state, such as $|000\ldots\rangle$; (3) long relevant decoherence times, much longer than the gate operation time; (4) a "universal" set of quantum gates; (5) a qubit-specific measurement capability; and two additional ones for communication; (6) the ability to interconvert stationary and flying qubits; (7) the ability faithfully to transmit flying qubits between specified locations. The rest of the criteria are self-explanatory, but (3) needs some explanation. Here, the decoherence refers to the loss of quantum coherence in quantum systems due to the influence from the environment. Suppose we have a state of superposition of two states $e^{-i\omega_a t-\gamma_a t}|a\rangle + e^{-i\omega_b t-\gamma_b t}|b\rangle$, where $\frac{1}{\gamma_a}$ and $\frac{1}{\gamma_b}$ are the decoherence times for states $|a\rangle$ and $|b\rangle$ and ω_a and ω_b are the frequencies. Our presentation follows [78].

10.6.1 Quantum bits or qubits

A classical computer relies on a binary bit, 0 or 1, to store, retrieve and transfer the information. This is realized in a binary circuit with transistors. A *single* circuit can have a high voltage or low voltage, thus, either 0 or 1.

Quantum computation is built upon an analogous concept, the quantum bit, or qubit for short. These qubits are quantum states, not a circuit like a classical computer. Switching is between quantum states. For this reason, a qubit must contain two or more states. A qubit that contains just one single state does not work, nor does a qubit that contains too many states to be selective. To this end, a *single* qubit consists of two states, $|0\rangle$ and $|1\rangle$. This is the first big difference between a classical bit, $\boxed{\text{either 0 or 1}}$, and a qubit, $\boxed{|0\rangle \text{ and } |1\rangle}$. Because a qubit has both $|0\rangle$ and $|1\rangle$, according to the superposition principle of a wavefunction (see Section 1.4.3), any possible combinations of $|0\rangle$ and $|1\rangle$ are a good wavefunction, or a good qubit.

We can represent these qubits through a general wavefunction,

$$|\psi(t)\rangle = \cos\left(\frac{\theta(t)}{2}\right)|0\rangle + e^{i\varphi(t)}\sin\left(\frac{\theta(t)}{2}\right)|1\rangle, \tag{10.6.1}$$

where $\theta(t)$ and $\varphi(t)$ are polar and azimuthal angles on the unit three-dimensional sphere as shown in Fig. 10.3. This sphere is often called the Bloch sphere. A qubit state is a unit vector in a two-dimensional complex vector space. It provides a useful means of visualizing the state of a single qubit and often serves as an excellent testbed for ideas about quantum computation. Because there are an infinite number of points on the unit sphere, in principle, it could store infinite information as far as we can operate on them.

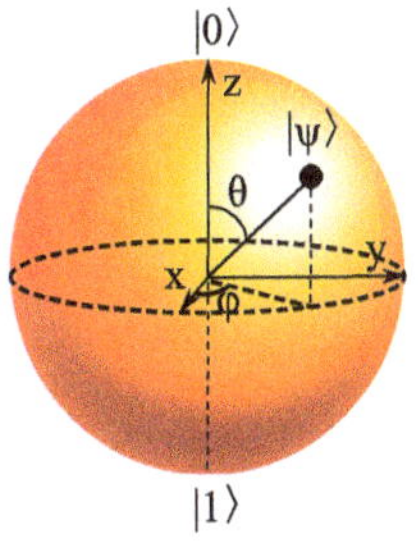

Figure 10.3: Bloch sphere representation of a qubit shows an infinitely possible number of qubits.

10.6.2 Single qubit and gates

A physical qubit is an atom, or a group of atoms. States can be nuclear, electronic, vibrational, photonic, or other. Consider a single qubit A that is a two-level system, with two computational basis states $|0\rangle$ and $|1\rangle$.[3] The system is in state $|\psi(t)\rangle$,

$$|\psi(t)\rangle = a(t)|0\rangle + b(t)|1\rangle, \tag{10.6.2}$$

where $a(t)$ and $b(t)$ are shorthand notations of coefficients in eq. (10.6.1). We carry out a measurement by left multiplying either $\langle 0|$ or $\langle 1|$; then, we find the probability is $|a(t)|^2$ in $|0\rangle$ and $|b(t)|^2$ in $|1\rangle$. Since the probability sum must be one, we have $|a(t)|^2+|b(t)|^2 = 1$.

Suppose we have an operator $\boxed{\mathrm{U}}$ that is applied to $|\psi(t)\rangle$. The resultant state is

$$|\psi'(t)\rangle = \boxed{\mathrm{U}}|\psi(t)\rangle.$$

Because any wavefunction must be normalized, this requires

$$\langle\psi'(t)|\psi'(t)\rangle = \langle\psi(t)|\boxed{\mathrm{U}}^{\dagger}\boxed{\mathrm{U}}|\psi(t)\rangle = 1.$$

Since we do not impose any restriction on ψ and ψ', this equation is generic and proves that $\boxed{\mathrm{U}}^{\dagger}\boxed{\mathrm{U}} = \boxed{\mathrm{I}}$ and $\boxed{\mathrm{U}}$ must be a 2×2 unitary matrix. $\boxed{\mathrm{I}}$ is a 2×2 identity matrix. This is the rotation matrix for a special unitary transformation group for two-dimensional vectors, SU(2) symmetry. Before we proceed, we highlight two cautions: (i) Here, the rotation has nothing to do with rotation of a physical qubit by some angles. Rather, we carry out a unitary transformation for basis states. Because it is easier for the reader to see the effect of the transformation, one usually adopts a geometrical view of transformation. (ii) Although one often invokes the similarity between a qubit and a spin,

3 The reason why we use the term "computational basis" is because $|0\rangle$ and $|1\rangle$ are the eigenstates of the initial $\hat{H}$. If $\hat{H}$ starts to change, $|0\rangle$ and $|1\rangle$ are only the basis functions of the eigenstates of the new $\hat{H}$. From the point of view of computational basis, the new eigenstates can be considered entangled.

we must ignore improper rotations, such as inversion and reflection, for qubits, except qubits themselves are spins because spins are axial vectors (see the Appendix).

We first rotate the "object" along the z-axis by γ, then along the y-axis by β, and finally along the z-axis by α, where α, β, and γ are the Eulerian angles. $\boxed{\mathrm{U}}$ is given by

$$\boxed{\mathrm{U}}(\alpha,\beta,\gamma) = \begin{pmatrix} e^{\frac{-i(\alpha+\gamma)}{2}}\cos\frac{\beta}{2} & -e^{\frac{-i(\alpha-\gamma)}{2}}\sin\frac{\beta}{2} \\ e^{\frac{i(\alpha-\gamma)}{2}}\sin\frac{\beta}{2} & e^{\frac{i(\alpha+\gamma)}{2}}\cos\frac{\beta}{2} \end{pmatrix}. \tag{10.6.3}$$

In principle, there is another common phase factor in $\boxed{\mathrm{U}}$, but it does not matter for a wavefunction since it is normalized anyway. Because α, β, γ are totally independent of each other, we have infinite possible $\boxed{\mathrm{U}}$ and $(a(t), b(t))$. One may already recognize some of them when we introduced the Pauli matrices (Section 6.1.2). In quantum computing, one uses different notations and introduces some extra matrices:

$$\boxed{\mathrm{I}} = \begin{pmatrix} 1 & 0 \\ 0 & 1 \end{pmatrix}, \quad \boxed{\mathrm{X}} = \begin{pmatrix} 0 & 1 \\ 1 & 0 \end{pmatrix}, \quad \boxed{\mathrm{Y}} = \begin{pmatrix} 0 & -i \\ i & 0 \end{pmatrix}, \quad \boxed{\mathrm{Z}} = \begin{pmatrix} 1 & 0 \\ 0 & -1 \end{pmatrix}, \tag{10.6.4}$$

$$\boxed{\mathrm{H}} = \frac{1}{\sqrt{2}}\begin{pmatrix} 1 & 1 \\ 1 & -1 \end{pmatrix}, \quad \boxed{\mathrm{S}} = \begin{pmatrix} 1 & 0 \\ 0 & i \end{pmatrix}, \quad \boxed{\mathrm{T}} = \sqrt{\boxed{\mathrm{S}}} = \begin{pmatrix} 1 & 0 \\ 0 & e^{i\pi/4} \end{pmatrix}. \tag{10.6.5}$$

Instead of naming them as a unitary matrix, in quantum computing, we call them gates. For instance, $\boxed{\mathrm{H}}$ and $\boxed{\mathrm{S}}$ are called the Hadamard[4] and phase gates, respectively. $\boxed{\mathrm{T}}$ is called $\pi/8$ gates because one can pull out $e^{i\pi/8}$ and then two nonzero elements are $e^{\pm i\pi/8}$. $\boxed{\mathrm{X}}$, $\boxed{\mathrm{Y}}$, $\boxed{\mathrm{Z}}$ are called Pauli-X, Pauli-Y, and Pauli-Z gates, where X, Y and Z do not necessarily correspond to Cartesian coordinates x, y, z.

Take a classical bit as an example. If we write classical bits 0 and 1 as column vectors $\binom{1}{0}$ and $\binom{0}{1}$, respectively, then the very operation of $\boxed{\mathrm{X}}$ is

$$\boxed{\mathrm{X}}\begin{pmatrix} 1 \\ 0 \end{pmatrix} = \begin{pmatrix} 0 \\ 1 \end{pmatrix}, \quad \boxed{\mathrm{X}}\begin{pmatrix} 0 \\ 1 \end{pmatrix} = \begin{pmatrix} 1 \\ 0 \end{pmatrix}. \tag{10.6.6}$$

What it does is to change 0 to 1, and 1 to 0. This is the NOT gate. For a classical computer, $\boxed{\mathrm{X}}$ is the only single-bit operation gate, and other gates need at least two bits.

As can be seen, for qubits, we have more gates even for a single qubit because a qubit has a phase and consists of both computational basis states, simultaneously. If we apply $\boxed{\mathrm{X}}$ to our qubit above, we find

$$\boxed{\mathrm{X}}\begin{pmatrix} \alpha(t) \\ \beta(t) \end{pmatrix} = \begin{pmatrix} \beta(t) \\ \alpha(t) \end{pmatrix}, \tag{10.6.7}$$

4 Named after the French mathematician Jacques Hadamard.

which shows that $\boxed{\mathrm{X}}$ exchanges the coefficients for two qubits, but does not switch qubits. This is the direct consequence of superposition. Naturally, one may ask whether this is indeed useful. The answer is affirmative and is part of ongoing research.

Example 6 (Hadamard gate). Suppose a qubit is in the state $|\psi\rangle = \frac{1}{\sqrt{2}}|0\rangle + \frac{1}{\sqrt{2}}|1\rangle$. (a) Find $\boxed{\mathrm{H}}|\psi\rangle$. (b) Find the percentage in $|0\rangle$ and $|1\rangle$.

(a) We first rewrite ψ in a matrix form, $\frac{1}{\sqrt{2}}\begin{pmatrix}1\\1\end{pmatrix}$. Then, we apply $\boxed{\mathrm{H}}$ to it, $\boxed{\mathrm{H}}|\psi\rangle$ to find $\begin{pmatrix}1\\0\end{pmatrix} = |0\rangle$. (b) We have 100 % in $|0\rangle$. This demonstrates that even our initial state has a 50 % chance in either $|0\rangle$ and $|1\rangle$, but the Hadamard gate can force the system to a definite $|0\rangle$. This provides the certainty needed for computing.

Exercise 10.6.2.

28. Suppose a qubit is in the state $|\psi\rangle = \frac{1}{\sqrt{2}}|0\rangle + \frac{1}{\sqrt{2}}|1\rangle$. (a) Find $\boxed{\mathrm{T}}|\psi\rangle$. (b) Find the percentage in $|0\rangle$ and $|1\rangle$.
29. Suppose a qubit is in the state $|\psi\rangle = |0\rangle$. (a) Find $\boxed{\mathrm{H}}|\psi\rangle$. (b) Find the percentage in $|0\rangle$ and $|1\rangle$.
30. Suppose a qubit is in the state $|\psi\rangle = \frac{1}{\sqrt{2}}|0\rangle + \frac{1}{\sqrt{2}}|1\rangle$. (a) Apply $\boxed{\mathrm{X}}$, $\boxed{\mathrm{H}}$, and $\boxed{\mathrm{Y}}$ consecutively to ψ. (b) Find the percentage in $|0\rangle$ and $|1\rangle$.

10.6.3 More than one qubit and entanglement

When we have more than one qubit, there are more gates. In fact, a classical computer already has many gates because one can individually address each bit. Quantum adds entanglement. To explain this clearly, let's assume that we have two qubits, A and B. One immediate question is: Are they indistinguishable just like identical particles? Yes and no. Yes, if we want to entangle or correlate them. No, if we want to control them.

We start with two qubits A and B, isolated from each other, with zero overlap between their respective *spatial* (not spin) wavefunctions, i. e., $\langle\phi_B|\phi_A\rangle = 0$, where ϕ_A and ϕ_B are wavefunctions localized at A and B, respectively.[5] Each can take $|0\rangle$ and $|1\rangle$ separately, so there are four possible eigenstates, $|\psi_0\rangle_n = |00\rangle, |01\rangle, |10\rangle, |11\rangle$, where n runs from 1 to 4. Here the first digit in $|ij\rangle$ belongs to A and the second to B. Because $\langle\phi_B|\phi_A\rangle = 0$, $|ij\rangle$ can be written as $|i\rangle \otimes |j\rangle$, where $\otimes$ means a direct product of two functions such as $\phi_A\phi_B$. These four states have *no entanglement*. If entangled, we remove the subscript 0 in $|\psi_0\rangle$.

To entangle two qubits, their spatial wavefunctions must overlap, or there is a coupling between them [79]. So, we move A and B closer until their wavefunctions overlap. From the view of point of the isolated case, A and B are entangled. The stated four basis states are no longer the eigenstate of the new composite system AB and instead serve

5 Even if $\langle\phi_B|\phi_A\rangle$ is zero at discrete single points in space, this is not considered as zero overlap. Zero overlap means that $\langle\phi_B|\phi_A\rangle$ is zero on a finite line segment.

as four computational basis states. The new wavefunction $\psi(t)$ is a superposition of the basis states,[6]

$$|\psi(t)\rangle = \underbrace{c_{00}(t)|00\rangle}_{1} + \underbrace{c_{01}(t)|01\rangle}_{2} + \underbrace{c_{10}(t)|10\rangle}_{3} + \underbrace{c_{11}(t)|11\rangle}_{4} . \tag{10.6.8}$$

It is no longer possible to write $|\psi\rangle$ as a product of two states, so $|\psi\rangle$ is an entangled state.[7] In addition, even basis states $|ij\rangle$ are also slightly different from those $|ij\rangle$ when A and B are isolated, but this is often ignored for simplicity.

Next, we operate on $|\psi\rangle$ using gates to generate different qubits. A single qubit gate is a 2×2 matrix. A two-qubit gate is a 4×4 matrix because the matrix size is 2^n,[8] where n is the number of qubits. In eq. (10.6.8) we specifically label those terms from 1 to 4. Here, a subtle physics problem, not a math problem, emerges which does not happen for a classical computer. For instance, the classical `control-NOT`, or CNOT, chooses a specific classical bit as the control and the other as the target bit to be controlled, without ambiguity. How do we know in advance which qubit is the target and which is the control in an entangled state? For this reason, we have to make A and B distinguishable so we can apply a gate. The only way we could do this is to move A and B slightly apart. Once we are done with gating, we move them back. Many researchers are using different nomenclatures for this, but basically this is what one is doing. Monroe and coworkers [79] used two different qubits: the hyperfine states of the Be ion as B, and the nuclear harmonic oscillation of the Be ion as A. Doing so loses many possible configurations (qubits) of $\psi(t)$, but there is no way out. Once one understands this, there is nothing fancy about two-qubit gates. The reliability of the CNOT-operation was measured to be on the order of 90 % in the beryllium ion trap in 1995 [79], and this is good enough for quantum computing.

With the stated caveat, we assume A is the control bit and B is the target bit. CNOT means that, if A is in $|0\rangle$, B keeps its state; if A is in $|1\rangle$, B flips its state, $|0\rangle \rightarrow |1\rangle$ or $|1\rangle \rightarrow |0\rangle$. Here, A never changes its state. We now develop a matrix for the $\boxed{\text{CNOT}}$ gate. If A is $|0\rangle$, by the definition of $\boxed{\text{CNOT}}$, B should not change its state, so we have $\boxed{\text{CNOT}}|00\rangle = |00\rangle$ and $\boxed{\text{CNOT}}|01\rangle = |01\rangle$. To this end, we get two nonzero elements, $\boxed{\text{CNOT}}(1,1) = 1$ and $\boxed{\text{CNOT}}(2,2) = 1$, where 1 and 2 are indices of $|00\rangle$ and $|01\rangle$ in eq. (10.6.8). Next, we apply $\boxed{\text{CNOT}}$ to $|10\rangle$, and we have $\boxed{\text{CNOT}}|10\rangle = |11\rangle$, so $\boxed{\text{CNOT}}(3,4) = 1$. This means that, if we apply $\boxed{\text{CNOT}}$ to the third state, we obtain the fourth state. Similarly, we can find $\boxed{\text{CNOT}}(4,3) = 1$. Only these four elements are nonzero, and the rest are zero. $\boxed{\text{CNOT}}$ in

6 Provided the four basis states are complete and can represent $\psi(t)$.

7 In many-body physics, an entangled state is called a correlated state; an uncorrelated one is called a Hartree–Fock state.

8 For those who are familiar with tensors, $c_{ijk\dots}$ is in fact a tensor. When we represent a tensor by a matrix, the matrix size grows as 2^n. This already happens in many-body physics, such as the Gutzwiller–Hubbard model, suitable for quantum computing.

matrix form is

$$\boxed{\text{CNOT}} = \begin{pmatrix} 1 & 0 & 0 & 0 \\ 0 & 1 & 0 & 0 \\ 0 & 0 & 0 & 1 \\ 0 & 0 & 1 & 0 \end{pmatrix}. \tag{10.6.9}$$

If we apply $\boxed{\text{CNOT}}$ to $|\psi(t)\rangle$ in eq. (10.6.8), we get $|\psi'(t)\rangle = c_{00}|00\rangle + c_{01}|01\rangle + c_{11}|10\rangle + c_{10}|11\rangle$. Other gates can be developed through $\boxed{\text{CNOT}}$ and single qubit gates [78].

Before we move on to the hardware, we wish to discuss the quantum information flow. Two qubits are particularly simple to demonstrate if we follow the idea of the density matrix renormalization group (DMRG) [80, 81] by writing $\psi(t) = \sum_{ij} c_{ij}|ij\rangle$. Now, suppose we are interested in qubit A. We define a density matrix (not to be confused with charge densities) as

$$\rho^A_{i1,i2} = \sum_j c^*_{i1,j} c_{i2,j}, \tag{10.6.10}$$

where the summation is over j (over environment B in DMRG). In DMRG, one diagonalizes this matrix and uses the eigenvectors with the largest eigenvalues to form a new operator for the next iteration. Here, the matrix is

$$\rho^A = \begin{pmatrix} |c_{00}|^2 + |c_{01}|^2 & c^*_{00}c_{10} + c^*_{01}c_{11} \\ c^*_{10}c_{00} + c^*_{11}c_{01} & |c_{10}|^2 + |c_{11}|^2 \end{pmatrix}. \tag{10.6.11}$$

Taking one of the Bell states $|\psi_1\rangle = \frac{1}{\sqrt{2}}(|00\rangle + |11\rangle)$ as an example, we find ρ^A only has diagonal elements of 1/2 and qubit A has a 50 % chance in the $|0\rangle$ and $|1\rangle$ states.

Exercise 10.6.3.

31. Where does the state entanglement come from?
32. Show that the trace of the density matrix in eq. (10.6.11) is 1.
33. Construct a controlled-Pauli-Z gate and find the matrix form for two qubits.

10.6.4 Hardware and quantum bits

A classical central processing unit (CPU) is based on a field transistor and a capacitor. Figure 10.4(a) shows an example. Similarly, a quantum processing unit (QPU) also needs hardware to operate [82], to store information, to correct errors, and to interface with a classical computer to carry out the instruction. At the core of hardware is to design and operate two quantum states. The recent quantum computing roadmap, currently hosted at `qist.lanl.gov`, identifies the following physical protocols: (1) nuclear magnetic resonance; (2) ion trap (Ba^+, Be^+, Ca^+, Cd^+, In^+, Mg^+, Sr^+, Yb^+); (3) neutral atoms;

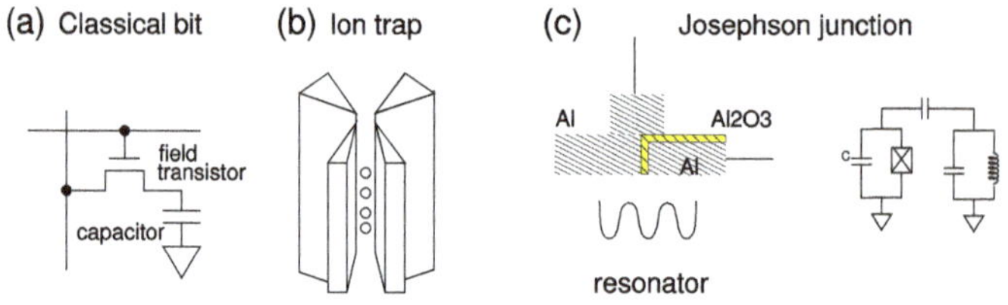

Figure 10.4: (a) A classical bit consists of a field transistor and a capacitor. (b) Quantum bits based on the Paul ion trap, where ions are trapped inside a harmonic potential well generated by four electrodes. Ions (circles) form a line. Two qubits are optical qubits derived from a ground state and an excited state and hyperfine qubits derived from the electronic ground state. (c) Josephson junction qubits. A resonator couples to the junction and reads out the state of qubits. Three types of qubits are phase qubit, flux qubit, and charge qubit.

(4) cavity quantum electrodynamics (QED) (Ba, Ca^+, Cs, Rb) (specifically, candidate systems include Rydberg atoms in microwave cavities, neutral atoms in optical cavities, and trapped ion cavities QED); (5) optics using photons; (6) the solid state has four systems: spins in confined structure, impurity spins, charged or excitonic systems, and mechanical systems (example systems include nuclear spin of P donors in Si, high-spin magnetic nanoparticles, shallow donors in Si, SiGe, or GaAs, paramagnetic ions in C_{60}, nitrogen vacancy (NV) diamond centers or rare-earth color centers, SiC, paramagnetic ions in carbon nanotubes, and high-Q nanocantilevers); (7) superconducting Josephson tunneling junctions.

We discuss only a few of them. Figure 10.4(b) shows QPU based on trapped ions. Radio-frequency electric and DC fields are employed to form a harmonic potential well for ions, called the Paul trap, where ions and atoms are confined and suspended [83]. The third dimension allows ions stacking on the top of each other. Due to the coulombic repulsion, ions are separated from each other. Qubits are two hyperfine electronic states.

Solid-state devices include superconducting Josephson tunneling junction (Fig. 10.4(c)). The Josephson junction is built from a superconductor and a thin insulating layer. Current tunneling through the insulator barrier has an intrinsic aharmonicity, where the ground and excited states are not evenly spaced and can be isolated from the rest. A microwave resonator is coupled to the junction to read qubits.

A viable qubit has four common features. First, quantum states must retain its coherence as long as possible, so one can operate on them. For this reason, one has to isolate qubits from the environment. Second, one has to be able to scale up multiple interacting qubits. The third is to manipulate qubits and interface with classical computers as cleanly as possible. However, the third requirement poses a fundamental contradiction to the first one. Fourth, qubits must have error correction.

It is probably the best to use an example to demonstrate what qubits look like. We are going to take the two lowest states of PIB (Fig. 10.5(a)), ϕ_1 and ϕ_2. Because the quantum number n in PIB cannot be zero, we will not use the quantum computing notation

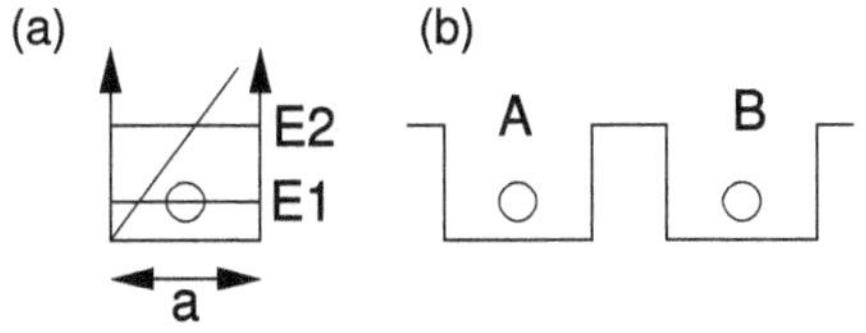

Figure 10.5: (a) Two eigenstates in an infinite quantum well of width a as a single qubit. The thin line denotes the perturbation or a gate, $\boxed{\mathtt{X}}$. (b) Two finite quantum wells form two qubits, A and B.

for states. Recall $\phi_n = \sqrt{\frac{2}{a}}\sin(\frac{n\pi x}{a})$ and $E_n = \frac{n^2\hbar^2\pi^2}{2ma^2}$. In the energy representation, the Hamiltonian for these two levels is diagonal and can be converted to $\boxed{\mathtt{I}}$ and $\boxed{\mathtt{Z}}$,

$$\hat{H}_0 = \begin{pmatrix} E_1 & 0 \\ 0 & E_2 \end{pmatrix} \to \hat{H}_0 = \frac{(E_1+E_2)}{2}\boxed{\mathtt{I}} + \frac{E_1-E_2}{2}\boxed{\mathtt{Z}}. \tag{10.6.12}$$

Since $\hat{H}_0$ is time-independent, the state evolves as $|\psi(t)\rangle = c_1 e^{-iE_1t/\hbar}|\phi_1\rangle + c_2 e^{-iE_2t/\hbar}|\phi_2\rangle$, where c_i is the coefficient and may depend on time t. $|\psi(t)\rangle$ is our qubit.

Next, we turn on a perturbation (the thin line in Fig. 10.5(a)) $\hat{H}_I = A(t)(\frac{x}{a}-\frac{1}{2})$, where $\frac{1}{2}$ is introduced to cancel the diagonal element $\langle\phi_n|x|\phi_n\rangle$, so only the off-diagonal elements $\langle\phi_1|x|\phi_2\rangle = -\frac{16a}{9\pi^2}$ are retained (see the problems on Section 2.8). $\hat{H}_I$ is

$$\hat{H}_I = -A(t)\frac{16a}{9\pi^2}\begin{pmatrix} 0 & 1 \\ 1 & 0 \end{pmatrix} = -A(t)\frac{16a}{9\pi^2}\boxed{\mathtt{X}}. \tag{10.6.13}$$

If we apply $\hat{H}_I$ to $\psi(t)$ at time t, we get a new qubit $|\psi'(t)\rangle$ as

$$|\psi'(t)\rangle = -A(t)\frac{16a}{9\pi^2}(c_1 e^{-iE_1t/\hbar}|\phi_2\rangle + c_2 e^{-iE_2t/\hbar}|\phi_1\rangle), \tag{10.6.14}$$

where $A(t)$ is a time-dependent amplitude. If $A(t)$ is a constant, then we only have a beating between these two states, no switching. If $A(t)$ is time-dependent, it must act during which coherence between these two states is active. In actual systems, excited states decay with time, i. e., $c_2(t) \propto e^{-\Gamma t}$, where Γ is positive. $|\psi'(t)\rangle$ in eq. (10.6.14) only sets up the initial state for the system. How $|\psi'(t)\rangle$ evolves with time must follow the time-dependent Schrödinger equation,

$$i\hbar\frac{\partial|\psi(t)\rangle}{\partial t} = (\hat{H}_0 + \hat{H}_I)|\psi(t)\rangle. \tag{10.6.15}$$

The application of $\boxed{\mathtt{X}}$ gate $\hat{H}_i(t)$ is just part of the story. Now, suppose that we want to develop a $\boxed{\mathtt{CNOT}}$ gate which needs two qubits, A and B. PIB does not work, because there is no coupling with other qubits. Figure 10.5(b) adopts two finite quantum wells, two qubits, A and B.[9] Because the potential barrier is low, A and B interact with each other. There are numerous challenges to make this model work as a CNOT gate. For instance, how could one apply a perturbation $\hat{H}_I$ on A (as the target qubit) without affecting B (as the control)? We have to end here. The future of quantum information science and technology belongs to those who are prepared to make it.

Exercise 10.6.4.

34. Starting from $|\psi(t)\rangle = c_1 e^{-iE_1 t/\hbar}|\phi_1\rangle + c_2 e^{-iE_2 t/\hbar}|\phi_2\rangle$, find $|\psi'(t)\rangle$ if we apply H to it.

35. What happens if we use two Hadamard gates to operate on $|\psi(t)\rangle$?

10.7 Berry phase

In general, phases in a wavefunction ψ are purely arbitrary and have been largely ignored for a good reason. If we compute the density, we find immediately that the phase drops out because of $\psi^*\psi$. However, as Berry pointed out [84], in some cases, the phase cannot be dropped, and more importantly, the phase can be detected. This section introduces a different phase, the Berry phase. This all starts from an apparent problem in the time-dependent Schrödinger equation where the time-derivative is a partial derivative.

10.7.1 Dynamical phase factor with time-independent Hamiltonian

We start from a time-independent Hamiltonian $\hat{H}(0)$. The eigenequation is,

$$\hat{H}(0)|n(0)\rangle = E_n|n(0)\rangle, \tag{10.7.1}$$

where E_n is the eigenvalue of the eigenstate $|n(0)\rangle$. At time t, the eigenstate evolves to

$$|n(t)\rangle = e^{-iE_n t/\hbar}|n(0)\rangle, \tag{10.7.2}$$

where $|n(0)\rangle$ is the eigenfunction at $t = 0$, and the exponential is the familiar dynamical phase factor. If we want to compute the expectation value of an operator $\hat{O}$, this phase factor does not change the result $\langle n(t)|\hat{O}|n(t)\rangle = \langle n(0)|\hat{O}|n(0)\rangle$. Therefore, for a time-independent Hamiltonian, $|n(t)\rangle$ is a stationary state. If the initial state is a superposition of several eigenstates, then each state carries a separate dynamical phase factor.

---

9 This is an oversimplified model. In real experiments, as shown by Monroe and coworkers [79], their two qubits consist of two hyperfine states and two vibrational harmonic states.

10.7.2 Time-dependent Hamiltonian and transition between states

However, if the Hamiltonian is time-dependent, the situation is different. A state evolves with time according to the time-dependent Schrödinger equation,

$$i\hbar\frac{\partial\psi}{\partial t} = \hat{H}(t)\psi. \tag{10.7.3}$$

As seen in Chapter 7, if $\hat{H}(t)$ changes with time rapidly, there will be transitions between eigenstates of the unperturbed Hamiltonian. Treating the partial time-derivative as a full-time derivative, we obtain eq. (7.3.1) as

$$C_l^{(1)}(t) = \frac{e\mathbf{r}_{ik}\cdot\mathbf{E}_0}{i\hbar}\frac{e^{i(\omega_l-\omega_k-\omega)t} - e^{i(\omega_l-\omega_k-\omega)t_0}}{i(\omega_l-\omega_k-\omega)}.$$

We see the phase change depends on $\omega_l-\omega_k-\omega$. If there is a resonant transition between states l and k, $\omega_l-\omega_k-\omega = 0$, the phase changes slowly. Otherwise, it changes rapidly.

10.7.3 Adiabatic approximation

In physics and chemistry, an adiabatic process means that a system evolves with time slowly, so slowly that there is no transition between eigenstates during the time propagation. The time evolution may be due to parameters in the system, such as the coordinates of a molecules and the direction of a magnetic field. We denote them as $\{\mathbf{R}(t)\}$, so the entire Hamiltonian is $\hat{H}(\mathbf{R}(t))$. Different from time-dependent perturbation theory, we do not separate the Hamiltonian into one unperturbed and the other external perturbed Hamiltonians. During the time evolution, an eigenstate, denoted as $|n(\mathbf{R}(t))\rangle$, remains an instantaneous eigenstate of an instantaneous Hamiltonian $\hat{H}(\mathbf{R}(t))$,[10]

$$\hat{H}(\mathbf{R}(t))|n(\mathbf{R}(t))\rangle = E_n(\mathbf{R}(t))|n(\mathbf{R}(t))\rangle, \tag{10.7.4}$$

where $E_n(\mathbf{R}(t))$ is the instantaneous eigenvalue obtained by diagonalizing $\hat{H}$. All the eigenstates are orthonormalized, $\langle n(\mathbf{R}(t))|m(\mathbf{R}(t))\rangle = \delta_{nm}$, at the same time t, but not at two different times t_1 and t_2 (see the following exercise).

In the meantime, regardless of whether we take an adiabatic approximation or not, the state must evolve according to the time-dependent Schrödinger equation,

$$i\hbar\frac{\partial|n(\mathbf{R}(t))\rangle}{\partial t} = \hat{H}(\mathbf{R}(t))|n(\mathbf{R}(t))\rangle, \tag{10.7.5}$$

From eqs. (10.7.4) and (10.7.5), we find

10 What we are doing here is to diagonalize the Hamiltonian at each time t.

$$i\hbar\frac{\partial|n(\mathbf{R}(t))\rangle}{\partial t}=E_n(\mathbf{R}(t))|n(\mathbf{R}(t))\rangle, \tag{10.7.6}$$

which can be integrated to yield

$$|n(\mathbf{R}(t))\rangle=\exp\left[-\frac{i}{\hbar}\int_0^t dt' E_n(\mathbf{R}(t'))\right]|n(\mathbf{R}(0))\rangle. \tag{10.7.7}$$

If we compare this expression with eq. (10.7.2), we see the dynamical phase factor is changed to the integration because the eigenenergy changes with time. However, going from eq. (10.7.6) to (10.7.7) explicitly assumes that one can convert the partial time derivative eq. (10.7.6) to a full time derivative, i. e., in essence, we have changed a partial differential equation to an ordinary differential equation. This, in general, is not possible, and is valid only if there is no other parameter changing.

We can make this clearly by seeking a general solution $|\psi(t)\rangle$ to the ordinary time-derivative Schrödinger equation and by expanding $|\psi(t)\rangle$ in $|n(\mathbf{R}(t))\rangle$, not $|n(\mathbf{R}(0))\rangle$,[11]

$$|\psi(t)\rangle=\sum_n|\psi_n(t)\rangle\equiv\sum_n a_n(t)|n(\mathbf{R}(t))\rangle\exp\left[-\frac{i}{\hbar}\int_0^t dt' E_n(\mathbf{R}(t'))\right], \tag{10.7.8}$$

where $|\psi_n(t)\rangle$ is defined through eq. (10.7.8) and $a_n(t)$ is a coefficient to be determined. We plug it into the ordinary time-differential Schrödinger equation,

$$i\hbar|\dot{\psi}(t)\rangle=\hat{H}(\mathbf{R}(t))|\psi(t)\rangle, \tag{10.7.9}$$

where the dot over $|\psi\rangle$ denotes the ordinary time derivative ($\frac{d}{dt}$), and then left multiply $\langle m(\mathbf{R}(t))|$ and integrate the resultant equation to get (see the exercises)

$$\dot{a}_m(t)=-\sum_n a_n(t)\langle m(\mathbf{R}(t))|\dot{n}(\mathbf{R}(t))\rangle\exp\left[-\frac{i}{\hbar}\int_0^t dt'(E_n(\mathbf{R}(t'))-E_m(\mathbf{R}(t')))\right], \tag{10.7.10}$$

which shows that a_m depends on $\langle m(\mathbf{R}(t))|\dot{n}(\mathbf{R}(t))\rangle$. If $\langle m(\mathbf{R}(t))|\dot{n}(\mathbf{R}(t))\rangle=0$, then a_m is simply a constant. This happens if $|n(\mathbf{R}(t))\rangle$ is real. But if it is a complex function, then $\langle m(\mathbf{R}(t))|\dot{n}(\mathbf{R}(t))\rangle$ may differ from zero, and we can estimate how large it is.

For $n\neq m$, from eq. (10.7.4) we can find

$$\langle m(\mathbf{R}(t))|\dot{n}(\mathbf{R}(t))\rangle=\frac{\langle m(\mathbf{R}(t))|\dot{\hat{H}}|n(\mathbf{R}(t))\rangle}{E_n(\mathbf{R}(t))-E_m(\mathbf{R}(t))}.$$

11 This is the key difference. If the Hamiltonian is time independent, we always use $|n(\mathbf{R}(0))\rangle$, but here the Hamiltonian changes with time. The completeness is fulfilled only at each time t. In general, states at different times are not even orthogonal.

If $E_n(\mathbf{R}(t)) - E_m(\mathbf{R}(t))$ is large, while the change in $\hat{H}$ is small, then the ratio is small. This is the condition for the adiabatic approximation. Then, we can ignore terms with $n \neq m$ and only retain terms with $n = m$. Doing so simplifies eq. (10.7.10) to

$$\dot{a}_m(t) = -a_m(t)\langle m(\mathbf{R}(t))|\dot{m}(\mathbf{R}(t))\rangle.$$

We integrate over time t to obtain

$$a_m(t) = a_m(0)\exp\left[-\int_0^t \langle m(\mathbf{R}(t'))|\dot{m}(\mathbf{R}(t'))\rangle dt'\right] \equiv a_m(0)e^{i\gamma_m(t)}, \tag{10.7.11}$$

where we have defined a phase γ_m through the exponential term. Equation (10.7.11) is true only if crossing terms are ignored, i. e., no transitions between states, which is the adiabatic approximation. From now on, we only focus on a single state. The key message here is that the phase γ_m appears because, once we have a time-dependent $\mathbf{R}(t)$, the physics goes beyond the partial time-differential Schrödinger equation. We need an ordinary time-differential Schrödinger equation. Going from a partial differential equation to an ordinary differential equation admits more roots.

Exercise 10.7.3.

36. (a) Show at two different times t_1 and t_2 that two states are not orthogonal, i. e., $\langle n(\mathbf{R}(t_1))|m(\mathbf{R}(t_2))\rangle \neq \delta_{nm}$. (b) Find a condition under which they are orthogonal.

37. Prove eq. (10.7.7) satisfies eq. (10.7.6).

38. Show that, if $|n(\mathbf{R}(t))\rangle$ is real, $\langle m(\mathbf{R}(t))|\dot{n}(\mathbf{R}(t))\rangle = 0$.

39. Starting from eq. (10.7.4), use the Hellmann–Feynman theorem to prove $\langle m(\mathbf{R}(t))|\dot{n}(\mathbf{R}(t))\rangle = \frac{\langle m(\mathbf{R}(t))|\dot{\hat{H}}|n(\mathbf{R}(t))\rangle}{E_n(\mathbf{R}(t))-E_m(\mathbf{R}(t))}$.

10.7.4 Geometric phase factor: a conceptual change

We pick up one state $|\psi_n(t)\rangle$ from the summation of eq. (10.7.8) and substitute $a_n(t)$ of eq. (10.7.11) into $|\psi_n(t)\rangle$ to find [84] (setting $a_n(0)$ to 1 in eq. (10.7.11)),

$$|\psi_n(t)\rangle = e^{i\gamma_n(t)}\exp\left[-\frac{i}{\hbar}\int_0^t dt' E_n(\mathbf{R}(t'))\right]|n(\mathbf{R}(t))\rangle. \tag{10.7.12}$$

At $t = 0$, $|\psi_n(0)\rangle = e^{i\gamma_n(0)}|n(\mathbf{R}(0))\rangle$ corresponds to the initial state. The extra phase $\gamma_n(t)$ can be found by requiring $|\psi_n\rangle$ to satisfy

$$i\hbar\frac{\partial|\psi_n(t)\rangle}{\partial t} = \hat{H}(t)|\psi_n(t)\rangle. \tag{10.7.13}$$

Doing so, we find

$$\frac{\partial \gamma_n(t)}{\partial t} = i\langle n(\mathbf{R}(t))|\frac{\partial}{\partial t}|n(\mathbf{R}(t))\rangle. \tag{10.7.14}$$

Before we move further, we pause to explain what we are doing. Equation (10.7.12) replaces $|n(\mathbf{R}(0))\rangle$ in eq. (10.7.7) with $e^{i\gamma_n(t)}|n(\mathbf{R}(t))\rangle$, so it goes beyond the dynamical phase factor. We can show this explicitly. We change the time derivative on $|n(\mathbf{R}(t))\rangle$ to the derivative on the parameters $\frac{\partial}{\partial t} = \nabla_{\mathbf{R}} \cdot \frac{d\mathbf{R}}{dt}$ (see the exercise). Because γ_n is a function of only t, $\frac{\partial}{\partial t}$ can be written as a full derivative, $\frac{d}{dt}$, so eq. (10.7.14) becomes

$$\frac{d\gamma_n(t)}{dt} = i\langle n(\mathbf{R})|\nabla_{\mathbf{R}}|n(\mathbf{R})\rangle \cdot \frac{d\mathbf{R}}{dt}. \tag{10.7.15}$$

Eliminating dt from both sides converts the time dependence to the parameter dependence,

$$d\gamma_n(t) = i\langle n(\mathbf{R})|\nabla_{\mathbf{R}}|n(\mathbf{R})\rangle \cdot d\mathbf{R},$$

and we integrate it along a path P in the parameter space (Fig. 10.6(a)).[12]

$$\gamma_n(\mathbf{R}_i \xrightarrow{P} \mathbf{R}_f) = \int_{\mathbf{R}_i}^{\mathbf{R}_f} i\langle n(\mathbf{R})|\nabla_{\mathbf{R}}|n(\mathbf{R})\rangle \cdot d\mathbf{R}. \tag{10.7.16}$$

Different from the dynamical phase factor (eq. (10.7.2)), $\gamma_n(t)$ only parametrically depends on t, where the initial and final times t_i and t_f enter $\gamma_n(t)$ through $\mathbf{R}_i$ and $\mathbf{R}_f$, respectively, with the geometry at the center of $\gamma_n(t)$, the geometrical phase factor.[13] Because $\gamma_n(t)$ appears in $e^{i\gamma_n(t)}$, it must be real; otherwise, it becomes unphysical. This requires that $\langle n(\mathbf{R})|\nabla_{\mathbf{R}}|n(\mathbf{R})\rangle$ must be imaginary. To simplify our notation, we define

$$\mathbf{A}_n(\mathbf{R}) = i\langle n(\mathbf{R})|\nabla_{\mathbf{R}}|n(\mathbf{R})\rangle, \tag{10.7.17}$$

where $\mathbf{A}_n(\mathbf{R})$ is called the Berry connection or Berry vector potential, already used in eq. (9.4.19) on Section 9.4.2.2. In this way, eq. (10.7.16) has a simpler form as

$$\gamma_n(t) = \int_P d\mathbf{R} \cdot \mathbf{A}_n(\mathbf{R}). \tag{10.7.18}$$

---

12 Because we have not specified what the parameters are, we can only choose a path. This is a path integral. We cannot use $\gamma_n(\mathbf{R}_f) - \gamma_n(\mathbf{R}_i)$ because the integral depends on the path.

13 We do not call it a spatial phase factor because $\mathbf{R}$ does not necessarily represent space as it can represent a magnetic field or crystal momentum.

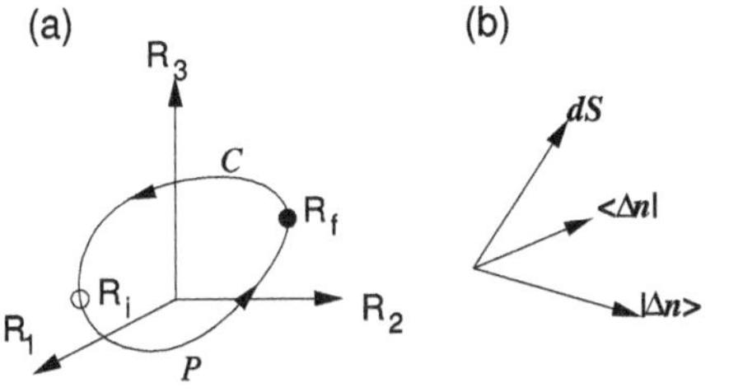

Figure 10.6: (a) Geometrical phase factor γ_n. (b) The triple product $d\mathbf{S} \cdot \langle \nabla_{\mathbf{R}} n(\mathbf{R})| \times |\nabla_{\mathbf{R}} n(\mathbf{R})\rangle$ represents a "volume".

Unfortunately, $\mathbf{A}_n(\mathbf{R})$ has an undesirable problem: It is not gauge invariant. Suppose we multiply $|n(\mathbf{R})\rangle$ by a phase factor $e^{i\zeta(\mathbf{R})}$, and we find $\mathbf{A}_n(\mathbf{R})$ changes as

$$\begin{aligned} |n(\mathbf{R})\rangle &\to e^{i\zeta(\mathbf{R})}|n(\mathbf{R})\rangle, \\ \mathbf{A}_n(\mathbf{R}) &\to \mathbf{A}_n(\mathbf{R}) - \nabla_{\mathbf{R}}\zeta(\mathbf{R}). \end{aligned} \tag{10.7.19}$$

This causes the problem for the geometrical phase $\gamma_n(t)$,

$$\begin{aligned} \gamma_n(t) &= \int_p d\mathbf{R} \cdot \mathbf{A}_n(\mathbf{R}) - \int_p d\mathbf{R} \cdot \nabla_{\mathbf{R}}\zeta(\mathbf{R}) \\ &= \int_p d\mathbf{R} \cdot \mathbf{A}_n(\mathbf{R}) + \zeta(\mathbf{R}(0)) - \zeta(\mathbf{R}(t)), \end{aligned} \tag{10.7.20}$$

where $\mathbf{R}(0)$ and $\mathbf{R}(t)$ are the initial and final parameters in the parameter space.

Example 7 (Berry connection for a two-dimensional planewave). A two-dimensional planewave has the wavefunction $\psi(x, y; k_x, k_y) = e^{ik_x x + ik_y y}/a$ normalized with a square $a \times a$, where the parameters are k_x and k_y. Find the Berry connection $\mathbf{A}$.

We take the partial derivatives with respect to k_x and k_y to find $\frac{\partial\psi}{\partial k_x} = ixe^{ik_x x + ik_y y}/a$ and $\frac{\partial\psi}{\partial k_y} = iye^{ik_x x + ik_y y}/a$. So,

$$\begin{aligned} A_x &= i \iint dxdy \frac{e^{-ik_x x - ik_y y}}{a} ix \frac{e^{ik_x x + ik_y y}}{a} = -a/2, \\ A_y &= i \iint dxdy \frac{e^{-ik_x x - ik_y y}}{a} iy \frac{e^{ik_x x + ik_y y}}{a} = -a/2. \end{aligned}$$

Finally, $\mathbf{A} = -\frac{a}{2}(\hat{x} + \hat{y})$.

Exercise 10.7.4.

40. Prove eq. (10.7.14).
41. Show $\frac{\partial}{\partial t}$ can be written as $\nabla_{\mathbf{R}} \cdot \frac{d\mathbf{R}}{dt}$.
42. (a) Prove $\langle n(\mathbf{R})|\nabla_{\mathbf{R}}|n(\mathbf{R})\rangle$ must be imaginary. (b) If $|n(\mathbf{R})\rangle$ is a real wavefunction, show the Berry connection must be zero.

43. Take the wavevector **k** as the parameter **R** to compute the Berry connections for a (unnormalized) planewave $\psi(r) = e^{i\mathbf{k}\cdot\mathbf{r}}$ in (a) a one-dimensional chain of length a and (b) a three-dimensional cube a^3.

10.7.5 Berry phase

The only way we can get rid of the extra term in eq. (10.7.20) is to choose a closed path C, so $\mathbf{R}(0) = \mathbf{R}(t)$. For ζ in eq. (10.7.19), this means

$$e^{i\zeta_n(\mathbf{R}(0))} = e^{i\zeta_n(\mathbf{R}(t))} \rightarrow \zeta(\mathbf{R}(0)) - \zeta(\mathbf{R}(t)) = 2\pi \times n, \tag{10.7.21}$$

where $n = 1, 2, 3, \ldots$ and γ_n can change by an integer multiple of 2π under the gauge transformation from eq. (10.7.19). This was first discovered by Berry [84].

This treatment is tantamount to assuming that, after a period T, the system returns to its original system, where $\hat{H}(T) = \hat{H}(0)$ and $\mathbf{R}(0) = \mathbf{R}(T)$. For a closed path C, γ_n is a gauge-invariant physical quantity,

$$\gamma_n(C) = \oint_C d\mathbf{R} \cdot \mathbf{A}_n(\mathbf{R}) = i \oint_C \langle n(\mathbf{R}) | \nabla_{\mathbf{R}} | n(\mathbf{R}) \rangle \cdot d\mathbf{R}, \tag{10.7.22}$$

which is called Berry's phase, a geometric phase in the parameter space. One may directly compute it with an eigenstate (see the exercise). We wish to make a few comments on the structure of eq. (10.7.22). Whenever the Berry connection is zero, the Berry phase is zero. Any real wavefunctions have a zero Berry phase. On the other hand, spinor wavefunctions can have a nonzero Berry phase.

Second, this integral has two separate integrations. One is over the spatial coordinate $d\mathbf{r}$ of the wavefunction $|n(\mathbf{R}, \mathbf{r})\rangle$, which is denoted by two angle brackets in the Dirac notation. The other is the integration over $\mathbf{R}$. We must compute the derivative $\nabla_{\mathbf{R}} |n(\mathbf{R})\rangle$ first. If $\{\mathbf{R}\}$ is one dimensional, then the integrations over $\mathbf{r}$ and $d\mathbf{R}$ can be swapped, and the entire integral is zero, or $\gamma_n(C) = 0$ (see the exercise).

Example 8 (Berry phase for a two-dimensional planewave). A two-dimensional planewave has the wavefunction $\psi(x, y; k_x, k_y) = e^{ik_x x + ik_y y}/a$ normalized with a square $a \times a$, where the parameters are k_x and k_y. Find the Berry phase γ_n over a loop starting from $(k_x, k_y) = (0, 0)\frac{2\pi}{a} \rightarrow (1, 0)\frac{2\pi}{a} \rightarrow (1, 1)\frac{2\pi}{a} \rightarrow (0, 1)\frac{2\pi}{a} \rightarrow (0, 0)\frac{2\pi}{a}$.

We start from the Berry connection on Section 10.7.4: $\mathbf{A}_n = -\frac{a}{2}(\hat{x} + \hat{y})$. There are four segments. From $(0, 0)\frac{2\pi}{a} \rightarrow (1, 0)\frac{2\pi}{a}$, $\int d\mathbf{R} \cdot \mathbf{A} = -\frac{a}{2} \int dk_x = -\pi$. For the next three segments, their geometrical phases are $-\pi$, $+\pi$, and $+\pi$. So, we have the Berry phase $\gamma_n(C) = -\pi - \pi + \pi + \pi = 0$. The phases in each segment cancel out.

Another way to calculate the Berry phase is to utilize Stokes' theorem to convert the loop integral in eq. (10.7.22) into a surface integral,

$$\gamma_n(C) = -\text{Im} \iint_C d\mathbf{S} \cdot [\nabla_{\mathbf{R}} \times \langle n(\mathbf{R})|\nabla_{\mathbf{R}}|n(\mathbf{R})\rangle] \tag{10.7.23}$$

$$\equiv -\text{Im} \iint_C d\mathbf{S} \cdot \mathbf{V}_n(\mathbf{R}) \tag{10.7.24}$$

$$= -\text{Im} \iint_C d\mathbf{S} \cdot \langle \nabla_{\mathbf{R}} n(\mathbf{R})| \times |\nabla_{\mathbf{R}} n(\mathbf{R})\rangle. \tag{10.7.25}$$

In eq. (10.7.24) $\mathbf{V}_n(\mathbf{R}) = \nabla_{\mathbf{R}} \times \mathbf{A}_n(\mathbf{R})$ is called the Berry curvature, the curl of the Berry connection $\mathbf{A}_n(\mathbf{R})$, and the integrand in eq. (10.7.25) is just like the enclosed volume by three vectors (Fig. 10.6(b)). For this reason, to have a nonzero $\gamma_n(C)$, $|\nabla_{\mathbf{R}} n(\mathbf{R})\rangle$ and $\langle \nabla_{\mathbf{R}} n(\mathbf{R})|$ must be two different vectors. Here is an example.

Example 9 (Berry phase for a two-dimensional planewave). A two-dimensional planewave has the wavefunction $\psi(x,y;k_x,k_y) = e^{ik_x x + ik_y y}/a$ normalized with a square $a \times a$, where the parameters are k_x and k_y. Find the Berry phase γ_n over a loop using eq. (10.7.25).

We first compute $\nabla_{\mathbf{R}}|\psi\rangle = ie^{ik_x x + ik_y y}(x\hat{x} + y\hat{y})/a$. Its conjugation is $\langle \nabla_{\mathbf{R}}\psi| = -ie^{-ik_x x - ik_y y}(x\hat{x} + y\hat{y})/a$. It is easy to see that these two vectors are parallel to each other. Therefore, eq. (10.7.25) yields $\gamma_n(C) = 0$.

In case the Hamiltonian $\hat{H}(\mathbf{R})$ of the system is given and its eigenstates $|n(\mathbf{R})\rangle$ can be computed, we can insert the completeness of the eigenstates into the cross-product in eq. (10.7.25), $\sum_{m\neq n}\langle \nabla_{\mathbf{R}} n(\mathbf{R})|m(\mathbf{R})\rangle \times \langle m(\mathbf{R})|\nabla_{\mathbf{R}} n(\mathbf{R})\rangle$, where we exclude $m = n$ because $\langle \nabla_{\mathbf{R}} n(\mathbf{R})|n(\mathbf{R})\rangle$ and $\langle n(\mathbf{R})|\nabla_{\mathbf{R}} n(\mathbf{R})\rangle$ are the same vector and their cross-product is zero. Starting from (10.7.4), we use the Hellmann–Feynman theorem (Section 5.7) to rewrite the derivative as

$$\langle m(\mathbf{R})|\nabla_{\mathbf{R}} n(\mathbf{R})\rangle = \langle m(\mathbf{R})|\nabla_{\mathbf{R}}\hat{H}|n(\mathbf{R})\rangle/(E_n - E_m), \quad m \neq n,$$

which translates the derivatives on the wavefunction to the Hamiltonian. Doing so leads to a more manageable expression for the Berry phase

$$\gamma_n(C) = -\text{Im} \iint_C d\mathbf{S} \cdot \mathbf{V}_n(\mathbf{R}), \tag{10.7.26}$$

where the Berry curvature $\mathbf{V}_n(\mathbf{R})$ is given by

$$\mathbf{V}_n(\mathbf{R}) = \sum_{m\neq n} \frac{\langle n(\mathbf{R})|\nabla_{\mathbf{R}}\hat{H}(\mathbf{R})|m(\mathbf{R})\rangle \times \langle m(\mathbf{R})|\nabla_{\mathbf{R}}\hat{H}(\mathbf{R})|n(\mathbf{R})\rangle}{(E_m(\mathbf{R}) - E_n(\mathbf{R}))^2}. \tag{10.7.27}$$

We present these two examples, but their Berry phases are all zero because the phase from one part of the closed path cancels that from the other part. Equation (10.7.27) reveals some interesting features to us. If we have a degeneracy, $E_m = E_n$ with $m \neq n$, then $\mathbf{V}$ can be very large. Next, we introduce an example with degeneracy that Berry used [84], and the problem at the end of this chapter uses a different method.

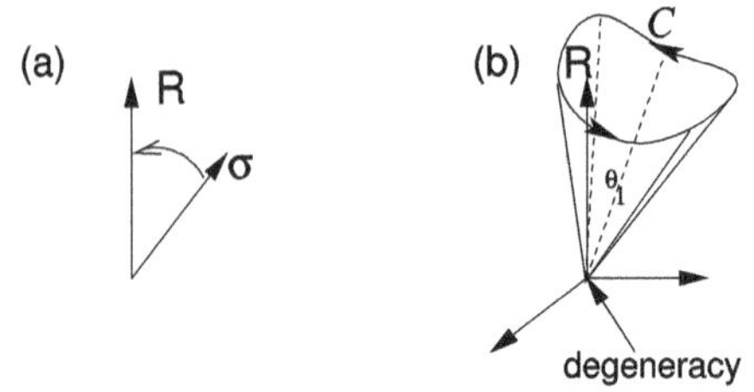

Figure 10.7: (a) We rotate $\hat{\boldsymbol{\sigma}}$ to **R** first and then back. (b) The closed path C is around **R**, with the polar angle of θ_1.

Example 10 (Berry phase for a spin). Suppose we have a Hamiltonian, $\hat{H}(\mathbf{R}) = \frac{1}{2}\begin{pmatrix} Z & X-iY \\ X+iY & -Z \end{pmatrix}$, where X, Y, and Z are the Cartesian coordinates of a vector **R**. Compute the Berry phase for the eigenstate with the largest eigenvalue over a circle C over the solid angle $\Omega(C)$, where C subtends at the degeneracy. In $\Omega(C)$, θ starts from 0 to θ_0.

We first find the eigenvalues of $\hat{H}$ by diagonalizing the Hamiltonian, $E_1 = \frac{1}{2}\sqrt{X^2+Y^2+Z^2} = \frac{1}{2}R$, and $E_2 = -\frac{1}{2}R$. The degeneracy occurs at $R = 0$. According to eq. (10.7.27), we need to find $\nabla_{\mathbf{R}}\hat{H}$. We note that the Hamiltonian can be written as $\hat{H} = \frac{1}{2}\mathbf{R}\cdot\hat{\boldsymbol{\sigma}}$, where $\hat{\boldsymbol{\sigma}}$ is the Pauli matrix of the spin, so $\nabla_{\mathbf{R}}\hat{H} = \frac{1}{2}\hat{\boldsymbol{\sigma}}$. Because $\hat{\boldsymbol{\sigma}}$'s quantization axis is not always along **R**, we temporally rotate the spin's z-axis along **R** (see Fig. 10.7(a)), so the eigenstates of $\hat{H}$ are simply $|1\rangle = |\alpha\rangle$ and $|2\rangle = |\beta\rangle$. This eases the calculation of **V** in eq. (10.7.27),

$$\mathbf{V}_1(\mathbf{R}) = \langle 1|\nabla_{\mathbf{R}}\hat{H}|2\rangle \times \langle 2|\nabla_{\mathbf{R}}\hat{H}|1\rangle/(E_1-E_2)^2 = \langle 1|\hat{\boldsymbol{\sigma}}|2\rangle \times \langle 2|\hat{\boldsymbol{\sigma}}|1\rangle/\left(4R^2\right),$$

which can be cast into the determinant

$$\mathbf{V}_1(\mathbf{R}) = \frac{1}{4R^2}\begin{vmatrix} \hat{x} & \hat{y} & \hat{z} \\ \langle 1|\hat{\sigma}_x|2\rangle & \langle 1|\hat{\sigma}_y|2\rangle & \langle 1|\hat{\sigma}_z|2\rangle \\ \langle 2|\hat{\sigma}_x|1\rangle & \langle 2|\hat{\sigma}_y|1\rangle & \langle 2|\hat{\sigma}_z|1\rangle \end{vmatrix}.$$

Three components are given by

$$\begin{aligned} V_{1x} &= \frac{1}{4R^2}\left(\langle 1|\hat{\sigma}_y|2\rangle\langle 2|\hat{\sigma}_z|1\rangle - \langle 1|\hat{\sigma}_z|2\rangle\langle 2|\hat{\sigma}_y|1\rangle\right), \\ V_{1y} &= \frac{1}{4R^2}\left(\langle 1|\hat{\sigma}_z|2\rangle\langle 2|\hat{\sigma}_x|1\rangle - \langle 1|\hat{\sigma}_x|2\rangle\langle 2|\hat{\sigma}_z|1\rangle\right), \\ V_{1z} &= \frac{1}{4R^2}\left(\langle 1|\hat{\sigma}_x|2\rangle\langle 2|\hat{\sigma}_y|1\rangle - \langle 1|\hat{\sigma}_y|2\rangle\langle 2|\hat{\sigma}_x|1\rangle\right). \end{aligned}$$

Because both $\langle 1|\hat{\sigma}_z|2\rangle$ and $\langle 2|\hat{\sigma}_z|1\rangle$ are zero, $V_{1x} = V_{1y} = 0$, where $|1\rangle$ and $2\rangle$ are just $|\alpha\rangle$ and $|\beta\rangle$. However, $\hat{\sigma}_x|\alpha(\beta)\rangle = |\beta(\alpha)\rangle$ and $\hat{\sigma}_y|\alpha(\beta)\rangle = +(-)i|\beta(\alpha)\rangle$, so $V_{1z} = \frac{1}{4R^2}(\langle 1|1\rangle\langle 2|i|2\rangle - \langle 2|2\rangle\langle 1|(-i)|1\rangle) = \frac{i}{2R^2}$. Next, we rotate $\mathbf{V}_1$ back to the unrotated axis,

$$\mathbf{V}_1 = \frac{i}{2R^2}\frac{\mathbf{R}}{R} = \frac{i\mathbf{R}}{2R^3},$$

which is plugged into eq. (10.7.27) to get

$$\gamma_1(C) = i\iint \mathbf{V}_1\cdot d\mathbf{S} = i\int \frac{i\mathbf{R}}{2R^3}\cdot\hat{\mathbf{R}}R^2\sin\theta d\theta d\phi = i\iint \frac{i}{2}\sin\theta d\theta d\phi = -\frac{1}{2}\iint d\Omega.$$

After integration, we have $\gamma_1(C) = -\frac{\Omega}{2}$, which shows the Berry phase is half the solid angle Ω that C subtends at the degeneracy. The degeneracy is just the origin of $\mathbf{R}$ (Fig. 10.7(b)). We now choose a circle around $\mathbf{R}$ with θ at θ_1, so the Berry phase is $\gamma_1(C) = -(1 - \cos\theta_1)\pi$. We see that, if $\theta_1 = 0$, $\gamma_1(C) = 0$. In this case, the Hamiltonian is real. If $\theta_1 \neq 0$, $\gamma_1(C)$ is nonzero. This means that the path encloses the degeneracy. It can be shown that the other eigenvalue has $\gamma_2(C) = -\gamma_1(C)$. They appear in pairs. □

In the beginning of this book, we emphasized that one needs both the wavefunction and operator to make a measurement. But the Berry phase changes this viewpoint. Here, everything is based on the wavefunction itself. There is no operator. Both Berry connection and curvature are the making of a wavefunction. Although most of the examples that have been used here are related to the spin, the Berry phase has found many applications, such as polarization [59].

Exercise 10.7.5.

44. What is the relationship between the geometrical phase and the Berry phase?
45. According to eq. (10.7.4), what can we say about the total energy of the system?
46. Show $\nabla_{\mathbf{R}} \times \langle n(\mathbf{R})|\nabla_{\mathbf{R}}|n(\mathbf{R})\rangle = \langle \nabla_{\mathbf{R}} n(\mathbf{R})| \times |\nabla_{\mathbf{R}} n(\mathbf{R})\rangle$.
47. Show $\gamma_n(C) = 0$ if the parameter space is one dimensional.
48. Compute the Berry phase for a (unnormalized) planewave $\psi(r) = e^{i\mathbf{k}\cdot\mathbf{r}}$ in a three-dimensional cube a^3 using a path $(0,0,0) \to (1,0,0) \to (1,1,0) \to (0,1,0) \to (0,0,0)$ in the unit of $\frac{2\pi}{a}$.

10.8 Problems

1. Two electrons are confined in $x \in (-\frac{a}{2}, \frac{a}{2})$. Suppose they are in a wavefunction $\psi(1,2) = \frac{1}{\sqrt{2}}(\phi_a(1)\phi_b(2) \pm \phi_b(1)\phi_a(2))$, where $\phi_a(x) = \cos(\frac{\pi x}{a})$ and $\phi_b = e^{-\frac{|x|}{a}}$. (a) Normalize both ϕ_a and ϕ_b and find $\langle\phi_a|\phi_b\rangle$. (b) Find the expectation values of x_1 and x_2. (c) Find the expectation value of x_1x_2 for the symmetric and antisymmetric wavefunctions and compare it with (b). What can you conclude?
2. Two identical quantum particles occupy two single-particle orbitals ψ_a and ψ_b. It is known that the density $\rho(1,2)$ includes following two terms: $\rho_a(1)\rho_b(2)$ and $\rho_{ab}(1)\rho_{ba}(2)$, where $\rho_a(1) = |\psi_a(1)|^2$, $\rho_b(2) = |\psi_b(2)|^2$, $\rho_{ab}(1) = \psi_a^*(1)\psi_b(1)$ and $\rho_{ba}(2) = \psi_b^*(2)\psi_a(2)$, but the remaining terms are unknown. (a) Because exchanging two identical particles should not change $\rho(1,2)$, write down all the possible expressions for $\rho(1,2)$. (b) Prove that, if $\psi_a = \psi_b$ and $\rho(1,2) = 0$, these particles must be a fermion. (c) Prove that, if $\psi_a = \psi_b$ and $\rho(1,2) \neq 0$, these particles must be a boson.
3. Two electrons in an atom have two different configurations: (1) One electron is in the $2p_x$ orbital and the other is in the $2p_z$ orbital and (2) one electron is in the $2p_x$ orbital and the other is in the $2p_y$ orbital. The spherical harmonics for $2p_x$, $2p_y$ and $2p_z$ are given in Table 4.1. (a) Use eq. (10.4.4) to compute the energies for (1) and (2). Since the radial integrals for (1) and (2) are exactly the same, only the integral over

the angle is needed. (b) According to the Hund second rule, (1) should have a lower Coulomb energy than (2). What can you conclude from your results?

4. Two qubits are in the state $|\Psi\rangle = \frac{1}{\sqrt{2}}(|00\rangle + |11\rangle)$, where the first bit belongs to qubit A and the second one belongs to B. (a) Suppose one applies the Hadamard gate on qubit A and $\boxed{\mathrm{T}}$ on qubit B. What is the final state? (b) Find the percentages of qubit A in $|0\rangle$ and $|1\rangle$.
5. Construct a controlled-Hadamard gate and find the matrix form for two qubits.
6. It is often customary that one writes four computational basis states as the Bell bases, $|\psi_1\rangle = \frac{1}{\sqrt{2}}(|00\rangle + |11\rangle)$, $|\psi_2\rangle = \frac{1}{\sqrt{2}}(|00\rangle - |11\rangle)$, $|\psi_3\rangle = \frac{1}{\sqrt{2}}(|01\rangle + |10\rangle)$, and $|\psi_4\rangle = \frac{1}{\sqrt{2}}(|01\rangle - |10\rangle)$. (a) Apply $\boxed{\mathrm{CNOT}}$ to each basis. (b) Set up the density matrices and diagonalize their eigenvalues and eigenvectors.
7. Starting from eq. (10.7.8), prove eq. (10.7.10).
8. Start from $\hat{H}(\mathbf{R}) = \frac{1}{2}\begin{pmatrix} Z & X-iY \\ X+iY & -Z \end{pmatrix}$, where X, Y and Z are the Cartesian coordinates of a vector $\mathbf{R}$. (a) Diagonalize $\hat{H}$ to find the eigenvalues and eigenvectors. (b) Find the Berry connection $\mathbf{A}$ for both eigenstates. (c) Compute the Berry phases for both eigenstates directly from eq. (10.7.21), without using eq. (10.7.27), and then compare the results on Section 10.7.5.
9. An electron is placed in a uniform magnetic field with $\mathbf{B}(t) = B[\sin(\theta(t))\cos(\phi(t))\hat{x} + \sin(\theta(t))\sin(\phi(t))\hat{y} + \cos(\theta(t))]\hat{z}$, where B is the amplitude of the magnetic field and ϕ and θ are the azimuthal and polar angles at time t. The Hamiltonian is $\hat{H}(t) = -\frac{g\mu_B}{\hbar}\mathbf{B}(t)\cdot\hat{\mathbf{S}}$. (a) Find the eigenvalues and eigenvectors. (b) Find the Berry connection $\mathbf{A}$ for both eigenstates. (c) Compute the Berry phases for both eigenstates. (d) If at $t = 0$ the system starts from a superposition state $|\psi(0) = (|\alpha\rangle + |\beta\rangle)/\sqrt{2}$, and after time T, $\hat{H}(t)$ returns to its initial form $\hat{H}(T) = \hat{H}(0)$, find the wavefunction $\psi(T)$ at $t = T$. (e) Find the expectation values of $\langle\psi(T)|\hat{\sigma}_x|\psi(T)\rangle$, $\langle\psi(T)|\hat{\sigma}_y|\psi(T)\rangle$, and $\langle\psi(T)|\hat{\sigma}_z|\psi(T)\rangle$, and compare their respective values at $t = 0$.
10. This problem is modeled on the Rabi Hamiltonian on Section 6.6.1. A system has three magnetic fields, B_x, B_y, B_z, $\hat{H} = \begin{pmatrix} -B_z & B_x+iB_y \\ B_x-iB_y & B_z \end{pmatrix}$, where B_x and B_y return to their original values after T. (a) Find the eigenvalues and eigenstates. (b) Find the Berry curvature. (c) Find the Berry phase. (d) If B_x and B_y are constant, but B_z returns to its initial value after T, what is the Berry phase?
11. The Berry phase can be defined around a discrete path [59], called Pancharatnam phase, instead of continuous paths given above. Suppose we have N number of states $\{\phi_i\}$ which are not orthogonal to each other. We go from ϕ_1 to ϕ_2 all the way to ϕ_N and then back to ϕ_1. The Berry phase $\gamma(C)$ is defined as $\gamma(C) = -\mathrm{Im}\ln[\langle\phi_1|\phi_2\rangle\langle\phi_2|\phi_3\rangle\cdots\langle\phi_{N-1}|\phi_N\rangle]$. Now suppose we start from the eigenfunctions of $\hat{S}_z$, the eigenfunctions of $\hat{S}_x$, the eigenfunctions of $\hat{S}_y$, back to the eigenfunctions of $\hat{S}_z$. Compute $\gamma(C)$.
12. Show around a closed path C, the energy conservation is violated in general.

13. Suppose $\psi_{\mathbf{k}}$ is a Bloch function (eq. (9.2.11)). Prove the lattice vector $\mathbf{R}$ can be calculated from the difference between two Berry connections

$$\mathbf{R} = -i\left(\left\langle \psi_{\mathbf{k}}(\mathbf{R}+\mathbf{r}) \middle| \frac{\partial \psi_{\mathbf{k}}(\mathbf{R}+\mathbf{r})}{\partial \mathbf{k}} \right\rangle - \left\langle \psi_{\mathbf{k}}(\mathbf{r}) \middle| \frac{\partial \psi_{\mathbf{k}}(\mathbf{r})}{\partial \mathbf{k}} \right\rangle\right).$$

This problem shows the connection between $\mathbf{k}$ and $\mathbf{R}$, which is independent of $\mathbf{r}$.

11 Appendix

11.1 Atomic units

Quantum mechanics uses atomic units since many quantities are extremely small in SI units. While the most used atomic units are broadly accepted in the community, two different energy units are used: Hartree and Rydberg units. They differ by a factor of 2. In this book, we will explicitly state which units we are using. The following table lists some of the most common physical quantities. The Planck constant is $h = 6.626 \times 10^{-34}$ J·s, and the reduced one is $\hbar = h/(2\pi)$. The Bohr radius is $a_0 = 0.529177210903$ Å. The electron mass is $m_e = 9.11 \times 10^{-31}$ kg. The charge is $e = 1.602 \times 10^{-19}$ C.

Quantity	Value
Time	$\hbar/E_h$
Length	a_0
Mass	m_e
Charge	e
Energy (Hartree)	$E_h = \hbar^2/(m_e a_0^2)$
Energy (Rydberg)	$E_r = \hbar^2/(2m_e a_0^2)$
Conductance	$G_0 = e^2/h$
Conductivity	$e^2/(\hbar a_0)$
Resistance	$R_0 = \hbar/e^2$
Resistivity	$\hbar a_0/e^2$
Velocity	$a_0 E_h/\hbar$
Angular momentum	$\hbar$
Momentum	$\hbar/a_0$
Force	E_h/a_0 or E_r/a_0
Electric current	$eE_h/\hbar$
Electric field	$E_h/(ea_0)$
Electric dipole moment	ea_0
Spin moment (Bohr magneton)	$\mu_B = e\hbar/(2m_e)$

11.2 Time-order operator

If we integrate Eq. (3.5.4) once, we have

$$\hat{U}(t, t_0) = \hat{U}(t_0, t_0) + (i\hbar)^{-1} \int_{t_0}^{t} \hat{H}_S \hat{U}(t', t_0) dt', \tag{11.1}$$

which means that $\hat{U}$ at time t depends on $\hat{U}$ at time t'. If we use the same equation repeatedly, we end up with terms within the integral on the right side like $(i\hbar)^{-n} \int_{t_0}^{t} \cdots \int_{t_0}^{t_n} dt_1 \cdots dt_n$. This means that their coefficient is just $(i\hbar)^{-n}$.

https://doi.org/10.1515/9783110672152-011

If we want to start from Eq. (3.5.8) and if $\hat{H}_S$ depends on time, $-i\hat{H}_S(t-t_0)/\hbar$ must be changed to an integral, $-i\int_{t_0}^{t}\hat{H}_S(t')dt'/\hbar$, so $\hat{U}=e^{-i\int_{t_0}^{t}\hat{H}_s(t')dt'/\hbar}$. However, this expression is different from Eq. (11.1). To see this clearly, we expand it to the second order,

$$e^{-i\int_{t_0}^{t}\hat{H}_s(t')dt'/\hbar}=1+\left(-i\int_{t_0}^{t}\hat{H}_s(t')dt'/\hbar\right)+\frac{1}{2!}\left(-i\int_{t_0}^{t}\hat{H}_s(t')dt'/\hbar\right)^2+\cdots, \tag{11.2}$$

where the second-order term has 1/2!, but (11.1) does not have. In addition, it has two time-integrals. In general, operators at different times are not commutable, $[\hat{H}_S(t'), \hat{H}_S(t'')]\neq 0$. Ignoring $(-i/\hbar)$, we have

$$\left(\int_{t_0}^{t}\hat{H}_S(t')dt'\right)^2=\int_{t_0}^{t}dt'\int_{t_0}^{t}dt''\hat{H}_S(t')\hat{H}_S(t''). \tag{11.3}$$

So, we have to know the time order t' and t'', i. e., which one occurs earlier. Freeman Dyson introduced the time order operator $\hat{T}$, so

$$\hat{T}\left(\int_{t_0}^{t}H_S(t')dt'\right)^2=2!\int_{t_0}^{t}dt'\int_{t_0}^{t}dt''\hat{H}_S(t')\hat{H}_S(t''), \tag{11.4}$$

where $t''<t'$. This 2! cancels $\frac{1}{2!}$ in Eq. (11.2), so all the terms in the expansion have no factorials in the denominator. $\hat{T}$ allows us to write $\hat{U}(t,t_0)$ more easily as

$$\hat{U}(t,t_0)=\hat{T}\exp\left(-\frac{i}{\hbar}\int_{t_0}^{t}\hat{H}(t')dt'\right). \tag{11.5}$$

11.3 Spin rotation

In physics, there are two types of vectors. One comprises the polar vectors, such as position and momentum, and the other the axial vectors, such as angular momentum and spin. Their main difference is that, if we invert a polar vector **A**, it changes from **A** to −**A**, but, if we invert an axial vector **A**, it does not change sign. We will explain the reason behind this by first reviewing the rotation for a polar vector **A** in a three-dimensional system.

11.3.1 Polar vectors

We denote R as a 3×3 rotation matrix. We first rotate the object along the z-axis by γ, then along the y-axis by β, and finally along the z-axis by α. α, β, and γ are the Eulerian angles. These three rotations are called the proper rotations because they do not involve inversion or mirror operations. $R_z(\alpha)$, $R_y(\beta)$, and $R_z(\gamma)$, respectively [27], are,

$$\begin{pmatrix} \cos\alpha & -\sin\alpha & 0 \\ \sin\alpha & \cos\alpha & 0 \\ 0 & 0 & 1 \end{pmatrix}, \quad \begin{pmatrix} \cos\beta & 0 & \sin\beta \\ 0 & 1 & 0 \\ -\sin\beta & 0 & \cos\beta \end{pmatrix}, \quad \begin{pmatrix} \cos\gamma & -\sin\gamma & 0 \\ \sin\gamma & \cos\gamma & 0 \\ 0 & 0 & 1 \end{pmatrix}. \tag{11.6}$$

One key feature of the proper rotation is that the determinant of these matrices is always 1. If we want to rotate a vector **A**, we can left multiply it by R, such as $\mathbf{A}' = R\mathbf{A}$.

Improper rotations contain mirror operations and inversion. The following first three matrices, denoted by Σ_{xy}, Σ_{yz}, and Σ_{zx}, represent the mirror operation with respect to the xy-, yz- and zx-planes, respectively. The last one, $\mathcal{I}$, is the inversion matrix.

$$\begin{pmatrix} 1 & 0 & 0 \\ 0 & 1 & 0 \\ 0 & 0 & -1 \end{pmatrix}, \quad \begin{pmatrix} -1 & 0 & 0 \\ 0 & 1 & 0 \\ 0 & 0 & 1 \end{pmatrix}, \quad \begin{pmatrix} 1 & 0 & 0 \\ 0 & -1 & 0 \\ 0 & 0 & 1 \end{pmatrix}, \quad \begin{pmatrix} -1 & 0 & 0 \\ 0 & -1 & 0 \\ 0 & 0 & -1 \end{pmatrix}, \tag{11.7}$$

These rotations form a group known as the $O(3)$ group.

11.3.2 Axial vectors: spin rotation and SU(2) rotation matrices

Axial vectors form a special group of vectors. If we apply an inversion $\mathcal{I}$ to them, they do not change signs. Take the angular momentum $\hat{\mathbf{L}} = \hat{\mathbf{r}} \times \hat{\mathbf{p}}$ as an example: $\mathcal{I}\hat{\mathbf{r}} = -\hat{\mathbf{r}}$ and $\mathcal{I}\hat{\mathbf{p}} = -\hat{\mathbf{p}}$, but $\mathcal{I}\hat{\mathbf{L}} = \hat{\mathbf{L}}$, no sign changes. So, $\hat{\mathbf{L}}$ is an axial vector. This is a vector in three-dimensional space, where each component is a scalar in CM and a scalar operator in QM.

Spin is also an axial vector, but, different from $\hat{\mathbf{L}}$, each component of spin is a 2×2 matrix, which cannot be operated by a 3×3 rotation matrix R. Instead, we must first find a rotation matrix that works in the two-dimension [24]. A set of these matrices form a group, called the SU(2) group, a special unitary transformation group for two-dimensional vectors. The respective rotation matrices about the z- and y-axes are [27],[1]

$$U_z(\alpha) = \begin{pmatrix} e^{-i\alpha/2} & 0 \\ 0 & e^{i\alpha/2} \end{pmatrix}, \quad U_y(\beta) = \begin{pmatrix} \cos(\beta/2) & -\sin(\beta/2) \\ \sin(\beta/2) & \cos(\beta/2) \end{pmatrix}.$$

1 Our matrices can be applied to spins directly and differ by a Hermite conjugation from those in [24]. The matrices given on page 105 of [24], pertain to the axis rotation, not the spin rotation, but the final $U(\alpha,\beta,\gamma)$ on page 109 of [24], which includes this Hermite conjugation, is identical to ours and those in [27].

For the same three rotations with Eulerian angles, α, β, and γ, the 2×2 rotation matrix is the product matrix $U(\alpha,\beta,\gamma) = U_z(\alpha)U_y(\beta)U_z(\gamma)$,

$$U(\alpha,\beta,\gamma) = \begin{pmatrix} e^{\frac{-i(\alpha+\gamma)}{2}} \cos\frac{\beta}{2} & -e^{\frac{-i(\alpha-\gamma)}{2}} \sin\frac{\beta}{2} \\ e^{\frac{i(\alpha-\gamma)}{2}} \sin\frac{\beta}{2} & e^{\frac{i(\alpha+\gamma)}{2}} \cos\frac{\beta}{2} \end{pmatrix}.$$

Because $\hat{S}_x$ is a matrix, the matrix rotation is different from the vector rotation. Suppose our rotation operator is U. To rotate $\hat{S}_x$, we must proceed as

$$\hat{S}'_x = U\hat{S}_x U^{-1},$$

where $\hat{S}_x$ is given in Eq. (6.1.7). Caution must be taken: *Only one component* can be manipulated at one time, and it is not possible to rotate all three components at the same time like $\hat{\mathbf{L}}$. To rotate $\hat{S}_y$ and $\hat{S}_z$, we have to carry out two different rotations. After a lengthy calculation, we find [27]

$$\begin{aligned}\hat{S}'_x = {} & (\cos\alpha\cos\beta\cos\gamma - \sin\alpha\sin\gamma)\hat{S}_x + (\cos\alpha\sin\gamma + \sin\alpha\cos\beta\cos\gamma)\hat{S}_y \\ & - \sin\beta\cos\gamma\hat{S}_z.\end{aligned} \tag{11.8}$$

This method takes care only of proper rotations, leaving out improper rotations.

11.3.3 $O(3)$ rotation matrices

A much simpler method was revealed to us by Rodriguez–Carvajal [85], Birss [86], and Chen [87], and employed in our own research [88]. Here, we introduce this much simpler procedure. To carry out the same three subsequent rotations of a spin $\hat{\mathbf{S}}$, $R(\alpha,\beta,\gamma) = R_1(\alpha)R_2(\beta)R_3(\gamma)$, we need only

$$\begin{pmatrix} \hat{S}'_x \\ \hat{S}'_y \\ \hat{S}'_z \end{pmatrix} = R(\alpha,\beta,\gamma)\mathrm{Det}(R) \begin{pmatrix} \hat{S}_x \\ \hat{S}_y \\ \hat{S}_z \end{pmatrix} = \mathrm{Det}(R) \begin{pmatrix} R_{11} & R_{12} & R_{13} \\ R_{21} & R_{22} & R_{23} \\ R_{31} & R_{32} & R_{33} \end{pmatrix} \begin{pmatrix} \hat{S}_x \\ \hat{S}_y \\ \hat{S}_z \end{pmatrix}.$$

For instance $\hat{S}'_x = (R_{11}\hat{S}_x + R_{12}\hat{S}_y + R_{13}\hat{S}_z)\mathrm{Det}(R)$. In this way, we can finish the job for all three spin components at once. The determinant $\mathrm{Det}(R)$ is the only difference between the spin and a regular vector. For this reason, spin is called an axial vector. The regular vector is renamed as a polar vector. When we apply an inversion $\mathcal{I}$ to a regular vector $\mathbf{A}$, it produces $\mathcal{I}\mathbf{A} = -\mathbf{A}$. But, if it applies to a spin $\hat{\mathbf{S}}$, $\mathrm{Det}(\mathcal{I})\mathcal{I}\hat{\mathbf{S}} = \hat{\mathbf{S}}$, so it does not change its direction. For a mirror operation, such as

$$\mathcal{M} = \begin{pmatrix} -1 & 0 & 0 \\ 0 & 1 & 0 \\ 0 & 0 & 1 \end{pmatrix},$$

it represents a reflection with respect to the yz-plane. We can do the same. For $\hat{S}_x$, we have

$$\begin{pmatrix} -1 & 0 & 0 \\ 0 & 1 & 0 \\ 0 & 0 & 1 \end{pmatrix} \mathrm{Det}(\mathcal{M}) \begin{pmatrix} \hat{S}_x \\ 0 \\ 0 \end{pmatrix} = \begin{pmatrix} \hat{S}_x \\ 0 \\ 0 \end{pmatrix},$$

which does not change the direction of $\hat{S}_x$. But for $\hat{S}_y$, the reflection does change:

$$\begin{pmatrix} -1 & 0 & 0 \\ 0 & 1 & 0 \\ 0 & 0 & 1 \end{pmatrix} \mathrm{Det}(\mathcal{M}) \begin{pmatrix} 0 \\ \hat{S}_y \\ 0 \end{pmatrix} = \begin{pmatrix} 0 \\ -\hat{S}_y \\ 0 \end{pmatrix}.$$

At no point, should we use U to rotate spin vectors.

11.3.4 What about spin eigenfunctions?

However, this method does not work for spin eigenfunctions because the spin eigenfunctions are spinors and double-valued. We have to use U. Suppose a spin state η whose rotated function is $U(\alpha, \beta, \nu)\eta$. Because U only has half-angles like $\frac{\alpha}{2}, \frac{\beta}{2}, \frac{\gamma}{2}$, a 2π-rotation, $e^{\frac{i2\pi}{2}} = -1$ produces a negative sign for the eigenfunction. For this reason, η are named as spinors. The reader may consult the books [86, 24] for a more advanced discussion.

11.4 Spin matrices for photon and useful math formulas

One key difference between electron spin and photon spin is that the photon spin matrices are 3×3 matrices, but the electron spin matrices are 2×2 matrices. They can be found in the same way as the electron spin. They are

$$\hat{\sigma}_x = \frac{1}{\sqrt{2}} \begin{pmatrix} 0 & 1 & 0 \\ 1 & 0 & 1 \\ 0 & 1 & 0 \end{pmatrix}, \quad \hat{\sigma}_y = \frac{i}{\sqrt{2}} \begin{pmatrix} 0 & -1 & 0 \\ 1 & 0 & -1 \\ 0 & 1 & 0 \end{pmatrix}, \quad \hat{\sigma}_z = \begin{pmatrix} 1 & 0 & 0 \\ 0 & 0 & 0 \\ 0 & 0 & -1 \end{pmatrix}. \tag{11.9}$$

The identity matrix I is a 3×3 matrix as well. Four Dirac matrices in Eq. (6.1.5) become

$$\hat{\gamma}_0 = \begin{pmatrix} I & 0 \\ 0 & -I \end{pmatrix}, \quad \hat{\gamma}_1 = \begin{pmatrix} 0 & \hat{\sigma}_x \\ -\hat{\sigma}_x & 0 \end{pmatrix}, \quad \hat{\gamma}_2 = \begin{pmatrix} 0 & \hat{\sigma}_y \\ -\hat{\sigma}_y & 0 \end{pmatrix}, \quad \hat{\gamma}_3 = \begin{pmatrix} 0 & \hat{\sigma}_z \\ -\hat{\sigma}_z & 0 \end{pmatrix}, \tag{11.10}$$

where $\hat{\sigma}_x$, $\hat{\sigma}_y$, $\hat{\sigma}_z$ are replaced by those in Eq. (11.9). In contrast to Ref. [12] which uses τ, we use $\hat{\sigma}$ instead. Three components of $\boldsymbol{\alpha}$ are $\alpha_i = \hat{\gamma}_0\hat{\gamma}_i$, where i runs from 1 to 3.

Table 11.1: The following expansions have been used in the coulomb potential $1/r$.

$(1 \pm x)^{-1/2} = 1 \mp \frac{1}{2}x + \frac{1}{2}\frac{3}{4}x^2 + \cdots$ where $x < 1$.

Table 11.2: Gradient, Divergence, and curl. f is a scalar function.

Cartesian $\nabla = \frac{\partial f}{\partial x}\hat{x} + \frac{\partial f}{\partial y}\hat{y} + \frac{\partial f}{\partial z}\hat{z}$
Cylindrical $\nabla = \frac{\partial f}{\partial r}\hat{r} + \frac{1}{r}\frac{\partial f}{\partial \phi}\hat{\phi} + \frac{\partial f}{\partial z}\hat{z}$
Spherical $\nabla = \frac{\partial f}{\partial r}\hat{r} + \frac{1}{r}\frac{\partial f}{\partial \theta}\hat{\theta} + \frac{1}{r\sin\theta}\frac{\partial f}{\partial \phi}\hat{\phi}$

Table 11.3: Divergence. **F** is a vector function.

Cartesian $\nabla \cdot \mathbf{F} = \frac{\partial F_x}{\partial x} + \frac{\partial F_y}{\partial y} + \frac{\partial F_z}{\partial z}$
Cylindrical $\nabla \cdot \mathbf{F} = \frac{1}{r}\frac{\partial (rF_r)}{\partial r} + \frac{1}{r}\frac{\partial F_\phi}{\partial \phi} + \frac{\partial F_z}{\partial z}$
Spherical $\nabla \cdot \mathbf{F} = \frac{1}{r^2}\frac{(r^2 F_r)}{\partial r} + \frac{1}{r\sin\theta}\frac{(\sin\theta F_\theta)}{\partial \theta} + \frac{1}{r\sin\theta}\frac{F_\phi}{\partial \phi}$

Table 11.4: Curl. **F** is a vector function.

Cartesian $\nabla \times \mathbf{F} = (\frac{\partial F_z}{\partial y} - \frac{\partial F_y}{\partial z})\hat{x} + (\frac{\partial F_x}{\partial z} - \frac{\partial F_z}{\partial x})\hat{y} + (\frac{\partial F_y}{\partial x} - \frac{\partial F_x}{\partial y})\hat{z}$
Cylindrical $\nabla \times \mathbf{F} = (\frac{1}{r}\frac{\partial F_z}{\partial \phi} - \frac{\partial F_\phi}{\partial z})\hat{r} + (\frac{\partial F_r}{\partial z} - \frac{\partial F_z}{\partial r})\hat{\phi} + \frac{1}{r}(\frac{\partial (rF_\phi)}{\partial r} - \frac{\partial F_r}{\partial \phi})\hat{z}$
Spherical $\nabla \times \mathbf{F} = \frac{1}{r\sin\theta}(\frac{\partial (F_\phi \sin\theta)}{\partial \theta} - \frac{F_\theta}{\partial \phi})\hat{r} + \frac{1}{r}(\frac{1}{\sin\theta}\frac{\partial F_r}{\partial \phi} - \frac{\partial (rF_\phi)}{\partial r})\hat{\theta} + \frac{1}{r}(\frac{\partial (rF_\theta)}{\partial r} - \frac{\partial F_r}{\partial \theta})\hat{\phi}$

Table 11.5: Laplacian. f is a scalar function.

Cartesian $\nabla^2 f = \frac{\partial^2 f}{\partial x^2} + \frac{\partial^2 f}{\partial y^2} + \frac{\partial^2 f}{\partial z^2}$
Cylindrical $\nabla^2 f = \frac{1}{r}\frac{\partial}{\partial r}(r\frac{\partial f}{\partial r}) + \frac{1}{r^2}\frac{\partial^2 f}{\partial \phi^2} + \frac{\partial^2 f}{\partial z^2}$
Spherical $\nabla^2 f = \frac{1}{r^2}\frac{\partial}{\partial r}(r^2\frac{\partial f}{\partial r}) + \frac{1}{r^2\sin\theta}\frac{\partial}{\partial \theta}(\sin\theta\frac{\partial f}{\partial \theta}) + \frac{1}{r^2\sin^2\theta}\frac{\partial^2 f}{\partial \phi^2}$

Table 11.6: Conversion of Cartesian coordinates to the spherical coordinates. We use $\cos\theta$ instead of θ and $\tan\phi$ instead of ϕ because it is simpler. They are used to compute partial derivatives such as $\partial r/\partial x$. To compute $\partial x/\partial r$, the second table should be used.

$r = \sqrt{x^2 + y^2 + z^2}$
$\cos\theta = z/\sqrt{x^2 + y^2 + z^2}$
$\tan\phi = y/x$

$x = r\sin\theta\cos\phi$
$y = r\sin\theta\sin\phi$
$z = \cos\theta$

Table 11.7: The first table: spherical unit vectors $(\hat{r}, \hat{\theta}, \hat{\phi})$ expressed in Cartesian unit vectors $(\hat{x}, \hat{y}, \hat{z})$. The second table: Cartesian unit vectors $(\hat{x}, \hat{y}, \hat{z})$ expressed in spherical unit vectors $(\hat{r}, \hat{\theta}, \hat{\phi})$.

$\hat{r} = (\sin\theta\cos\phi)\hat{x} + (\sin\theta\sin\phi)\hat{y} + (\cos\theta)\hat{z}$
$\hat{\theta} = (\cos\theta\cos\phi)\hat{x} + (\cos\theta\sin\phi)\hat{y} - (\sin\theta)\hat{z}$
$\hat{\phi} = -(\sin\phi)\hat{x} + (\cos\phi)\hat{y}$

$\hat{x} = (\sin\theta\cos\phi)\hat{r} + (\cos\theta\cos\phi)\hat{\theta} - (\sin\phi)\hat{\phi}$
$\hat{y} = (\sin\theta\sin\phi)\hat{r} + (\cos\theta\sin\phi)\hat{\theta} - (\cos\phi)\hat{\phi}$
$\hat{z} = (\cos\theta)\hat{r} - (\sin\theta)\hat{\theta}$

Table 11.8: The first table: cylindrical unit vectors $(\hat{r}, \hat{\theta}, \hat{\phi})$ expressed in Cartesian unit vectors $(\hat{x}, \hat{y}, \hat{z})$. The second table: Cartesian unit vectors $(\hat{x}, \hat{y}, \hat{z})$ expressed in spherical unit vectors $(\hat{r}, \hat{\theta}, \hat{\phi})$.

$\hat{r} = (\sin\theta\cos\phi)\hat{x} + (\sin\theta\sin\phi)\hat{y} + (\cos\theta)\hat{z}$
$\hat{\theta} = (\cos\theta\cos\phi)\hat{x} + (\cos\theta\sin\phi)\hat{y} - (\sin\theta)\hat{z}$
$\hat{\phi} = -(\sin\phi)\hat{x} + (\cos\phi)\hat{y}$

$\hat{x} = (\sin\theta\cos\phi)\hat{r} + (\cos\theta\cos\phi)\hat{\theta} - (\sin\phi)\hat{\phi}$
$\hat{y} = (\sin\theta\sin\phi)\hat{r} + (\cos\theta\sin\phi)\hat{\theta} - (\cos\phi)\hat{\phi}$
$\hat{z} = (\cos\theta)\hat{r} - (\sin\theta)\hat{\theta}$

Table 11.9: Trigonometry.

$\sin^2(a) + \cos^2(a) = 1$, $\cos^2(a) - \sin^2(a) = \cos(2a)$, $1 + \tan^2(a) = \sec^2(a)$, $1 + \cot^2(a) = \csc^2(a)$
$\sin(a \pm b) = \sin(a)\cos(b) \pm \cos(a)\sin(b)$, $\cos(a \pm b) = \cos(a)\cos(b) \mp \sin(a)\sin(b)$
$\sin(a)\sin(b) = \frac{-1}{2}[\cos(a+b) - \cos(a-b)]$, $\cos(a)\cos(b) = \frac{1}{2}[\cos(a+b) + \cos(a-b)]$,
$\sin(a)\cos(b) = \frac{1}{2}[\sin(a+b) + \sin(a-b)]$
$\sin a + \sin b = 2\sin(\frac{a+b}{2})\cos(\frac{a-b}{2})$, $\sin a - \sin b = 2\sin(\frac{a-b}{2})\cos(\frac{a+b}{2})$
$\cos a + \cos b = 2\cos(\frac{a+b}{2})\cos(\frac{a-b}{2})$, $\cos a - \cos b = -2\sin(\frac{a+b}{2})\sin(\frac{a-b}{2})$

Table 11.10: Integration.

$\int x\sin(ax)dx = -\frac{x}{a}\cos(ax) + \frac{1}{a^2}\sin(ax)$
$\int x^2\sin(ax)dx = \frac{2x}{a^2}\sin(ax) + \frac{2-a^2x^2}{a^3}\cos(ax)$
$\int x\cos(ax)dx = \frac{x}{a}\sin(ax) + \frac{1}{a^2}\cos(ax)$
$\int x^2\cos(ax)dx = \frac{2x}{a^2}\cos(ax) + \frac{a^2x^2-2}{a^3}\sin(ax)$
$\int \sin(ax)\cos(bx)dx = -\frac{\cos[(a-b)x]}{2(a-b)} - \frac{\cos[(a+b)x]}{2(a+b)}$ $\lvert a\rvert \neq \lvert b\rvert$
$\int x\sin(ax)\sin(bx)dx = \frac{1}{2}\Big[\frac{x}{a-b}\sin[(a-b)x] + \frac{1}{(a-b)^2}\cos[(a-b)x]$
$\quad - \frac{x}{a+b}\sin[(a+b)x] - \frac{1}{(a+b)^2}\cos[(a+b)x]\Big]$
$\int \frac{1}{b+c\exp(ax)} = \frac{x}{b} - \frac{1}{ab}\ln[b + c\exp(ax)]$
$\int \exp(ax)xdx = \frac{ax-1}{a^2}\exp(ax)$
$\int \exp(ax)x^2dx = (\frac{x^2}{a} - \frac{2x}{a^2} + \frac{2}{a^3})\exp(ax)$
$\int_{-\infty}^{\infty}\exp(-ax^2 + bx)dx = \sqrt{\frac{\pi}{a}}\exp(\frac{b^2}{4a})$, $\int_{-\infty}^{\infty}x\exp(-ax^2 + bx)dx = \frac{\sqrt{\pi}b}{2a^{3/2}}\exp(\frac{b^2}{4a})$
$\int_{-\infty}^{\infty}x^2\exp(-ax^2 + bx)dx = \frac{\sqrt{\pi}(2a+b^2)}{4a^{5/2}}\exp(\frac{b^2}{4a})$

11.5 Computer codes

Listing 11.1: Completeness.

```
!       this code checks the completeness of eigenfunctions for PIB
        implicit double precision (a-h,o-z)
        pi=dacos(-1d0)
        x=0.5465d0
        a=10d0
        open(1,file='completeness')
        n=7
        one=0d0
        dx=1d-2*a
        do ix=1,100
           x=ix*dx
           psi=dsqrt(2/a)*dsin(n*pi*x/a)
           one=one+psi*psi
        enddo
!        write(*,*)one*dx
!       stop
        mesh=300
        numberofstates=300
        do ix=1,mesh
           x=ix*a/mesh
           one=0d0
           do n=1,numberofstates
              psi=dsqrt(2/a)*dsin(n*pi*x/a)
              one=one+psi*psi
           enddo
           write(1,*)x,one/(numberofstates/a)
        enddo
        close(1)
        end
```

Listing 11.2: Even roots of finite square well.

```
!       This code computes the roots of
!       Equation: eta = +/- omega cos(eta)
!       written on Aug, 2021,
!       revised on Sept. 27, 2021

!       to Run: compile the code: ifort evenroot.f or gfortran evenroot.f
!       then: ./a.out
!       press enter if the code produces warnings.
!       Results: Final results are stored in  evenroot.eta-omegaxcoseta,
!
        implicit double precision (a-h,o-z)
        double precision root(100),m
        integer debug
        debug=0
        open(10,file='evenroot.eta-omegaxcoseta')
        write(10,927)'#Omega (pi),eta, eta,zero,zero, xi (positive root)'
```

```
927   format(a)
      open(20,file='evenroot.eta+omegaxcoseta')
      write(20,927)'#Omega (pi),eta,eta,zero,zero, xi (negative root)'
      pi=dacos(-1d0)
!     this is tolerance of my solution
      tol=1d-12
!     this runs over omega. For each omega, I found a multiple roots for
!     eta
      do iw=1,80
         w=0.1d0*iw*pi
         d=0.1
!         goto 123
         icount=0
         open(1,file='tmp')
         do i=1,10000
            q=i*d
            write(1,*)q,fp(q,w)
            if(fp(q,w)*fp(q+d,w).lt.0d0)then
               icount=icount+1
               root(icount)=q
            endif
         enddo
         close(1)
         gap=1d0                      !must a small one
         do i=1,icount
            write(*,*)i,root(i),'positive possible root',w
            if(i.ne.1)then
               if(root(i)-root(i-1).lt.gap)then
                  gap=root(i)-root(i-1)
               endif
            endif
         enddo
!         read(*,*)
         do i=1,icount
            b=root(i)+gap/5d0
            a=root(i)-gap/5d0
            write(*,*)'initial a and b',a,b
            do while((b-a)>tol)
               m=a+(b-a)/2d0
               if(fp(a,w)*fp(m,w).gt.0d0) then
                  a=m
               else
                  b=m
               endif
               if(debug.eq.1) write(*,*)a,b
            enddo
103         xi=a*dtan(a)
            write(*,*)'w',w/pi,a/pi,b,fp(a,w),fp(b,w),xi
            if(dabs(fp(a,w)).gt.5d-12)then
            write(*,*)'positive root not accurate,increase gap,press..'
               gap=2*gap
               b=root(i)+gap/5d0
```

```
            a=root(i)-gap/5d0
            write(*,*)'initial a and b',a,b
            do while((b-a)>tol)
               m=a+(b-a)/2d0
               if(fp(a,w)*fp(m,w).gt.0d0) then
                  a=m
               else
                  b=m
               endif
             if(debug.eq.1)  write(*,*)a,b
            enddo
            read(*,*)
            goto 103
         endif
!         read(*,*)
         if(xi.gt.0d0)then
            write(10,*)w/pi,a/pi,b/pi,fp(a,w),fp(b,w),xi/pi
         endif
      enddo
!    second equation starts here eta = - w cos(eta)
 123  icount=0
!     -------------------------------------------------
!    /     the following is to get approximate roots. \
      open(1,file='tmp')
      do i=1,10000
         q=i*d
         write(1,*)q,fn(q,w)
!    if two values of function fn at q and q+d change sign, this means
!    it crosses a root.
         if(fn(q,w)*fn(q+d,w).lt.0d0)then
!    count how many roots are found
            icount=icount+1
            if(icount.gt.100)stop'increase dimension for root'
!    root records the actual roots
            root(icount)=q
         endif
      enddo
      close(1)
!    \_____end_________________________________________/

! in order to avoid the code missing a root, I need to find the minimum
! gap that I can choose. The default is 1d0
      gap=1d0
      do i=1,icount
         write(*,*)i,root(i),'negative possible root',w/pi
         if(i.ne.1)then
            if(root(i)-root(i-1).lt.gap)then
               gap=root(i)-root(i-1)
            endif
         endif
      enddo
      if(debug.eq.1)read(*,*)
```

```
!      stop
          tol=1d-12
!--accurate root finder----from Michael Heath's book: Middle point method.
          do i=1,icount
             b=root(i)+gap/5d0   !b is chosen slightly higher than root
!by gap/5
             a=root(i)-gap/5d0   !a is chosen slightly lower than root by
!gap/5
             if(debug.eq.1)then
                write(*,*)'initial a b',a,b
             endif
             do while((b-a)>tol)
                m=a+(b-a)/2d0
!      this is from Michael Heath's book: Middle point method.
                if(fn(a,w)*fn(m,w).gt.0d0) then
                   a=m
                else
                   b=m
                endif
             enddo
!      here a is the root for eta. From eta, we can find xi. xi=eta\tan(eta)
 104         xi=a*dtan(a)
             write(*,*)'w',w/pi,a/pi,b,fn(a,w),fn(b,w),xi
             if(debug.eq.1)read(*,*)
             if(dabs(fn(a,w)).gt.1d-12)then
                write(*,*)'This root not accurate, increase gap, press..'
                gap=2*gap
                b=root(i)+gap/5d0
                a=root(i)-gap/5d0
                do while((b-a)>tol)
                   m=a+(b-a)/2d0
!      this is from Michael Heath's book: Middle point method.
                   if(fn(a,w)*fn(m,w).gt.0d0) then
                      a=m
                   else
                      b=m
                   endif
                enddo
                read(*,*)
                goto 104
             endif
!      we only accept roots with both xi and eta positive.
             if(xi.gt.0d0)then
                write(20,*)w/pi,a/pi,b/pi,fn(a,w),fn(b,w),xi/pi
             endif
          enddo
       enddo
       close(10)
       close(20)
       end

       double precision function fp(q,w)
```

```
      implicit double precision (a-h,o-z)
      fp=q/w-dcos(q)             !positive root
      end

      double precision function fn(q,w)
      implicit double precision (a-h,o-z)
      fn=q/w+dcos(q)             !negative root
      end
```

Listing 11.3: Even wavefunction of finite square well

```
!     This code computes finite.well.wavefunction.
!     input: even.w.eta.used. This file is selectively copied from
!     evenroot.eta-omegaxcoseta or evenroot.eta+omegaxcoseta, depending
!     on whether you want to have both roots. In general, you do. If so,
!     you need to copy the root for the same w into even.w.eta.used.
!     Selectively means: you have to choose a w (in pi).

!     Sept. 29, 2021 written.
!     Sept. 30, I convert it to AI

      program even_wavefunction
      implicit double precision (a-h,o-z)
      double precision norm1,norm2
      w=1.0000000000000000
      eta=0.73908513321475766
      w=0.31415926535897931
      eta=0.30011694986751536
      pi=dacos(-1d0)
!     Quantum well width
      a=10d0
!     Grid mesh size
      dx=0.02d0
      open(1,file='even.w.eta.used') !this includes a few root for one w
!     file: w.eta.even is taken from the output file of
!     evenroot.f. There are two files: root.eta-omegaxcoseta and
!     root.eta+omegaxcoseta
!
!     here is an example
!    5.0000000000000000          0.47003254934310262
!    5.0000000000000000           1.4090642703195637
!    5.0000000000000000           2.3446403463001162
!    5.0000000000000000           3.2594932830920658
!    5.0000000000000000           4.1843795598082147
 10   read(1,*,end=1,err=1)w,eta
      w=w*pi
      eta=eta*pi
!     find xi from eta and w
      xi=dsqrt(w**2-eta**2)
!     the following is from Eq. XXX
      denom=a/2*(1d0+dcos(eta)**2/xi+dsin(2d0*eta)/(2d0*eta))
      denom=dsqrt(denom)
```

```
      AII=1d0/denom
      AI=dcos(eta)*dexp(xi)*AII
      AIII=AI
!     the following formula is from Eq. on page 14. AI=. Here BI and AI
!     are the same thing. I want to check whether using AII or AI has
!     any difference. Finding: There is no difference.
      BI=2d0/a*dexp(xi*2)/(1d0/xi+(2d0*eta+dsin(2d0*eta))/(2d0*eta
     c     *dcos(eta)**2))
      BI=dsqrt(BI)
      write(*,*)AI,BI
!      read(*,*)
      write(*,*)'AI AII and AIII',AI,AII,AIII,'denom',denom
      open(2,file='even.wavefunction',position='append')
      open(20,file='even.eigenvalue',position='append')
      write(20,*)eta**2/a**2
      close(20)
      do i=-100,100
         x=dx*i*a
         if(x.lt.-a/2)then       !wavefunction from - infinity to -a/2
            write(2,*)x,AI*dexp(xi*2d0*x/a)
         else if(x.ge.-a/2.and.x.lt.a/2)then  !wavefunction [-a/2, a/2]
            write(2,*)x,AII*dcos(eta*2*x/a)
         else  if(x.ge.a/2)then
            write(2,*)x,AIII*dexp(-xi*2d0*x/a) ![a/2,infinity]
         endif
      enddo
      write(2,*)'&'
      close(2)
      norm1=a/2*AI**2*dexp(-2*xi)/xi
      norm2=a/2*AII**2*(1d0+dsin(2d0*eta)/eta/2d0)
      write(*,*)'Plotting wavefunction',w,eta,norm1,norm2,norm1+norm2
      read(*,*)
      goto 10
 1    close(1)
      end
```

Listing 11.4: Odd roots of Finite square well.

```
!     This gives the odd root.
!     This code computes the roots of Equation: eta = +/- omega sin(eta)
!     written on Aug, 2021, revised on Sept. 27, 2021

!     Oct. 1, 2021. I found a bug in eta = +/- omega sin(eta). Solution:
!     eta must give a negative ctan(eta) to be accepted as a root since
!     xi=-eta tan(eta) and xi and eta must be both positive.

      implicit double precision (a-h,o-z)
      double precision root(100),m
      open(10,file='oddroot.eta-omegaxsineta')
      write(10,927)'#Omega (pi),eta, eta,zero,zero, xi (positive root)'
 927  format(a)
      open(20,file='oddroot.eta+omegaxsineta')
```

```
      write(20,927)'#Omega (pi),eta,eta,zero,zero, xi (negative root)'
      pi=dacos(-1d0)
!     this is tolerance of my solution
      tol=1d-12
!     this runs over omega. For each omega, I found a multiple roots for
!     eta
      do iw=1,80
         w=0.1d0*iw*pi
         d=0.1
!         goto 123
         icount=0
         open(1,file='tmp')
         do i=1,10000
            q=i*d
            write(1,*)q,fp(q,w)
            if(fp(q,w)*fp(q+d,w).lt.0d0)then
               icount=icount+1
               root(icount)=q
            endif
         enddo
         close(1)
         gap=1d0                    !this is the initial gap
         do i=1,icount
            write(*,*)i,root(i),'positive part possible root'
            if(i.ne.1)then
               if(root(i)-root(i-1).lt.gap)then
                  gap=root(i)-root(i-1)
               endif
            endif
         enddo
 103     do i=1,icount
            b=root(i)+gap/5d0
            a=root(i)-gap/5d0
            write(*,*)'my initial a and b',a,b
            do while((b-a)>tol)
               m=a+(b-a)/2d0
               if(fp(a,w)*fp(m,w).gt.0d0) then
                  a=m
               else
                  b=m
               endif
!               write(*,*)'a b',a,b
            enddo
            xi=-a/dtan(a)        !this is xi=-eta ctan(a)
            write(*,*)'w +',iw,w,a,b,fp(a,w),fp(b,w),xi
            if(dabs(fp(a,w)).gt.1d-12)then
            write(*,*)'This root is not accurate, increase gap, press..'
               gap=5*gap
               read(*,*)
               goto 103
            endif
!            if(iw.le.10)read(*,*)
```

```
            if(xi.gt.1d-12)then !10.1.2021
               write(10,*)w/pi,a/pi,b/pi,fp(a,w),fp(b,w),xi/pi
            endif
         enddo
!      second equation starts here eta = - w sin(eta)
 123     icount=0
         write(*,*)'negative root iw',iw
!       ------------------------------------------------
!      /     the following is to get approximate roots. \
         open(1,file='tmp')
         do i=1,10000
            q=i*d
            write(1,*)q,fn(q,w)
!      if two values of function fn at q and q+d change sign, this means
!      it crosses a root.
            if(fn(q,w)*fn(q+d,w).lt.0d0)then
!      count how many roots are found
               icount=icount+1
               if(icount.gt.100)stop'increase dimension for root'
!      root records the actual roots
               root(icount)=q
            endif
         enddo
         close(1)
!      \_____end________________________________________/

! in order to avoid the code misses a root, I need to find the minimum
! gap that I can choose. The default is 1d0. do not choose a big one.
         gap=1d0
         do i=1,icount
            write(*,*)i,root(i),'possible root'
            if(i.ne.1)then
               if(root(i)-root(i-1).lt.gap)then
                  gap=root(i)-root(i-1)
               endif
            endif
         enddo
!      stop
         tol=1d-12
!---------accurate root finder-----------------\
 104     do i=1,icount
            b=root(i)+gap/5d0   !b is chosen slightly higher than root
!by gap/5
            a=root(i)-gap/5d0   !a is chosen slightly lower than root by
!gap/5
            do while((b-a)>tol)
               m=a+(b-a)/2d0
!      this is from Michael Heath's book: Middle point method.
               if(fn(a,w)*fn(m,w).gt.0d0) then
                  a=m
               else
                  b=m
```

```
            endif
         enddo
!     here a is the root for eta. From eta, we can find xi. xi=eta\tan(eta)
         xi=-a/dtan(a)         !10.1.2021
         write(*,*)'w -',iw, w,a,b,fn(a,w),fn(b,w),xi
         if(dabs(fn(a,w)).gt.1d-12)then
       write(*,*)'This root is not accurate, increase gap, press..'
            gap=5*gap
            read(*,*)
            goto 104
         endif

!          read(*,*)
!          if(iw.le.10)read(*,*)
!     we only accept roots with both xi and eta positive.
         if(xi.gt.1d-12)then
            write(*,*)'success'
            write(20,*)w/pi,a/pi,b/pi,fn(a,w),fn(b,w),xi/pi
         endif
      enddo
      enddo
      close(10)
      close(20)
      end

      double precision function fp(q,w)
      implicit double precision (a-h,o-z)
      fp=q/w-dsin(q)              !positive root
      end

      double precision function fn(q,w)
      implicit double precision (a-h,o-z)
      fn=q/w+dsin(q)             !negative root
      end
```

Listing 11.5: Odd wavefunction of finite square well

```
!     Quantum Mechanics by Guo-ping Zhang, Mingsu Si, Thomas F. George.
!     Finite.well.wavefunction.
!     Sept. 29, 2021 written.
!     Sept. 30, 2021. Revised to compute the odd wavefunction.
      implicit double precision (a-h,o-z)
      double precision norm1,norm2
      pi=dacos(-1d0)
!
!     Quantum well width
      a=10d0
!     Grid mesh size
      dx=0.02d0
      open(1,file='odd.w.eta.used') !this includes a few  root for one w
 10   read(1,*,end=1,err=1)w,eta
      w=w*pi
```

```
      eta=eta*pi
!     find xi from eta and w
      xi=dsqrt(w**2-eta**2)
!     from page 20

      den=a/2*dexp(-xi*2d0)*(1d0/xi+(2*eta-dsin(2*eta))/
     c  (2*eta*dsin(eta)**2))
      AI=1/dsqrt(den)
      AIII=-AI                        !page 19
      AII=-AI*dexp(-xi)/dsin(eta)
      write(*,*)'AI AII and AIII',AI,AII,AIII,'denom',denom
      open(2,file='odd.wavefunction',position='append')
      open(20,file='odd.eigenenergy',position='append')
      write(20,*)eta**2/a**2
      close(20)
      do i=-100,100
         x=dx*i*a
         if(x.lt.-a/2)then        !wavefunction from - infinity to -a/2
            write(2,*)x,AI*dexp(xi*2d0*x/a)
         else if(x.ge.-a/2.and.x.lt.a/2)then  !wavefunction [-a/2, a/2]
            write(2,*)x,AII*dsin(eta*2*x/a)
         else  if(x.ge.a/2)then
            write(2,*)x,AIII*dexp(-xi*2d0*x/a) ![a/2,infinity]
         endif
      enddo
      write(2,*)'&'
      close(2)
      norm1=a/2*AI**2*dexp(-2*xi)/xi
      norm2=a/2*AII**2*(1d0-dsin(2d0*eta)/eta/2d0)
      write(*,*)'Plotting wavefunction',w,eta,norm1,norm2,norm1+norm2
      read(*,*)
      goto 10
 1    close(1)
      end
```

Listing 11.6: Transmission coefficient

```
!     written in 2021.
!     this code computes the transmission coefficient, reflection
!     coefficient.
      implicit double precision (a-h,o-z)
      double precision me
      echarge=1.602d-19
      me=9.11d-31
      h=6.626d-34
      pi=dacos(-1d0)
      hbar=h/(2*pi)
!     potential energy barrier V0= 2eV
      V0=2d0*echarge                    !eV
!     barrier width[0,a] in units of angstrom.
      a=3d-10
      open(1,file='transmission.versus.E.with.a.is.3Angstrom')
```

```
      do ie=1,99
         E=ie/50d0*echarge
         qa=dsqrt(2*me*(V0-E))*a/hbar
         T=((dsqrt((V0-E)/E)-dsqrt(E/(V0-E)))*dsinh(qa))**2/4
         T=T+(dcosh(qa))**2
         T=1/T
         write(1,*)e/echarge,T!,ie,qa
      enddo
      close(1)
      open(1,file='transmission.versus.a.E.is.1eV')
      E=1d0*echarge
      do ie=1,10
         a=ie*1d-10
         qa=dsqrt(2*me*(V0-E))*a/hbar
         T=((dsqrt((V0-E)/E)-dsqrt(E/(V0-E)))*dsinh(qa))**2/4
         T=T+(dcosh(qa))**2
         T=1/T
         write(1,*)a/1d-10,T!,ie,qa
      enddo
      close(1)
      end
```

Listing 11.7: Tunneling wavefunction

```
!     Fri Dec 17 09:36:39 EST 2021
!     this code computes the scattering wavefunctions.

!     E: energy for the incident electron, E=alpha*V0
!     alpha: must be between 0 to 1. this is the ratio
!     aa: the barrier width in Angstrom
!     implicit double precision (a-h,o-z)
      implicit none
      double precision qa,ka,q,k,alpha,m
      double complex t,r,a,b,i,dem,wave,cxp
      double precision pi,hbar,aa,v0,x,step,echarge
      i=dcmplx(0.d0,1d0)
      pi=dacos(-1d0)
      hbar=6.626d-34/(2d0*pi)    !reduced planck constant
      m=9.11d-31                 !electron mass
      echarge=1.602d-19          !electron charge
      alpha=0.6d0                !ratio
      aa=10d0                    !barrier width
      V0=1d0                     !potential barrier height
      k=dsqrt(2d0*m*alpha*V0*echarge)/hbar*1d-10
      q=dsqrt(2d0*m*(1d0-alpha)*V0*echarge)/hbar*1d-10
      ka=k*aa
      qa=q*aa
      write(*,*)'wavevectors k q ',k,q
      dem=(1d0-ka**2/qa**2)*sinh(qa)-2*i*ka/qa*cosh(qa)
      t=dcmplx(dcos(ka),-dsin(ka))*(-2*i*ka/qa)/dem
      write(*,*)'Transmitted t',t,t*dconjg(t)
      dem=sinh(qa)*(ka**2-qa**2)+2*i*ka*qa*cosh(qa)
```

```
      r=(ka**2+qa**2)*sinh(qa)/dem
      write(*,*)'Reflected r',r,r*dconjg(r)
      write(*,*)'T+R',t*dconjg(t)+r*dconjg(r)
      a=0.5d0*((1d0+i*ka/qa)+r*(1-i*ka/qa)) !Coefficient A of wavefunction
                                            !in Region II
      b=0.5d0*((1d0-i*ka/qa)+r*(1+i*ka/qa)) !Coefficient B of
                                            !wavefunction in Region II
      write(*,*)'Coefficient A and B',a,b
      step=0.01d0
      open(1,file='tunnel.wave')
      do x=-100d0,100d0,step
         if(x.lt.0d0) then
            wave=1d0*cxp(k*x)+1d0*r*cxp(-k*x)
         endif
         if(x.ge.aa) then
            wave=t*cxp(k*x)
         endif
         if(x.ge.0d0.and.x.le.aa) then
            wave=a*dexp(q*x)+b*dexp(-q*x)
         endif
         write(1,1)x,wave
      enddo
1     format(d10.3,2(d20.10,1x))
      close(1)
      end
      double precision function sinh(x)
      double precision x
      sinh=(dexp(x)-dexp(-x))/2d0
      end
      double precision function cosh(x)
      double precision x
      coshh=(dexp(x)+dexp(-x))/2d0
      end
      double  complex function cxp(x)
      double precision x
      cxp=dcmplx(dcos(x),dsin(x))
      end
```

Listing 11.8: Hermite function.

```
!     this code generates Hermite polynomials and wavefunction for
!     harmonic oscillator
      implicit double precision (a-h,o-z)
      character*40 wf
      pi=dacos(-1d0)
      do n=2,10
         write(wf,'(a,I2.2)')'wavefunction',n
         write(*,*)wf
         open(n+50,file=wf)
      enddo
      do ieta=-100,100
         eta=ieta*0.1d0
```

```
          h0=1d0
          h1=2*eta
          m=1
          do n=2,10
!             write(wf,'(a,I2.2)')'wavefunction',n
!             write(*,*)wf
!             read(*,*)
             m=m*n
             factor=dexp(-eta**2/2)/dsqrt(2d0**n*m)
             h2=h(h0,h1,n,eta)
             h0=h1
             h1=h2
             psi=h2*factor
!       classical x0 (dimensionless). this is the classical maximum,
!       i.e. amplitude of the pendulum. x0 is computed from
!       E=(1/2)mw^2x0^2=(n+1/2)hbar w.

             x0=dsqrt(2d0*n+1d0)
             if(dabs(x0).gt.dabs(eta))then !if eta < x0, classical has value
                p=1d0/(pi*dsqrt(x0**2-eta**2))
             else
                p=0d0
             endif
!       if(n.eq.10)write(*,*)eta,psi**2,p !*eta
 50          format(f12.5,3x,3(e16.10,1x))
             write(n+50,50)eta,psi,psi**2,p !*eta
             !if(n.eq.2)write(*,*)eta,psi**2,p !*eta
          enddo
       enddo
       do n=2,10
          close(n+50)
       enddo
       end
       double precision function h(h0,h1,n,eta)
       implicit double precision (a-h,o-z)
       double precision h0,h1
       h=2d0*eta*h1-2*(n-1)*h0
       end
```

Listing 11.9: Atomic orbitals.

```
!       This code allows one to plot, s, p, d, f, g, h and m orbitals, one at a time
       program legendre
       implicit double precision (a-h,o-z)
       double precision neutral,charged
       pi=dacos(-1d0)
       open(1,file='orbital')
       do theta=0,2*pi,0.001*pi
          u=dcos(theta)
          func=1d0                  !s orbital
          func=dcos(theta)          !p orbital
          goto 1
```

```
!         move this line to your desired location. Currently, it is p orbital
          func=0.5d0*(3*dcos(theta)**2-1d0) ! d orbital
          func=0.5d0*(5*u**3-3*u) ! f orbital
          func=(35*u**4-30*u**2+3)/8 ! g orbital
          func=(63*u**5-70*u**3+15*u)/8! h orbital
          func=(46189*u**10-109395*u**8+90090*u**6-30030*u**4+3465*u**2
     c         -63)/256 !m orbital
 1        x=dabs(func        )*dcos(theta)
!         x=     (dcos(theta))*dcos(theta)
          y=dabs(func        )*dsin(theta)
!         y=     (dcos(theta))*dsin(theta)
          write(1,*)x,y
       enddo
       close(1)
       end
```

Listing 11.10: Hydrogen wavefunction.

```
!      Thu Mar 17 15:14:36 EDT 2022
c      ab initio calculation for hydrogen atom, and other spherical
c      potentials. Atomic units are used. The potential is defined in the
c      potential subroutine.
       implicit double precision  (a-h,o-z)
!      nm: dimension of the matrix. You want to be large enough
!      dr: this is Delta r, your mesh grid size. Length in unit of bohr.
       parameter(nm=400,dr=0.06d0)
!      a: Hamiltonian matrix
!      e: eigenvalues: to get the true value. we need to divide it by 2,
!      but the equation has 2 in it.
!      f: used internal for the diagonalization routine.
!      z: wavefunction
       dimension a(nm,nm),e(nm),f(nm),z(nm,nm),h_kin(nm,nm)
       do i=1,nm
          do j=1,nm
             a(i,j)=0.d0
          enddo
       enddo
       write(*,*)'Input angular quantum number'
       read(*,*)l
       do i=1,nm
          r=i*dr
          a(i,i)=v(r,l)+2/(dr**2)
       enddo

       do i=1,nm-1
          a(i,i+1)=-1/dr**2
          a(i+1,i)=-1/dr**2
       enddo

       CALL TRED2(nm,nm,a,E,F,Z)!diagonalization code
       CALL TQL2(nm,nm,E,F,Z,IERR)!we obtain from the LAPACK
```

```
      open(1,file='eigenvalue')
      do i=1,nm
         write(1,*)i,e(i)
      enddo
      close(1)
      open(1,file='transitionmatrix.incorrect')
                                   !since the transition is forbidden
                                   !for the same quantum number l
                                   !that is why they are not correct here

                                   !to calculate the s-->p, you need
                                   !calculate at least twice, one for S
                                   !wave, the other for P wave, then you
                                   !can calculate the transition matrix
                                   !elements.

      do i=1,nm
         do j=1,nm
            tr=0.d0
            do k=1,nm
               dipole=k*dr
               tr=tr+z(k,i)*z(k,j)*dipole
            enddo
            write(1,10)i,j,tr
         enddo
      enddo
      close(1)
 10   format(1x,2I5,1x,f10.3)
c     kinetic energy operator is a matrix with off-diagonal elements
c     equal to -1/dr**2, diagal equal +2/dr**2
      open(1,file='kineticenergy')
      do i=1,nm
         do j=1,nm
            H_kin(i,j)=0.d0
         enddo
      enddo
      do i=1,nm
         H_kin(i,i)=2.d0/dr**2! this is diagonal
      enddo
      do i=1,nm-1
         H_kin(i+1,i)=-1.d0/dr**2 !off-diagonal
         H_kin(i,i+1)=-1.d0/dr**2 !off-diagonal
      enddo
      do n=1,nm
         e_kin=0.d0
         do i=1,nm
            do j=1,nm
               e_kin=e_kin+z(i,n)*h_kin(i,j)*z(j,n)
            enddo
         enddo
         write(1,11)n,e_kin
      enddo
```

```
      close(1)
11    format(1x,I4,d16.8)
1     write(*,*)'input which state you want to view'
      read(*,*)kk !choose main quantum number
      open(2,file='wf')
      do i=1,nm
c         write(*,*)i,e(i)
        write(2,*)i*dr,z(i,kk)! z is the wavefunction, i is mesh grid index
      enddo
      close(2)
      goto 1
      end

      double precision function v(r,l)
      implicit double precision (a-h,o-z)
!     v=-2.d0   /r+(l+1)*l/r**2 !this is a hydrogen
      v=-2.d0   /r+(l+1)*l/r**2 !this is a hydrogen
!this simulates the well potential, you can uncomment them, if you want to use.
!
!                      ______________V=5
!       |         |
!       |         |
!       |_________| V=-5
!     r=0        r=2

c      if(r.lt.2.d0)then
c         v=-5.d0
c      else
c         v=5.d0
c      endif
      end
```

Listing 11.11: Molecular orbital.

```
      implicit double precision (a-h,o-z)
      c1=1d0/sqrt(2d0)
!      c1=1d0/sqrt(3d0)
      c2=-dsqrt(1d0-c1**2)
      xA=-1d0
      xB=1d0
      a0=0.529d0
      do ix=-50,50
        x=ix*0.1d0
        p1=dexp(-dabs(x-xA)/a0)
        p2=dexp(-dabs(x-xB)/a0)
        psi=c1*p1+c2*p2
        rho=psi**2
        write(*,*)x,rho
      enddo
      end
```

Listing 11.12: Hermite function.

```
      implicit double precision (a-h,o-z)
      c1=1d0/sqrt(2d0)
!      c1=1d0/sqrt(3d0)
      c2=-dsqrt(1d0-c1**2)
      xA=-1d0
      xB=1d0
      a0=0.529d0
      do ix=-50,50
         x=ix*0.1d0
         p1=dexp(-dabs(x-xA)/a0)
         p2=dexp(-dabs(x-xB)/a0)
         psi=c1*p1+c2*p2
         rho=psi**2
         write(*,*)x,rho
      enddo
      end
```

Listing 11.13: flugge.f.

```
!     this code plots, the energy as a function of distance R  between two protons
!     this code finds the energy minimum for several lambdas
!     use the formula from our book, more accurate than Zeng's and Flugge's books.
      implicit none
      double precision dx,rho,L,S,D,X,energy,emin,rhomin,bohr
      double precision K,E,Es
      integer i,iL
      dx=0.01d0
      bohr=5.29177210903d-1
      rhomin=0d0
      open(1,file='flugge.energy.versus.rho')
      open(2,file='flugge.lambda.energy')
      open(3,file='flugge.energy.versus.R')
      open(4,file='flugge.S.D.X.versus.R')
      do iL=0,500
         L=1d0+IL*dx*0.1d0              !this is lambda
         emin=100d0
         do i=1,1000
            rho=dx*i
            S=(1d0+rho+rho**2/3d0)*dexp(-rho) !overlap matrix
            K=1/rho*(1d0-(1d0+rho)*dexp(-2*rho))
            E=(1+rho)*dexp(-rho)
            Es=-L**2/2d0+(L*(L-1d0)-L*K+L*(L-2)*E)/(1d0+S)
            energy=Es+L/rho !the second term is proton-proton repulsion
            write(1,*)rho,energy
            if(dabs(L-1.238d0).lt.1d-6)then
               write(3,*)rho/L*bohr,energy,Es,L/rho
               write(4,*)rho/L*bohr,S,K,E
            endif
            if(energy.le.emin)then
               emin=energy
               rhomin=rho
```

```
            endif
         enddo
         write(1,*)'&',L
         write(*,*)'Lambda',L,'E_min',emin,'  rho_min',rhomin/L!*bohr
         write(2,*)L,emin,rhomin/L*bohr
      enddo
      close(1)
      close(2)
      close(3)
      close(4)
      end
```

Listing 11.14: Kronig-Penney model.

```
      implicit double precision (a-h,o-z)
      double precision root(100),m
      double precision k
      pi=dacos(-1d0)
      tol=1d-12
      A=10d0
      R=0.2d0
      open(1,file='F.versus.epsilon')
      do i=0,300
         e=0.01d0*i+1d-10
         write(1,*)e,f0(A,R,E)
      enddo
      close(1)
!       stop
      delta=1d-2
      Emax=5d0                   !this is the maximum energy that I want
      E=1d-5                     !this is starting E
      Ebandtop=E                 !I set the maximum energy for the last band
      bandtop=E                  !this is to collect the band top
      open(11,file='allowed.band')
 2    do icoska=-100,100         !cos(ka) runs from -1 to 1.
! 2    do icoska=0,100           !This runs from -1 to 1.
         coska=icoska*0.01d0
         E=EBandtop+delta               !this is starting E, I have to add
                                        !a small amount to skip the top of
                                        !the last band
         sign=1d0
 10      do while (sign>0d0)     !test whether F-coska crosses zero. If
                                 !so, sign must be negative.
            sign=f(A,R,E,coska)*f(A,R,E+Delta,coska)
            E=E+delta
            write(*,*)'E',E,sign
!            read(*,*)
            if(E.gt.Emax) goto 1
         enddo
         if(sign>0)stop'wrong sign'
!      start to search for the exact value of the root.
         bb=E
```

```
         aa=E-delta
!      write(*,*)'aa bb line 30',aa,bb,f(A,R,aa,coska),f(A,R,bb,coska)
!      read(*,*)
!      this method is from  Michael T. Heath's Scientific Computing book, page 224.
         do while((bb-aa)>tol)
            m=aa+(bb-aa)/2d0
!      write(*,*)'>------',ff(A,R,aa),ff(A,R,m)
            if(f(A,R,aa,coska)*f(A,R,m,coska).gt.0d0) then
               aa=m
            else
               bb=m
            endif
         enddo
         write(*,*)'aa',aa,'  bb',bb,coska,dacos(coska)
!         read(*,*)
         write(*,*)f(A,R,aa,coska),f(A,R,bb,coska),aa,bb
!note that bands are plotted as Energy versus ka, not coska.
!If plotted versus coska, it is pretty much a straight line.
         write(11,*)dacos(coska)/pi,aa
         bandtop=max(bandtop,aa)
         write(*,*)'bandtop',bandtop
      enddo                         !icoska
      Ebandtop=bandtop              !I want to get ready for next band.
      if(Ebandtop.gt.Emax) goto 1 !I am done
      write(11,*)'&'
      write(*,*)'Next band starts'
!       read(*,*)
      goto 2
 1    continue
      close(11)
      stop
      end
      double precision function f0(A,R,E)!this is F function in the book.
      implicit double precision (a-h,o-z)
      if(E.lt.1d0)then
         f0=(1d0-2d0*e)/(2d0*dsqrt(E*(1d0-E)))*dsinh(r*A*dsqrt(1d0-E))
         f0=f0*dsin(A*dsqrt(E))
         f0=f0+dcosh(r*A*dsqrt(1d0-E))*dcos(A*dsqrt(E))
      else if (E.eq.1d0)then
         f0=dcos(A)-R*A*dsin(A)/2d0
      else if (E.gt.1d0)then
        f0=(1d0-2d0*e)/(2d0*dsqrt(E*(-1d0+E)))*dsin(r*A*dsqrt(-1d0+E))
         f0=f0*dsin(A*dsqrt(E))
         f0=f0+dcos(r*A*dsqrt(-1d0+E))*dcos(A*dsqrt(E))
      else
         stop'wrong'
      endif
      end
      double precision function f(A,R,E,coska)
      implicit double precision (a-h,o-z)
      f=0d0
      if(E.lt.1d0)then
```

```
      f=(1d0-2d0*e)/(2d0*dsqrt(E*(1d0-E)))*dsinh(r*A*dsqrt(1d0-E))
      f=f*dsin(A*dsqrt(E))
      f=f+dcosh(r*A*dsqrt(1d0-E))*dcos(A*dsqrt(E))
   else if (E.eq.1d0)then
      f=dcos(A)-R*A*dsin(A)/2d0
   else if (E.gt.1d0)then
      f=(1d0-2d0*e)/(2d0*dsqrt(E*(-1d0+E)))*dsin(r*A*dsqrt(-1d0+E))
      f=f*dsin(A*dsqrt(E))
      f=f+dcos(r*A*dsqrt(-1d0+E))*dcos(A*dsqrt(E))
   else
      stop'wrong'
   endif
   f=f-coska
   end
```

Listing 11.15: Fermi surface.

```
!     this code computes the Fermi surface in two-dimensional system.
      implicit none
      integer N,i,ix,iy
      double precision pi,dx,E,x,y
      N=1000
      pi=dacos(-1d0)
      dx=0.001d0*pi
      E=1.d0                        !this the energy that I set
      goto 1
      do ix=-n,n
         x=dx*ix
         write(*,*)x,dcos(x)
      enddo
      stop
 1    continue
      do i=-n,0
         x=i*dx
         y=-E-dcos(x)
         if(dabs(y).le.1d0)then
            y=dacos(y)
            write(*,*)x/pi,y/pi
         endif
      enddo
      write(*,*)'&'
      do i=1,n
         x=i*dx
         y=-E-dcos(x)
         if(dabs(y).le.1d0)then
            y=dacos(y)
            write(*,*)x/pi,y/pi
         endif
      enddo
!     stop
      write(*,*)'&'
      do i=n,0,-1
```

```
      x=i*dx
      y=-E-dcos(x)
      if(dabs(y).le.1d0)then
         y=dacos(y)
         write(*,*)x/pi,-y/pi
      endif
   enddo
   write(*,*)'&'
   do i=0,-n,-1
      x=i*dx
      y=-E-dcos(x)
      if(dabs(y).le.1d0)then
         y=dacos(y)
         write(*,*)x/pi,-y/pi
      endif
   enddo
   end
```

Listing 11.16: fermi200.f.

```
!     this code computers the chemical potential as a function of kt for
!     a fixed number of electrons.
      implicit none
      double precision a,b,fermi,m,tol,kt
      integer ikt
      tol=1d-14
      open(3,file='chemical.versus.kt')
      do ikt=1,18
         kt=0.01d0*ikt
         a=-10d0
         b=10d0
         do while ((b-a)>tol)
            m=a+(b-a)/2d0
            if(fermi(a,kt)*fermi(m,kt).gt.0d0) then
               a=m
            else
               b=m
            endif
            write(*,*)a,b,'a,b'
         enddo
         write(*,*)'Fermi energy',a,b
         write(3,*)kt,a
      enddo
      close(3)
      end
      double precision function fermi(ef,kt)
      implicit none
      integer ktot,ne, n,nk1,nk2,nk3,m,i
      double precision ek,ef,sum,kt,f
      double precision pi,lambda,h,e,emass,hbar,number
      parameter(NE=40,Ktot=100)
      double precision E0        !this is the site energy
```

```
      double precision t          !hopping integral
      double precision ka         !the product k and a (lattice constant),
                                  !starts from -pi to pi
      double precision dk         !k mesh
      pi=dacos(-1d0)
      E0=-2d0                     !eV
      t=1d0                       !eV
!     kt=0.12d0                 !eV
      number=0d0
      dk=2*pi/ktot
      open(1,file='out.fermi200')
      open(2,file='out1.fermi200')
      do i=-ktot/2,ktot/2         !ka from -pi to pi
         ka=i*dk
         ek=E0-t*dcos(ka)
         f=dexp((ek-ef)/kt)+1d0
         f=1d0/f
         number=number+f
         write(1,*)ek,f,number
         write(2,*)ka/pi,ek
      enddo
      close(1)
      close(2)
!      write(*,*)'number',number/ktot,number
!      goto 1
      fermi=number-NE
      write(*,*)'number',number
!     read(*,*)
      end
```

Bibliography

[1] G. N. Lewis, The conservation of photons, Nature **118**, 874–875 (1926).
[2] A. H. Compton, A quantum theory of the scattering of X-rays by light elements, Phys. Rev. **21**, 483 (1923).
[3] W. Elsasser, Naturwissenschaften **13**, 711 (1925).
[4] C. J. Davisson and L. H. Germer, Diffraction of electrons by a crystal of nickel, Phys. Rev. **30**, 705 (1927).
[5] M. Arndt, O. Nairz, J. Voss-Andreae, C. Keller, G. van der Zouw, and A. Zeilinger, Wave-particle duality of C_{60}, Nature **401**, 680 (1999).
[6] B. Brezger, L. Hackermüller, S. Uttenthaler, J. Petschinka, M. Arndt, and A. Zeilinger, Matter-wave interferometer for large molecules, Phys. Rev. Lett. **88**, 100404 (2002).
[7] C. Brand, S. Troyer, C. Knobloch, R. Cheshnovsky, and M. Arndt, Single-, double-, and triple-slit diffraction of molecular matter waves, Am. J. Phys. **89**, 1132 (2021).
[8] W. D. Phillips, Nobel Lecture: laser cooling and trapping of neutral atoms, Rev. Mod. Phys. **70**, 721 (1998).
[9] M. F. Crommie, C. P. Lutz, and D. M. Eigler, Confinement of electrons to quantum corrals on a metal surface, Science **262**, 218 (1993).
[10] E. J. Galvez, Charles H. Holbrow, M. J. Pysher, J. W. Martin, N. Courtemanche, L. Heilig, and J. Spencer, Interference with correlated photons: five quantum mechanics experiments for undergraduates, Am. J. Phys. **73**, 127 (2005).
[11] P. A. M. Dirac, The Principles of Quantum Mechanics, Oxford University Press (1958).
[12] P. J. Mohr, Solutions of the Maxwell equations and photon wavefunctions, Ann. Phys. **325**, 607–663 (2010).
[13] S. M. Barnett, Resolution of the Abraham–Minkowski dilemma, Phys. Rev. Lett. **104**, 070401 (2010).
[14] E. Noether, Nachr. d. König. Gesellsch. d. Wiss. zu Göttingen, Math-phys. Klasse, 235–257 (1918). Invariant Variation Problems, M. A. Tavel's English translation of Noether's Theorems in Transport Theory and Statistical Physics **1**: 186–207 (1971). Accessible at https://doi.org/10.48550/arXiv.physics/0503066.
[15] P. Fulde, Electron Correlations in Molecules and Solids, Springer Berlin, Heidelberg (1991).
[16] I.-H. Tan, G. L. Snider, L. D. Chang, and E. L. Hu, A self-consistent solution of Schrödinger–Poisson equations using a nonuniform mesh, J. Appl. Phys. **68**, 4071 (1990).
[17] D. J. Griffiths and D. F. Schroeter, Introduction to Quantum Mechanics, 3rd edition, Cambridge University Press (2018).
[18] D. H. McIntyre, Quantum Mechanics, Pearson (2012).
[19] J. Y. Zeng, Quantum Mechanics, in Chinese, Scientific Publishing, Beijing (1984).
[20] H. C. Ohanian, Modern Physics, Prentice Hall, Upper Saddle River, NJ 07458 (1995).
[21] F. Constantinescu and E. Magyari, Problems in Quantum Mechanics, Pergamon Press, Oxford (1971).
[22] C. Kittel, Introduction to Solid State Physics, 7th edition, John Wiley & Sons, Inc., New York (1996).
[23] B. G. Moore, Orbital plots using Gnuplot, J. Chem. Educ. **77**, 785 (2000).
[24] M. Tinkham, Group Theory and Quantum Mechanics, McGraw-Hill Book Company, New York (1964).
[25] M. Lax, Symmetry Principles in Solid State and Molecular Physics, John Wiley & Sons, Inc., New York (1974).
[26] P. H. E. Meijer and E. Bauer, Group Theory: the Application to Quantum Mechanics, North-Holland Publishing Company, Amsterdam (1962).
[27] X. D. Xie, P. Jiang, and F. Lu, Group Theory and Its Applications in Physics, China Science Publishing and Media Ltd. Beijing (1984).
[28] J. S. Huebner, Comment on the Franck–Hertz experiment, Am. J. Phys. **44**, 302 (1976).
[29] E. Fermi, Notes on Quantum mechanics, The University of Chicago Press (1961).
[30] J. J. Sakurai, Modern Quantum Mechanics, ed. by S. F. Tuan, Pearson (2005).
[31] W. Gerlach and O. Stern, Z. Phys. **9**, 353 (1922).

https://doi.org/10.1515/9783110672152-012

[32] L. I. Schiff, Quantum Mechanics, McGraw-Hill, New York (1955).
[33] L. J. Norrby, Why is mercury liquid?, J. Chem. Educ. **68**, 110 (1991).
[34] S. E. Harris, J. E. Field, and A. Imamoglu, Phys. Rev. Lett. **64**, 1107 (1990).
[35] M. Fleischhauer, A. Imamoglu, and J. P. Marangos, Rev. Mod. Phys. **77**, 633 (2005).
[36] M. O. Scully and M. Suhail Zubairy, Quantum Optics, Cambridge Press (2006).
[37] N. Bloembergen, E. M. Purcell, and R. V. Pound, Phys. Rev. **73**, 679 (1948).
[38] J.-L. Basdevant and J. Dalibard, The Quantum Mechanics Solver, Springer-Verlag, Berlin, Heidelberg (2006).
[39] G. P. Zhang, Hidden identity in a generic λ system: applications to coherent population trapping, J. Phys. B, At. Mol. Opt. Phys. **46**, 035504 (2013).
[40] Y. R. Shen, The Principles of Nonlinear Optics, John Wiley & Sons, Inc., Hoboken, New Jersey (2003).
[41] R. W. Boyd, Nonlinear Optics, Academic Press (1991).
[42] P. N. Butcher and D. Cotter, The Elements of Nonlinear Optics, Cambridge University Press (1990).
[43] S. Mukamel, Principles of Nonlinear Optical Spectroscopy, Oxford University Press, New York (1995).
[44] M. D. Fayer, Ultrafast Infrared and Raman Spectroscopy, Marcel Dekker, Inc., New York and Basel (2001).
[45] S. Flügge, Practical Quantum Mechanics (1974).
[46] E. D. Palik, Handbook of Optical Constants of Solids, Academic (1985).
[47] T. H. Maiman, Stimulated optical radiation in ruby, Nature **187**, 493 (1960).
[48] T. H. Maiman, Optical and microwave-optical experiments in ruby, Phys. Rev. Lett. **4**, 493 (1960).
[49] K. Gottfried and T.-M. Yan, Quantum Mechanics: Fundamentals, 2nd edition, Springer (July 15, 2004).
[50] X. Sun, Solitons and Polarons in Conjugated Polymers, in Chinese, Sichuan Publishing House (1987).
[51] D. A. Papaconstantopoulous, Handbook of the Band Structure of Elemental Solids, 2nd edition, Springer, New York (2015).
[52] J. P., Lowe, Quantum Chemistry, Academic Press, Inc., San Diego (1993).
[53] H. Shirakawa, Nobel Lecture: the discovery of polyacetylene film – the dawning of an era of conducting polymers, Rev. Mod. Phys. **73**, 713 (2000).
[54] Alan G. MacDiarmid, Nobel Lecture: synthetic metals: a novel role for organic polymers, Rev. Mod. Phys. **73**, 701 (2001).
[55] Alan J. Heeger, Nobel Lecture: semiconducting and metallic polymers: the fourth generation of polymeric materials, Rev. Mod. Phys. **73**, 681 (2001).
[56] R. de L. Kronig and W. G. Penney, Quantum mechanics of electrons in crystal lattices, Proc. R. Soc. Lond. A **130**, 499–513 (1931).
[57] D. A. McQuarrie, The Kronig–Penney model: a single lecture illustrating the band structure of solids, Chem. Educ. **1**, 1–10 (1996).
[58] N. W. Ashcroft and N. D. Mermin, Solid State Physics, 1st edition, Cengage Learning (January 2, 1976).
[59] D. Vanderbilt, Berry Phases in Electronic Structure Theory, Cambridge University Press, Cambridge (2018).
[60] I. Blount, Formalisms of band theory, Solid State Phys. **13**, 305 (1962).
[61] J. Callaway, Quantum Theory of the Solid State, Academic Press, Inc., New York (1974). See page 467.
[62] N. Marzari, A. A. Mostofi, J. R. Yates, I. Souza, and D. Vanderbilt, Maximally localized Wannier functions: theory and applications, Rev. Mod. Phys. **84**, 1419 (2012).
[63] S. Datta, Quantum Transport: Atom to Transistor, Cambridge University Press (2005).
[64] L. Ye and X. Y. Peng, Selected Topics in Condensed Matter Physics, Fudan University Press, Shanghai, China (2003).
[65] S. Franck, P. Poncharal, Z. L. Wang, and W. A. de Heer, Carbon nanotube quantum resistors, Science **280**, 1744 (1998).
[66] D. M. Gillingham, I. Linington, C. Müller, and J. A. C. Bland, e^2/h quantization of the conduction in Cu nanowires, J. Appl. Phys. **93**, 7388 (2003).
[67] R. Landauer, IBM J. Res. Dev. **1**, 223 (1957); IBM J. Res. Dev. **32**, 306 (1957).

[68] G. Binnig and H. Rohrer, Scanning tunneling microscopy – from birth to adolescence, Rev. Mod. Phys. **59**, 615 (1987).

[69] M. S. Dresselhaus, G. Dresselhaus, and P. C. Eklund, Science of Fullerenes and Carbon Nanotubes, Academic Press, San Diego, California (1996).

[70] J. Menendez and J. B. Page, Vibrational spectroscopy of C_{60}, in Light Scattering in Solids VIII Fullerenes, Semiconductor Surfaces, Coherent Phonons, edited by M. Cardona and G. Güntherodt, Springer (2000).

[71] W. M. You, C. L. Wang, F. C. Zhang, and Z. B. Su, Application of a Su–Schrieffer–Heeger-like model to the intramolecular electron-phonon coupling in C_{60} clusters, Phys. Rev. B **47**, 4765 (1993).

[72] P. Hohenberg and W. Kohn, Inhomogeneous electron gas, Phys. Rev. **136** B864 (1964).

[73] W. Kohn and L. J. Sham, Self-consistent equations including exchange and correlation effects, Phys. Rev. **140**, A1133 (1965).

[74] G. Racah, Theory of complex spectra. II, Phys. Rev. **62**, 438 (1942).

[75] J. Stöhr and H. C. Siegmann, Magnetism: From Fundamentals to Nanoscale Dynamics, Springer-Verlag, Berlin/Heidelberg, New York (2006).

[76] J. M. D. Coey, Magnetism and Magnetic Materials, Cambridge University Press (2010).

[77] D. P. DiVincenzo, The physical implementation of quantum computation, Fortschr. Phys. **48**, 771–783 (2000).

[78] M. A. Nielsen and I. L. Chuang, Quantum Computation and Quantum Information, Cambridge University Press (2000).

[79] C. Monroe, D. M. Meekhof, B. E. King, W. M. Itano, and D. J. Wineland, Demonstration of a fundamental quantum logic gate, Phys. Rev. Lett. **75**, 4714 (1995).

[80] S. R. White, Density matrix formulation for quantum renormalization groups, Phys. Rev. Lett. **69**, 2863 (1992).

[81] G. P. Zhang, Ground-state phase diagram of the one-dimensional extended Hubbard model: a density-matrix renormalization-group approach, Phys. Rev. B **56**, 9189 (1997).

[82] H. Ball, M. Biercuk, and M. Hush, Quantum firmware and the quantum computing stack, Phys. Today, March (2021), page 28.

[83] B. B. Blinov, D. Leibfried, C. Monroe, and D. J. Wineland, Quantum computing with trapped ion hyperfine qubits, Quantum Inf. Process. **3** (2004).

[84] M. V. Berry, Quantal phase factors accompanying adiabatic changes, Proc. R. Soc. Lond. A **392**, 45 (1984).

[85] Juan Rodriguez-Carvajal, Introduction to magnetic space groups, Lecture notes.

[86] R. R. Birss, Symmetry and Magnetism, North-Holland Publishing Company, Amsterdam (1964).

[87] J.-Q. Chen, Group Representation Theory for Physicists, World Scientific, Singapore (1989).

[88] G. P. Zhang, M. S. Si, M. Murakami, Y. H. Bai, and T. F. George, Generating high-order optical and spin harmonics from ferromagnetic monolayers, Nat. Commun. **9**, 3031 (2018).

Index

CNOT 316, 319
H 314
I 314, 319
S 314
T 314
X 314, 319
Y 314
Z 314, 319

absorption 219
adiabatic 297
adiabatic approximation 321, 323
Alzheimer's disease 126
amplification 224
angular momentum 110
anticommutation 167
anticommutator 60
AO 231
aufbau principle 125
axial vector 314

band structure 245
basis function 9, 10, 30, 143, 239
Beer–Lambert law 221
Bell state 317
Berry connection 271, 324, 325, 331
Berry curvature 327
Berry phase 320, 326
Berry vector potential 324
Binnig–Rohrer factor 287
blackbody radiation 1, 81
Bloch equation 203
Bloch sphere 312
Bloch sum 259
Bloch theorem 241
Bloch wavefunction 255
body-centered cubic 251
bonding state 263
Born statistics 7
Bose–Einstein condensation 292
Bose–Einstein distribution 292
boson 300
bound state 32, 40, 41, 48, 242
boundary condition 32, 33, 41, 43
Bravais lattice 252
Brillouin zone 247, 290, 291

canonical momentum 205
Cd lamp 187
central field potential 115
central potential 118
channel 278
chemical potential 276
classical bit 312, 314, 318
closed system 25
coherence 312
coherent population trapping 198
coherent state 75, 76
commutation 38, 59
commutator 27
completeness 29, 34, 111
complex conjugate 24, 26
computational basis 313, 314
conduction band 273
configuration 301
conjugate 63
conjugated molecular orbitals 231
conjugation 4
conservation of angular momentum 117
continuity equation 25
continuous state 32, 48, 49, 242
Coulomb gauge 206
Coulomb integral 308, 309
CPU 317
crossing term 306
crystal momentum 254
current density 25

ddσ-orbital 230
de Broglie wavelength 3, 44
degeneracy 117
degenerate perturbation 262, 266
density matrix 27, 317
density matrix renormalization group 317
density of states (DOS) 284
dipole selection rule 145, 208, 213
Dirac matrix 168
Dirac notation 5
distinguishable 298
Divicenzo criteria 312
dpσ-orbital 230
drain 278
dynamical phase factor 29, 320, 324

https://doi.org/10.1515/9783110672152-013

effective atomic number 125
effective mass 264
Ehrenfest's theorem 81
eigenenergy 16, 29, 33, 44
eigenequation 32
eigenfunction 16, 28, 60, 253
eigenstate 16, 29, 34
eigenvector 15
electromagnetically induced transparency 198
emission 219
energy band 241, 269
energy band gap 273
energy representation 211
energy-time uncertainty 213, 214
entangled state 316
entanglement 299, 303, 315
equipartition theorem 83
Eulerian angle 314
exchange charge 301
exchange integral 308, 309
exchange interaction 309
expansion coefficient 30
expectation values 13

face-centered cubic 251
Fermi energy 272
Fermi surface 274
Fermi velocity 276
Fermi wavevector 273, 274
Fermi–Dirac distribution 276
fine structure constant 96
finite square well 40
first-order energy correction 258
force 16
Franck–Hertz experiment 127
Freeman Dyson 333
function space 10

gauge invariant 325
Gaussian integrals 338
geometric phase factor 324
group momentum 60, 263
group velocity 263
Grover search scheme 312
gyromagnetic ratio 178
gyromagnetic ratio (nuclei) 199

Hadamard gate 314, 315
Hamiltonian operator 15
hardware 317
Hartree 121
Hartree–Fock state 316
Heisenberg equation of motion 79, 199
Heisenberg model 309
Heisenberg picture 77
Heisenberg uncertainty principle 60, 68
Hellmann–Feynman theorem 227, 323, 327
Hermite polynomial 69
Hermitian conjugate 27
Hermitian operator 14, 167
Hermitian transpose 5
hexagonal lattice 251
Hilbert space 10, 34, 73
hole 89, 272
HOMO 236
hopping integral 235
Hubbard model 316
Hund's rule 190
hydrogenic ions 125
hyperfine spectral line 96
hyperfine states 316

I-V curve 18
identical particle 306
identity operator 59
infinite quantum well 319
inner product 5

jj coupling 190
Josephson tunneling junction 318

kinetic energy expectation value 307
kinetic energy operator 15
Kramers–Kronig relation 226
Kronecker delta 34

ladder operator 63, 70
Lagrange multiplier 150
Laguerre polynomial 120
Landauer theory 281
LAPACK 121
Laplacian 118
Larmor frequency 195, 199
lattice 251
lattice constant 239, 252
lattice point 251
law of conservation 115
LCAO 153, 231, 265

Lennard–Jones potential 65
lifetime 218
light-emitting diodes 18
Liouville equation 26
LS coupling 190
LUMO 236

magnetic angular momentum quantum number 103
magnetism 309
magnon 311
many-body wavefunction 236
matrix mechanics 5, 59, 121, 143, 154, 169, 231, 235
matrix representation 112
Maxwell–Boltzmann distribution 82
Maxwell–Boltzmann statistics 7
measurement 13, 17
mechanical momentum 205
mesh grid 122
mesh grid point 5
metastable 224
microwave cavity 318
mixed state 38
molecular wavefunction 230
momentum operator 14

nearly free-electron model 256
Noether's theorem 22
norm 61
normal mode 289
normalization 24, 29, 34
number of nodes 34
numerical solution 121

observables 13
operators 13
orbital angular momentum 95, 100, 178
orbital moment 178
orbital-dependent potential 122
ordinary time-differential Schrödinger equation 322, 323
orthogonalization 29
outer product 5
overlap integral 306

p-representation 4, 14
parameter 321
parametric process 219
parity 36

Paul ion trap 318
Pauli exclusion principle 236, 272, 297, 301
Pauli matrix 314
permutation 299
phase difference 291
phonons 292
photoelectric effect 87
photon 2
photon Hamiltonian 16
PIB 319
Planck's law 85
polyacetylene 239
population inversion 224
position operator 14
potential energy operator 15
$pp\sigma$-orbital 230
precession frequency 195
primitive unit cell 250
probability 47
probability amplitude 6, 11
probability density 10, 17, 61, 64
projection 111
pure state 38

QPU 317
quantization 1
quantization condition 33
quantum fluctuation 72
quantum information 311
quantum state 4, 6
qubit 312

r-representation 4, 14, 15, 65, 100
Rabi frequency 196
recursion relation 69
reflected wave 52
relativistic Hamiltonian 16
representation 111
Ritz method 153
Rodrigues formula 69
rotating wave approximation 195
rotation of a wavefunction 116
rotation of an operator 116
rotational symmetry 103, 115
Russell–Saunders coupling 190
RWA 195
Rydberg 121

Schrödinger picture 73

Schrödinger representation 73
selection rule 36, 193, 208
semiconductor quantum well 40
separation of time and space variables 28
Shor factoring method 311
simple cubic 251
single slit 18
singlet 189
singlet spatial wavefunction 304
site energy 234
Slater determinant 236, 301
solar cells 87
source 278
spσ-orbital 231
spatial quantization 108
spatial wavefunction 315
specific heat 292
spectroscopic term notation 173
spherical harmonics 103, 105
spherical symmetry 118
spin and spatial entanglement 305
spin angular momentum 170
spin moment 178
spin singlet 302
spin triplet 302
spin waves 311
spin-orbit coupling 174
spinor 172
spinors 170
ssσ-orbital 229
stationary state 37
stationary state evolution 29
SU(2) symmetry 313
superposition 10, 23, 50, 312
susceptibility 220

targeted state 136, 137, 143
tensor 316
Thomas–Fermi kinetic energy functional 152
tight binding model 265
tight-binding approximation 234, 235, 239, 267
time derivative 321
time order operator 333
time translational symmetry 23
time-dependent perturbation theory 204
time-independent Schrödinger equation 27
transitions 97
translation 250
translational operator 253
translational symmetry 253
transpose 26
traveling wave 289
triplet 189

unbound state 49
uncorrelated wavefunction 297
unentangled wavefunction 297
unit cell 250
unitary matrix 116, 167
unitary operator 78, 313

valence band 273
variational principle 150
velocity operator 25, 80
virial theorem 95, 153, 164
von Klitzing constant 279

wave mechanics 4
wave-particle duality 3, 6
wavefunction 4, 6
wavepacket 3, 60, 76
work function 89

x-ray edges 176
XMCD 97

Zeeman disk 187
Zeeman effect 187
Zeeman energy 179
zero-point energy 68, 72
zero-point fluctuation 160
zero-point vibration 157

www.ingramcontent.com/pod-product-compliance
Lightning Source LLC
Chambersburg PA
CBHW080705220326
41598CB00033B/5319

* 9 7 8 3 1 1 0 6 7 2 1 2 1 *